STUDENT'S SOLUTIONS MANUAL

JEFFERY A. COLE

Anoka-Ramsey Community College

PREALGEBRA: AN INTEGRATED APPROACH

Margaret L. Lial

American River College

Diana L. Hestwood

Minneapolis Community and Technical College

Boston San Francisco New York
London Toronto Sydney Tokyo Singapore Madrid
Mexico City Munich Paris Cape Town Hong Kong Montreal

Reproduced by Pearson Addison-Wesley from electronic files supplied by the author.

Publishing as Pearson Addison-Wesley, 75 Arlington Street, Boston, MA 02116.

ISBN-13: 978-0-321-44163-8
ISBN-10: 0-321-44163-X

1 2 3 4 5 6 BB 10 09 08 07

Preface

This *Student's Solutions Manual* contains solutions to selected exercises in the text *Prealgebra: An Integrated Approach* by Margaret L. Lial and Diana L. Hestwood. It contains solutions to the odd-numbered exercises in each section, all Relating Concepts exercises, as well as solutions to all the exercises in the review sections, the chapter tests, and the cumulative review sections.

This manual is a text supplement and should be read along *with* the text. You should read all exercise solutions in this manual because many concept explanations are given and then used in subsequent solutions. All concepts necessary to solve a particular problem are not reviewed for every exercise. If you are having difficulty with a previously covered concept, refer back to the section where it was covered for more complete help.

A significant number of today's students are involved in various outside activities, and find it difficult, if not impossible, to attend all class sessions; this manual should help meet the needs of these students. In addition, it is my hope that this manual's solutions will enhance the understanding of all readers of the material and provide insights to solving other exercises.

I appreciate feedback concerning errors, solution correctness or style, and manual style. Any comments may be sent directly to me at the address below, at jeff.cole@anokaramsey.edu, or in care of the publisher, Pearson Addison-Wesley.

I would like to thank Ken Grace, of Anoka-Ramsey Community College, and Jeannine Grace, for typesetting the manuscript and providing assistance with many features of the manual; Marv Riedesel and Mary Johnson, for their careful accuracy checking and valuable suggestions; Jim McLaughlin, for his help with the entire art package; and the authors and Maureen O'Connor and Lauren Morse, of Pearson Addison-Wesley, for entrusting me with this project.

Jeffery A. Cole
Anoka-Ramsey Community College
11200 Mississippi Blvd. NW
Coon Rapids, MN 55433

Table of Contents

CHAPTER 1 INTRODUCTION TO ALGEBRA: INTEGERS

1.1 Place Value

1.1 Section Exercises

1. The whole numbers are: 15; 0; 83,001

3. The whole numbers are: 7; 362,049

5. The 2 in 61,284 is in the hundreds place.

7. The 2 in 284,100 is in the hundred-thousands place.

9. The 2 in 725,837,166 is in the ten-millions place.

11. The 2 in 253,045,701,000 is in the hundred-billions place.

13. Name the place value for each zero in

    302,016,450,098,570.

    From left to right: ten-trillions, hundred-billions, millions, hundred-thousands, and ones.

15. 8421 in words: eight thousand, four hundred twenty-one.

17. 46,205 in words: forty-six thousand, two hundred five.

19. 3,064,801 in words: three million, sixty-four thousand, eight hundred one.

21. 840,111,003 in words: eight hundred forty million, one hundred eleven thousand, three.

23. 51,006,888,321 in words: fifty-one billion, six million, eight hundred eighty-eight thousand, three hundred twenty-one.

25. 3,000,712,000,000 in words: three trillion, seven hundred twelve million.

27. Forty-six thousand, eight hundred five
    The first group name is *thousand*, so you need to fill *two groups* of three digits.

    $\underline{0}\,\underline{4}\,\underline{6},\underline{8}\,\underline{0}\,\underline{5} = 46{,}805$

29. Five million, six hundred thousand, eighty-two
    The first group name is *million,* so you need to fill *three groups* of three digits.

    $\underline{0}\,\underline{0}\,\underline{5},\underline{6}\,\underline{0}\,\underline{0},\underline{0}\,\underline{8}\,\underline{2} = 5{,}600{,}082$

31. Two hundred seventy-one million, nine hundred thousand
    The first group name is *million,* so you need to fill *three groups* of three digits.

    $\underline{2}\,\underline{7}\,\underline{1},\underline{9}\,\underline{0}\,\underline{0},\underline{0}\,\underline{0}\,\underline{0} = 271{,}900{,}000$

33. Twelve billion, four hundred seventeen million, six hundred twenty-five thousand, three hundred ten
    The first group name is *billion,* so you need to fill *four groups* of three digits.

    $\underline{0}\,\underline{1}\,\underline{2},\underline{4}\,\underline{1}\,\underline{7},\underline{6}\,\underline{2}\,\underline{5},\underline{3}\,\underline{1}\,\underline{0} = 12{,}417{,}625{,}310$

35. Six hundred trillion, seventy-one million, four hundred
    The first group name is *trillion,* so you need to fill *five groups* of three digits.

    $\underline{6}\,\underline{0}\,\underline{0},\underline{0}\,\underline{0}\,\underline{0},\underline{0}\,\underline{7}\,\underline{1},\underline{0}\,\underline{0}\,\underline{0},\underline{4}\,\underline{0}\,\underline{0} =$
    600,000,071,000,400

37. 6375 in words: six thousand, three hundred seventy-five

39. One hundred one million, two hundred eighty thousand
    The first group is *millions,* so fill *three groups* of three digits.

    $\underline{1}\,\underline{0}\,\underline{1},\underline{2}\,\underline{8}\,\underline{0},\underline{0}\,\underline{0}\,\underline{0} = 101{,}280{,}000$

41. \$93,972,602 in words: ninety-three million, nine hundred seventy-two thousand, six hundred two dollars

43. Fifty-five million, eight hundred
    The first group is *millions,* so you need to fill *three groups* of three digits.

    $\underline{0}\,\underline{5}\,\underline{5},\underline{0}\,\underline{0}\,\underline{0},\underline{8}\,\underline{0}\,\underline{0} = 55{,}000{,}800$

45. 6,400,000 in words: six million, four hundred thousand each day. 2,336,000,000 in words: two billion, three hundred thirty-six million in one year.

47. To make the largest possible whole number, arrange the digits from largest to smallest.

    97651100 → 97,651,100

    In words: ninety-seven million, six hundred fifty-one thousand, one hundred.

    To make the smallest possible whole number, arrange the numbers from smallest to largest with one exception: because we must use all the digits, start with the smallest nonzero digit.

    10015679 → 10,015,679

    In words: ten million, fifteen thousand, six hundred seventy-nine.

48. Answers will vary.

49.

sixty-fours	thirty-twos	sixteens	eights	fours	twos	ones
1,	1	1	1,	1	1	1

(a) $5 = 4 + 1 =$ binary 101

(b) $10 = 8 + 2 =$ binary 1010

(c) $15 = 8 + 4 + 2 + 1 =$ binary 1111

50. **(a)** Answers will vary but should mention that the location or place in which a digit is written gives it a different value.

(b) $8 = 5 + 3 =$ VIII
$38 = 30 + 5 + 3 =$ XXXVIII
$275 = 200 + 50 + 20 + 5 =$ CCLXXV
$3322 = 3000 + 300 + 20 + 2 =$ MMMCCCXXII

(c) The Roman system is *not* a place value system because no matter what place it's in, M = 1000, C = 100, etc. One disadvantage is that it takes much more space to write many large numbers; another is that there is no symbol for zero.

1.2 Introduction to Signed Numbers

1.2 Section Exercises

1. "Above sea level" implies a positive number.

$^+29{,}035$ feet or 29,035 feet

3. "Below zero" implies a negative number.

$^-128.6$ degrees

5. "Lost a total of 18 yards" implies a negative number.

$^-18$ yards

7. "Won \$100" implies a positive number.

$^+\$100$ or \$100

9. "Lost $6\frac{1}{2}$ pounds" implies a negative number.

$^-6\frac{1}{2}$ pounds

11. Graph $^-3, 3, 0, ^-5$

$^-5$ $^-4$ $^-3$ $^-2$ $^-1$ 0 1 2 3 4 5

13. Graph $^-1, 4, ^-2, 5$

$^-5$ $^-4$ $^-3$ $^-2$ $^-1$ 0 1 2 3 4 5

15. Graph $^-4\frac{1}{2}, \frac{1}{2}, 0, ^-8$

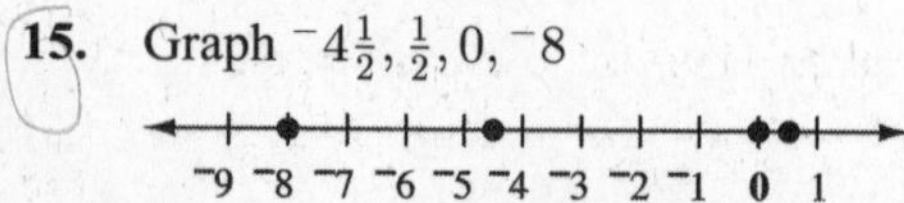

17. 10 is to the *right* of 2 on the number line, so 10 is *greater than* 2. Write $10 > 2$.

19. $^-1$ is to the *left* of 0 on the number line, so $^-1$ is *less than* 0. Write $^-1 < 0$.

21. $^-10$ is to the *left* of 2 on the number line, so $^-10$ is *less than* 2. Write $^-10 < 2$.

23. $^-3$ is to the *right* of $^-6$ on the number line, so $^-3$ is *greater than* $^-6$. Write $^-3 > {}^-6$.

25. $^-10$ is to the *left* of $^-2$ on the number line, so $^-10$ is *less than* $^-2$. Write $^-10 < {}^-2$.

27. 0 is to the *right* of $^-8$ on the number line, so 0 is *greater than* $^-8$. Write $0 > {}^-8$.

29. 10 is to the *right* of $^-2$ on the number line, so 10 is *greater than* $^-2$. Write $10 > {}^-2$.

31. $^-4$ is to the *left* of 4 on the number line, so $^-4$ is *less than* 4. Write $^-4 < 4$.

33. $|15| = 15$ because the distance from 0 to 15 on the number line is 15 spaces.

35. $|{}^-3| = 3$ because the distance between 0 and $^-3$ on the number line is 3 spaces.

37. $|0| = 0$ because the distance from 0 to 0 on the number line is 0 spaces.

39. $|200| = 200$ because the distance between 0 and 200 on the number line is 200 spaces.

41. $|{}^-75| = 75$ because the distance between 0 and $^-75$ on the number line is 75 spaces.

43. $|{}^-8042| = 8042$ because the distance between 0 and $^-8042$ on the number line is 8042 spaces.

45. Graph $^-1.5$ as A, 0.5 as B, $^-1$ as C, and 0 as D.

A C D B
$^-2$ $^-1$ 0 1 2

46. From Exercise 45, in order from lowest to highest: $^-1.5, ^-1, 0, 0.5$

47. A: $^-1.5$ is in the Below $^-1$ range. This patient may be at risk.

B: 0.5 is in the Above 0 range. This patient is above normal.

C: $^-1$ is in the 0 to $^-1$ range. This patient is normal.

D: 0 is in the 0 to $^-1$ range. This patient is normal.

48. **(a)** A patient who did not understand the importance of the negative sign would think the interpretation of $^-1.5$ was "above normal" (range Above 0) and wouldn't get treatment.

(b) For Patient D's score of 0, the sign plays no role. Zero is neither positive nor negative.

1.3 Adding Integers

1.3 Section Exercises

1. $^-2 + 5 = {^+3}$ or 3

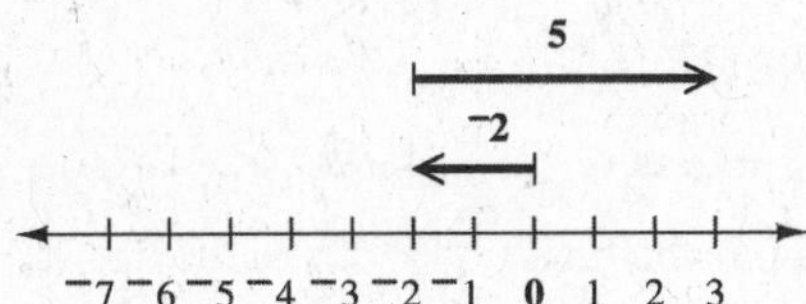

3. $^-5 + {^-2} = {^-7}$

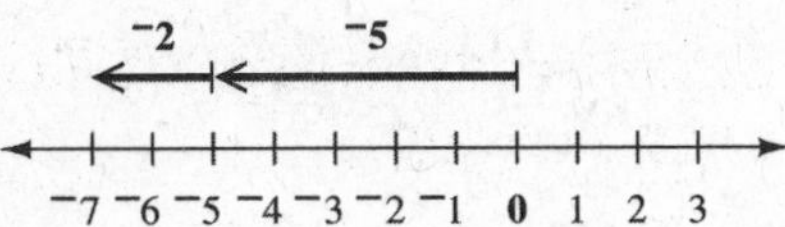

5. $3 + {^-4} = {^-1}$

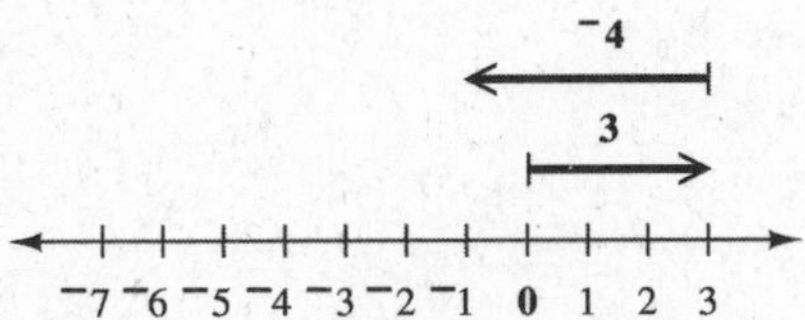

7. **(a)** $^-5 + {^-5} = {^-10}$ Adding *like* signed integers

Step 1: Add the absolute values.

$|^-5| = 5$

Add $5 + 5$ to get 10.

Step 2: Both integers are negative, so the sum is negative.

$^-5 + {^-5} = {^-10}$

(b) $5 + 5 = 10$ Adding *like* signed integers Both addends are positive, so the sum is positive.

9. **(a)** $7 + 5 = 12$ Adding *like* signed integers Both addends are positive, so the sum is positive.

(b) $^-7 + {^-5} = {^-12}$ Adding *like* signed integers

Step 1: Add the absolute values.

$|^-7| = 7; |^-5| = 5$

Add $7 + 5$ to get 12.

Step 2: Both integers are negative, so the sum is negative.

$^-7 + {^-5} = {^-12}$

11. **(a)** $^-25 + {^-25} = {^-50}$ Adding *like* signed integers

Step 1: Add the absolute values.

$|^-25| = 25$

Add $25 + 25$ to get 50.

Step 2: Both integers are negative, so the sum is negative.

$^-25 + {^-25} = {^-50}$

(b) $25 + 25 = 50$ Adding *like* signed integers Both addends are positive, so the sum is positive.

13. **(a)** $48 + 110 = 158$ Adding *like* signed integers Both addends are positive, so the sum is positive.

(b) $^-48 + {^-110} = {^-158}$ Adding *like* signed integers

Step 1: Add the absolute values.

$|^-48| = 48; |^-110| = 110$

Add $48 + 110$ to get 158.

Step 2: Both numbers are negative, so the sum is negative.

$^-48 + {^-110} = {^-158}$

15. The absolute values are the same in each pair of answers, so the only difference in the sums is the common sign.

17. **(a)** $^-6 + 8$ Adding *unlike* signed integers

Step 1: $|^-6| = 6; |8| = 8$

Subtract $8 - 6$ to get 2.

Step 2: 8 has the larger absolute value and is positive, so the sum is positive.

$^-6 + 8 = {^+2}$ or 2

(b) $6 + {^-8}$ Adding *unlike* signed integers

Step 1: $|6| = 6; |^-8| = 8$

Subtract $8 - 6$ to get 2.

Step 2: $^-8$ has the larger absolute value and is negative, so the sum is negative.

$6 + {^-8} = {^-2}$

19. **(a)** $^-9 + 2$ Adding *unlike* signed integers

Step 1: $|^-9| = 9; |2| = 2$

Subtract $9 - 2$ to get 7.

Step 2: $^-9$ has the larger absolute value and is negative, so the sum is negative.

$^-9 + 2 = {^-7}$

(b) $9 + {^-2}$ Adding *unlike* signed integers

Step 1: $|9| = 9; |^-2| = 2$

Subtract $9 - 2$ to get 7.

Step 2: 9 has the larger absolute value and is positive, so the sum is positive.

$9 + {^-2} = {^+7}$ or 7

21. **(a)** $20 + {}^{-}25$ Adding *unlike* signed integers

Step 1: $|20| = 20; |{}^{-}25| = 25$

Subtract $25 - 20$ to get 5.

Step 2: ${}^{-}25$ has the larger absolute value and is negative, so the sum is negative.

$20 + {}^{-}25 = {}^{-}5$

(b) ${}^{-}20 + 25$ Adding *unlike* signed integers

Step 1: $|{}^{-}20| = 20; |25| = 25$

Subtract $25 - 20$ to get 5.

Step 2: 25 has the larger absolute value and is positive, so the sum is positive.

${}^{-}20 + 25 = {}^{+}5$ or 5

23. **(a)** $200 + {}^{-}50$ Adding *unlike* signed integers

Step 1: $|200| = 200; |{}^{-}50| = 50$

Subtract $200 - 50$ to get 150.

Step 2: 200 has the larger absolute value and is positive, so the sum is positive.

$200 + {}^{-}50 = {}^{+}150$ or 150

(b) ${}^{-}200 + 50$ Adding *unlike* signed integers

Step 1: $|{}^{-}200| = 200; |50| = 50$

Subtract $200 - 50$ to get 150.

Step 2: ${}^{-}200$ has the larger absolute value and is negative, so the sum is negative.

${}^{-}200 + 50 = {}^{-}150$

25. Each pair of answers differs only in the sign of the answer. This occurs because the signs of the addends are reversed.

27. ${}^{-}8 + 5$ Adding *unlike* signed integers

Step 1: $|{}^{-}8| = 8; |5| = 5$

Subtract $8 - 5$ to get 3.

Step 2: ${}^{-}8$ has the larger absolute value and is negative, so the sum is negative.

${}^{-}8 + 5 = {}^{-}3$

29. ${}^{-}1 + 8$ Adding *unlike* signed integers

Step 1: $|{}^{-}1| = 1; |8| = 8$

Subtract $8 - 1$ to get 7.

Step 2: 8 has the larger absolute value and is positive, so the sum is positive.

${}^{-}1 + 8 = {}^{+}7$ or 7

31. ${}^{-}2 + {}^{-}5$ Adding *like* signed integers

Step 1: $|{}^{-}2| = 2; |{}^{-}5| = 5$

Add $2 + 5$ to get 7.

Step 2: Both integers are negative, so the sum is negative.

${}^{-}2 + {}^{-}5 = {}^{-}7$

33. $6 + {}^{-}5$ Adding *unlike* signed integers

Step 1: $|6| = 6; |{}^{-}5| = 5$

Subtract $6 - 5$ to get 1.

Step 2: 6 has the larger absolute value and is positive, so the sum is positive.

$6 + {}^{-}5 = {}^{+}1$ or 1

35. $4 + {}^{-}12$ Adding *unlike* signed integers

Step 1: $|4| = 4; |{}^{-}12| = 12$

Subtract $12 - 4$ to get 8.

Step 2: ${}^{-}12$ has the larger absolute value and is negative, so the sum is negative.

$4 + {}^{-}12 = {}^{-}8$

37. ${}^{-}10 + {}^{-}10$ Adding *like* signed integers

Step 1: $|{}^{-}10| = 10; |{}^{-}10| = 10$

Add $10 + 10$ to get 20.

Step 2: Both integers are negative, so the sum is negative.

${}^{-}10 + {}^{-}10 = {}^{-}20$

39. ${}^{-}17 + 0 = {}^{-}17$

Adding zero to any number leaves the number unchanged.

41. $1 + {}^{-}23$ Adding *unlike* signed integers

Step 1: $|1| = 1; |{}^{-}23| = 23$

Subtract $23 - 1$ to get 22.

Step 2: ${}^{-}23$ has the larger absolute value and is negative, so the sum is negative.

$1 + {}^{-}23 = {}^{-}22$

43. ${}^{-}2 + {}^{-}12 + {}^{-}5$ *Add left to right.*
$= {}^{-}14 + {}^{-}5$
$= {}^{-}19$

45. $8 + 6 + {}^{-}8$ *Commute addends.*
$= 8 + {}^{-}8 + 6$ *Add left to right.*
$= 0 + 6$
$= 6$

47. ${}^{-}7 + 6 + {}^{-}4$ *Add left to right.*
$= {}^{-}1 + {}^{-}4$
$= {}^{-}5$

49. ${}^{-}3 + {}^{-}11 + 14$ *Add left to right.*
$= {}^{-}14 + 14$
$= 0$

51. $10 + {}^-6 + {}^-3 + 4$ *Add left to right.*
$= 4 + {}^-3 + 4$
$= 1 + 4$
$= 5$

53. ${}^-7 + 28 + {}^-56 + 3$ *Add left to right.*
$= 21 + {}^-56 + 3$
$= {}^-35 + 3$
$= {}^-32$

55. "Yards gained" are positive (${}^+13$), and "yards lost" are negative (${}^-17$).

$13 + {}^-17 = {}^-4$ yards

The team lost 4 yards.

57. The overdrawn amount is negative (${}^-\$62$), and the deposit amount is positive (${}^+\$50$).

${}^-\$62 + \$50 = {}^-\$12$

Nick is \$12 overdrawn.

59. \$88 stolen implies a loss of money or ${}^-\$88$.

Jay received \$35 back implies a gain of money or ${}^+\$35$.

${}^-\$88 + \$35 = {}^-\$53$

Jay's net loss was \$53.

61. Jeff: ${}^-20 + 75 + {}^-55$ *Add left to right.*
$= 55 + {}^-55$
$= 0$ points

Terry: $42 + {}^-15 + 20$ *Add left to right.*
$= 27 + 20$
$= 47$ points

63. ${}^-6 + {}^-5 + {}^-1 + {}^-2$ *Add left to right.*
$= {}^-11 + {}^-1 + {}^-2$
$= {}^-12 + {}^-2$
$= {}^-14$

Tiger Woods' total score was ${}^-14$.

65. ${}^-3 + 0 + 4 + {}^-1$ *Add left to right.*
$= {}^-3 + 4 + {}^-1$
$= 1 + {}^-1$
$= 0$

Miguel Jimenez's total score was 0 (even par).

67. $\underbrace{{}^-18 + {}^-5}_{{}^-23} = \underbrace{{}^-5 + {}^-18}_{{}^-23}$ *Commutative property*

Both sums are ${}^-23$.

69. $\underbrace{{}^-4 + 15}_{{}^+11} = \underbrace{15 + {}^-4}_{{}^+11}$ *Commutative property*

Both sums are ${}^+11$ or 11.

71. $6 + {}^-14 + 14$

Option 1: $(6 + {}^-14) + 14 = {}^-8 + 14$
$= 6$

Option 2: $6 + ({}^-14 + 14) = 6 + 0$
$= 6$

Option 2 is easier.

73. ${}^-14 + {}^-6 + {}^-7$

Option 1: $({}^-14 + {}^-6) + {}^-7 = {}^-20 + {}^-7$
$= {}^-27$

Option 2: ${}^-14 + ({}^-6 + {}^-7) = {}^-14 + {}^-13$
$= {}^-27$

Option 1 might seem easier.

75. Some possibilities are:
${}^-6 + 0 = {}^-6$; $10 + 0 = 10$; $0 + 3 = 3$

77. Be sure to use the *negative* key as opposed to the *subtraction* key.

${}^-7081 + 2965 = {}^-4116$

79. ${}^-179 + {}^-61 + 8926 = 8686$

81. $86 + {}^-99{,}000 + 0 + 2837 = {}^-96{,}077$

1.4 Subtracting Integers

1.4 Section Exercises

1. The opposite of 6 is ${}^-6$. $6 + {}^-6 = 0$

3. The opposite of ${}^-13$ is 13. ${}^-13 + 13 = 0$

5. The opposite of 0 is 0. $0 + 0 = 0$

7. $19 - 5$ *Change subtraction to addition. Change 5 to* ${}^-5$.
$= 19 + {}^-5$
$= 14$

9. $10 - 12$ *Change subtraction to addition. Change 12 to* ${}^-12$.
$= 10 + {}^-12$
$= {}^-2$

11. $7 - 19$ *Change subtraction to addition. Change 19 to* ${}^-19$.
$= 7 + {}^-19$
$= {}^-12$

13. ${}^-15 - 10$ *Change subtraction to addition. Change 10 to* ${}^-10$.
$= {}^-15 + {}^-10$
$= {}^-25$

15. ${}^-9 - 14$ *Change subtraction to addition. Change 14 to* ${}^-14$.
$= {}^-9 + {}^-14$
$= {}^-23$

17. $^-3 - {}^-8$ *Change subtraction to addition. Change $^-8$ to $^+8$.*

$= {}^-3 + {}^+8$
$= 5$

19. $6 - {}^-14$ *Change subtraction to addition. Change $^-14$ to $^+14$.*

$= 6 + {}^+14$
$= 20$

21. $1 - {}^-10$ *Change subtraction to addition. Change $^-10$ to $^+10$.*

$= 1 + {}^+10$
$= 11$

23. $^-30 - 30$ *Change subtraction to addition. Change 30 to $^-30$.*

$= {}^-30 + {}^-30$
$= {}^-60$

25. $^-16 - {}^-16$ *Change subtraction to addition. Change $^-16$ to $^+16$.*

$= {}^-16 + {}^+16$
$= 0$

27. $13 - 13$ *Change subtraction to addition. Change 13 to $^-13$.*

$= 13 + {}^-13$
$= 0$

29. $0 - 6$ *Change subtraction to addition. Change 6 to $^-6$.*

$= 0 + {}^-6$
$= {}^-6$

31. **(a)** $3 - {}^-5$ *Change subtraction to addition. Change $^-5$ to $^+5$.*

$= 3 + {}^+5$
$= 8$

(b) $3 - 5$ *Change subtraction to addition. Change 5 to $^-5$.*

$= 3 + {}^-5$
$= {}^-2$

(c) $^-3 - {}^-5$ *Change subtraction to addition. Change $^-5$ to $^+5$.*

$= {}^-3 + {}^+5$
$= 2$

(d) $^-3 - 5$ *Change subtraction to addition. Change 5 to $^-5$.*

$= {}^-3 + {}^-5$
$= {}^-8$

33. **(a)** $4 - 7$ *Change subtraction to addition. Change 7 to $^-7$.*

$= 4 + {}^-7$
$= {}^-3$

(b) $4 - {}^-7$ *Change subtraction to addition. Change $^-7$ to $^+7$.*

$= 4 + {}^+7$
$= 11$

(c) $^-4 - 7$ *Change subtraction to addition. Change 7 to $^-7$.*

$= {}^-4 + {}^-7$
$= {}^-11$

(d) $^-4 - {}^-7$ *Change subtraction to addition. Change $^-7$ to $^+7$.*

$= {}^-4 + {}^+7$
$= 3$

35. $^-2 - 2 - 2$ *Change all subtractions to additions. Change 2 to $^-2$.*

$= {}^-2 + {}^-2 + {}^-2$ *Add left to right.*
$= {}^-4 + {}^-2$
$= {}^-6$

37. $9 - 6 - 3 - 5$ *Change all subtractions to additions. Change 6 to $^-6$, 3 to $^-3$, and 5 to $^-5$.*

$= 9 + {}^-6 + {}^-3 + {}^-5$ *Add left to right.*
$= 3 + {}^-3 + {}^-5$
$= 0 + {}^-5$
$= {}^-5$

39. $3 - {}^-3 - 10 - {}^-7$ *Change all subtractions to additions. Change $^-3$ to $^+3$, 10 to $^-10$, and $^-7$ to $^+7$.*

$= 3 + {}^+3 + {}^-10 + {}^+7$ *Add left to right.*
$= 6 + {}^-10 + {}^+7$
$= {}^-4 + {}^+7$
$= 3$

41. $^-2 + {}^-11 - {}^-3$ *Change subtraction to addition. Change $^-3$ to $^+3$.*

$= {}^-2 + {}^-11 + {}^+3$ *Add left to right.*
$= {}^-13 + {}^+3$
$= {}^-10$

43. $4 - {}^-13 + {}^-5$ *Change subtraction to addition. Change $^-13$ to $^+13$.*

$= 4 + {}^+13 + {}^-5$ *Add left to right.*
$= 17 + {}^-5$
$= 12$

45. $6 + 0 - 12 + 1$ *Change subtraction to addition. Change 12 to* $^-12$.
$= 6 + 0 + {}^-12 + 1$ *Add left to right.*
$= 6 + {}^-12 + 1$
$= {}^-6 + 1$
$= {}^-5$

47. **(a)** The 30°F column and the 10 mph wind row intersect at 21°F. The difference between the actual temperature and the wind chill temperature is $30 - 21 = 30 + {}^-21 = 9$ degrees.

(b) The 15°F column and the 15 mph wind row intersect at 0°F. The difference between the actual temperature and the wind chill temperature is $15 - 0 = 15$ degrees.

(c) The 5°F column and the 25 mph wind row intersect at $^-17$°F. The difference between the actual temperature and the wind chill temperature is $5 - {}^-17 = 5 + {}^+17 = 22$ degrees.

(d) The $^-10$°F column and the 35 mph wind row intersect at $^-41$°F. The difference between the actual temperature and the wind chill temperature is $^-10 - {}^-41 = {}^-10 + {}^+41 = 31$ degrees.

49. The student forgot to change 6 to its opposite, $^-6$.

Correct Method:
$^-6 - 6$ *Change subtraction to addition. Change 6 to* $^-6$.
$= {}^-6 + {}^-6$ *Add.*
$= {}^-12$

51. $^-2 + {}^-11 + |{}^-2|$ *$|{}^-2| = 2$ because the distance from 0 to $^-2$ is 2 units.*
$= {}^-2 + {}^-11 + 2$ *Add left to right.*
$= {}^-13 + 2$
$= {}^-11$

53. $0 - |{}^-7 + 2|$ *Simplify the sum within the absolute value bars first.*
$= 0 - |{}^-5|$ *$|{}^-5| = 5$ because the distance from 0 to $^-5$ is 5 units.*
$= 0 - 5$ *Change subtraction to addition. Change 5 to* $^-5$.
$= 0 + {}^-5$ *Add.*
$= {}^-5$

55. $^-3 - ({}^-2 + 4) + {}^-5$ *Simplify the sum within the parentheses first.*
$= {}^-3 - 2 + {}^-5$ *Change subtraction to addition and change 2 to* $^-2$.
$= {}^-3 + {}^-2 + {}^-5$ *Add left to right.*
$= {}^-5 + {}^-5$
$= {}^-10$

57. $^-3 - 5 = {}^-3 + {}^-5 = {}^-8$
$5 - {}^-3 = 5 + 3 = 8$

$^-4 - {}^-3 = {}^-4 + 3 = {}^-1$
$^-3 - {}^-4 = {}^-3 + 4 = 1$

Subtraction is *not* commutative; the absolute value of the answer is the same, but the sign changes.

58. Subtracting 0 from a number does *not* change the number. For example, $^-5 - 0 = {}^-5$. But subtracting a number from 0 *does* change the number to its opposite. For example, $0 - {}^-5 = 5$.

1.5 Problem Solving: Rounding and Estimating

1.5 Section Exercises

1. 42 (nearest ten)

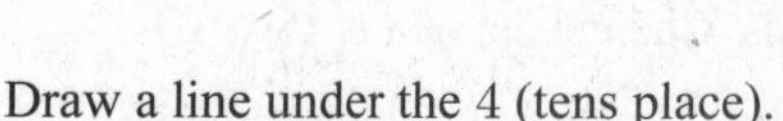

Draw a line under the 4 (tens place).
$\underline{4}2$ is closer to 40.

3. $^-691$ (nearest hundred)

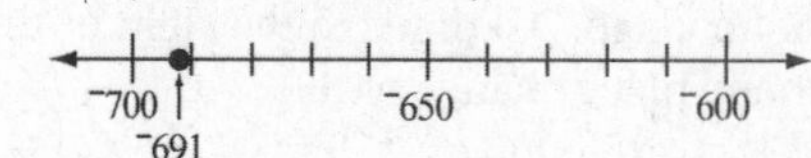

Draw a line under the 6 (hundreds place).
$^-\underline{6}91$ is closer to $^-700$.

5. 54,402 (nearest thousand)

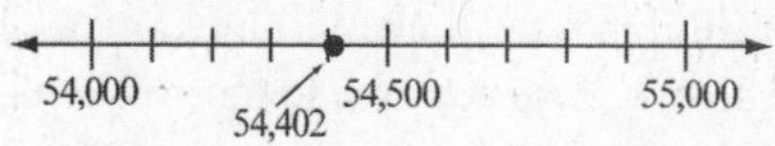

Draw a line under the first 4 (thousands place).
5$\underline{4}$,402 is closer to 54,000.

7. $6\underline{2}5 \approx 630$ (nearest ten)
Next digit is 5 or more. Tens place changes. Add 1 to 2. Change 5 to 0.

9. $^-10\underline{8}3 \approx {}^-1080$ (nearest ten)
Next digit is 4 or less. Tens place remains 8. Change 3 to 0.

11. $7\underline{8}62 \approx 7900$ (nearest hundred)
Next digit is 5 or more. Hundreds place changes. Add 1 to 8. Change 6 and 2 to 0 .

13. $^-86{,}\underline{8}13 \approx {}^-86{,}800$ (nearest hundred)
Next digit is 4 or less. Hundreds place remains 8. Change 1 and 3 to 0.

15. $42{,}\underline{4}95 \approx 42{,}500$ (nearest hundred)
Next digit is 5 or more. Hundreds place changes. Add 1 to 4. Change 9 and 5 to 0.

17. $^-5\underline{9}96 \approx {}^-6000$ (nearest hundred)
Next digit is 5 or more. Hundreds place changes. Add 1 to 9 and regroup 1 to thousands. Change 9 and 6 to 0. (**Regrouping** is also called *carrying*.)

19. $15{,}\underline{7}58 \approx 15{,}800$ (nearest hundred)
Next digit is 5 or more. Hundreds place changes. Add 1 to 7. Change 5 and 8 to 0.

21. $^-7\underline{8}{,}499 \approx {}^-78{,}000$ (nearest thousand)
Next digit is 4 or less. Thousands place remains 8. Change 4, 9, and 9 to 0.

23. $\underline{5}847 \approx 6000$ (nearest thousand)
Next digit is 5 or more. Thousands place changes. Add 1 to 5. Change 8, 4, and 7 to 0.

25. $5\underline{3}{,}182 \approx 53{,}000$ (nearest thousand)
Next digit is 4 or less. Thousands place remains 3. Change 1, 8, and 2 to 0.

27. $5\underline{9}5{,}008 \approx 600{,}000$ (nearest ten-thousand)
Next digit is 5 or more. Ten-thousands place changes. Add 1 to 9 and regroup 1 to hundred thousands. Change 5 and 8 to 0.

29. $5{,}\underline{4}44{,}000 \approx 5{,}400{,}000$ (nearest hundred-thousand)
Next digit is 4 or less. Hundred-thousands place does not change. All digits to the right of the underlined place change to 0.

31. $^-\underline{8}{,}906{,}422 \approx {}^-9{,}000{,}000$ (nearest million)
Next digit is 5 or more. Millions place changes. Add 1 to 8. Change other digits to 0.

33. $13\underline{9}{,}610{,}000 \approx 140{,}000{,}000$ (nearest million)
Next digit is 5 or more. Millions place changes. Add 1 to 9. Regroup one to ten-millions. Change 6 and 1 to zeros.

35. $2\underline{9}2{,}987{,}333 \approx 290{,}000{,}000$ (nearest ten-million)
Next digit is 4 or less. Ten-millions place does not change. All digits to the right of the underlined place change to 0.

37. $19{,}\underline{9}51{,}880{,}500 \approx 20{,}000{,}000{,}000$ (nearest hundred-million)
Next digit is 5 or more. Hundred-millions place changes. Add 1 to 9. Write 0 and regroup 1 to the ten-billions place. All digits to the right of the underlined place change to 0.

39. $\underline{8}{,}608{,}200{,}100 \approx 9{,}000{,}000{,}000$ (nearest billion)
Next digit is 5 or more. Billions place changes. Add 1 to 8. All digits to the right of the underlined place change to 0.

41. $\underline{3}1{,}500 \approx 30{,}000$ miles
Next digit is 4 or less. Leave 3 as 3. Change 1 and 5 to 0. 31,500 is closer to 30,000 than 40,000.

43. $^-\underline{5}6 \approx {}^-60$ degrees
Next digit is 5 or more. Change 5 to 6 and change 6 to 0. $^-56$ is closer to $^-60$ than $^-50$.

45. $\$\underline{9}942 \approx \$10{,}000$
Next digit is 5 or more. Add 1 to 9 and regroup 1 to the ten-thousands place. Change 9, 4, and 2 to 0. \$9942 is closer to \$10,000 than \$9000.

47. $\underline{6}0{,}950{,}000 \approx 60{,}000{,}000$ Americans
Next digit is 4 or less. Leave 6 as 6. Change 9 and 5 to 0. 60,950,000 is closer to 60,000,000 than 70,000,000.

49. $^-\underline{2}55 \approx {}^-300$ feet
Next digit is 5 or more. Change 2 to 3. Change 5s to 0. $^-255$ is closer to $^-300$ than $^-200$.

51. $\underline{6}35{,}478 \approx 600{,}000$ people in Alaska
Next digit is 4 or less. Leave 6 as 6. Change all other digits to zeros. 635,478 is closer to 600,000 than 700,000.

$\underline{3}4{,}501{,}994 \approx 30{,}000{,}000$ people in California
Next digit is 4 or less. Leave 3 as 3. Change all other digits to zeros. 34,501,994 is closer to 30,000,000 than 40,000,000.

53. Answers will vary but should mention looking only at the second digit, rounding first digit up when second digit is 5 or more, leaving first digit unchanged when second digit is 4 or less. Examples will vary. Some possibilities are $\underline{2}7 \approx 30$, $\underline{6}41 \approx 600$.

55. $^-42 + 89$
$^-\underline{4}2$ is closer to $^-40$ than $^-50$.
$\underline{8}9$ is closer to 90 than 80.

Estimate: $^-40 + 90 = 50$;
Exact: $^-42 + 89 = 47$

57. $16 + {}^-97$
$\underline{1}6$ is closer to 20 than 10.
$^-\underline{9}7$ is closer to $^-100$ than $^-90$.

Estimate: $20 + {}^-100 = {}^-80$;
Exact: $16 + {}^-97 = {}^-81$

59. $^-273 + {}^-399$
$^-\underline{2}73$ is closer to $^-300$ than $^-200$.
$^-\underline{3}99$ is closer to $^-400$ than $^-300$.

Estimate: $^-300 + {}^-400 = {}^-700$;
Exact: $^-273 + {}^-399 = {}^-672$

61. $3081 + 6826$
$\underline{3}081$ is closer to 3000 than 4000.
$\underline{6}826$ is closer to 7000 than 6000.

Estimate: $3000 + 7000 = 10{,}000$;
Exact: $3081 + 6826 = 9907$

63. $23 - 81$ *Change subtraction to addition.*
$23 + {}^-81$ *Change 81 to ⁻81.*

$\underline{2}3$ is closer to 20 than 30.
${}^-\underline{8}1$ is closer to ${}^-80$ than ${}^-90$.

Estimate: $20 + {}^-80 = {}^-60$;
Exact: $23 - 81 = 23 + {}^-81 = {}^-58$

65. ${}^-39 - 39$ *Change subtraction to addition.*
${}^-39 + {}^-39$ *Change 39 to ⁻39.*

${}^-\underline{3}9$ is closer to ${}^-40$ than ${}^-30$.

Estimate: ${}^-40 + {}^-40 = {}^-80$;
Exact: ${}^-39 - 39 = {}^-39 + {}^-39 = {}^-78$

67. ${}^-106 + 34 - {}^-72$ *Change subtraction to addition of the opposite.*
${}^-106 + 34 + {}^+72$

${}^-\underline{1}06 \approx {}^-100$; $\underline{3}4 \approx 30$; $\underline{7}2 \approx 70$

Estimate: ${}^-100 + 30 + {}^+70 = 0$;
Exact:
${}^-106 + 34 - {}^-72 = {}^-106 + 34 + {}^+72 = 0$

69. Already raised: $52,882 ≈ $50,000
Amount needed: $78,650 ≈ $80,000
Amount that still needs to be collected:

Estimate: 80,000 − 50,000 = $30,000;
Exact: 78,650 − 52,882 = $25,768

71. Estimate Dorene's expenses.
Rent: $685 ≈ $700
Food: $325 ≈ $300
Childcare: $320 ≈ $300
Transportation: $182 ≈ $200
Other: $150 ≈ $200

Estimate: Dorene's total expenses:
$700 + $300 + $300 + $200 + $200 = $1700.
Estimate Dorene's monthly take home pay.
$1920 ≈ $2000
Subtract Dorene's expenses from her take home pay to estimate her monthly savings.
$2000 − $1700 = $300.

Exact:
Total Expenses = $685 + $325 + $320 + $182 + $150 = $1662
Monthly savings = $1920 − $1662 = $258.

73. The final temperature equals the initial temperature plus the two increases.

${}^-102 \approx {}^-100$; $37 \approx 40$; $52 \approx 50$

Estimate: ${}^-100 + 40 + 50 = {}^-10$ degrees
Exact: ${}^-102 + 37 + 52 = {}^-13$ degrees

75. $\underline{4}12 \approx 400$ doors
$\underline{1}47 \approx 100$ windows
Total number of doors and windows:
Estimate: $400 + 100 = 500$ doors and windows
Exact: $412 + 147 = 559$ doors and windows

1.6 Multiplying Integers

1.6 Section Exercises

1. **(a)** $9 \cdot 7 = 63$ Factors have the *same* sign, so the product is *positive.*

(b) ${}^-9 \cdot {}^-7 = 63$ Factors have the *same* sign, so the product is *positive.*

(c) ${}^-9 \cdot 7 = {}^-63$ Factors have *different* signs, so the product is *negative.*

(d) $9 \cdot {}^-7 = {}^-63$ Factors have *different* signs, so the product is *negative.*

3. **(a)** $7({}^-8) = {}^-56$ Factors have *different* signs, so the product is *negative.*

(b) ${}^-7(8) = {}^-56$ Factors have *different* signs, so the product is *negative.*

(c) $7(8) = 56$ Factors have the *same* sign, so the product is *positive.*

(d) ${}^-7({}^-8) = 56$ Factors have the *same* sign, so the product is *positive.*

5. ${}^-5 \cdot 7 = {}^-35$ (*different* signs, product is *negative*)

7. $({}^-5)(9) = {}^-45$
(*different* signs, product is *negative*)

9. $3({}^-6) = {}^-18$ (*different* signs, product is *negative*)

11. $10({}^-5) = {}^-50$
(*different* signs, product is *negative*)

13. $({}^-1)(40) = {}^-40$
(*different* signs, product is *negative*)

15. ${}^-56 \cdot 1 = {}^-56$; multiplication property of 1

17. ${}^-8({}^-4) = 32$ (*same* signs, product is *positive*)

19. $11 \cdot 7 = 77$ (*same* signs, product is *positive*)

21. $25 \cdot 0 = 0$; multiplication property of 0

23. ${}^-19({}^-7) = 133$ (*same* signs, product is *positive*)

25. ${}^-13({}^-1) = 13$ (*same* signs, product is *positive*)

27. $(0)({}^-25) = 0$; multiplication property of 0

29. $^-4 \cdot {}^-6 \cdot 2$ *Multiply from left to right.*
$= 24 \cdot 2$
$= 48$

31. $(^-4)(^-2)(^-7)$ *Multiply from left to right.*
$= 8(^-7)$
$= {}^-56$

33. $5(^-8)(4)$ *Multiply from left to right.*
$= {}^-40(4)$
$= {}^-160$

35. $(^-3)(5) = {}^-15$ (negative product, different signs)

37. $^-3 \cdot 10 = {}^-30$ (negative product, different signs)

39. $^-17 = 17(^-1)$ (negative product, different signs)

41. $(0)(^-350) = 0$ (multiplication property of 0)

43. $5 \cdot {}^-4 \cdot 5 = {}^-100$ because
$^-20 \cdot 5 = {}^-100$ (negative product, different signs)

45. $(^-4)(^-5)(^-2) = {}^-40$ because
$(^-4)(10) = {}^-40$ (negative product, different signs)

47. Commutative property: changing the order of the factors does not change the product.
Associative property: changing the *grouping* of the factors does not change the product.
Examples will vary. Some possibilities are
$^-4 \cdot 7 = 7 \cdot {}^-4 = {}^-28$,
$(2 \cdot 5) \cdot {}^-8 = 2 \cdot (5 \cdot {}^-8) = {}^-80$.

49. Examples will vary. Some possibilities are:

(a) $6 \cdot {}^-1 = {}^-6$; $2 \cdot {}^-1 = {}^-2$; $15 \cdot {}^-1 = {}^-15$

(b) $^-6 \cdot {}^-1 = 6$; $^-2 \cdot {}^-1 = 2$; $^-15 \cdot {}^-1 = 15$

The result of multiplying any nonzero number times $^-1$ is the number with the opposite sign.

50.
$^-2 \cdot {}^-2 = 4$
$^-2 \cdot {}^-2 \cdot {}^-2 = {}^-8$
$^-2 \cdot {}^-2 \cdot {}^-2 \cdot {}^-2 = 16$
$^-2 \cdot {}^-2 \cdot {}^-2 \cdot {}^-2 \cdot {}^-2 = {}^-32$

The absolute value doubles each time and the sign changes. The next three products are
$^-2 \cdot {}^-32 = 64$, $^-2 \cdot 64 = {}^-128$, and
$^-2 \cdot {}^-128 = 256$.

51. $9(^-3 + 5)$ rewritten by using the distributive property is $9 \cdot {}^-3 + 9 \cdot 5$.

$9(^-3 + 5) = 9 \cdot {}^-3 + 9 \cdot 5$
$9(2) = {}^-27 + 45$
$18 = 18$

53. $25 \cdot 8$ rewritten by using the commutative property is $8 \cdot 25$.

$25 \cdot 8 = 8 \cdot 25$
$200 = 200$

55. $^-3 \cdot (2 \cdot 5)$ rewritten using the associative property is $(^-3 \cdot 2) \cdot 5$.

$^-3 \cdot (2 \cdot 5) = (^-3 \cdot 2) \cdot 5$
$^-3 \cdot 10 = {}^-6 \cdot 5$
$^-30 = {}^-30$

57. Income: $\$324 \approx \$300$
52 weeks ≈ 50

Estimate: $\$300 \cdot 50 = \$15{,}000$
Exact: $\$324 \cdot 52 = \$16{,}848$

59. Monthly loss: $^-\$9950 \approx {}^-\$10{,}000$
12 months ≈ 10

Estimate: $^-\$10{,}000 \cdot 10 = {}^-\$100{,}000$
Exact: $^-\$9950 \cdot 12 = {}^-\$119{,}400$

61. Tuition: \$182 per credit $\approx$ \$200 per credit
13 credits $\approx$ 10 credits

Estimate: $\$200 \cdot 10 = \$2000$
Exact: $\$182 \cdot 13 = \$2366$

63. Hours: $24 \approx 20$
365 days ≈ 400

Estimate: $20 \cdot 400 = 8000$ hours
Exact: $24 \cdot 365 = 8760$ hours

65. $^-8 \cdot |{}^-8 \cdot 8|$ Multiply the factors within the absolute value (different signs, negative product).
$= {}^-8 \cdot |{}^-64|$ $^-64$ is 64 units from 0, so $|{}^-64| = 64$.
$= {}^-8 \cdot 64$ Multiply.
$= {}^-512$ (different signs, negative product)

67. $(^-37)(^-1)(85)(0) = 0$;
multiplication property of 0

69. $|6 - 7| \cdot {}^-355{,}299$ Subtract within the absolute value first.
$= |6 + {}^-7| \cdot {}^-355{,}299$
$= |{}^-1| \cdot {}^-355{,}299$ $^-1$ is 1 unit from 0, so $|{}^-1| = 1$
$= 1 \cdot {}^-355{,}299$ multiplication property of 1
$= {}^-355{,}299$

71. The charge for each cat's shots will be $\$24 + \$29 = \$53$.
The total for all four cats will be four times that, plus the office visit charge.

$4 \cdot \$53 + \$35$ Do the multiplication first.
$= \$212 + \$35$
$= \$247$

73. The temperature drops 3 degrees ($^-3$ degrees) for every 1000 feet climbed into the air. An altitude of 24,000 feet would require 24 increases of 1000 feet each.

$^-3 \cdot 24 = {}^-72$ degrees

The temperature at 24,000 feet is $50 + {}^-72 = {}^-22$ degrees.

75. Points possible on tests:
$6(100) = 600$ points
Bonus points possible on tests:
$6(4) = 24$ points
Points possible on quizzes:
$8(6) = 48$ points
Points possible on homework assignments:
$20(5) = 100$ points
Total points possible:
$600 + 24 + 48 + 100 = 772$ points

1.7 Dividing Integers

1.7 Section Exercises

1. **(a)** $14 \div 2 = 7$ (*same* signs, quotient is *positive*)

(b) $^-14 \div {}^-2 = 7$
(*same* signs, quotient is *positive*)

(c) $14 \div {}^-2 = {}^-7$
(*different* signs, quotient is *negative*)

(d) $^-14 \div 2 = {}^-7$
(*different* signs, quotient is *negative*)

3. **(a)** $^-42 \div 6 = {}^-7$
(*different* signs, quotient is *negative*)

(b) $^-42 \div {}^-6 = 7$
(*same* signs, quotient is *positive*)

(c) $42 \div {}^-6 = {}^-7$
(*different* signs, quotient is *negative*)

(d) $42 \div 6 = 7$ (*same* signs, quotient is *positive*)

5. **(a)** $\frac{35}{35} = 1$; any nonzero number divided by itself is 1.

(b) $\frac{35}{1} = 35$; any number divided by 1 is the number.

(c) $\frac{^-13}{1} = {}^-13$; any number divided by 1 is the number.

(d) $\frac{^-13}{^-13} = 1$; any nonzero number divided by itself is 1.

7. **(a)** $\frac{0}{50} = 0$; zero divided by any nonzero number is 0.

(b) $\frac{50}{0}$ is undefined; division by zero is undefined.

(c) $\frac{^-11}{0}$ is undefined; division by zero is undefined.

(d) $\frac{0}{^-11} = 0$; zero divided by any nonzero number is 0.

9. $\frac{^-8}{2} = {}^-4$ (*different* signs, quotient is *negative*)

11. $\frac{21}{^-7} = {}^-3$ (*different* signs, quotient is *negative*)

13. $\frac{^-54}{^-9} = 6$ (*same* signs, quotient is *positive*)

15. $\frac{55}{^-5} = {}^-11$ (*different* signs, quotient is *negative*)

17. $\frac{^-28}{0}$ is undefined. Division by zero is undefined.

19. $\frac{14}{^-1} = {}^-14$ (*different* signs, quotient is *negative*)

21. $\frac{^-20}{^-2} = 10$ (*same* signs, quotient is *positive*)

23. $\frac{^-48}{^-12} = 4$ (*same* signs, quotient is *positive*)

25. $\frac{^-18}{18} = {}^-1$ (*different* signs, quotient is *negative*)

27. $\frac{0}{^-9} = 0$;
zero divided by any nonzero number is 0.

29. $\frac{^-573}{^-3} = 191$ (*same* signs, quotient is *positive*)

31. $\frac{163{,}672}{^-328} = {}^-499$
(*different* signs, quotient is *negative*)

33. $^-60 \div 10 \div {}^-3$ No parentheses, so start at the left.
$= {}^-6 \div {}^-3$
$= 2$

35. $^-64 \div {}^-8 \div {}^-2$ No parentheses, so start at the left.
$= 8 \div {}^-2$
$= {}^-4$

37. $100 \div {}^-5({}^-2)$ Start at the left.
$= {}^-20({}^-2)$
$= 40$

39. $48 \div 3 \cdot (12 \div {}^-4)$ Start inside the parentheses. Now work from the left.

$= 48 \div 3 \cdot {}^-3$
$= 16 \cdot {}^-3$
$= {}^-48$

41. ${}^-5 \div {}^-5({}^-10) \div {}^-2$ Start at the left.
$= 1({}^-10) \div {}^-2$
$= {}^-10 \div {}^-2$
$= 5$

43. $64 \cdot 0 \div {}^-8(10)$ Start at the left.
$= 0 \div {}^-8(10)$
$= 0(10)$
$= 0$

45. $2 \div 1 = 2$ but $1 \div 2 = 0.5$, so division is not commutative.

46. $\underbrace{(12 \div 6)}_{\underbrace{2 \div 2}_{1}} \div 2$ different quotients $12 \div \underbrace{(6 \div 2)}_{\underbrace{12 \div 3}_{4}}$

Division is not associative.

47. Similar: If the signs match, the result is positive. If the signs are different, the result is negative. Different: Multiplication is commutative, division is not. You can multiply by 0, but dividing by 0 is undefined.

48. Examples will vary. The properties are: Any nonzero number divided by itself is 1(example: $4 \div 4 = 1$). Any number divided by 1 is the number (example: ${}^-7 \div 1 = {}^-7$). Division by 0 is undefined (example: ${}^-5 \div 0$ is undefined). Zero divided by any other number (except 0) is 0 (example: $0 \div 9 = 0$).

49. Examples will vary.

(a) $\frac{{}^-6}{{}^-1} = 6; \frac{{}^-2}{{}^-1} = 2; \frac{{}^-15}{{}^-1} = 15$

(b) $\frac{6}{{}^-1} = {}^-6; \frac{2}{{}^-1} = {}^-2; \frac{15}{{}^-1} = {}^-15$

When dividing by ${}^-1$, change the sign of the dividend to its opposite to get the quotient.

50. Division is not commutative. $\frac{0}{{}^-3} = 0$ because $0 \cdot {}^-3 = 0$. But $\frac{{}^-3}{0}$ is undefined because when $\frac{{}^-3}{0} = ?$ is rewritten as $? \cdot 0 = {}^-3$, no number can replace ? and make a true statement.

51. Depth below sea level implies a negative, ${}^-35{,}836$ feet. Use division to find the size of each step.
${}^-35{,}836 \approx {}^-40{,}000$; $17 \approx 20$

Estimate: ${}^-40{,}000 \div 20 = {}^-2000$ feet
Exact: ${}^-35{,}836 \div 17 = {}^-2108$ feet

53. Overdrawn implies a negative, ${}^-\$238$. Transfer of money into the account, implies a positive, \$450.
${}^-238 \approx {}^-200$, $450 \approx 500$

Estimate: ${}^-200 + 500 = \$300$
Exact: ${}^-238 + 450 = \$212$

55. The number of non-foggy days equals the total number of days in a year minus the number of foggy days. $365 \approx 400$, $106 \approx 100$

Estimate: $400 - 100 = 300$ days
Exact: $365 - 106 = 259$ days

57. Descending implies a negative, ${}^-730$ feet each minute. ${}^-730 \approx {}^-700$
Because the plane took $37 \approx 40$ minutes to land, use multiplication to find how far the plane descended.

Estimate: ${}^-700 \cdot 40 = {}^-28{,}000$ feet
Exact: ${}^-730 \cdot 37 = {}^-27{,}010$ feet

59. Use division to find how many miles were covered in each hour. $315 \approx 300$, $5 \approx 5$

Estimate: $300 \div 5 = 60$ miles
Exact: $315 \div 5 = 63$ miles

61. To find the average, add all the scores and divide by the number of scores.
Sum of scores: $143 + 190 + 162 + 177 = 672$
Number of scores: 4 scores given
Average score $= \frac{672}{4} = 168$

63. Calculate the total weight from the data on the back, then use subtraction to find the difference between that and the figure on the front.

Total weight from back:
$13 \cdot 40$ grams $= 520$ grams

Difference from front: $520 - 510 = 10$ grams

The back claims 10 more grams than the front.

65. The \$302 already in Stephanie's account and her \$347 paycheck are positives. The money she paid for day care, \$116, and rent, \$548, are negatives.

$$\$302 + {}^-\$116 + {}^-\$548 + \$347 = {}^-\$15$$

Stephanie's balance is ${}^-\$15$. She is overdrawn by \$15.

67. Use division to convert minutes to hours.

$\dfrac{1000 \text{ minutes}}{60 \text{ minutes per hour}}$

$$\begin{array}{r} 16 \\ 60\overline{)1000} \\ \underline{60} \\ 400 \\ \underline{360} \\ 40 \end{array}$$

1000 minutes = 16 hours, with 40 minutes left over.

69. Use division to find the number of rooms.

$\dfrac{163 \text{ people}}{5 \text{ people per room}}$

$$\begin{array}{r} 32 \\ 5\overline{)163} \\ \underline{15} \\ 13 \\ \underline{10} \\ 3 \end{array}$$

32 rooms will be full, and that leaves 3 people. So 33 rooms are needed, with space for 2 people $(5 - 3 = 2)$ unused.

71. $|^-8| \div {^-4} \cdot |^-5| \cdot |1|$ Simplify the absolute values first.

$= 8 \div {^-4} \cdot 5 \cdot 1$ No parentheses, so start from the left.

$= {^-2} \cdot 5 \cdot 1$
$= {^-10} \cdot 1$
$= {^-10}$

73. $^-6(^-8) \div (^-5 - {^-5})$ Start inside the parentheses.

$= {^-6} \cdot {^-8} \div 0$ Division by zero is undefined.

75. Start by entering 1 000 000 000.
Divide by 60. ($\approx$ 16,666,667 minutes)
Divide by 60. ($\approx$ 277,778 hours)
Divide by 24. ($\approx$ 11,574 days)
Divide by 365. ($\approx$ 31.7 years)
31.70979198 rounds to 32 years to receive one billion dollars.

Summary Exercises on Operations with Integers

1. $2 - 8$ Add the opposite.
$= 2 + {^-8}$
$= {^-6}$

3. $^-14 - {^-7}$ Add the opposite.
$= {^-14} + 7$
$= {^-7}$

5. $^-9(^-7)$ (*same* signs, product is *positive*)
$= 63$

7. $(1)(^-56)$ Multiplication property of 1
$= {^-56}$

9. $5 - {^-7}$ Add the opposite.
$= 5 + 7$
$= 12$

11. $^-18 + 5 = {^-13}$

13. $^-40 - {^-40}$ Add the opposite.
$= {^-40} + {^+40}$ Addition of opposites is 0.
$= 0$

15. $8(^-6)$ (*different* signs, product is *negative*)
$= {^-48}$

17. $^-5(10)$ (*different* signs, product is *negative*)
$= {^-50}$

19. $0 - 14$ Add the opposite.
$= 0 + {^-14}$ Addition property of 0.
$= {^-14}$

21. $^-13 + 13$ Addition of opposites is 0.
$= 0$

23. $20 - 50$ Add the opposite.
$= 20 + {^-50}$
$= {^-30}$

25. $(^-4)(^-6)(2)$ Multiply from left to right.
$= (24)(2)$
$= 48$

27. $^-60 \div 10 \div {^-3}$ Divide from left to right.
$= {^-6} \div {^-3}$
$= 2$

29. $64(0) \div {^-8}$ Multiply.
$= 0 \div {^-8}$ Divide.
$= 0$

31. $^-9 + 8 + {^-2}$ Add from left to right.
$= {^-1} + {^-2}$
$= {^-3}$

33. $8 + 6 + {^-8}$ Add from left to right.
$= 14 + {^-8}$
$= 6$

35. $^-25 \div {^-1} \div {^-5}$ Divide from left to right.
$= 25 \div {^-5}$
$= {^-5}$

37. $^-72 \div {^-9} \div {^-4}$ Divide from left to right.
$= 8 \div {^-4}$
$= {^-2}$

39. $9 - 6 - 3 - 5$ Add the opposite.
$= 9 + {}^-6 + {}^-3 + {}^-5$ Add from left to right.
$= 3 + {}^-3 + {}^-5$
$= 0 + {}^-5$
$= {}^-5$

41. ${}^-1(9732)({}^-1)({}^-1)$ Multiply from left to right.
$= ({}^-9732)({}^-1)({}^-1)$
$= (9732)({}^-1)$
$= {}^-9732$

43. ${}^-10 - 4 + 0 + 18$ Add the opposite.
$= {}^-10 + {}^-4 + 0 + 18$ Add from left to right.
$= {}^-14 + 0 + 18$
$= {}^-14 + 18$
$= 4$

45. $5 - |{}^-3| + 3$ Absolute value first
$= 5 - 3 + 3$ Add the opposite.
$= 5 + {}^-3 + 3$ Add from left to right.
$= 2 + 3$
$= 5$

47. ${}^-3 - ({}^-2 + 4) - 5$ Parentheses first
$= {}^-3 - 2 - 5$ Add the opposites.
$= {}^-3 + {}^-2 + {}^-5$ Add from left to right.
$= {}^-5 + {}^-5$
$= {}^-10$

49. **(a)** If zero is divided by a nonzero number, the quotient is 0.

(b) If any number is multiplied by 0, the product is 0.

(c) If a nonzero number is divided by itself, the quotient is 1.

1.8 Exponents and Order of Operations

1.8 Section Exercises

1. Exponential Form: 4^3
Factored Form: $4 \cdot 4 \cdot 4$
Simplified: 64
Read as: 4 cubed or 4 to the third power

3. Exponential Form: 2^7
Factored Form: $2 \cdot 2 \cdot 2 \cdot 2 \cdot 2 \cdot 2 \cdot 2$
Simplified: 128
Read as: 2 to the seventh power

5. Exponential Form: 5^4
Factored Form: $5 \cdot 5 \cdot 5 \cdot 5$
Simplified: 625
Read as: 5 to the fourth power

7. Exponential Form: 7^2
Factored Form: $7 \cdot 7$
Simplified: 49
Read as: 7 squared

9. Exponential Form: 10^1
Factored Form: 10
Simplified: 10
Read as: 10 to the first power

11. **(a)** $10^1 = 10$

(b) $10^2 = 10 \cdot 10 = 100$

(c) $10^3 = 10 \cdot 10 \cdot 10$
$= 100 \cdot 10$
$= 1000$

(d) $10^4 = 10 \cdot 10 \cdot 10 \cdot 10$
$= 100 \cdot 10 \cdot 10$
$= 1000 \cdot 10$
$= 10{,}000$

13. **(a)** $4^1 = 4$

(b) $4^2 = 4 \cdot 4 = 16$

(c) $4^3 = 4 \cdot 4 \cdot 4$
$= 16 \cdot 4$
$= 64$

(d) $4^4 = 4 \cdot 4 \cdot 4 \cdot 4$
$= 16 \cdot 4 \cdot 4$
$= 64 \cdot 4$
$= 256$

15. 5^{10} on the calculator is 9,765,625.

17. 2^{12} on the calculator is 4096.

19. $({}^-2)^2 = ({}^-2)({}^-2)$
$= 4$

21. $({}^-5)^2 = ({}^-5)({}^-5)$
$= 25$

23. $({}^-4)^3 = ({}^-4)({}^-4)({}^-4)$
$= 16({}^-4)$
$= {}^-64$

25. $({}^-3)^4 = ({}^-3)({}^-3)({}^-3)({}^-3)$
$= 9({}^-3)({}^-3)$
$= {}^-27({}^-3)$
$= 81$

27. $({}^-10)^3 = ({}^-10)({}^-10)({}^-10)$
$= 100({}^-10)$
$= {}^-1000$

29. $1^4 = 1$; 1 times itself any number of times equals 1.

31. $3^3 \cdot 2^2$ Apply the exponents first.
$3^3 = 3 \cdot 3 \cdot 3 = 27$
$2^2 = 2 \cdot 2 = 4$
$= 27 \cdot 4$
$= 108$

33. $2^3(^-5)^2$ Apply the exponents first.
$2^3 = 2 \cdot 2 \cdot 2 = 8$
$(^-5)^2 = {^-5} \cdot {^-5} = 25$
$= 8(25)$
$= 200$

35. $6^1(^-5)^3$ Apply the exponents first.
$6^1 = 6$
$(^-5)^3 = (^-5)(^-5)(^-5) = {^-125}$
$= 6(^-125)$
$= {^-750}$

37. $(^-2)(^-2)^4$ Apply the exponent first.
$(^-2)^4 = (^-2)(^-2)(^-2)(^-2) = 16$
$= {^-2}(16)$
$= {^-32}$

39. $(^-2)^2 = \underline{4}$ $(^-2)^6 = \underline{64}$
$(^-2)^3 = \underline{^-8}$ $(^-2)^7 = \underline{^-128}$
$(^-2)^4 = \underline{16}$ $(^-2)^8 = \underline{256}$
$(^-2)^5 = \underline{^-32}$ $(^-2)^9 = \underline{^-512}$

(a) When a negative number is raised to an even power, the answer is positive; when raised to an odd power, the answer is negative.

(b) 15 is odd, so the sign of $(^-2)^{15}$ is negative. 24 is even, so the sign of $(^-2)^{24}$ is positive.

41. $12 \div 6(^-3)$ Divide.
$= 2(^-3)$ Multiply.
$= {^-6}$

43. ${^-1} + 15 - 7 - 7$ Add left to right.
$= 14 - 7 - 7$ Add the opposite.
$= 14 + {^-7} + {^-7}$ Add left to right.
$= 7 + {^-7}$ Add.
$= 0$

45. $10 - 7^2$ Exponent: $7^2 = 7 \cdot 7$
$= 10 - 49$ Add the opposite.
$= 10 + {^-49}$ Add.
$= {^-39}$

47. $2 - {^-5} + 3^2$ Exponent: $3^2 = 3 \cdot 3$
$= 2 - {^-5} + 9$ Add the opposite.
$= 2 + 5 + 9$ Add from left to right.
$= 7 + 9$ Add.
$= 16$

49. $3 + 5(6 - 2)$ Add the opposite.
$= 3 + 5(6 + {^-2})$ Parentheses
$= 3 + 5(4)$ Multiply.
$= 3 + 20$ Add.
$= 23$

51. ${^-7} + 6(8 - 14)$ Add the opposite.
$= {^-7} + 6(8 + {^-14})$ Parentheses
$= {^-7} + 6(^-6)$ Multiply.
$= {^-7} + {^-36}$ Add.
$= {^-43}$

53. $2(^-3 + 5) - (9 - 12)$ Add the opposite.
$= 2(^-3 + 5) - (9 + {^-12})$ Parentheses
$= 2(2) - (^-3)$ Multiply.
$= 4 - (^-3)$ Add the opposite.
$= 4 + {^+3}$ Add.
$= 7$

55. ${^-5}(7 - 13) \div {^-10}$ Add the opposite.
$= {^-5}(7 + {^-13}) \div {^-10}$ Parentheses
$= {^-5}(^-6) \div {^-10}$ Multiply.
$= 30 \div {^-10}$ Divide.
$= {^-3}$

57. $9 \div (^-3)^2 + {^-1}$ Exponent first
$= 9 \div 9 + {^-1}$ Divide.
$= 1 + {^-1}$ Add.
$= 0$

59. $2 - {^-5}(^-2)^3$ Exponent first
$= 2 - {^-5}(^-8)$ Multiply.
$= 2 - 40$ Add the opposite.
$= 2 + {^-40}$ Add.
$= {^-38}$

61. ${^-2}(^-7) + 3(9)$ Multiply.
$= 14 + 3(9)$ Multiply.
$= 14 + 27$ Add.
$= 41$

63. $30 \div {^-5} - 36 \div {^-9}$ Divide.
$= {^-6} - 36 \div {^-9}$ Divide.
$= {^-6} - {^-4}$ Add the opposite.
$= {^-6} + {^+4}$ Add.
$= {^-2}$

65. $2(5) - 3(4) + 5(3)$ Multiply.
$= 10 - 3(4) + 5(3)$ Multiply.
$= 10 - 12 + 5(3)$ Multiply.
$= 10 - 12 + 15$ Add the opposite.
$= 10 + {^-12} + 15$ Add.
$= {^-2} + 15$ Add.
$= 13$

67. $4(3^2) + 7(3 + 9) - {^-6}$ Parentheses first
$= 4(3^2) + 7(12) - {^-6}$ Exponent next
$= 4(9) + 7(12) - {^-6}$ Multiply.
$= 36 + 7(12) - {^-6}$ Multiply.
$= 36 + 84 - {^-6}$ Add the opposite.
$= 36 + 84 + 6$ Add.
$= 120 + 6$ Add.
$= 126$

69. $(^-4)^2 \cdot (7-9)^2 \div 2^3$ Parentheses first
$= (^-4)^2 \cdot (^-2)^2 \div 2^3$ Exponents next
$= 16 \cdot 4 \div 8$ Multiply.
$= 64 \div 8$ Divide.
$= 8$

71. $\dfrac{^-1+5^2-{}^-3}{^-6-9+12}$

Numerator:
$^-1+5^2-{}^-3$ Exponents first
$= {}^-1+25-{}^-3$ Add the opposite.
$= {}^-1+25+3$ Add left to right.
$= 24+3$
$= 27$

Denominator:
$^-6-9+12$ Add the opposite.
$= {}^-6+{}^-9+12$ Add left to right.
$= {}^-15+12$
$= {}^-3$

Last step is division: $\dfrac{27}{^-3} = {}^-9$

73. $\dfrac{^-2(4^2)-4(6-2)}{^-4(8-13)\div{}^-5}$

Numerator:
$^-2(4^2)-4(6-2)$ Parentheses first
$= {}^-2(4^2)-4(4)$ Exponent
$= {}^-2(16)-4(4)$ Multiply from left to right.
$= {}^-32-16$ Add the opposite.
$= {}^-32+{}^-16$ Add.
$= {}^-48$

Denominator:
$^-4(8-13)\div{}^-5$ Parentheses first
$= {}^-4(^-5)\div{}^-5$ Multiply.
$= 20\div{}^-5$ Divide.
$= {}^-4$

Last step is division: $\dfrac{^-48}{^-4} = 12$

75. $\dfrac{2^3\cdot(^-2-5)+4(^-1)}{4+5(^-6\cdot 2)+(5\cdot 11)}$

Numerator:
$2^3\cdot(^-2-5)+4(^-1)$ Parentheses: $^-2-5={}^-2+{}^-5={}^-7$
$= 2^3\cdot(^-7)+4(^-1)$ Exponent
$= 8\cdot(^-7)+4(^-1)$ Multiply.
$= {}^-56+4(^-1)$ Multiply.
$= {}^-56+{}^-4$ Add.
$= {}^-60$

Denominator:
$4+5(^-6\cdot 2)+(5\cdot 11)$ Parentheses
$= 4+5(^-12)+(5\cdot 11)$ Parentheses
$= 4+5(^-12)+55$ Multiply.
$= 4+{}^-60+55$ Add left to right.
$= {}^-56+55$
$= {}^-1$

Last step is division: $\dfrac{^-60}{^-1} = 60$

77. $5^2(9-11)(^-3)(^-3)^3$ Parentheses first
$= 5^2(^-2)(^-3)(^-3)^3$ Exponents
$= 25(^-2)(^-3)(^-27)$ Multiply left to right.
$= {}^-50(^-3)(^-27)$
$= 150(^-27)$
$= {}^-4050$

79. $|^-12|\div 4+2\cdot|(^-2)^3|\div 4$ Work inside the absolute value signs first.
$= |^-12|\div 4+2\cdot|^-8|\div 4$ Absolute values
$= 12\div 4+2\cdot 8\div 4$ Divide.
$= 3+2\cdot 8\div 4$ Multiply.
$= 3+16\div 4$ Divide.
$= 3+4$ Add.
$= 7$

81. $\dfrac{^-9+18\div{}^-3(^-6)}{32-4(12)\div 3(2)}$

Numerator:
$^-9+18\div{}^-3(^-6)$ Divide.
$= {}^-9+{}^-6(^-6)$ Multiply.
$= {}^-9+36$ Add.
$= 27$

Denominator:
$32-4(12)\div 3(2)$ Multiply.
$= 32-48\div 3(2)$ Divide.
$= 32-16(2)$ Multiply.
$= 32-32$ Add the opposite.
$= 32+{}^-32$ Add.
$= 0$

Last step is division: $\dfrac{27}{0}$ is undefined.

Chapter 1 Review Exercises

1. The whole numbers are: 86; 0; 35,600

2. 806 in words: eight hundred six

3. 319,012 in words: three hundred nineteen thousand, twelve

4. 60,003,200 in words: sixty million, three thousand, two hundred

5. 15,749,000,000,006 in words: fifteen trillion, seven hundred forty-nine billion, six

6. Five hundred four thousand, one hundred
The first group name is *thousand,* so you need to fill *two groups* of three digits.

$\underline{5}\,\underline{0}\,\underline{4}, \underline{1}\,\underline{0}\,\underline{0} = 504{,}100$

7. Six hundred twenty million, eighty thousand
The first group name is *million,* so you need to fill *three groups* of three digits.

$\underline{6}\,\underline{2}\,\underline{0}, \underline{0}\,\underline{8}\,\underline{0}, \underline{0}\,\underline{0}\,\underline{0} = 620{,}080{,}000$

8. Ninety-nine billion, seven million, three hundred fifty-six
The first group name is *billion,* so you need to fill *four groups* of three digits.

$\underline{0}\,\underline{9}\,\underline{9}, \underline{0}\,\underline{0}\,\underline{7}, \underline{0}\,\underline{0}\,\underline{0}, \underline{3}\,\underline{5}\,\underline{6} = 99{,}007{,}000{,}356$

9. Graph $^-3\frac{1}{2}, 2, ^-5, 0$

$^-5\ ^-4\ ^-3\ ^-2\ ^-1\ 0\ 1\ 2\ 3\ 4\ 5$

10. 0 is to the *right* of $^-4$ on the number line, so 0 is *greater than* $^-4$. Write $0 > {}^-4$.

11. $^-3$ is to the *left* of $^-1$ on the number line, so $^-3$ is *less than* $^-1$. Write $^-3 < {}^-1$.

12. 2 is to the *right* of $^-2$ on the number line, so 2 is *greater than* $^-2$. Write $2 > {}^-2$.

13. $^-2$ is to the *left* of 1 on the number line, so $^-2$ is *less than* 1. Write $^-2 < 1$.

14. $|{}^-5| = 5$ because the distance from 0 to $^-5$ on the number line is 5 spaces.

15. $|9| = 9$ because the distance from 0 to 9 on the number line is 9 spaces.

16. $|0| = 0$ because the distance from 0 to 0 on the number line is 0 spaces.

17. $|{}^-125| = 125$ because the distance from 0 to $^-125$ on the number line is 125 spaces.

18. $^-9 + 8 = {}^-1$

Add *unlike* signed integers.

$|{}^-9| = 9$; $|8| = 8$; Subtract $9 - 8$ to get 1.

$^-9$ has the larger absolute value and is negative, so the sum is negative.

19. $^-8 + {}^-5 = {}^-13$

Add *like* signed integers.

$|{}^-8| = 8$; $|{}^-5| = 5$; Add $8 + 5$ to get 13.

Both numbers are negative, so the sum is negative.

20. $16 + {}^-19 = {}^-3$

Add *unlike* signed integers.

$|16| = 16$; $|{}^-19| = 19$; Subtract $19 - 16$ to get 3.

$^-19$ has the larger absolute value and is negative, so the sum is negative.

21. $^-4 + 4 = 0$
Addition of opposites is always zero.

22. $6 + {}^-5 = {}^+1$ or 1

23. $^-12 + {}^-12 = {}^-24$

24. $0 + {}^-7 = {}^-7$

25. $^-16 + 19 = {}^+3$ or 3

26. $9 + {}^-4 + {}^-8 + 3$ *Add from left to right.*
$= 5 + {}^-8 + 3$
$= {}^-3 + 3$
$= 0$

27. $^-11 + {}^-7 + 5 + {}^-4$ *Add from left to right.*
$= {}^-18 + 5 + {}^-4$
$= {}^-13 + {}^-4$
$= {}^-17$

28. The opposite of $^-5$ is 5. $^-5 + 5 = 0$

29. The opposite of 18 is $^-18$. $18 + {}^-18 = 0$

30. $5 - 12$ *Change subtraction to addition. Change 12 to $^-12$.*
$= 5 + {}^-12$
$= {}^-7$

31. $24 - 7$ *Add the opposite.*
$= 24 + {}^-7$
$= 17$

32. $^-12 - 4$ *Add the opposite.*
$= {}^-12 + {}^-4$
$= {}^-16$

33. $4 - {}^-9$ *Add the opposite.*
$= 4 + {}^+9$
$= 13$

34. $^-12 - {}^-30$ *Add the opposite.*
$= {}^-12 + {}^+30$
$= 18$

35. $^-8 - 14$ *Add the opposite.*
$= {}^-8 + {}^-14$
$= {}^-22$

36. $^-6 - {}^-6$ *Add the opposite.*
$= {}^-6 + {}^+6$
$= 0$

37. $^{-}10 - 10$ *Add the opposite.*
$= {}^{-}10 + {}^{-}10$
$= {}^{-}20$

38. $^{-}8 - {}^{-}7$ *Add the opposite.*
$= {}^{-}8 + {}^{+}7$
$= {}^{-}1$

39. $0 - 3$ *Add the opposite.*
$= 0 + {}^{-}3$
$= {}^{-}3$

40. $1 - {}^{-}13$ *Add the opposite.*
$= 1 + {}^{+}13$
$= 14$

41. $15 - 0$ *Add the opposite.*
$= 15 + 0$
$= 15$

42. $3 - 12 - 7$ *Add the opposites.*
$= 3 + {}^{-}12 + {}^{-}7$ *Add left to right.*
$= {}^{-}9 + {}^{-}7$
$= {}^{-}16$

43. $^{-}7 - {}^{-}3 + 7$ *Add the opposite.*
$= {}^{-}7 + {}^{+}3 + 7$
$= {}^{-}7 + 7 + {}^{+}3$ *Commutative property*
$= 0 + {}^{+}3$
$= 3$

44. $4 + {}^{-}2 - 0 - 10$ *Add the opposites.*
$= 4 + {}^{-}2 + 0 + {}^{-}10$ *Add left to right.*
$= 2 + 0 + {}^{-}10$
$= 2 + {}^{-}10$
$= {}^{-}8$

45. $^{-}12 - 12 + 20 - {}^{-}4$ *Add the opposites.*
$= {}^{-}12 + {}^{-}12 + 20 + {}^{+}4$ *Add left to right.*
$= {}^{-}24 + 20 + 4$
$= {}^{-}4 + 4$
$= 0$

46. $2\underline{0}5 \approx 210$

Underline the tens place. The next digit is 5 or more. Add 1 to 0. Change 5 to 0.

47. $5\underline{9},499 \approx 59,000$

Underline the thousands place. The next digit is 4 or less. Leave 9 as 9. Change all digits to the right of the underlined place to zeros.

48. $8\underline{5},066,000 \approx 85,000,000$

Underline the millions place. The next digit is 4 or less. Leave 5 as 5. Change all digits to the right of the underlined place to zeros.

49. $^{-}2\underline{9}63 \approx {}^{-}3000$

Underline the hundreds place. The next digit is 5 or more. Add 1 to 9. Write 0 and regroup the one to the thousands place. Change all digits to the right of the underlined place to zeros.

50. $^{-}7,0\underline{6}3,885 \approx {}^{-}7,060,000$

Underline the ten-thousands place. The next digit is 4 or less. Leave 6 as 6. Change all digits to the right of the underlined place to zeros.

51. $39\underline{9},712 \approx 400,000$

Underline the thousands place. The next digit is 5 or more. Add 1 to 9. Write 0 and regroup the one to the ten-thousands place: $9 + 1 = 10$. Write 0 and regroup the one to the hundred-thousands place: $3 + 1 = 4$. Change all digits to the right of the underlined place to zeros.

52. Weight loss implies a negative number:

$^{-}\underline{1}97$ pounds $\approx {}^{-}200$ pounds

Underline the first digit. The next digit is 5 or more. Add 1 to 1. Change all digits to the right of the underlined place to zeros.

53. Below sea level implies a negative number:

$^{-}\underline{1}312 \approx {}^{-}1000$ feet

Underline the first digit. The next digit is 4 or less. Leave 1 as 1. Change all digits to the right of the underlined place to zeros.

54. $\underline{3}62,000,000$ directories $\approx 400,000,000$ directories

Underline the first digit. The next digit is 5 or more. Add 1 to 3. Change all digits to the right of the underlined place to zeros.

55. $\underline{9},104,000,000$ people $\approx 9,000,000,000$ people

Underline the first digit. The next digit is 4 or less. Leave 9 as 9. Change all digits to the right of the underlined place to zeros.

56. $^{-}6(9) = {}^{-}54$ (*different* signs, product is *negative*)

57. $({}^{-}7)({}^{-}8) = 56$ (*same* signs, product is *positive*)

58. $10({}^{-}10) = {}^{-}100$
(*different* signs, product is *negative*)

59. $^{-}45 \cdot 0 = 0$; multiplication property of 0

60. $^{-}1({}^{-}24) = 24$ (*same* signs, product is *positive*)

61. $17 \cdot 1 = 17$; multiplication property of 1

62. $4({}^{-}12) = {}^{-}48$
(*different* signs, product is *negative*)

63. $({}^{-}5)({}^{-}25) = 125$ (*same* signs, product is *positive*)

64. $^-3(^-4)(^-3)$ *Multiply from left to right.*
$= 12 \cdot {^-3}$
$= {^-36}$

65. $^-5(2)(^-5)$ *Multiply from left to right.*
$= {^-10} \cdot {^-5}$
$= 50$

66. $(^-8)(^-1)(^-9)$ *Multiply from left to right.*
$= 8(^-9)$
$= {^-72}$

67. $\frac{^-63}{^-7} = 9$ (*same* signs, quotient is *positive*)

68. $\frac{70}{^-10} = {^-7}$ (*different* signs, quotient is *negative*)

69. $\frac{^-15}{0}$ is undefined. Division by zero is undefined.

70. $^-100 \div {^-20} = 5$ (*same* signs, quotient is *positive*)

71. $18 \div {^-1} = {^-18}$
(*different* signs, quotient is *negative*)

72. $\frac{0}{12} = 0$; 0 divided by any nonzero number is 0.

73. $\frac{^-30}{^-2} = 15$ (*same* signs, quotient is *positive*)

74. $\frac{^-35}{35} = {^-1}$ (*different* signs, quotient is *negative*)

75. $^-40 \div {^-4} \div {^-2}$ *Divide from left to right.*
$= 10 \div {^-2}$
$= {^-5}$

76. $^-18 \div 3(^-3)$ *Divide.*
$= {^-6}(^-3)$ *Multiply.*
$= 18$

77. $0 \div {^-10}(5) \div 5$ *Divide.*
$= 0(5) \div 5$ *Multiply.*
$= 0 \div 5$ *Divide.*
$= 0$

78. 1250 hours are separated into 8-hour days.

$$\begin{array}{r} 156 \\ 8\overline{)1250} \\ \underline{8} \\ 45 \\ \underline{40} \\ 50 \\ \underline{48} \\ 2 \end{array}$$

It took 156 work days of 8 hours each, plus 2 extra hours.

79. $10^4 = 10 \cdot 10 \cdot 10 \cdot 10 = 10{,}000$

80. $2^5 = 2 \cdot 2 \cdot 2 \cdot 2 \cdot 2 = 32$

81. $3^3 = 3 \cdot 3 \cdot 3 = 27$

82. $(^-4)^2 = (^-4)(^-4) = 16$

83. $(^-5)^3 = (^-5)(^-5)(^-5)$
$= (25)(^-5)$
$= {^-125}$

84. $8^1 = 8$

85. $6^2 \cdot 3^2$ *Exponents*
$= 36 \cdot 9$
$= 324$

86. $5^2(^-2)^3$ *Exponents*
$= 25(^-8)$
$= {^-200}$

87. $^-30 \div 6 - 4(5)$ *Divide.*
$= {^-5} - 4(5)$ *Multiply.*
$= {^-5} - 20$ *Add the opposite.*
$= {^-5} + {^-20}$ *Add.*
$= {^-25}$

88. $6 + 8(2 - 3)$ *Parentheses first* $2 - 3 = 2 + {^-3}$
$= 6 + 8(^-1)$ *Multiply.*
$= 6 + {^-8}$ *Add.*
$= {^-2}$

89. $16 \div 4^2 + (^-6 + 9)^2$ *Parentheses first*
$= 16 \div 4^2 + (3)^2$ *Exponents*
$= 16 \div 16 + 9$ *Divide.*
$= 1 + 9$ *Add.*
$= 10$

90. $^-3(4) - 2(5) + 3(^-2)$ *Multiply left to right.*
$= {^-12} - 10 + {^-6}$ *Add the opposite.*
$= {^-12} + {^-10} + {^-6}$ *Add left to right.*
$= {^-22} + {^-6}$
$= {^-28}$

91. $\frac{^-10 + 3^2 - {^-9}}{3 - 10 - 1}$

Numerator:
$^-10 + 3^2 - {^-9}$ *Exponent*
$= {^-10} + 9 - {^-9}$ *Add the opposite.*
$= {^-10} + 9 + {^+9}$ *Add from left to right.*
$= {^-1} + {^+9}$
$= 8$

Denominator:
$3 - 10 - 1$ *Add the opposites.*
$= 3 + {^-10} + {^-1}$ *Add from left to right.*
$= {^-7} + {^-1}$
$= {^-8}$

Last step is division: $\frac{8}{^-8} = {^-1}$

92. $\dfrac{^-1(1-3)^3+12\div 4}{^-5+24\div 8\cdot 2(6-6)+5}$

Numerator:

$^-1(1-3)^3+12\div 4$ *Parentheses first* $1-3=1+{^-3}={^-2}$

$={^-1}({^-2})^3+12\div 4$ *Exponent*

$={^-1}({^-8})+12\div 4$ *Multiply.*

$=8+12\div 4$ *Divide.*

$=8+3$ *Add.*

$=11$

Denominator:

$^-5+24\div 8\cdot 2(6-6)+5$ *Parentheses first*

$={^-5}+24\div 8\cdot 2(0)+5$ *Divide.*

$={^-5}+3\cdot 2(0)+5$ *Multiply.*

$={^-5}+6(0)+5$ *Multiply.*

$={^-5}+0+5$ *Add.*

$={^-5}+5$ *Add.*

$=0$

Last step is division: $\dfrac{11}{0}$ is undefined.

93. **[1.3]** Associative property of addition

94. **[1.6]** Commutative property of multiplication

95. **[1.3]** Addition property of 0

96. **[1.6]** Multiplication property of 0

97. **[1.6]** Distributive property

98. **[1.6]** Associative property of multiplication

99. **[1.6]** 192 rounds to 200.
$11,900 rounds to $10,000.

Estimate: $\$10{,}000\cdot 200=\$2{,}000{,}000$ total value
Exact: $\$11{,}900\cdot 192=\$2{,}284{,}800$ total value

100. **[1.3]** Account balance of $185 rounds to $200.
The deposit of $428 rounds to $400.
The check for $706 rounds to $700.

Estimate: $\$200+\$400+{^-\$700}={^-\$100}$
Exact: $\$185+\$428+{^-\$706}={^-\$93}$

101. **[1.7]** 24 gallons rounds to 20 gallons.
840 miles rounds to 800 miles.
Divide miles by gallons to get miles per gallon.

Estimate: $800\div 20=40$ miles for each gallon
Exact: $840\div 24=35$ miles for each gallon

102. **[1.6]** 19 calculators rounds to 20.
12 modems rounds to 10.
$39 rounds to $40. $85 rounds to $90.

Estimate:
$(\$40\cdot 20)+(\$90\cdot 10)=\$800+\$900=\$1700$
Exact:
$(\$39\cdot 19)+(\$85\cdot 12)=\$741+\$1020=\$1761$

103. **[1.3]** Expenses imply negative numbers.

Jan. $\$2400+{^-\$3100}={^-\$700}$ (loss)
Feb. $\$1900+{^-\$2000}={^-\$100}$ (loss)
Mar. $\$2500+{^-\$1800}=\$700$ (profit)
Apr. $\$2300+{^-\$1400}=\$900$ (profit)
May $\$1600+{^-\$1600}=\$0$ (neither)
June $\$1900+{^-\$1200}=\$700$ (profit)

104. **[1.3]** January had the greatest loss.
April had the greatest profit.

105. **[1.7]** To find her average monthly income, add her income from each month and divide by the number of months.

$$\frac{\$2400+\$1900+\$2500+\$2300+\$1600+\$1900}{6}$$
$$=\frac{\$12{,}600}{6}=\$2100$$

106. **[1.7]** To find her average monthly expenses, add her expenses from each month and divide by the number of months.

$$\frac{\$3100+\$2000+\$1800+\$1400+\$1600+\$1200}{6}$$
$$=\frac{\$11{,}100}{6}=\$1850$$

(Since this is an expense, we could write it as $^-\$1850$.)

Chapter 1 Test

1. 20,008,307 in words: twenty million, eight thousand, three hundred seven

2. Thirty billion, seven hundred thousand, five

The first group name is *billion*, so you need to fill *four groups* of three digits.

$\underline{030},\underline{000},\underline{700},\underline{005}=30{,}000{,}700{,}005$

3. Graph $3, {^-2}, 0, {^-\frac{1}{2}}$

$^-3$ $^-2$ $^-1$ 0 1 2 3

4. 0 is to the *right* of $^-3$ on the number line, so 0 is *greater than* $^-3$. Write $0>{^-3}$.

$^-2$ is to the *left* of $^-1$ on the number line, so $^-2$ is *less than* $^-1$. Write $^-2<{^-1}$.

5. $|10|=10$ because the distance from 0 to 10 on the number line is 10 spaces.

$|{^-14}|=14$ because the distance from 0 to $^-14$ on the number line is 14 spaces.

6. $3-9$ *Add the opposite.*
$=3+{^-9}$
$={^-6}$

7. $^-12+7={^-5}$

8. $\dfrac{^-28}{^-4} = 7$ (*same* signs, quotient is *positive*)

9. $^-1(40) = {^-40}$
(*different* signs, product is *negative*)

10. $^-5 - {^-15}$ *Change subtraction to addition.*
$= {^-5} + {^+15}$ *Add.*
$= 10$

11. $(^-8)(^-8) = 64$ (*same* signs, product is *positive*)

12. $^-25 + {^-25}$ *Add.*
$= {^-50}$

13. $\frac{17}{0}$ is undefined.

14. $^-30 - 30$ *Change subtraction to addition.*
$= {^-30} + {^-30}$ *Add.*
$= {^-60}$

15. $\dfrac{50}{^-10} = {^-5}$ (*different* signs, quotient is *negative*)

16. $5(^-9) = {^-45}$ (*different* signs, product is *negative*)

17. $0 - {^-6}$ *Change subtraction to addition.*
$= 0 + {^+6}$ *Addition property of zero*
$= 6$

18. $^-35 \div 7(-5)$ *Divide.*
$= {^-5}(^-5)$ *Multiply.*
$= 25$

19. $^-15 - {^-8} + 7$ *Change subtraction to addition.*
$= {^-15} + {^+8} + 7$ *Add from left to right.*
$= {^-7} + 7$
$= 0$

20. $3 - 7(^-2) - 8$ *Multiply.*
$= 3 - {^-14} - 8$ *Change subtraction to addition.*
$= 3 + {^+14} + {^-8}$ *Add from left to right.*
$= 17 + {^-8}$
$= 9$

21. $(^-4)^2 \cdot 2^3$ *Exponents*
$= 16 \cdot 8$
$= 128$

22. $\dfrac{5^2 - 3^2}{(4)(^-2)}$

Numerator:
$5^2 - 3^2$ *Exponents*
$= 25 - 9$ *Add the opposite.*
$= 25 + {^-9}$ *Add.*
$= 16$

Denominator:
$(4)(^-2)$ *Multiply.*
$= {^-8}$

Last step is division: $\dfrac{16}{^-8} = {^-2}$

23. $^-2(^-4 + 10) + 5(4)$ *Parentheses*
$= {^-2}(6) + 5(4)$ *Multiply.*
$= {^-12} + 5(4)$ *Multiply.*
$= {^-12} + 20$ *Add.*
$= 8$

24. $^-3 + (^-7 - {^-10}) + 4(6 - 10)$ *Parentheses*
$= {^-3} + (^-7 + {^+10}) + 4(6 + {^-10})$
$= {^-3} + 3 + 4(^-4)$ *Multiply.*
$= {^-3} + 3 + {^-16}$ *Add.*
$= 0 + {^-16}$ *Add.*
$= {^-16}$

25. An exponent shows how many times to use a factor in repeated multiplication. Examples will vary. Some possibilities are
$(2)^4 = 2 \cdot 2 \cdot 2 \cdot 2 = 16$ and
$(^-3)^2 = (^-3)(^-3) = 9$.

26. Commutative property: changing the order of addends does not change the sum.

One possible example: $2 + 5 = 5 + 2$

Associative property: changing the *grouping* of addends does not change the sum.

One possible example:
$(^-1 + 4) + 2 = {^-1} + (4 + 2)$

27. $\underline{8}51 \approx 900$

Underline the hundreds place. The next digit is 5 or more. Add 1 to 8. Change all digits to the right of the underlined place to zeros.

28. $36{,}42\underline{0}{,}498{,}725 \approx 36{,}420{,}000{,}000$

Underline the millions place. The next digit is 4 or less. Leave 0 as 0. Change all digits to the right of the underlined place to zeros.

29. $34\underline{9},812 \approx 350,000$

Underline the thousands place. The next digit is 5 or more. Add 1 to 9. Write 0 and regroup 1 to the ten-thousands place. Change all digits to the right of the underlined place to zeros.

30. \$184 account balance rounds to \$200.
The \$293 deposit rounds to \$300.
The \$506 tuition check rounds to \$500.

Balance in her account:
Estimate: $\$200 + \$300 + {}^-\$500 = \$0$
Exact: $\$184 + \$293 + {}^-\$506 = {}^-\$29$

31. Loss of 1140 yards rounds to 1000 yds.
12 games rounds to 10 games.

Average loss in each game:
Estimate: ${}^-1000 \div 10 = {}^-100$ yards
Exact: ${}^-1140 \div 12 = {}^-95$ yards

32. Cereal 1: 220 calories rounds to 200
Cereal 2: 110 calories rounds to 100
31 days rounds to 30

Estimate of calories saved by eating Cereal 2:
$200 - 100 = 100$ calories per day
$30 \cdot 100 = 3000$ calories in the month

Exact calories saved by eating Cereal 2:
$220 - 110 = 110$ calories per day
$31 \cdot 110 = 3410$ calories in the month

33. The difference between the high and low temperatures on Mars is:

${}^-10 - {}^-100$ *Add the opposite.*
$= {}^-10 + {}^+100$
$= 90$ degrees difference

34. 1276 books separated into cartons holding 48 books each.

```
      2 6
48 | 1 2 7 6
       9 6
      -----
       3 1 6
       2 8 8
      ------
         2 8
```

Anthony will need 27 cartons to ship all of the books. He will have 26 full boxes and one box with only 28 books in it.

CHAPTER 2 UNDERSTANDING VARIABLES AND SOLVING EQUATIONS

2.1 Introduction to Variables

2.1 Section Exercises

1. $c + 4$ c is a variable.
   4 is a constant.

3. $^-3 + m$ m is a variable.
   $^-3$ is a constant.

5. $5h$ h is a variable.
   5 is a coefficient.

7. $2c - 10$ c is a variable.
   2 is a coefficient.
   10 is a constant.

9. $x - y$ Both x and y are variables.

11. $^-6g + 9$ g is a variable.
    $^-6$ is a coefficient.
    9 is a constant.

13. Expression (rule) for ordering robes: $g + 10$

    (a) Evaluate the expression when there are 654 graduates.

    $g + 10$ Replace g with 654.
    $\underbrace{654 + 10}$ Follow the rule and add.
    664 robes must be ordered.

    (b) Evaluate the expression when there are 208 graduates.

    $g + 10$ Replace g with 208.
    $\underbrace{208 + 10}$ Follow the rule and add.
    218 robes must be ordered.

    (c) Evaluate the expression when there are 95 graduates.

    $g + 10$ Replace g with 95.
    $\underbrace{95 + 10}$ Follow the rule and add.
    105 robes must be ordered.

15. Expression (rule) for finding perimeter of equilateral triangle of side length s: $3s$

    (a) Evaluate the expression when s, the side length, is 11 inches.

    $3s$ Replace s with 11.
    $\underbrace{3 \cdot 11}$ Follow the rule and multiply.
    33 inches is the perimeter.

    (b) Evaluate the expression when s, the side length, is 3 feet.

    $3s$ Replace s with 3.
    $\underbrace{3 \cdot 3}$ Follow the rule and multiply.
    9 feet is the perimeter.

17. Expression (rule) for ordering brushes: $3c - 5$

    (a) Evaluate the expression when c, the class size, is 12.

    $3c - 5$ Replace c with 12.
    $\underbrace{3 \cdot 12} - 5$ Multiply before subtracting.
    $\underbrace{36 - 5}$
    31 brushes must be ordered.

    (b) Evaluate the expression when c, the class size, is 16.

    $3c - 5$ Replace c with 16.
    $\underbrace{3 \cdot 16} - 5$ Multiply before subtracting.
    $\underbrace{48 - 5}$
    43 brushes must be ordered.

19. Expression (rule) for average test score, where p is the total points and t is the number of tests: p/t

    (a) Evaluate the expression when p, the total points, is 332 and t, the number of tests, is 4.

    $\frac{p}{t}$ Replace p with 332 and t with 4.
    $\frac{332}{4}$ Follow the rule and divide.
    83 points is the average test score.

    (b) Evaluate the expression when p, the total points, is 637 and t, the number of tests, is 7.

    $\frac{p}{t}$ Replace p with 637 and t with 7.
    $\frac{637}{7}$ Follow the rule and divide.
    91 points is the average test score.

21.

Value of x	Expression $x + x + x + x$	Expression $4x$
12	$12 + 12 + 12 + 12$ is 48	$4 \cdot 12$ is 48
0	$0 + 0 + 0 + 0$ is 0	$4 \cdot 0$ is 0
$^-5$	$^-5 + {}^-5 + {}^-5 + {}^-5$ is $^-20$	$4 \cdot {}^-5$ is $^-20$

23.

Value of x	Value of y	Expression $^-2x + y$
$^-4$	5	$^-2(^-4) + 5$ is $8 + 5$ is 13
$^-6$	$^-2$	$^-2(^-6) + {}^-2$ is $12 + {}^-2$ is 10
0	$^-8$	$^-2(0) + {}^-8$ is $0 + {}^-8$ is $^-8$

25. A variable is a letter that represents the part of a rule that varies or changes depending on the situation. An expression expresses, or tells, the rule for doing something. For example, $c + 5$ is an expression, and c is the variable.

27. Let b represent "a number." Multiplying a number by 1 leaves the number unchanged.

$b \cdot 1 = b$ or $1 \cdot b = b$

29. Let b represent "any number." Any number divided by 0 is undefined.

$\frac{b}{0}$ is undefined or $b \div 0$ is undefined.

31. c^6 written without exponents is

$c \cdot c \cdot c \cdot c \cdot c \cdot c$

33. x^4y^3 written without exponents is

$x \cdot x \cdot x \cdot x \cdot y \cdot y \cdot y$

35. $^-3a^3b$ can be written as $^-3 \cdot a \cdot a \cdot a \cdot b$. The exponent 3 applies only to the base a.

37. $9xy^2$ can be written as $9 \cdot x \cdot y \cdot y$. The exponent 2 applies only to the base y.

39. $^-2c^5d$ can be written as $^-2 \cdot c \cdot c \cdot c \cdot c \cdot c \cdot d$. The exponent 5 applies only to the base c.

41. a^3bc^2 can be written as $a \cdot a \cdot a \cdot b \cdot c \cdot c$. The exponent 3 applies only to the base a. The exponent 2 applies only to the base c.

43. Evaluate t^2 when t is $^-4$.

t^2 means

$t \cdot t$ Replace t with $^-4$.

$\underbrace{^-4 \cdot {}^-4}$ Multiply.

16

45. Evaluate rs^3 when r is $^-3$ and s is 2.

rs^3 means

$r \cdot s \cdot s \cdot s$ Replace r with $^-3$ and s with 2.

$\underbrace{^-3 \cdot 2} \cdot 2 \cdot 2$ Multiply left to right.

$\underbrace{^-6 \cdot 2} \cdot 2$

$\underbrace{^-12 \cdot 2}$

$^-24$

47. Evaluate $3rs$ when r is $^-3$ and s is 2.

$3rs$ means

$3 \cdot r \cdot s$ Replace r with $^-3$ and s with 2.

$\underbrace{3 \cdot {}^-3} \cdot 2$ Multiply left to right.

$\underbrace{^-9 \cdot 2}$

$^-18$

49. Evaluate $^-2s^2t^2$ when s is 2 and t is $^-4$.

$^-2s^2t^2$ means

$^-2 \cdot s \cdot s \cdot t \cdot t$ Replace s with 2 and t with $^-4$.

$\underbrace{^-2 \cdot 2} \cdot 2 \cdot {}^-4 \cdot {}^-4$ Multiply left to right.

$\underbrace{^-4 \cdot 2} \cdot {}^-4 \cdot {}^-4$

$\underbrace{^-8 \cdot {}^-4} \cdot {}^-4$

$\underbrace{32 \cdot {}^-4}$

$^-128$

51. Evaluate $r^2s^5t^3$ when r is $^-3$, s is 2, and t is $^-4$, using a calculator.

$r^2s^5t^3$ Replace r with $^-3$, s with 2, and t with $^-4$.

$\underbrace{(^-3)^2(2)^5(^-4)^3}$ Use the y^x key.

$\underbrace{(9)(32)(^-64)}$ Multiply left to right.

$\underbrace{(288)(^-64)}$

$^-18{,}432$

53. Evaluate $^-10r^5s^7$ when r is $^-3$ and s is 2, using a calculator.

$^-10r^5s^7$ Replace r with $^-3$ and s with 2.

$^-10\ \underbrace{(^-3)^5}\ \underbrace{(2)^7}$ Use the y^x key.

$\underbrace{^-10(^-243)}\ (128)$ Multiply left to right.

$\underbrace{2430(128)}$

$311{,}040$

55. Evaluate $|xy| + |xyz|$ when x is 4, y is $^-2$, and z is $^-6$.

$|xy| + |xyz|$ Replace x with 4, y with $^-2$, and z with $^-6$.

$|\underbrace{4 \cdot {}^-2}| + |\underbrace{4 \cdot {}^-2} \cdot {}^-6|$ Multiply left to right within the absolute value signs.

$|^-8| + |\underbrace{^-8 \cdot {}^-6}|$

$|^-8| + |48|$ Evaluate the absolute values.

$\underbrace{8 + 48}$ Add.

56

57. Evaluate $\frac{z^2}{^-3y + z}$ when z is $^-6$ and y is $^-2$.

$\frac{z^2}{^-3y + z}$ Replace z with $^-6$ and y with $^-2$.

$\frac{(^-6)^2}{^-3(^-2) + {}^-6}$ Follow the order of operations.

$\frac{36}{0}$ Numerator: $(^-6)^2 = {}^-6 \cdot {}^-6 = 36$
Denominator: $^-3(^-2) + {}^-6 = 6 + {}^-6$

Undefined Division by 0 is undefined.

59. **(a)** Evaluate $\frac{s}{5}$ when s is 15 seconds.

$\frac{s}{5}$ Replace s with 15.

$\frac{15}{5}$ Divide.

3 miles

(b) Evaluate $\frac{s}{5}$ when s is 10 seconds.

$\frac{s}{5}$ Replace s with 10.

$\frac{10}{5}$ Divide.

2 miles

(c) Evaluate $\frac{s}{5}$ when s is 5 seconds.

$\frac{s}{5}$ Replace s with 5.

$\frac{5}{5}$ Divide.

1 mile

60. **(a)** Using part (c) of Exercise 59, the distance covered in $2\frac{1}{2}$ seconds is half of the distance covered in 5 seconds, or $\frac{1}{2}$ mile.

(b) Using part (a) of Exercise 59, the time to cover $1\frac{1}{2}$ miles is half the time to cover 3 miles, or $7\frac{1}{2}$ seconds. Or, using parts (b) and (c), find the number halfway between 5 seconds and 10 seconds.

(c) Using parts (a) and (b) of Exercise 59, find the number halfway between 10 seconds and 15 seconds; that is $12\frac{1}{2}$ seconds.

2.2 Simplifying Expressions

2.2 Section Exercises

1. $2b^2$ and b^2 are the only like terms in the expression. The variable parts match; both are b^2. The coefficients are 2 and 1.

3. ^-xy and $2xy$ are the like terms in the expression. The variable parts match; both are xy. The coefficients are $^-1$ and 2.

5. $7, 3,$ and $^-4$ are like terms. There are no variable parts; constants are considered like terms.

7. $6r + 6r$ These are like terms. Add the coefficients.

$(6+6)r$

$12r$ The variable part, r, stays the same.

9. $x^2 + 5x^2$ These are like terms. Rewrite x^2 as $1x^2$.

$1x^2 + 5x^2$ Add the coefficients.

$(1+5)x^2$

$6x^2$ The variable part, x^2, stays the same.

11. $p - 5p$ These are like terms. Rewrite p as $1p$.

$1p - 5p$ Change subtraction to adding the opposite.

$1p + {}^-5p$ Add the coefficients.

$(1 + {}^-5)p$

^-4p The variable part, p, stays the same.

13. $^-2a^3 - a^3$ These are like terms. Rewrite a^3 as $1a^3$.

$^-2a^3 - 1a^3$ Change subtraction to adding the opposite.

$^-2a^3 + {}^-1a^3$ Add the coefficients.

$({}^-2 + {}^-1)a^3$

$^-3a^3$ The variable part, a^3, stays the same.

15. $\underbrace{c - c}$

0 Any number minus itself is 0.

17. $9xy + xy - 9xy$ These are like terms. Rewrite xy as $1xy$.

$9xy + 1xy - 9xy$ Change subtraction to adding the opposite.

$9xy + 1xy + {}^-9xy$ Add the coefficients.

$(9 + 1 + {}^-9)xy$

$1xy$ or xy The variable part, xy, stays the same.

19. $5t^4 + 7t^4 - 6t^4$ These are like terms. Change subtraction to adding the opposite.

$5t^4 + 7t^4 + {}^-6t^4$ Add the coefficients.

$(5 + 7 + {}^-6)t^4$

$6t^4$ The variable part, t^4, stays the same.

21. $y^2 + y^2 + y^2 + y^2$ These are like terms. Write in the understood coefficient of 1.

$1y^2 + 1y^2 + 1y^2 + 1y^2$

$(1+1+1+1)y^2$ Add the coefficients.

$4y^2$ The variable part, y^2, stays the same.

23. $^-x - 6x - x$ These are like terms. Rewrite ^-x as ^-1x and x as $1x$.

$^-1x - 6x - 1x$ Change subtraction to adding the opposite.

$^-1x + ^-6x + ^-1x$ Add the coefficients.

$(^-1 + ^-6 + ^-1)x$

^-8x The variable part, x, stays the same.

25. $8a + 4b + 4a$ Use the commutative property to rewrite the expression so that like terms are next to each other.

$8a + 4a + 4b$ Add the coefficients of like terms.

$(8 + 4)a + 4b$

$12a + 4b$ The variable part, a, stays the same.

27. $6 + 8 + 7rs$ Use the commutative property to put the constants at the end.

$7rs + 6 + 8$ Add the coefficients of like terms.

$7rs + 14$ The only like terms are constants.

29. $a + ab^2 + ab^2$ Write in the understood coefficient of 1.

$1a + 1ab^2 + 1ab^2$ Add the coefficients of like terms.

$1a + (1 + 1)ab^2$

$1a + 2ab^2$ The variable part, ab^2, stays the same.

or

$a + 2ab^2$

31. $6x + y - 8x + y$ Write in the understood coefficients of 1. Change subtraction to adding the opposite.

$6x + 1y + ^-8x + 1y$ Rewrite using the commutative property.

$6x + ^-8x + 1y + 1y$ Add the coefficients of like terms.

$\underbrace{(6 + ^-8)}x + \underbrace{(1 + 1)}y$

$^-2x \quad + \quad ^+2y$

$^-2x + 2y$

33. $8b^2 - a^2 - b^2 + a^2$ Write in the understood coefficient of 1.

$8b^2 - 1a^2 - 1b^2 + 1a^2$ Change subtraction to adding the opposite.

$8b^2 + ^-1a^2 + ^-1b^2 + 1a^2$ Rewrite using the commutative property.

$8b^2 + ^-1b^2 + ^-1a^2 + 1a^2$ Add the coefficients of like terms.

$\underbrace{(8 + ^-1)b^2} + \underbrace{(^-1 + 1)a^2}$

$7b^2 \quad + \quad 0 \cdot a^2$

$7b^2 + 0$

$7b^2$

35. $^-x^3 + 3x - 3x^2 + 2$

There are no like terms. The expression cannot be simplified.

37. $^-9r + 6t - s - 5r + s + t - 6t + 5s - r$

Write in the understood coefficient of 1.

Change subtraction to adding the opposite.

$^-9r + 6t + ^-1s + ^-5r + 1s + 1t + ^-6t + 5s + ^-1r$

Rewrite using the commutative property.

$^-9r + ^-5r + ^-1r + ^-1s + 1s + 5s + 6t + 1t + ^-6t$

Add the coefficients of like terms.

$\underbrace{(^-9 + ^-5 + ^-1)r} + \underbrace{(^-1 + 1 + 5)s} + \underbrace{(6 + 1 + ^-6)t}$

$^-15r \quad + \quad ^+5s \quad + \quad ^+1t$

$^-15r + 5s + t$

39. By using the associative property, we can write $3(10a)$ as

$$(3 \cdot 10) \cdot a = 30 \cdot a = 30a.$$

So, $3(10a)$ simplifies to $30a$.

41. By using the associative property, we can write $^-4(2x^2)$ as

$$(^-4 \cdot 2) \cdot x^2 = ^-8 \cdot x^2 = ^-8x^2.$$

So, $^-4(2x^2)$ simplifies to $^-8x^2$.

43. By using the associative property, we can write $5(^-4y^3)$ as

$$(5 \cdot ^-4) \cdot y^3 = ^-20 \cdot y^3 = ^-20y^3.$$

So, $5(^-4y^3)$ simplifies to $^-20y^3$.

45. By using the associative property, we can write $^-9(^-2cd)$ as

$$(^-9 \cdot ^-2) \cdot c \cdot d = 18 \cdot c \cdot d = 18cd.$$

So, $^-9(^-2cd)$ simplifies to $18cd$.

47. By using the associative property, we can write $7(3a^2bc)$ as

$$(7 \cdot 3) \cdot a^2 \cdot b \cdot c = 21 \cdot a^2 \cdot b \cdot c = 21a^2bc.$$

So, $7(3a^2bc)$ simplifies to $21a^2bc$.

49. $^-12(^-w)$ Write in the understood coefficient of $^-1$.

$^-12(^-1w)$ Rewrite using the associative property.

$(^-12 \cdot ^-1)w$

$12 \cdot w$

$12w$

51. $6(b+6)$ Distributive property
$6 \cdot b + 6 \cdot 6$
$6b + 36$

53. $7(x-1)$ Distributive property
$7 \cdot x - 7 \cdot 1$
$7x - 7$

55. $3(7t+1)$ Distributive property
$3 \cdot 7t + 3 \cdot 1$
$21t + 3$

57. ${}^-2(5r+3)$ Distributive property
${}^-2 \cdot 5r + {}^-2 \cdot 3$
${}^-10r + {}^-6$ Change addition to subtraction of the opposite.
${}^-10r - 6$

59. ${}^-9(k+4)$ Distributive property
${}^-9 \cdot k + {}^-9 \cdot 4$
${}^-9k + {}^-36$ Change addition to subtraction of the opposite.
${}^-9k - 36$

61. $50(m-6)$ Distributive property
$50 \cdot m - 50 \cdot 6$
$50m - 300$

63. $10 + 2(4y+3)$ Distributive property
$10 + 2 \cdot 4y + 2 \cdot 3$
$10 + 8y + 6$ Rewrite using the commutative property.
$8y + 10 + 6$ Combine like terms.
$8y + 16$

65. $6(a^2 - 2) + 15$ Distributive property
$6 \cdot a^2 - 6 \cdot 2 + 15$
$6a^2 - 12 + 15$ Combine like terms.
$6a^2 + 3$

67. $2 + 9(m-4)$ Distributive property
$2 + 9 \cdot m - 9 \cdot 4$
$2 + 9m - 36$ Change subtraction to adding the opposite.
$2 + 9m + {}^-36$ Rewrite using the commutative property.
$9m + 2 + {}^-36$ Add the coefficients of like terms.
$9m + {}^-34$ Change addition to subtraction of the opposite.
$9m - 34$

69. ${}^-5(k+5) + 5k$ Distributive property
${}^-5 \cdot k + {}^-5 \cdot 5 + 5k$
${}^-5k + {}^-25 + 5k$ Rewrite using the commutative property.
${}^-5k + 5k + {}^-25$ Add the coefficients of like terms.
$\underbrace{({}^-5+5)}k + {}^-25$
$0k + {}^-25$ Zero times any number is 0.
$\underbrace{0 + {}^-25}$ Zero added to any number is the number
${}^-25$

71. $4(6x-3) + 12$ Distributive property
$4 \cdot 6x - 4 \cdot 3 + 12$
$24x - 12 + 12$ Change subtraction to adding the opposite.
$24x + {}^-12 + 12$ Combine like terms. Any number plus its opposite is 0.
$24x + 0$
$24x$

73. $5 + 2(3n+4) - n$ Distributive property
$5 + 2 \cdot 3n + 2 \cdot 4 - n$ Rewrite n as $1n$.
$5 + 6n + 8 - 1n$ Change subtraction to adding the opposite.
$5 + 6n + 8 + {}^-1n$ Rewrite using the commutative property.
$5 + 8 + 6n + {}^-1n$ Add the coefficients of like terms.
$(5+8) + (6 + {}^-1)n$
$13 + 5n$ or $5n + 13$

75. ${}^-p + 6(2p-1) + 5$ Distributive property
${}^-p + 6 \cdot 2p - 6 \cdot 1 + 5$
${}^-p + 12p - 6 + 5$ Rewrite ${}^-p$ as ${}^-1p$. Change subtraction to adding the opposite.
${}^-1p + 12p + {}^-6 + 5$ Add the coefficients of like terms.
$({}^-1 + 12)p + ({}^-6 + 5)$
$11p + {}^-1$ Change addition to subtraction of the opposite.
$11p - 1$

77. A simplified expression still has variables, but is written in a simpler way. When evaluating an expression, the variables are all replaced by specific numbers and the final result is a numerical answer.

79. Like terms have matching variable parts, that is, matching letters and exponents. The coefficients do not have to match. Examples will vary. Possible examples: In $^-6x + 9 + x$, the terms ^-6x and x are like terms. In $4k + 3 - 8k^2 + 10$, the terms 3 and 10 are like terms.

81. $\underbrace{^-2x + 7x} + 8$

$5x + 8$

Keep the variable part unchanged when combining like terms.

83. $^-4(3y) - 5 + 2(5y + 7)$ Distributive property

$^-4 \cdot 3y - 5 + 2 \cdot 5y + 2 \cdot 7$

$^-12y - 5 + 10y + 14$ Change subtraction to adding the opposite.

$^-12y + {}^-5 + 10y + 14$ Group like terms and add the coefficients.

$\underbrace{^-12y + 10y} + \underbrace{^-5 + 14}$

$^-2y \quad + \quad 9$

$^-2y + 9$

85. $^-10 + 4(^-3b + 3) + 2(6b - 1)$

Distributive property

$^-10 + 4 \cdot {}^-3b + 4 \cdot 3 + 2 \cdot 6b - 2 \cdot 1$

$^-10 + {}^-12b + 12 + 12b - 2$

Change subtraction to adding the opposite.

$^-10 + {}^-12b + 12 + 12b + {}^-2$

Group like terms and add the coefficients.

$\underbrace{^-12b + 12b} + \underbrace{^-10 + 12 + {}^-2}$

$0b \quad + \quad 0$

0

87. $^-5(^-x + 2) + 8(^-x) + 3(^-2x - 2) + 16$

Distributive property

$^-5 \cdot {}^-x + {}^-5 \cdot 2 + 8 \cdot {}^-x + 3 \cdot {}^-2x - 3 \cdot 2 + 16$

$5x + {}^-10 + {}^-8x + {}^-6x - 6 + 16$

Change subtraction to adding the opposite.

$5x + {}^-10 + {}^-8x + {}^-6x + {}^-6 + 16$

Group like terms and add the coefficients.

$\underbrace{5x + {}^-8x + {}^-6x} + \underbrace{^-10 + {}^-6 + 16}$

$^-9x \quad + \quad 0$

^-9x

Summary Exercises on Variables and Expressions

1. Expression (rule) for finding the perimeter of an octagon of side length s: $8s$

(a) Evaluate the expression when s, the side length, is 4 yards.

$8s$ Replace s with 4.

$\underbrace{8 \cdot 4}$ Follow the rule and multiply.

32 yards is the perimeter.

(b) Evaluate the expression when s, the side length, is 15 inches.

$8s$ Replace s with 15.

$\underbrace{8 \cdot 15}$ Follow the rule and multiply.

120 inches is the perimeter.

3. wxy

Replace w with 5, x with $^-2$, and y with $^-6$.

$\underbrace{5 \cdot {}^-2} \cdot {}^-6$ Multiply left to right.

$\underbrace{^-10 \cdot {}^-6}$

60

5. ^-4wy

Replace w with 5 and y with $^-6$.

$\underbrace{^-4 \cdot 5} \cdot {}^-6$ Multiply left to right.

$\underbrace{^-20 \cdot {}^-6}$

120

7. $w^2x^5 = w \cdot w \cdot x \cdot x \cdot x \cdot x \cdot x$

Replace w with 5 and x with $^-2$.

$\underbrace{5 \cdot 5} \cdot {}^-2 \cdot {}^-2 \cdot {}^-2 \cdot {}^-2 \cdot {}^-2$ Multiply left to right.

$\underbrace{25 \cdot {}^-2} \cdot {}^-2 \cdot {}^-2 \cdot {}^-2 \cdot {}^-2$

$\underbrace{^-50 \cdot {}^-2} \cdot {}^-2 \cdot {}^-2 \cdot {}^-2$

$\underbrace{100 \cdot {}^-2} \cdot {}^-2 \cdot {}^-2$

$\underbrace{^-200 \cdot {}^-2} \cdot {}^-2$

$\underbrace{400 \cdot {}^-2}$

$^-800$

9. $10b + 4b + 10b = (10 + 4 + 10)b$
$= 24b$

11. $^-8(c + 4) = {}^-8 \cdot c + {}^-8 \cdot 4$
$= {}^-8c + {}^-32$
or $^-8c - 32$

13. $^-4(^-3c^2d) = (^-4 \cdot {}^-3) \cdot c^2d$
$= 12c^2d$

15. $2(3w + 4) = 2 \cdot 3w + 2 \cdot 4$
$= (2 \cdot 3)w + 2 \cdot 4$
$= 6w + 8$

17. $^-10(^-5x^3y^2) = (^-10 \cdot {}^-5) \cdot x^3y^2$
$= 50x^3y^2$

19. $21 + 7(h^2 - 3) = 21 + 7 \cdot h^2 - 7 \cdot 3$
$= 21 + 7h^2 - 21$
$= 7h^2$

21. $^-4(8y-5)+5 = {}^-4\cdot 8y - {}^-4\cdot 5+5$
$= ({}^-4\cdot 8)\cdot y - {}^-20+5$
$= {}^-32y+20+5$
$= {}^-32y+25$

23. $^-n+5(4n-2)+11 = {}^-n+5\cdot 4n-5\cdot 2+11$
$= {}^-n+(5\cdot 4)\cdot n-10+11$
$= {}^-n+20n+{}^-10+11$
$= ({}^-1+20)\cdot n+1$
$= 19n+1$

2.3 Solving Equations Using Addition

2.3 Section Exercises

1. $n-50=8$ Given equation
$58-50 \stackrel{?}{=} 8$ Replace n with 58.
$8=8$
Yes, 58 is the solution.
(No need to check 42 and 60.)

3. $^-6=y+10$ Given equation
$^-6 \stackrel{?}{=} {}^-4+10$ Replace y with $^-4$.
$^-6 \neq 6$
No, $^-4$ is not the solution.
$^-6 \stackrel{?}{=} {}^-16+10$ Replace y with $^-16$.
$^-6={}^-6$
Yes, $^-16$ is the solution.

5. $t+12=0$ Given equation
$0+12 \stackrel{?}{=} 0$ Replace t with 0.
$12 \neq 0$
No, 0 is not the solution.
$^-12+12 \stackrel{?}{=} 0$ Replace t with $^-12$.
$0=0$
Yes, $^-12$ is the solution.

7. $p+5 = 9$ Add the opposite of 5, $^-5$, to both sides.
$\underline{{}^-5 \quad\; {}^-5}$
$p+0 = 4$
$p = 4$ The solution is 4.

Check: $p+5=9$ Replace p with 4.
$4+5=9$
$9=9$ Balances

9. $8 = r-2$ Add the opposite.
$8 = r+{}^-2$ Add the opposite of $^-2$, 2, to both sides.
$\underline{2 \quad\; 2}$
$10 = r+0$
$10 = r$ The solution is 10.

Check: $8=r-2$ Replace r with 10.
$8=10-2$
$8=8$ Balances

11. $^-5 = n+3$ Add the opposite of 3, $^-3$, to both sides.
$\underline{{}^-3 \quad\; {}^-3}$
$^-8 = n+0$
$^-8 = n$ The solution is $^-8$.

Check: $^-5=n+3$ Replace n with $^-8$.
$^-5={}^-8+3$
$^-5={}^-5$ Balances

13. $^-4+k = 14$ Add the opposite of $^-4$, 4, to both sides.
$\underline{4 \quad\; 4}$
$0+k = 18$
$k = 18$ The solution is 18.

Check: $^-4+k=14$ Replace k with 18.
$^-4+18=14$
$14=14$ Balances

15. $y-6 = 0$ Add the opposite.
$y+{}^-6 = 0$ Add the opposite of $^-6$, 6, to both sides.
$\underline{6 \quad\; 6}$
$y+0 = 6$
$y = 6$ The solution is 6.

Check: $y-6=0$ Replace y with 6.
$6-6=0$ Add the opposite.
$6+{}^-6=0$
$0=0$ Balances

17. $7 = r+13$ Add the opposite of 13, $^-13$, to both sides.
$\underline{{}^-13 \quad\; {}^-13}$
$^-6 = r+0$
$^-6 = r$ The solution is $^-6$.

Check: $7=r+13$ Replace r with $^-6$.
$7={}^-6+13$
$7=7$ Balances

19. $x-12 = {}^-1$ Add the opposite.
$x+{}^-12 = {}^-1$ Add the opposite of $^-12$, 12, to both sides.
$\underline{12 \quad\; 12}$
$x+0 = 11$
$x = 11$ The solution is 11.

Check: $x-12={}^-1$ Replace x with 11.
$11-12={}^-1$ Add the opposite.
$11+{}^-12={}^-1$
$^-1={}^-1$ Balances

21.
$$\begin{aligned} {}^-5 &= {}^-2 + t && \text{Add the opposite of } {}^-2, \\ & && \text{2, to both sides.} \\ \underline{2} & \quad \underline{2} \\ {}^-3 &= 0 + t \\ {}^-3 &= t && \text{The solution is } {}^-3. \end{aligned}$$

Check:
$$\begin{aligned} {}^-5 &= {}^-2 + t && \text{Replace } t \text{ with } {}^-3. \\ {}^-5 &= {}^-2 + {}^-3 \\ {}^-5 &= {}^-5 && \text{Balances} \end{aligned}$$

23. The given solution is $^-2$.

Check:
$$\begin{aligned} z - 5 &= 3 && \text{Replace } z \text{ with } {}^-2. \\ {}^-2 - 5 &= 3 && \text{Add the opposite.} \\ {}^-2 + {}^-5 &= 3 \\ {}^-7 &\neq 3 && \text{Does not balance} \end{aligned}$$

Correct solution:

$$\begin{aligned} z - 5 &= 3 && \text{Add the opposite.} \\ z + {}^-5 &= 3 && \text{Add the opposite of } {}^-5, \\ & && \text{5, to both sides.} \\ \underline{5} & \quad \underline{5} \\ z + 0 &= 8 \\ z &= 8 && \text{The solution is 8.} \end{aligned}$$

Check:
$$\begin{aligned} z - 5 &= 3 && \text{Replace } z \text{ with 8.} \\ 8 - 5 &= 3 && \text{Add the opposite.} \\ 8 + {}^-5 &= 3 \\ 3 &= 3 && \text{Balances} \end{aligned}$$

25. The given solution is $^-18$.

Check:
$$\begin{aligned} 7 + x &= {}^-11 && \text{Replace } x \text{ with } {}^-18. \\ 7 + {}^-18 &= {}^-11 \\ {}^-11 &= {}^-11 && \text{Balances} \end{aligned}$$

$^-18$ is the correct solution.

27. The given solution is 10.

Check:
$$\begin{aligned} {}^-10 &= {}^-10 + b && \text{Replace } b \text{ with 10.} \\ {}^-10 &= {}^-10 + 10 \\ {}^-10 &\neq 0 && \text{Does not balance} \end{aligned}$$

Correct solution:

$$\begin{aligned} {}^-10 &= {}^-10 + b && \text{Add the opposite of } {}^-10, \\ & && \text{10, to both sides.} \\ \underline{10} & \quad \underline{10} \\ 0 &= 0 + b \\ 0 &= b && \text{The solution is 0.} \end{aligned}$$

Check:
$$\begin{aligned} {}^-10 &= {}^-10 + b && \text{Replace } b \text{ with 0.} \\ {}^-10 &= {}^-10 + 0 \\ {}^-10 &= {}^-10 && \text{Balances} \end{aligned}$$

29.
$$\begin{aligned} c - 4 &= {}^-8 + 10 && \text{Simplify the right side.} \\ c - 4 &= 2 && \text{Add the opposite.} \\ c + {}^-4 &= 2 && \text{Add 4 to both sides.} \\ \underline{4} & \quad \underline{4} \\ c + 0 &= 6 \\ c &= 6 && \text{The solution is 6.} \end{aligned}$$

Check:
$$\begin{aligned} c - 4 &= {}^-8 + 10 && \text{Replace } c \text{ with 6.} \\ 6 - 4 &= {}^-8 + 10 \\ 2 &= 2 && \text{Balances} \end{aligned}$$

31.
$$\begin{aligned} {}^-1 + 4 &= y - 2 && \text{Simplify the left side.} \\ 3 &= y - 2 && \text{Add the opposite.} \\ 3 &= y + {}^-2 && \text{Add 2 to both sides.} \\ \underline{2} & \quad \underline{2} \\ 5 &= y + 0 \\ 5 &= y && \text{The solution is 5.} \end{aligned}$$

Check:
$$\begin{aligned} {}^-1 + 4 &= y - 2 && \text{Replace } y \text{ with 5.} \\ {}^-1 + 4 &= 5 - 2 && \text{Add the opposite.} \\ {}^-1 + 4 &= 5 + {}^-2 \\ 3 &= 3 && \text{Balances} \end{aligned}$$

33.
$$\begin{aligned} 10 + b &= {}^-14 - 6 && \text{Add the opposite.} \\ 10 + b &= {}^-14 + {}^-6 && \text{Add.} \\ 10 + b &= {}^-20 && \text{Add } {}^-10 \\ \underline{{}^-10} & \quad \underline{{}^-10} && \text{to both sides.} \\ 0 + b &= {}^-30 \\ b &= {}^-30 && \text{The solution is } {}^-30. \end{aligned}$$

Check:
$$\begin{aligned} 10 + b &= {}^-14 - 6 && \text{Replace } b \text{ with } {}^-30. \\ 10 + {}^-30 &= {}^-14 + {}^-6 \\ {}^-20 &= {}^-20 && \text{Balances} \end{aligned}$$

35.
$$\begin{aligned} t - 2 &= 3 - 5 && \text{Add the opposites.} \\ t + {}^-2 &= 3 + {}^-5 && \text{Simplify the right side.} \\ t + {}^-2 &= {}^-2 && \text{Add 2 to both sides.} \\ \underline{2} & \quad \underline{2} \\ t + 0 &= 0 \\ t &= 0 && \text{The solution is 0.} \end{aligned}$$

Check:
$$\begin{aligned} t - 2 &= 3 - 5 && \text{Replace } t \text{ with 0.} \\ 0 - 2 &= 3 - 5 && \text{Add the opposites.} \\ 0 + {}^-2 &= 3 + {}^-5 \\ {}^-2 &= {}^-2 && \text{Balances} \end{aligned}$$

37.
$$\begin{aligned} 10z - 9z &= {}^-15 + 8 && \text{Add the opposite.} \\ 10z + {}^-9z &= {}^-15 + 8 && \text{Combine like terms.} \\ 1z &= {}^-7 && 1z \text{ is the same as } z. \\ z &= {}^-7 && \text{The solution is } {}^-7. \end{aligned}$$

Check:
$$\begin{aligned} 10z - 9z &= {}^-15 + 8 && \text{Replace } z \text{ with } {}^-7. \\ 10 \cdot {}^-7 - 9 \cdot {}^-7 &= {}^-15 + 8 \\ {}^-70 - {}^-63 &= {}^-7 && \text{Add the opposite.} \\ {}^-70 + 63 &= {}^-7 \\ {}^-7 &= {}^-7 && \text{Balances} \end{aligned}$$

39. $^-5w + 2 + 6w = {}^-4 + 9$ Rearrange and combine like terms.

$\underbrace{^-5w + 6w} + 2 = \underbrace{^-4 + 9}$

$1w \quad + 2 = \quad 5$ Add $^-2$ to both sides.

$\quad\;\; ^-2 \quad\; ^-2$

$1w \quad + 0 = 3$

$w = 3$ The solution is 3.

Check:

$^-5w + 2 + 6w = {}^-4 + 9$ Replace w with 3.

$^-5 \cdot 3 + 2 + 6 \cdot 3 = {}^-4 + 9$

$^-15 + 2 + 18 = {}^-4 + 9$

$5 = 5$ Balances

41. $^-3 - 3 = 4 - 3x + 4x$ Add the opposites.

$\underbrace{^-3 + {}^-3} = 4 + \underbrace{^-3x + 4x}$ Combine like terms.

$^-6 = 4 + 1x$ Add $^-4$ to both sides.

$^-4 \quad\; ^-4$

$^-10 = 0 + 1x$

$^-10 = 1x$ $1x$ is the same as x.

$^-10 = x$ The solution is $^-10$.

43. $^-3 + 7 - 4 = {}^-2a + 3a$ Add the opposite.

$^-3 + 7 + {}^-4 = {}^-2a + 3a$ Combine like terms.

$0 = a$ The solution is 0.

45. $y - 75 = {}^-100$ Add the opposite.

$y + {}^-75 = {}^-100$ Add 75 to both sides.

$75 \quad\; 75$

$y + 0 = {}^-25$

$y = {}^-25$ The solution is $^-25$.

47. $^-x + 3 + 2x = 18$ Rearrange and combine like terms.

$x + 3 = 18$ Add $^-3$ to both sides.

$^-3 \quad ^-3$

$x + 0 = 15$

$x = 15$ The solution is 15.

49. $82 = {}^-31 + k$ Add 31 to both sides.

$31 \quad\; 31$

$113 = 0 + k$

$113 = k$ The solution is 113.

51. $^-2 + 11 = 2b - 9 - b$ Add the opposite.

$^-2 + 11 = 2b + {}^-9 + {}^-b$ Rearrange and combine like terms.

$9 = b + {}^-9$ Add 9 to both sides.

$9 \quad\; 9$

$18 = b + 0$

$18 = b$ The solution is 18.

53. $r - 6 = 7 - 10 - 8$ Add the opposites.

$r + {}^-6 = 7 + {}^-10 + {}^-8$ Combine like terms.

$r + {}^-6 = {}^-11$ Add 6 to both sides.

$6 \quad\; 6$

$r + 0 = {}^-5$

$r = {}^-5$ The solution is $^-5$.

55. $^-14 = n + 91$ Add $^-91$ to both sides.

$^-91 \quad ^-91$

$^-105 = n + 0$

$^-105 = n$ The solution is $^-105$.

57. $^-9 + 9 = 5 + h$ Combine like terms.

$0 = 5 + h$ Add $^-5$ to both sides.

$^-5 \quad ^-5$

$^-5 = 0 + h$

$^-5 = h$ The solution is $^-5$.

59. No, the solution is $^-14$, the number used to replace x in the original equation.

61. $g + 10 = 305$ Add the opposite of $10, {}^-10$, to both sides.

$^-10 \quad ^-10$

$g + 0 = 295$

$g = 295$

There were 295 graduates this year.

63. $92 = c + 37$ Add $^-37$ to both sides.

$^-37 \quad ^-37$

$55 = c + 0$

$55 = c$

When the temperature is 92 degrees, a field cricket chirps 55 times (in 15 seconds).

65. $p - 65 = 45$ Add the opposite.

$p + {}^-65 = 45$ Add 65 to both sides.

$65 \quad 65$

$p + 0 = 110$

$p = 110$

Ernesto's parking fees average $110 per month in winter.

67. $^-17 - 1 + 26 - 38 = {}^-3 - m - 8 + 2m$

Change all subtractions to adding the opposite. Write the understood coefficient of 1.

$^-17 + {}^-1 + 26 + {}^-38 = {}^-3 + {}^-1m + {}^-8 + 2m$

Use the commutative property to group like terms on the right side.

$^-17 + {}^-1 + 26 + {}^-38 = {}^-3 + {}^-8 + {}^-1m + 2m$

continued

Combine like terms on each side.
$^-30 = {}^-11 + 1m$
To get m by itself, add the opposite of $^-11$, 11, to both sides.

$$\begin{aligned} ^-30 &= {}^-11 + 1m \\ 11 &\quad\; 11 \\ \hline ^-19 &= 1m \\ ^-19 &= m \end{aligned}$$
The solution is $^-19$.

69. $^-6x + 2x + 6 + 5x = |0 - 9| - |^-6 + 5|$
Change subtraction within absolute value to adding the opposite and rearrange the terms.
$^-6x + 2x + 5x + 6 = |0 + {}^-9| - |^-6 + 5|$
Simplify inside absolute value signs. Collect like terms.
$x + 6 = |^-9| - |^-1|$
Evaluate absolute values.
$x + 6 = 9 - 1$
Change subtraction to adding the opposite.

$$\begin{aligned} x + 6 &= 9 + {}^-1 \\ x + 6 &= 8 \qquad \text{Add } ^-6 \text{ to both sides.} \\ ^-6 &\quad\; ^-6 \\ \hline x + 0 &= 2 \\ x &= 2 \end{aligned}$$
The solution is 2.

71. **(a)** Equations will vary. Some possibilities are:

$$\begin{aligned} n - 1 &= {}^-3 \quad \text{Add the opposite.} \\ n + {}^-1 &= {}^-3 \quad \text{Add 1 to both sides.} \\ 1 &\quad\; 1 \\ \hline n + 0 &= {}^-2 \\ n &= {}^-2 \end{aligned}$$
The solution is $^-2$.

$$\begin{aligned} 8 &= x + 10 \quad \text{Add the opposite of } 10, {}^-10, \text{ to both sides.} \\ ^-10 &\quad\; ^-10 \\ \hline ^-2 &= x + 0 \\ ^-2 &= x \end{aligned}$$
The solution is $^-2$.

(b) Equations will vary. Some possibilities are:

$$\begin{aligned} y + 6 &= 6 \quad \text{Add the opposite of } 6, {}^-6, \text{ to both sides.} \\ ^-6 &\quad\; ^-6 \\ \hline y + 0 &= 0 \\ y &= 0 \end{aligned}$$
The solution is 0.

$$\begin{aligned} ^-5 &= {}^-5 + b \quad \text{Add the opposite of } {}^-5, 5, \text{ to both sides.} \\ 5 &\quad\; 5 \\ \hline 0 &= 0 + b \\ 0 &= b \end{aligned}$$
The solution is 0.

72. **(a)**

$$\begin{aligned} x + 1 &= 1\tfrac{1}{2} \quad \text{Add the opposite of } 1, {}^-1, \text{ to both sides.} \\ ^-1 &\quad\; ^-1 \\ \hline x + 0 &= \tfrac{1}{2} \\ x &= \tfrac{1}{2} \end{aligned}$$
The solution is $\frac{1}{2}$.

(b)

$$\begin{aligned} \tfrac{1}{4} &= y - 1 \quad \text{Add the opposite.} \\ \tfrac{1}{4} &= y + {}^-1 \quad \text{Add the opposite of } {}^-1, 1, \text{ to both sides.} \\ 1 &\quad\; 1 \\ \hline 1\tfrac{1}{4} &= y + 0 \\ 1\tfrac{1}{4} &= y \quad \text{or} \quad y = \tfrac{5}{4} \end{aligned}$$
The solution is $\frac{5}{4}$.

(c)

$$\begin{aligned} \$2.50 + n &= \$3.35 \quad \text{Add the opposite of } \$2.50, {}^-\$2.50, \text{ to both sides.} \\ ^-\$2.50 &\quad\; ^-\$2.50 \\ \hline \$0 + n &= \$0.85 \\ n &= \$0.85 \end{aligned}$$
The solution is \$0.85.

(d) Equations will vary. Some possibilities are:
$x - 2 = 1\frac{3}{5}$ The solution is $3\frac{3}{5}$.
$a - \$7.32 = \$9.16$ The solution is \$16.48.
$5c - \$11.20 = 4c - \$2.00$ The solution is \$9.20.

2.4 Solving Equations Using Division

2.4 Section Exercises

1. $6z = 12$ Divide *both* sides by 6.
$$\frac{6z}{6} = \frac{12}{6}$$
$z = 2$ The solution is 2.

Check: $6z = 12$ Replace z with 2.
$6 \cdot 2 = 12$
$12 = 12$ Balances

3. $48 = 12r$ Divide *both* sides by 12.
$$\frac{48}{12} = \frac{12r}{12}$$
$4 = r$ The solution is 4.

Check: $48 = 12r$ Replace r with 4.
$48 = 12 \cdot 4$
$48 = 48$ Balances

5. $3y = 0$ Divide *both* sides by 3.
$$\frac{3y}{3} = \frac{0}{3}$$
$y = 0$ The solution is 0.

Check: $3y = 0$ Replace y with 0.
$3 \cdot 0 = 0$
$0 = 0$ Balances

7. $^-7k = 70$ Divide *both* sides by $^-7$.
$$\frac{^-7k}{^-7} = \frac{70}{^-7}$$
$k = {}^-10$ The solution is $^-10$.

Check: $^-7k = 70$ Replace k with $^-10$.
$^-7 \cdot {}^-10 = 70$
$70 = 70$ Balances

9. $^-54 = {}^-9r$ Divide *both* sides by $^-9$.
$\frac{^-54}{^-9} = \frac{^-9r}{^-9}$
$6 = r$ The solution is 6.

Check: $^-54 = {}^-9r$ Replace r with 6.
$^-54 = {}^-9 \cdot 6$
$^-54 = {}^-54$ Balances

11. $^-25 = 5b$ Divide *both* sides by 5.
$\frac{^-25}{5} = \frac{5b}{5}$
$^-5 = b$ The solution is $^-5$.

Check: $^-25 = 5b$ Replace b with $^-5$.
$^-25 = 5 \cdot {}^-5$
$^-25 = {}^-25$ Balances

13. $2r = {}^-7 + 13$ Combine like terms.
$2r = 6$ Divide *both* sides by 2.
$\frac{2r}{2} = \frac{6}{2}$
$r = 3$ The solution is 3.

Check: $2r = {}^-7 + 13$ Replace r with 3.
$2 \cdot 3 = {}^-7 + 13$
$6 = 6$ Balances

15. $^-12 = 5p - p$ Add the opposite.
$^-12 = 5p + {}^-p$ Rewrite ^-p as ^-1p.
$^-12 = 5p + {}^-1p$ Combine like terms.
$^-12 = 4p$ Divide *both* sides by 4.
$\frac{^-12}{4} = \frac{4p}{4}$
$^-3 = p$ The solution is $^-3$.

Check: $^-12 = 5p - p$ Replace p with $^-3$.
$^-12 = 5 \cdot {}^-3 - {}^-3$
$^-12 = {}^-15 - {}^-3$ Add the opposite.
$^-12 = {}^-15 + {}^+3$
$^-12 = {}^-12$ Balances

17. $3 - 28 = 5a$ Add the opposite.
$3 + {}^-28 = 5a$ Combine like terms.
$\frac{^-25}{5} = \frac{5a}{5}$ Divide *both* sides by 5.
$^-5 = a$ The solution is $^-5$.

19. $x - 9x = 80$ Add the opposite.
$x + {}^-9x = 80$ Rewrite x as $1x$.
$1x + {}^-9x = 80$ Combine like terms.
$^-8x = 80$ Divide *both* sides by $^-8$.
$\frac{^-8x}{^-8} = \frac{80}{^-8}$
$x = {}^-10$ The solution is $^-10$.

21. $13 - 13 = 2w - w$ Add the opposite.
$13 + {}^-13 = 2w + {}^-w$ Rewrite ^-w as ^-1w.
$13 + {}^-13 = 2w + {}^-1w$ Combine like terms.
$0 = 1w$ $1w$ is the same as w.
$0 = w$ The solution is 0.

23. $3t + 9t = 20 - 10 + 26$ Add the opposite.
$3t + 9t = 20 + {}^-10 + 26$ Combine like terms.
$12t = 36$ Divide *both* sides by 12.
$\frac{12t}{12} = \frac{36}{12}$
$t = 3$ The solution is 3.

25. $0 = {}^-9t$ Divide *both* sides by $^-9$.
$\frac{0}{^-9} = \frac{^-9t}{^-9}$
$0 = t$ The solution is 0.

27. $^-14m + 8m = 6 - 60$ Add the opposite.
$^-14m + 8m = 6 + {}^-60$ Combine like terms.
$^-6m = {}^-54$ Divide *both* sides by $^-6$.
$\frac{^-6m}{^-6} = \frac{^-54}{^-6}$
$m = 9$ The solution is 9.

29. $100 - 96 = 31y - 35y$ Add the opposite.
$100 + {}^-96 = 31y + {}^-35y$ Combine like terms.
$4 = {}^-4y$ Divide *both* sides by $^-4$.
$\frac{4}{^-4} = \frac{^-4y}{^-4}$
$^-1 = y$ The solution is $^-1$.

31. $3(2z) = {}^-30$ To multiply on the left, use the associative property.
$(3 \cdot 2) \cdot z = {}^-30$
$6z = {}^-30$ Divide *both* sides by 6.
$\frac{6z}{6} = \frac{^-30}{6}$
$z = {}^-5$ The solution is $^-5$.

33. $50 = {}^-5(5p)$ To multiply on the right, use the associative property.
$50 = ({}^-5 \cdot 5) \cdot p$
$50 = {}^-25p$ Divide *both* sides by $^-25$.
$\frac{50}{^-25} = \frac{^-25p}{^-25}$
$^-2 = p$ The solution is $^-2$.

35. $^-2({}^-4k) = 56$ Associative property
$({}^-2 \cdot {}^-4) \cdot k = 56$
$8k = 56$ Divide *both* sides by 8.
$\frac{8k}{8} = \frac{56}{8}$
$k = 7$ The solution is 7.

37. $^-90 = {}^-10({}^-3b)$ Associative property
$^-90 = ({}^-10 \cdot {}^-3) \cdot b$
$^-90 = 30b$ Divide *both* sides by 30.
$\frac{^-90}{30} = \frac{30b}{30}$
$^-3 = b$ The solution is $^-3$.

39. $^-x = 32$ Write in the understood $^-1$.
$^-1x = 32$ Divide *both* sides by $^-1$.
$\frac{^-1x}{^-1} = \frac{32}{^-1}$
$x = {}^-32$ The solution is $^-32$.

41. $^-2 = {}^-w$ Write in the understood $^-1$.
$^-2 = {}^-1w$ Divide *both* sides by $^-1$.
$\frac{^-2}{^-1} = \frac{^-1w}{^-1}$
$2 = w$ The solution is 2.

43. $^-n = {}^-50$ Write in the understood $^-1$.
$^-1n = {}^-50$ Divide *both* sides by $^-1$.
$\frac{^-1n}{^-1} = \frac{^-50}{^-1}$
$n = 50$ The solution is 50.

45. $10 = {}^-p$ Write in the understood $^-1$.
$10 = {}^-1p$ Divide *both* sides by $^-1$.
$\frac{10}{^-1} = \frac{^-1p}{^-1}$
$^-10 = p$ The solution is $^-10$.

47. Each solution is the opposite of the number in the equation. So the rule is: When you change the variable from negative to positive, then change the number in the equation to its opposite.
In $^-x = 5$, the opposite of 5 is $^-5$, so $x = {}^-5$.

49. Divide by the coefficient of x, which is 3, *not* by the opposite of 3.

$3x = 16 - 1$
$3x = 15$
$\frac{3x}{3} = \frac{15}{3}$
$x = 5$ The correct solution is 5.

51. $3s = 45$ Divide *both* sides by 3.
$\frac{3s}{3} = \frac{45}{3}$
$s = 15$

The length of one side is 15 feet.

53. $120 = 5s$ Divide *both* sides by 5.
$\frac{120}{5} = \frac{5s}{5}$
$24 = s$

The length of one side is 24 meters.

55. $89 - 116 = {}^-4({}^-4y) - 9(2y) + y$

Use the associative property to simplify the products.

$89 - 116 = ({}^-4 \cdot {}^-4) \cdot y - (9 \cdot 2) \cdot y + y$
$89 - 116 = 16y - 18y + y$

Change subtraction to adding the opposite. Combine like terms.

$89 + {}^-116 = 16y + {}^-18y + y$
$^-27 = {}^-1y$

Divide both sides by the coefficient, $^-1$, to get y by itself.

$\frac{^-27}{^-1} = \frac{^-1y}{^-1}$
$27 = y$ The solution is 27.

57. $^-37(14x) + 28(21x) = |72 - 72| + |{}^-166 + 96|$

Simplify within the absolute values.

$^-37(14x) + 28(21x) = |0| + |{}^-70|$

Simplify the absolute values.

$^-37(14x) + 28(21x) = 0 + 70$

Use the associative property to multiply on the left.

$({}^-37 \cdot 14) \cdot x + (28 \cdot 21) \cdot x = 0 + 70$
$^-518x + 588x = 70$

Combine like terms.

$70x = 70$

Divide both sides by the coefficient, 70, to get x by itself.

$\frac{70x}{70} = \frac{70}{70}$
$x = 1$ The solution is 1.

2.5 Solving Equations with Several Steps

2.5 Section Exercises

1. $7p + 5 = 12$ To get $7p$ by itself, add
$\underline{\;{}^-5 \quad {}^-5\;}$ $^-5$ to both sides.
$7p + 0 = 7$
$7p = 7$ Divide *both* sides by 7.
$\frac{7p}{7} = \frac{7}{7}$
$p = 1$ The solution is 1.

Check: $7p + 5 = 12$
$7(1) + 5 = 12$
$7 + 5 = 12$
$12 = 12$ Balances

3. $2 = 8y - 6$ Add the opposite.
$2 = 8y + {}^-6$ To get $8y$ by itself, add
$\underline{6} \qquad \underline{\quad 6}$ 6 to both sides.
$8 = 8y + 0$ Divide *both* sides by 8.
$\frac{8}{8} = \frac{8y}{8}$
$1 = y$ The solution is 1.

Check: $2 = 8y - 6$
$2 = 8(1) - 6$
$2 = 8 - 6$
$2 = 2$ Balances

5. ${}^-3m + 1 = 1$ To get ${}^-3m$ by itself, add
$\underline{{}^-1} \qquad \underline{{}^-1}$ ${}^-1$ to both sides.
${}^-3m + 0 = 0$
${}^-3m = 0$ Divide *both* sides by ${}^-3$.
$\frac{{}^-3m}{{}^-3} = \frac{0}{{}^-3}$
$m = 0$ The solution is 0.

Check: ${}^-3m + 1 = 1$
${}^-3(0) + 1 = 1$
$0 + 1 = 1$
$1 = 1$ Balances

7. $28 = {}^-9a + 10$ To get ${}^-9a$ by itself, add
$\underline{{}^-10} \qquad \underline{{}^-10}$ ${}^-10$ to both sides.
$18 = {}^-9a + 0$
$18 = {}^-9a$ Divide *both* sides by ${}^-9$.
$\frac{18}{{}^-9} = \frac{{}^-9a}{{}^-9}$
${}^-2 = a$ The solution is ${}^-2$.

Check: $28 = {}^-9a + 10$
$28 = {}^-9({}^-2) + 10$
$28 = 18 + 10$
$28 = 28$ Balances

9. ${}^-5x - 4 = 16$ Add the opposite.
${}^-5x + {}^-4 = 16$ To get ${}^-5x$ by itself, add
$\underline{4} \qquad \underline{4}$ 4 to both sides.
${}^-5x + 0 = 20$
${}^-5x = 20$ Divide *both* sides by ${}^-5$.
$\frac{{}^-5x}{{}^-5} = \frac{20}{{}^-5}$
$x = {}^-4$ The solution is ${}^-4$.

Check: ${}^-5x - 4 = 16$
${}^-5({}^-4) - 4 = 16$
$20 - 4 = 16$
$16 = 16$ Balances

11. Solve with the variable on the *left* side.

$6p - 2 = 4p + 6$ Add the opposite.
$6p + {}^-2 = 4p + 6$ Add ${}^-4p$ to both sides.
$\underline{{}^-4p} \qquad \underline{{}^-4p}$
$2p + {}^-2 = 0 + 6$
$2p + {}^-2 = 6$ Add 2 to both sides.
$\underline{2} \qquad \underline{2}$
$2p + 0 = 8$
$2p = 8$ Divide *both* sides by 2.
$\frac{2p}{2} = \frac{8}{2}$
$p = 4$ The solution is 4.

Solve with the variable on the *right* side.

$6p - 2 = 4p + 6$ Add the opposite.
$6p + {}^-2 = 4p + 6$ Add ${}^-6p$ to both sides.
$\underline{{}^-6p} \qquad \underline{{}^-6p}$
$0 + {}^-2 = {}^-2p + 6$
${}^-2 = {}^-2p + 6$ Add ${}^-6$ to both sides.
$\underline{{}^-6} \qquad \underline{{}^-6}$
${}^-8 = {}^-2p + 0$
${}^-8 = {}^-2p$ Divide *both* sides by ${}^-2$.
$\frac{{}^-8}{{}^-2} = \frac{{}^-2p}{{}^-2}$
$4 = p$ The solution is 4.

Check: $6p - 2 = 4p + 6$
$6(4) - 2 = 4(4) + 6$
$24 - 2 = 16 + 6$
$22 = 22$ Balances

13. Solve with the variable on the *left* side.

${}^-2k - 6 = 6k + 10$ Add the opposite.
${}^-2k + {}^-6 = 6k + 10$ Add ${}^-6k$ to both sides.
$\underline{{}^-6k} \qquad \underline{{}^-6k}$
${}^-8k + {}^-6 = 0 + 10$
${}^-8k + {}^-6 = 10$ Add 6 to both sides.
$\underline{6} \qquad \underline{6}$
${}^-8k + 0 = 16$ Divide *both* sides by ${}^-8$.
$\frac{{}^-8k}{{}^-8} = \frac{16}{{}^-8}$
$k = {}^-2$ The solution is ${}^-2$.

continued

Solve with the variable on the *right* side.

$$\begin{aligned}
{}^-2k - 6 &= 6k + 10 && \text{Add the opposite.}\\
{}^-2k + {}^-6 &= 6k + 10 && \text{Add } 2k \text{ to both sides.}\\
\underline{2k \qquad} &\quad \underline{\quad 2k \qquad}\\
0 + {}^-6 &= 8k + 10\\
{}^-6 &= 8k + 10 && \text{Add } {}^-10 \text{ to both sides.}\\
\underline{{}^-10 \qquad} &\quad \underline{\quad {}^-10 \qquad}\\
{}^-16 &= 8k + 0 && \text{Divide } \textit{both} \text{ sides by 8.}\\
\frac{{}^-16}{8} &= \frac{8k}{8}\\
{}^-2 &= k && \text{The solution is } {}^-2.
\end{aligned}$$

Check:

$$\begin{aligned}
{}^-2k - 6 &= 6k + 10\\
{}^-2({}^-2) - 6 &= 6({}^-2) + 10\\
4 + {}^-6 &= {}^-12 + 10\\
{}^-2 &= {}^-2 && \text{Balances}
\end{aligned}$$

15. Solve with the variable on the *left* side.

$$\begin{aligned}
{}^-18 + 7a &= 2a + 7 && \text{Add } {}^-2a \text{ to both sides.}\\
\underline{\quad {}^-2a} &\quad \underline{\quad {}^-2a \quad}\\
{}^-18 + 5a &= 0 + 7\\
{}^-18 + 5a &= 7 && \text{Add 18 to both sides.}\\
\underline{18 \qquad} &\quad \underline{\quad 18 \quad}\\
0 + 5a &= 25\\
5a &= 25 && \text{Divide both sides by 5.}\\
\frac{5a}{5} &= \frac{25}{5}\\
a &= 5 && \text{The solution is 5.}
\end{aligned}$$

Solve with the variable on the *right* side.

$$\begin{aligned}
{}^-18 + 7a &= 2a + 7 && \text{Add } {}^-7a \text{ to both sides.}\\
\underline{\quad {}^-7a} &\quad \underline{\quad {}^-7a \quad}\\
{}^-18 + 0 &= {}^-5a + 7\\
{}^-18 &= {}^-5a + 7 && \text{Add } {}^-7 \text{ to both sides.}\\
\underline{\quad {}^-7} &\quad \underline{\qquad {}^-7}\\
{}^-25 &= {}^-5a + 0\\
{}^-25 &= {}^-5a && \text{Divide both sides by } {}^-5.\\
\frac{{}^-25}{{}^-5} &= \frac{{}^-5a}{{}^-5}\\
5 &= a && \text{The solution is 5.}
\end{aligned}$$

Check:

$$\begin{aligned}
{}^-18 + 7a &= 2a + 7\\
{}^-18 + 7(5) &= 2(5) + 7\\
{}^-18 + 35 &= 10 + 7\\
17 &= 17 && \text{Balances}
\end{aligned}$$

17.

$$\begin{aligned}
8(w - 2) &= 32 && \text{Distribute.}\\
8w - 16 &= 32 && \text{Add the opposite.}\\
8w + {}^-16 &= 32 && \text{Add 16 to both sides.}\\
\underline{\qquad 16} &\quad \underline{\quad 16 \quad}\\
8w + 0 &= 48\\
8w &= 48 && \text{Divide } \textit{both} \text{ sides by 8.}\\
\frac{8w}{8} &= \frac{48}{8}\\
w &= 6 && \text{The solution is 6.}
\end{aligned}$$

19.

$$\begin{aligned}
{}^-10 &= 2(y + 4) && \text{Distribute.}\\
{}^-10 &= 2y + 8 && \text{Add } {}^-8 \text{ to both sides.}\\
\underline{\quad {}^-8} &\quad \underline{\qquad {}^-8 \quad}\\
{}^-18 &= 2y + 0\\
{}^-18 &= 2y && \text{Divide } \textit{both} \text{ sides by 2.}\\
\frac{{}^-18}{2} &= \frac{2y}{2}\\
{}^-9 &= y && \text{The solution is } {}^-9.
\end{aligned}$$

21.

$$\begin{aligned}
{}^-4(t + 2) &= 12 && \text{Distribute.}\\
{}^-4t + {}^-8 &= 12 && \text{Add 8 to both sides.}\\
\underline{\qquad 8} &\quad \underline{\quad 8 \quad}\\
{}^-4t + 0 &= 20\\
{}^-4t &= 20 && \text{Divide } \textit{both} \text{ sides by } {}^-4.\\
\frac{{}^-4t}{{}^-4} &= \frac{20}{{}^-4}\\
t &= {}^-5 && \text{The solution is } {}^-5.
\end{aligned}$$

23.

$$\begin{aligned}
6(x - 5) &= {}^-30 && \text{Distribute.}\\
6x - 30 &= {}^-30 && \text{Add the opposite.}\\
6x + {}^-30 &= {}^-30 && \text{Add 30 to both sides.}\\
\underline{\qquad 30} &\quad \underline{\quad 30 \quad}\\
6x + 0 &= 0\\
6x &= 0 && \text{Divide } \textit{both} \text{ sides by 6.}\\
\frac{6x}{6} &= \frac{0}{6}\\
x &= 0 && \text{The solution is 0.}
\end{aligned}$$

25.

$$\begin{aligned}
{}^-12 &= 12(h - 2) && \text{Distribute.}\\
{}^-12 &= 12h - 24 && \text{Add the opposite.}\\
{}^-12 &= 12h + {}^-24 && \text{Add 24 to both sides.}\\
\underline{24} &\quad \underline{\qquad\quad 24}\\
12 &= 12h + 0\\
12 &= 12h && \text{Divide } \textit{both} \text{ sides by 12.}\\
\frac{12}{12} &= \frac{12h}{12}\\
1 &= h && \text{The solution is 1.}
\end{aligned}$$

27.

$$\begin{aligned}
0 &= {}^-2(y + 2) && \text{Distribute.}\\
0 &= {}^-2y + {}^-4 && \text{Add 4 to both sides.}\\
\underline{4} &\quad \underline{\qquad\quad 4}\\
4 &= {}^-2y + 0\\
4 &= {}^-2y && \text{Divide } \textit{both} \text{ sides by } {}^-2.\\
\frac{4}{{}^-2} &= \frac{{}^-2y}{{}^-2}\\
{}^-2 &= y && \text{The solution is } {}^-2.
\end{aligned}$$

29.
$$\begin{array}{rcll}
6m + 18 &=& 0 & \text{To get } 6m \text{ by itself, add} \\
^{-}18 && ^{-}18 & ^{-}18 \text{ to both sides.} \\
\hline
6m + 0 &=& ^{-}18 & \\
6m &=& ^{-}18 & \text{Divide } \textit{both} \text{ sides by 6.} \\
\dfrac{6m}{6} &=& \dfrac{^{-}18}{6} & \\
m &=& ^{-}3 & \text{The solution is } ^{-}3.
\end{array}$$

31.
$$\begin{array}{rcll}
6 &=& 9w - 12 & \text{Add the opposite.} \\
6 &=& 9w + {}^{-}12 & \text{Add 12 to both sides.} \\
12 && 12 & \\
\hline
18 &=& 9w + 0 & \\
18 &=& 9w & \text{Divide } \textit{both} \text{ sides by 9.} \\
\dfrac{18}{9} &=& \dfrac{9w}{9} & \\
2 &=& w & \text{The solution is 2.}
\end{array}$$

33.
$$\begin{array}{rcll}
5x &=& 3x + 10 & \text{Add } ^{-}3x \text{ to both sides} \\
&&& \text{to get the variable} \\
&&& \text{term on one side.} \\
^{-}3x && ^{-}3x & \\
\hline
2x &=& 0 + 10 & \\
2x &=& 10 & \text{Divide } \textit{both} \text{ sides by 2.} \\
\dfrac{2x}{2} &=& \dfrac{10}{2} & \\
x &=& 5 & \text{The solution is 5.}
\end{array}$$

35.
$$\begin{array}{rcll}
2a + 11 &=& 8a - 7 & \text{Add the opposite.} \\
2a + 11 &=& 8a + {}^{-}7 & \text{Add } ^{-}2a \text{ to both sides.} \\
^{-}2a && ^{-}2a & \\
\hline
0 + 11 &=& 6a + {}^{-}7 & \\
11 &=& 6a + {}^{-}7 & \text{Add 7 to both sides.} \\
7 && 7 & \\
\hline
18 &=& 6a + 0 & \\
18 &=& 6a & \text{Divide } \textit{both} \text{ sides by 6.} \\
\dfrac{18}{6} &=& \dfrac{6a}{6} & \\
3 &=& a & \text{The solution is 3.}
\end{array}$$

37.
$$\begin{array}{rcll}
7 - 5b &=& 28 + 2b & \text{Add the opposite.} \\
7 + {}^{-}5b &=& 28 + 2b & \text{Add } 5b \text{ to both sides.} \\
5b && 5b & \\
\hline
7 + 0 &=& 28 + 7b & \\
7 &=& 28 + 7b & \text{Add } ^{-}28 \text{ to both sides.} \\
^{-}28 && ^{-}28 & \\
\hline
^{-}21 &=& 0 + 7b & \\
^{-}21 &=& 7b & \text{Divide } \textit{both} \text{ sides by 7.} \\
\dfrac{^{-}21}{7} &=& \dfrac{7b}{7} & \\
^{-}3 &=& b & \text{The solution is } ^{-}3.
\end{array}$$

39.
$$\begin{array}{rcll}
^{-}20 + 2k &=& k - 4k & \text{Add the opposite.} \\
^{-}20 + 2k &=& k + {}^{-}4k & \text{Combine like terms.} \\
^{-}20 + 2k &=& ^{-}3k & \text{Add } ^{-}2k \text{ to both sides.} \\
^{-}2k && ^{-}2k & \\
\hline
^{-}20 + 0 &=& ^{-}5k & \\
^{-}20 &=& ^{-}5k & \text{Divide } \textit{both} \text{ sides by } ^{-}5. \\
\dfrac{^{-}20}{^{-}5} &=& \dfrac{^{-}5k}{^{-}5} & \\
4 &=& k & \text{The solution is 4.}
\end{array}$$

41.
$$\begin{array}{rcll}
10(c - 6) + 4 &=& 2 + c - 58 & \text{Distribute.} \\
10c - 60 + 4 &=& 2 + c - 58 & \text{Add the opposites.} \\
10c + {}^{-}60 + 4 &=& 2 + c + {}^{-}58 & \text{Group like terms.} \\
10c + {}^{-}60 + 4 &=& 2 + {}^{-}58 + c & \text{Combine like terms.} \\
10c + {}^{-}56 &=& ^{-}56 + c & \text{Add } ^{-}c \text{ to both} \\
^{-}c && ^{-}c & \text{sides.} \\
\hline
9c + {}^{-}56 &=& ^{-}56 + 0 & \\
9c + {}^{-}56 &=& ^{-}56 & \text{Add 56 to both sides.} \\
56 && 56 & \\
\hline
9c + 0 &=& 0 & \text{Divide } \textit{both} \text{ sides by 9.} \\
\dfrac{9c}{9} &=& \dfrac{0}{9} & \\
c &=& 0 & \text{The solution is 0.}
\end{array}$$

43.
$$\begin{array}{rcll}
^{-}18 + 13y + 3 &=& 3(5y - 1) - 2 & \text{Distribute.} \\
^{-}18 + 13y + 3 &=& 15y - 3 - 2 & \text{Add the opposites.} \\
^{-}18 + 13y + 3 &=& 15y + {}^{-}3 + {}^{-}2 & \text{Group like terms.} \\
13y + {}^{-}18 + 3 &=& 15y + {}^{-}3 + {}^{-}2 & \text{Combine like terms.} \\
13y + {}^{-}15 &=& 15y + {}^{-}5 & \text{Add } ^{-}13y \\
^{-}13y && ^{-}13y & \text{to both sides.} \\
\hline
0 + {}^{-}15 &=& 2y + {}^{-}5 & \\
^{-}15 &=& 2y + {}^{-}5 & \text{Add 5} \\
5 && 5 & \text{to both sides.} \\
\hline
^{-}10 &=& 2y + 0 & \\
^{-}10 &=& 2y & \text{Divide both sides by 2.} \\
\dfrac{^{-}10}{2} &=& \dfrac{2y}{2} & \\
^{-}5 &=& y & \text{The solution is } ^{-}5.
\end{array}$$

45.
$$\begin{array}{rcll}
6 - 4n + 3n &=& 20 - 35 & \text{Add the opposites.} \\
6 + {}^{-}4n + 3n &=& 20 + {}^{-}35 & \text{Combine like terms.} \\
6 + {}^{-}1n &=& ^{-}15 & \text{Add } ^{-}6 \text{ to} \\
^{-}6 && ^{-}6 & \text{both sides.} \\
\hline
0 + {}^{-}1n &=& ^{-}21 & \text{Divide both sides by } ^{-}1. \\
\dfrac{^{-}1n}{^{-}1} &=& \dfrac{^{-}21}{^{-}1} & \\
n &=& 21 & \text{The solution is 21.}
\end{array}$$

47.
$$\begin{aligned}
6(c-2) &= 7(c-6) && \text{Distribute.}\\
6c-12 &= 7c-42 && \text{Add the opposites.}\\
6c+{}^{-}12 &= 7c+{}^{-}42 && \text{Add } {}^{-}6c \text{ to}\\
{}^{-}6c \quad & \quad\ {}^{-}6c && \text{both sides.}\\
0+{}^{-}12 &= c+{}^{-}42 && \\
{}^{-}12 &= c+{}^{-}42 && \text{Add 42 to both sides.}\\
42 \quad & \quad\ 42 && \\
30 &= c+0 && \\
30 &= c && \text{The solution is 30.}
\end{aligned}$$

49.
$$\begin{aligned}
{}^{-}5(2p+2)-7 &= 3(2p+5) && \text{Distribute.}\\
{}^{-}10p+{}^{-}10-7 &= 6p+15 && \text{Add the opposite.}\\
{}^{-}10p+{}^{-}10+{}^{-}7 &= 6p+15 && \text{Combine like terms.}\\
{}^{-}10p+{}^{-}17 &= 6p+15 && \text{Add } {}^{-}6p \text{ to both}\\
{}^{-}6p \quad & \quad\ {}^{-}6p && \text{sides.}\\
{}^{-}16p+{}^{-}17 &= 0+15 && \\
{}^{-}16p+{}^{-}17 &= 15 && \text{Add 17 to both}\\
17 \quad & \quad\ 17 && \text{sides.}\\
{}^{-}16p+0 &= 32 && \text{Divide both sides by } {}^{-}16.\\
\frac{{}^{-}16p}{{}^{-}16} &= \frac{32}{{}^{-}16} && \\
p &= {}^{-}2 && \text{The solution is } {}^{-}2.
\end{aligned}$$

51.
$$\begin{aligned}
{}^{-}6b-4b+7b &= 10-b+3b && \text{Add the opposites.}\\
{}^{-}6b+{}^{-}4b+7b &= 10+{}^{-}b+3b && \text{Combine like terms.}\\
{}^{-}3b &= 10+2b && \text{Add } {}^{-}2b \text{ to}\\
{}^{-}2b \quad & \quad\ {}^{-}2b && \text{both sides.}\\
{}^{-}5b &= 10+0 && \\
{}^{-}5b &= 10 && \text{Divide both sides by } {}^{-}5.\\
\frac{{}^{-}5b}{{}^{-}5} &= \frac{10}{{}^{-}5} && \\
b &= {}^{-}2 && \text{The solution is } {}^{-}2.
\end{aligned}$$

53. The series of steps may vary. One possibility is:

$$\begin{aligned}
{}^{-}2t-10 &= 3t+5 && \text{Change subtraction to adding the opposite.}\\
{}^{-}2t+{}^{-}10 &= 3t+5 && \text{Add } 2t \text{ to both sides}\\
2t \quad & \quad\ 2t && \text{(addition property).}\\
0+{}^{-}10 &= 5t+5 && \text{Add } {}^{-}5 \text{ to both sides}\\
{}^{-}5 \quad & \quad\ {}^{-}5 && \text{(addition property).}\\
\frac{{}^{-}15}{5} &= \frac{5t}{5} && \text{Divide both sides by 5 (division property).}\\
{}^{-}3 &= t &&
\end{aligned}$$

The solution is $^{-}3$.

55. *Check:*
$$\begin{aligned}
{}^{-}8+4a &= 2a+2\\
{}^{-}8+4(3) &= 2(3)+2\\
{}^{-}8+12 &= 6+2\\
4 &\neq 8
\end{aligned}$$

The check does not balance, so 3 is not the correct solution. The student added $^{-}2a$ to $^{-}8$ on the left side, instead of adding $^{-}2a$ to $4a$. The correct solution, obtained using $^{-}8+2a=2, 2a=10$, is $a=5$.

57. **(a)** It must be negative, because the sum of two positive numbers is always positive.

(b) The sum of x and a positive number is negative, so x must be negative.

58. **(a)** It must be positive, because the sum of two negative numbers is always negative.

(b) The sum of d and a negative number is positive, so d must be positive.

59. **(a)** It must be positive. When the signs are the same, the product is positive, and when the signs are different, the product is negative.

(b) The product of n and a negative number is negative, so n must be positive.

60. **(a)** It must be negative also. When the signs are different, the product is negative, and when the signs match, the product is positive.

(b) The product of y and a negative number is positive, so y must be negative.

Chapter 2 Review Exercises

1. **(a)** In the expression $^{-}3+4k$, k is the variable, 4 is the coefficient, and $^{-}3$ is the constant term.

(b) The term that has 20 as the constant term and $^{-}9$ as the coefficient is $^{-}9y+20$.

2. **(a)** Evaluate $4c+10$ when c is 15.

$$\begin{aligned}
&4c+10 && \text{Replace } c \text{ with 15.}\\
&\underbrace{4\cdot 15}+10 && \\
&\underbrace{60+10} && \\
&70 && \text{Order 70 test tubes.}
\end{aligned}$$

(b) Evaluate $4c+10$ when c is 24.

$$\begin{aligned}
&4c+10 && \text{Replace } c \text{ with 24.}\\
&\underbrace{4\cdot 24}+10 && \\
&\underbrace{96+10} && \\
&106 && \text{Order 106 test tubes.}
\end{aligned}$$

3. **(a)** x^2y^4 means $x\cdot x\cdot y\cdot y\cdot y\cdot y$

(b) $5ab^3$ means $5\cdot a\cdot b\cdot b\cdot b$

4. **(a)** n^2 means

$n \cdot n$ Replace n with $^{-}3$.

$\underbrace{^{-}3 \cdot {^{-}3}}$

9

(b) n^3 means

$n \cdot n \cdot n$ Replace n with $^{-}3$.

$\underbrace{^{-}3 \cdot {^{-}3}} \cdot {^{-}3}$

$\underbrace{9 \cdot {^{-}3}}$

$^{-}27$

(c) $^{-}4mp^2$ means

$^{-}4 \cdot m \cdot p \cdot p$ Replace m with 2 and p with 4.

$\underbrace{^{-}4 \cdot 2} \cdot 4 \cdot 4$

$\underbrace{^{-}8 \cdot 4} \cdot 4$

$\underbrace{^{-}32 \cdot 4}$

$^{-}128$

(d) $5m^4n^2$ means

$5 \cdot m \cdot m \cdot m \cdot m \cdot n \cdot n$ Replace m with 2 and n with $^{-}3$.

$\underbrace{5 \cdot 2} \cdot 2 \cdot 2 \cdot 2 \cdot {^{-}3} \cdot {^{-}3}$

$\underbrace{10 \cdot 2} \cdot 2 \cdot 2 \cdot {^{-}3} \cdot {^{-}3}$

$\underbrace{20 \cdot 2} \cdot 2 \cdot {^{-}3} \cdot {^{-}3}$

$\underbrace{40 \cdot 2} \cdot {^{-}3} \cdot {^{-}3}$

$\underbrace{80 \cdot {^{-}3}} \cdot {^{-}3}$

$\underbrace{^{-}240 \cdot {^{-}3}}$

720

5. $ab + ab^2 + 2ab$

$\underline{1ab} + ab^2 + \underline{2ab}$ Combine like terms.

$3ab + ab^2$ or $ab^2 + 3ab$

6. $^{-}3x + 2y - x - 7$ Rewrite x as $1x$.

$^{-}3x + 2y - 1x - 7$ Add the opposite.

$\underline{^{-}3x} + 2y + \underline{^{-}1x} + {^{-}7}$ Combine like terms.

$^{-}4x + 2y + {^{-}7}$ or $^{-}4x + 2y - 7$

7. $^{-}8({^{-}2}g^3)$ Associative property

$({^{-}8} \cdot {^{-}2}) \cdot g^3$

$16 \cdot g^3$

$16g^3$

8. $4(3r^2t)$ Associative property

$(4 \cdot 3) \cdot r^2t$

$12 \cdot r^2t$

$12r^2t$

9. $5(k + 2)$ Distribute.

$5 \cdot k + 5 \cdot 2$

$5k + 10$

10. $^{-}2(3b + 4)$ Distribute.

$^{-}2 \cdot 3b + {^{-}2} \cdot 4$

$^{-}6b + {^{-}8}$ or $^{-}6b - 8$

11. $3(2y - 4) + 12$ Distribute.

$\underbrace{3 \cdot 2y} - \underbrace{3 \cdot 4} + 12$

$6y - \quad 12 + 12$

$6y + {^{-}12} + 12$

$6y + 0$

$6y$

12. $^{-}4 + 6(4x + 1) - 4x$ Distribute.

$^{-}4 + 24x + 6 - 4x$

$^{-}4 + 24x + 6 + {^{-}4x}$

$2 + 20x$ or $20x + 2$

13. Expressions will vary. One possibility is $6a^3 + a^2 + 3a - 6$.

14.

$$\begin{aligned} 16 + n &= 5 \quad \text{Add } {^{-}16} \text{ to both sides.} \\ \underline{^{-}16} \quad &\quad \underline{^{-}16} \\ 0 + n &= {^{-}11} \\ n &= {^{-}11} \quad \text{The solution is } {^{-}11}. \end{aligned}$$

Check: $16 + n = 5$

$16 + {^{-}11} = 5$

$5 = 5$ Balances

15.

$$\begin{aligned} ^{-}4 + 2 &= 2a - 6 - a \\ ^{-}4 + 2 &= 2a + {^{-}6} + {^{-}1a} \\ ^{-}2 &= 1a + {^{-}6} \\ \underline{6} \quad &\quad \underline{6} \\ 4 &= 1a + 0 \\ 4 &= a \end{aligned}$$

Check: $^{-}4 + 2 = 2a - 6 - a$

$^{-}4 + 2 = 2(4) - 6 - 4$

$^{-}2 = 8 + {^{-}6} + {^{-}4}$

$^{-}2 = 2 + {^{-}4}$

$^{-}2 = {^{-}2}$ Balances

The solution is 4.

16. $48 = {^{-}6m}$ Divide *both* sides by $^{-}6$.

$$\frac{48}{^{-}6} = \frac{^{-}6m}{^{-}6}$$

$^{-}8 = m$ The solution is $^{-}8$.

17.

$$\begin{aligned} k - 5k &= {^{-}40} \\ 1k - 5k &= {^{-}40} \\ 1k + {^{-}5k} &= {^{-}40} \quad \text{Combine like terms.} \\ ^{-}4k &= {^{-}40} \quad \text{Divide } \textit{both} \text{ sides by } {^{-}4}. \\ \frac{^{-}4k}{^{-}4} &= \frac{^{-}40}{^{-}4} \\ k &= 10 \quad \text{The solution is 10.} \end{aligned}$$

18. $\underbrace{^{-}17+11+6}=7t$

$0 = 7t$ Divide *both* sides by 7.

$\frac{0}{7}=\frac{7t}{7}$

$0=t$ The solution is 0.

19. $^{-}2p+5p=3-21$

$^{-}2p+5p=3+{}^{-}21$

$3p={}^{-}18$ Divide *both* sides by 3.

$\frac{3p}{3}=\frac{^{-}18}{3}$

$p={}^{-}6$ The solution is $^{-}6$.

20. $^{-}30=3(^{-}5r)$

$^{-}30={}^{-}15r$ Divide *both* sides by $^{-}15$.

$\frac{^{-}30}{^{-}15}=\frac{^{-}15r}{^{-}15}$

$2=r$ The solution is 2.

21. $12={}^{-}h$

$12={}^{-}1h$

$\frac{12}{^{-}1}=\frac{^{-}1h}{^{-}1}$

$^{-}12=h$ The solution is $^{-}12$.

22. $12w-4 = 8w+12$

$12w+{}^{-}4 = 8w+12$

$\underline{^{-}8w} \quad \underline{^{-}8w}$

$4w+{}^{-}4 = 0+12$

$4w+{}^{-}4 = 12$ Add 4 to both sides.

$\underline{4} \quad \underline{4}$

$4w+0 = 16$

$4w = 16$ Divide *both* sides by 4.

$\frac{4w}{4}=\frac{16}{4}$

$w = 4$ The solution is 4.

23. $0={}^{-}4(c+2)$ Distribute.

$0={}^{-}4\cdot c+{}^{-}4\cdot 2$

$0={}^{-}4c+{}^{-}8$

$8 \quad \underline{8}$ Add 8 to both sides.

$8={}^{-}4c+0$

$8={}^{-}4c$ Divide *both* sides by $^{-}4$.

$\frac{8}{^{-}4}=\frac{^{-}4c}{^{-}4}$

$^{-}2=c$ The solution is $^{-}2$.

24. $34 = 2n+4$ Add $^{-}4$ to both sides.

$\underline{^{-}4} \quad \underline{^{-}4}$

$30 = 2n+0$

$30 = 2n$ Divide *both* sides by 2.

$\frac{30}{2}=\frac{2n}{2}$

$15 = n$

The number of employees is 15.

25. [2.5] $12+7a = 4a-3$ Add $^{-}4a$ to both sides.

$\underline{^{-}4a} \quad \underline{^{-}4a}$

$12+3a = 0-3$

$12+3a = {}^{-}3$

$\underline{^{-}12} \quad \underline{^{-}12}$

$0+3a = {}^{-}15$ Divide *both* sides by 3.

$\frac{3a}{3}=\frac{^{-}15}{3}$

$a = {}^{-}5$ The solution is $^{-}5$.

26. [2.5] $^{-}2(p-3) = {}^{-}14$ Distribute.

$^{-}2p+6 = {}^{-}14$ Add $^{-}6$ to both sides.

$\underline{^{-}6} \quad \underline{^{-}6}$

$^{-}2p+0 = {}^{-}20$

$\frac{^{-}2p}{^{-}2}=\frac{^{-}20}{^{-}2}$

$p = 10$ The solution is 10.

27. [2.5] $10y = 6y+20$ Add $^{-}6y$ to both sides.

$\underline{^{-}6y} \quad \underline{^{-}6y}$

$4y = 0+20$

$\frac{4y}{4}=\frac{20}{4}$

$y = 5$ The solution is 5.

28. [2.5] $2m-7m=5-20$ Add the opposites.

$2m+{}^{-}7m=5+{}^{-}20$ Combine like terms.

$^{-}5m={}^{-}15$ Divide *both* sides by $^{-}5$.

$\frac{^{-}5m}{^{-}5}=\frac{^{-}15}{^{-}5}$

$m=3$ The solution is 3.

29. [2.5]

$$\begin{aligned} 20 &= 3x - 7 \\ 20 &= 3x + {}^{-}7 \\ \underline{7} \quad & \quad \underline{\quad\quad 7} \\ 27 &= 3x + 0 \\ \frac{27}{3} &= \frac{3x}{3} \\ 9 &= x \end{aligned}$$

The solution is 9.

30. [2.5]

$$\begin{aligned} b + 6 &= 3b - 8 \\ \underline{{}^{-}3b} \quad & \quad \underline{{}^{-}3b} \\ {}^{-}2b + 6 &= 0 - 8 \\ {}^{-}2b + 6 &= {}^{-}8 \\ \underline{{}^{-}6} \quad & \quad \underline{{}^{-}6} \\ {}^{-}2b + 0 &= {}^{-}14 \\ \frac{{}^{-}2b}{{}^{-}2} &= \frac{{}^{-}14}{{}^{-}2} \\ b &= 7 \end{aligned}$$

The solution is 7.

31. [2.3]

$$\begin{aligned} z + 3 &= 0 \\ \underline{{}^{-}3} \quad & \quad \underline{{}^{-}3} \\ z + 0 &= {}^{-}3 \\ z &= {}^{-}3 \end{aligned}$$

The solution is $^{-}3$.

32. [2.5]

$$\begin{aligned} 3(2n - 1) &= 3(n + 3) && \text{Distribute.} \\ 6n - 3 &= 3n + 9 \\ \underline{{}^{-}3n} \quad & \quad \underline{{}^{-}3n} \\ 3n - 3 &= 0 + 9 \\ 3n - 3 &= 9 && \text{Add 3 to both sides.} \\ \underline{3} \quad & \quad \underline{3} \\ 3n + 0 &= 12 \\ \frac{3n}{3} &= \frac{12}{3} \\ n &= 4 \end{aligned}$$

The solution is 4.

33. [2.5]

$$\begin{aligned} {}^{-}4 + 46 &= 7({}^{-}3t + 6) && \text{Distribute.} \\ {}^{-}4 + 46 &= {}^{-}21t + 42 \\ 42 &= {}^{-}21t + 42 && \text{Add } {}^{-}42 \text{ to both sides.} \\ \underline{{}^{-}42} \quad & \quad \underline{{}^{-}42} \\ 0 &= {}^{-}21t + 0 \\ \frac{0}{{}^{-}21} &= \frac{{}^{-}21t}{{}^{-}21} \\ 0 &= t \end{aligned}$$

The solution is 0.

34. [2.5]

$$\begin{aligned} 6 + 10d - 19 &= 2(3d + 4) - 1 && \text{Distribute.} \\ 6 + 10d + {}^{-}19 &= 6d + 8 - 1 \\ {}^{-}13 + 10d &= 6d + 7 \\ \underline{{}^{-}6d} \quad & \quad \underline{{}^{-}6d} \\ {}^{-}13 + 4d &= 0 + 7 \\ {}^{-}13 + 4d &= 7 && \text{Add 13 to both sides.} \\ \underline{13} \quad & \quad \underline{13} \\ 0 + 4d &= 20 \\ \frac{4d}{4} &= \frac{20}{4} \\ d &= 5 \end{aligned}$$

The solution is 5.

35. [2.5]

$$\begin{aligned} {}^{-}4(3b + 9) &= 24 + 3(2b - 8) && \text{Distribute.} \\ {}^{-}12b - 36 &= 24 + 6b - 24 \\ {}^{-}12b + {}^{-}36 &= 24 + 6b + {}^{-}24 \\ {}^{-}12b + {}^{-}36 &= 6b && \text{Add } 12b \text{ to both sides.} \\ \underline{12b} \quad & \quad \underline{12b} \\ 0 + {}^{-}36 &= 18b \\ \frac{{}^{-}36}{18} &= \frac{18b}{18} \\ {}^{-}2 &= b \end{aligned}$$

The solution is $^{-}2$.

Chapter 2 Test

1. In the expression $^{-}7w + 6$, $^{-}7$ is the coefficient, w is the variable, and 6 is the constant term.

2. Evaluate the expression $3a + 2c$ when a is 45 and c is 21.

$$\begin{gathered} 3a + 2c \\ 3 \cdot 45 + 2 \cdot 21 \\ 135 + 42 \\ 177 \end{gathered}$$

Buy 177 hot dogs.

3. x^5y^3 means $x \cdot x \cdot x \cdot x \cdot x \cdot y \cdot y \cdot y$

4. $4ab^4$ means $4 \cdot a \cdot b \cdot b \cdot b \cdot b$

5. $^{-}2s^2t$ means

$$\begin{gathered} {}^{-}2 \cdot s \cdot s \cdot t \\ \underbrace{{}^{-}2 \cdot {}^{-}5} \cdot {}^{-}5 \cdot 4 \\ \underbrace{10 \cdot {}^{-}5} \cdot 4 \\ \underbrace{{}^{-}50 \cdot 4} \\ {}^{-}200 \end{gathered}$$

Replace s with $^{-}5$ and t with 4.

6.

$$\begin{gathered} 3w^3 - 8w^3 + w^3 \\ 3w^3 - 8w^3 + 1w^3 \\ \underbrace{3w^3 + {}^{-}8w^3} + 1w^3 \\ \underbrace{{}^{-}5w^3 + 1w^3} \\ {}^{-}4w^3 \end{gathered}$$

7.
$$xy - xy$$
$$1xy - 1xy$$
$$(1-1)xy$$
$$0xy$$
$$0$$

8.
$$^{-}6c - 5 + 7c + 5$$
$$^{-}6c + {}^{-}5 + 7c + 5$$
$$\underbrace{^{-}6c + 7c} + \underbrace{^{-}5 + 5}$$
$$1c \quad + \quad 0$$
$$1c \text{ or } c$$

9. $3m^2 - 3m + 3mn$
There are no like terms.
The expression cannot be simplified.

10. $^{-}10(4b^2)$ Associative property of multiplication
$$(^{-}10 \cdot 4) \cdot b^2$$
$$^{-}40b^2$$

11. $^{-}5(^{-}3k)$ Associative property of multiplication
$$(^{-}5 \cdot {}^{-}3) \cdot k$$
$$15k$$

12. $7(3t + 4)$ Distributive property
$$7(3t) + 7(4)$$
$$21t + 28$$

13. $^{-}4(a + 6)$ Distributive property
$$^{-}4 \cdot a + {}^{-}4 \cdot 6$$
$$^{-}4a + {}^{-}24$$
$$^{-}4a - 24$$

14.
$$^{-}8 + 6(x - 2) + 5$$
$$^{-}8 + 6x - 12 + 5$$
$$^{-}8 + 6x + {}^{-}12 + 5 \quad \text{Combine like terms.}$$
$$6x + {}^{-}15 \text{ or } 6x - 15$$

15.
$$^{-}9b - c - 3 + 9 + 2c$$
$$^{-}9b - 1c - 3 + 9 + 2c$$
$$^{-}9b + {}^{-}1c + {}^{-}3 + 9 + 2c$$
$$^{-}9b + c + 6$$

16.
$$^{-}4 = x - 9$$
$$\underline{9 \qquad\quad 9}$$
$$5 = x + 0$$
$$5 = x$$

Check:
$$^{-}4 = x - 9$$
$$^{-}4 = 5 - 9$$
$$^{-}4 = {}^{-}4$$
Balances

The solution is 5.

17.
$$^{-}7w = 77$$
$$\frac{^{-}7w}{^{-}7} = \frac{77}{^{-}7}$$
$$w = {}^{-}11$$

Check:
$$^{-}7w = 77$$
$$^{-}7 \cdot {}^{-}11 = 77$$
$$77 = 77$$
Balances

The solution is $^{-}11$.

18.
$$^{-}p = 14$$
$$^{-}1p = 14$$
$$\frac{^{-}1p}{^{-}1} = \frac{14}{^{-}1}$$
$$p = {}^{-}14$$

Check:
$$^{-}p = 14$$
$$^{-}1p = 14$$
$$^{-}1 \cdot {}^{-}14 = 14$$
$$14 = 14$$
Balances

The solution is $^{-}14$.

19.
$$^{-}15 = {}^{-}3(a + 2)$$
$$^{-}15 = {}^{-}3a - 6$$
$$\underline{6 \qquad\qquad 6}$$
$$^{-}9 = {}^{-}3a$$
$$\frac{^{-}9}{^{-}3} = \frac{^{-}3a}{^{-}3}$$
$$3 = a$$

Check:
$$^{-}15 = {}^{-}3(a + 2)$$
$$^{-}15 = {}^{-}3(3 + 2)$$
$$^{-}15 = {}^{-}3(5)$$
$$^{-}15 = {}^{-}15$$
Balances

The solution is 3.

20.
$$6n + 8 - 5n = {}^{-}4 + 4$$
$$6n + 8 + {}^{-}5n = 0$$
$$n + 8 = 0$$
$$\underline{\quad {}^{-}8 \qquad {}^{-}8}$$
$$n = {}^{-}8$$

The solution is $^{-}8$.

21.
$$5 - 20 = 2m - 3m$$
$$5 + {}^{-}20 = 2m + {}^{-}3m$$
$$^{-}15 = {}^{-}1m$$
$$\frac{^{-}15}{^{-}1} = \frac{^{-}1m}{^{-}1}$$
$$15 = m$$

The solution is 15.

22.
$$^{-}2x + 2 = 5x + 9 \quad \text{Add } 2x \text{ to both sides.}$$
$$\underline{2x \qquad\qquad 2x}$$
$$2 = 7x + 9$$
$$\underline{^{-}9 \qquad\qquad {}^{-}9}$$
$$\frac{^{-}7}{7} = \frac{7x}{7}$$
$$^{-}1 = x$$

The solution is $^{-}1$.

23.
$$3m - 5 = 7m - 13$$
$$\underline{^{-}3m \qquad\qquad {}^{-}3m}$$
$$0 - 5 = 4m - 13$$
$$^{-}5 = 4m - 13 \quad \text{Add 13 to both sides.}$$
$$\underline{13 \qquad\qquad 13}$$
$$\frac{8}{4} = \frac{4m}{4}$$
$$2 = m$$

The solution is 2.

24.
$$\begin{aligned} 2 + 7b - 44 &= {}^-3b + 12 + 9b \\ 7b - 42 &= 6b + 12 \\ {}^-6b \quad & \quad {}^-6b \\ \hline 1b - 42 &= 12 \\ 42 \quad & \quad 42 \\ \hline 1b &= 54 \\ b &= 54 \end{aligned}$$

Add 42 to both sides.

The solution is 54.

25.
$$\begin{aligned} 3c - 24 &= 6(c - 4) \\ 3c - 24 &= 6c - 24 \\ {}^-3c \quad & \quad {}^-3c \\ \hline {}^-24 &= 3c - 24 \\ 24 \quad & \quad 24 \\ \hline 0 &= 3c \\ \frac{0}{3} &= \frac{3c}{3} \\ 0 &= c \end{aligned}$$

Distribute.

Add 24 to both sides.

The solution is 0.

26. Equations will vary. Two possibilities are $x - 5 = {}^-9$ and ${}^-24 = 6y$.

Solving:

$$\begin{aligned} x - 5 &= {}^-9 \\ {}^+5 \quad & \quad {}^+5 \\ \hline x &= {}^-4 \end{aligned} \qquad \begin{aligned} {}^-24 &= 6y \\ \frac{{}^-24}{6} &= \frac{6y}{6} \\ {}^-4 &= y \end{aligned}$$

Cumulative Review Exercises (Chapters 1–2)

1. 306,000,004,210 in words is three hundred six billion, four thousand, two hundred ten.

2. eight hundred million, sixty-six thousand: 800,066,000

3. **(a)** ${}^-3$ lies to the *right* of ${}^-10$ on the number line, so ${}^-3 > {}^-10$.

(b) ${}^-1$ lies to the *left* of 0 on the number line, so ${}^-1 < 0$.

4. **(a)** ${}^-6 + 2 = 2 + {}^-6$

Commutative property of addition:

Changing the order of the addends does not change the sum.

(b) $0 \cdot 25 = 0$

Multiplication property of zero: Multiplying any number by 0 gives a product of 0.

(c) $5({}^-6 + 4) = 5 \cdot {}^-6 + 5 \cdot 4$

Distributive property:

Multiplication distributes over addition.

5. **(a)** $9047 \approx 9000$

Underline the hundreds place: $9\underline{0}47$

The next digit is 4 or less, so leave 0 as 0. Change 4 and 7 to 0.

(b) $289{,}610 \approx 290{,}000$

Underline the thousands place: $28\underline{9}{,}610$

The next digit is 5 or more, so add 1 to 9, write the 0 and add 1 to the ten-thousands place. Change 6 and 1 to 0.

6. $0 - 8$ Change subtraction to adding the opposite.
$= 0 + {}^-8$
$= {}^-8$

7. $|{}^-6| + |4|$ ${}^-6$ is 6 units from 0. 4 is 4 units from 0.
$= 6 + 4$
$= 10$

8. ${}^-3({}^-10)$ Same sign, positive product
$= 30$

9. $({}^-5)^2$
$= {}^-5 \cdot {}^-5$ Same sign, positive product
$= 25$

10. $\frac{{}^-42}{{}^-6}$ Same sign, positive quotient
$= 7$

11. ${}^-19 + 19$ Addition of a number and its opposite is zero.
$= 0$

12. $({}^-4)^3$ Exponent
$\underbrace{{}^-4 \cdot {}^-4} \cdot {}^-4$ Multiply left to right.
$\underbrace{16 \cdot {}^-4}$
${}^-64$

13. $\frac{{}^-14}{0}$ is undefined. Division by 0 is undefined.

14. ${}^-5 \cdot 12$ Different signs, negative product
$= {}^-60$

15. ${}^-20 - 20$ Change subtraction to adding the opposite.
$= {}^-20 + {}^-20$
$= {}^-40$

16. $\frac{45}{{}^-5} = {}^-9$ Different signs, negative quotient

17. ${}^{-}50 + 25 = {}^{-}25$

18. ${}^{-}10 + 6(4 - 7)$ Distribute.

$${}^{-}10 + \underbrace{6 \cdot 4} - \underbrace{6 \cdot 7}$$
$${}^{-}10 + 24 - 42$$
$$\underbrace{{}^{-}10 + 24} + {}^{-}42$$
$$\underbrace{14 + {}^{-}42}$$
$${}^{-}28$$

19. $$\frac{{}^{-}20 - 3({}^{-}5) + 16}{({}^{-}4)^2 - 3^3}$$

Numerator:

${}^{-}20 - 3({}^{-}5) + 16$ Multiply.
${}^{-}20 - {}^{-}15 + 16$ Add the opposite.
$\underbrace{{}^{-}20 + 15} + 16$ Add left to right.
$\underbrace{{}^{-}5 + 16}$
11

Denominator:

$$({}^{-}4)^2 - 3^3$$
$$\underbrace{({}^{-}4)({}^{-}4)} - \underbrace{3 \cdot 3 \cdot 3}$$
$$16 \quad - \quad 27$$
$$\underbrace{16 + {}^{-}27}$$
$${}^{-}11$$

Last step is division: $\dfrac{11}{{}^{-}11} = {}^{-}1$

20. 22 days rounds to 20.
616 miles rounds to 600.
Average distance "per" day implies division.

Estimate: $\dfrac{600 \text{ miles}}{20 \text{ days}} = 30$ miles per day

Exact: $\dfrac{616 \text{ miles}}{22 \text{ days}} = 28$ miles per day

21. ${}^{-}48$ degrees rounds to ${}^{-}50$.
"Rise" of 23 degrees rounds to 20.

A start temperature of ${}^{-}48$ degrees followed by a rise of 23 degrees implies addition.

Estimate: ${}^{-}50 + 20 = {}^{-}30$ degrees
Exact: ${}^{-}48 + 23 = {}^{-}25$ degrees

22. 52 shares rounds to 50.
\$2132 rounds to \$2000.
\$8 rounds to \$10.

Each stock dropped in value by \$8 and Doug owned 52 shares. Multiply to find out how much money he lost. Then, subtract this amount from the original total value.

Estimate: $\$2000 - (50 \cdot 10) = \$1500$
Exact: $\$2132 - (52 \cdot 8) = \$1716$

23. \$552 rounds to \$600.
\$35 rounds to \$40.
12 months (in one year) rounds to 10.

Estimate: $10(\$600 + \$40)$
$10(\$640) = \$6400$

Exact: $12(\$552 + \$35) = 12(\$587) = \$7044$

24. ${}^{-}4ab^3c^2$ means ${}^{-}4 \cdot a \cdot b \cdot b \cdot b \cdot c \cdot c$

25. $3xy^3$ means

$3 \cdot x \cdot y \cdot y \cdot y$ Replace x with ${}^{-}5$ and y with ${}^{-}2$.
$\underbrace{3 \cdot {}^{-}5} \cdot {}^{-}2 \cdot {}^{-}2 \cdot {}^{-}2$ Multiply left to right.
$\underbrace{{}^{-}15 \cdot {}^{-}2} \cdot {}^{-}2 \cdot {}^{-}2$
$\underbrace{30 \cdot {}^{-}2} \cdot {}^{-}2$
$\underbrace{{}^{-}60 \cdot {}^{-}2}$
120

26. $3h - 7h + 5h$ Change subtraction to adding the opposite.
$\underbrace{3h + {}^{-}7h} + 5h$ Combine like terms.
$\underbrace{{}^{-}4h + 5h}$
$1h$ or h

27. $c^2d - c^2d$ Write with the understood coefficients of 1.
$= 1c^2d - 1c^2d$ Change subtraction to adding the opposite.
$= 1c^2d + {}^{-}1c^2d$ Combine like terms.
$= (1 + {}^{-}1)c^2d$
$= 0 \cdot c^2d$
$= 0$

28. $$4n^2 - 4n + 6 - 8 + n^2$$
$$4n^2 + {}^{-}4n + 6 + {}^{-}8 + n^2$$
$$\underbrace{4n^2 + n^2} + {}^{-}4n + \underbrace{6 + {}^{-}8}$$
$$5n^2 \quad + {}^{-}4n + \quad {}^{-}2$$
or $5n^2 - 4n - 2$

29. ${}^{-}10(3b^2)$ Associative property
$\underbrace{({}^{-}10 \cdot 3)}b^2$
${}^{-}30b^2$

30. $7(4p - 4)$ Distribute.
$\underbrace{7 \cdot 4p} - \underbrace{7 \cdot 4}$
$28p - 28$

31. $$3 + 5({}^{-}2w^2 - 3) + w^2$$
$$3 + {}^{-}10w^2 - 15 + w^2$$
$$3 + {}^{-}10w^2 + {}^{-}15 + w^2$$
$${}^{-}9w^2 + {}^{-}12 \text{ or } {}^{-}9w^2 - 12$$

32.
$$\begin{aligned} 3x &= x - 8 \\ {}^{-}x & \quad {}^{-}x \\ 2x &= 0 - 8 \\ 2x &= {}^{-}8 \\ \frac{2x}{2} &= \frac{{}^{-}8}{2} \\ x &= {}^{-}4 \end{aligned}$$

Check:
$$\begin{aligned} 3x &= x - 8 \\ 3({}^{-}4) &= {}^{-}4 - 8 \\ {}^{-}12 &= {}^{-}4 + {}^{-}8 \\ {}^{-}12 &= {}^{-}12 \end{aligned}$$
Balances

The solution is $^{-}4$.

33.
$$\begin{aligned} {}^{-}44 &= {}^{-}2 + 7y \\ 2 & \quad 2 \\ {}^{-}42 &= 0 + 7y \\ {}^{-}42 &= 7y \\ \frac{{}^{-}42}{7} &= \frac{7y}{7} \\ {}^{-}6 &= y \end{aligned}$$

Check:
$$\begin{aligned} {}^{-}44 &= {}^{-}2 + 7y \\ {}^{-}44 &= {}^{-}2 + 7({}^{-}6) \\ {}^{-}44 &= {}^{-}2 + {}^{-}42 \\ {}^{-}44 &= {}^{-}44 \end{aligned}$$
Balances

The solution is $^{-}6$.

34.
$$\begin{aligned} 2k - 5k &= {}^{-}21 \\ 2k + {}^{-}5k &= {}^{-}21 \\ {}^{-}3k &= {}^{-}21 \\ \frac{{}^{-}3k}{{}^{-}3} &= \frac{{}^{-}21}{{}^{-}3} \\ k &= 7 \end{aligned}$$

Check:
$$\begin{aligned} 2k - 5k &= {}^{-}21 \\ 2(7) - 5(7) &= {}^{-}21 \\ 14 - 35 &= {}^{-}21 \\ 14 + {}^{-}35 &= {}^{-}21 \\ {}^{-}21 &= {}^{-}21 \end{aligned}$$
Balances

The solution is 7.

35.
$$\begin{aligned} m - 6 &= {}^{-}2m + 6 \\ 2m & \quad 2m \\ 3m - 6 &= 0 + 6 \\ 3m - 6 &= 6 \\ 6 & \quad 6 \\ 3m + 0 &= 12 \\ \frac{3m}{3} &= \frac{12}{3} \\ m &= 4 \end{aligned}$$

The solution is 4.

Check:
$$\begin{aligned} m - 6 &= {}^{-}2m + 6 \\ 4 - 6 &= {}^{-}2(4) + 6 \\ 4 + {}^{-}6 &= {}^{-}8 + 6 \\ {}^{-}2 &= {}^{-}2 \end{aligned}$$
Balances

36.
$$\begin{aligned} 4 - 4x &= 18 + 10x \\ 4x & \quad 4x \\ 4 + 0 &= 18 + 14x \\ 4 &= 18 + 14x \\ {}^{-}18 & \quad {}^{-}18 \\ {}^{-}14 &= 0 + 14x \\ \frac{{}^{-}14}{14} &= \frac{14x}{14} \\ {}^{-}1 &= x \end{aligned}$$

Add $4x$ to both sides.

The solution is $^{-}1$.

37.
$$\begin{aligned} 18 &= {}^{-}r \\ 18 &= {}^{-}1r \\ \frac{18}{{}^{-}1} &= \frac{{}^{-}1r}{{}^{-}1} \\ {}^{-}18 &= r \end{aligned}$$

The solution is $^{-}18$.

38.
$$\begin{aligned} {}^{-}8b - 11 + 7b &= b - 1 \\ {}^{-}1b - 11 &= 1b - 1 \\ 1b & \quad 1b \\ 0b - 11 &= 2b - 1 \\ {}^{-}11 &= 2b - 1 \\ 1 & \quad 1 \\ {}^{-}10 &= 2b + 0 \\ \frac{{}^{-}10}{2} &= \frac{2b}{2} \\ {}^{-}5 &= b \end{aligned}$$

Add $1b$ to both sides.

Add 1 to both sides.

The solution is $^{-}5$.

39.
$$\begin{aligned} {}^{-}2(t + 1) &= 4(1 - 2t) \\ {}^{-}2t + {}^{-}2 &= 4 - 8t \\ 2 & \quad 2 \\ {}^{-}2t + 0 &= 6 - 8t \\ {}^{-}2t &= 6 - 8t \\ 8t & \quad 8t \\ 6t &= 6 + 0 \\ \frac{6t}{6} &= \frac{6}{6} \\ t &= 1 \end{aligned}$$

Add 2 to both sides.

Add $8t$ to both sides.

The solution is 1.

40.
$$\begin{aligned} 5 + 6y - 23 &= 5(2y + 8) - 10 \\ 5 + 6y + {}^{-}23 &= 10y + 40 + {}^{-}10 \\ 6y + {}^{-}18 &= 10y + 30 \\ 18 & \quad 18 \\ 6y + 0 &= 10y + 48 \\ 6y &= 10y + 48 \\ {}^{-}10y & \quad {}^{-}10y \\ {}^{-}4y &= 0 + 48 \\ \frac{{}^{-}4y}{{}^{-}4} &= \frac{48}{{}^{-}4} \\ y &= {}^{-}12 \end{aligned}$$

Add 18 to both sides.

The solution is $^{-}12$.

CHAPTER 3 SOLVING APPLICATION PROBLEMS

3.1 Problem Solving: Perimeter

3.1 Section Exercises

1. The perimeter formula for a square is $P = 4s$.

$P = 4s$ Replace s with 9 cm.
$P = 4 \cdot 9$ cm
$P = 36$ cm

The perimeter of the square is 36 cm.

3. The perimeter formula for a square is $P = 4s$.

$P = 4s$ Replace s with 25 in.
$P = 4 \cdot 25$ in.
$P = 100$ in.

The perimeter of the square is 100 in.

5. A square park measuring 1 mile (mi) on each side

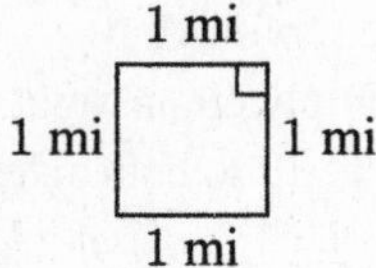

$P = 4s$ Replace s with 1 mi.
$P = 4 \cdot 1$ mi
$P = 4$ mi

The perimeter of the park is 4 miles.

7. A 22 mm square postage stamp

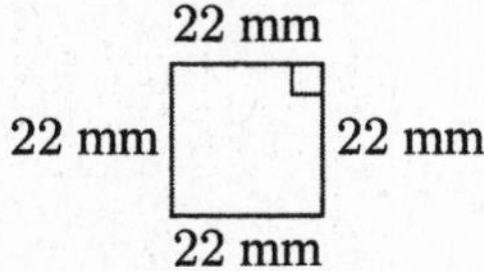

$P = 4s$ Replace s with 22 mm.
$P = 4 \cdot 22$ mm
$P = 88$ mm

The perimeter of the stamp is 88 mm.

9. The perimeter of a square is 120 ft.

$P = 4s$ Replace P with 120 ft.
$120 \text{ ft} = 4s$ Divide both sides by 4.
$\frac{120 \text{ ft}}{4} = \frac{4s}{4}$
$30 \text{ ft} = s$

The length of one side of the square is 30 ft.

11. The perimeter of a square is 4 millimeters (mm).

$P = 4s$ Replace P with 4 mm.
$4 \text{ mm} = 4s$ Divide both sides by 4.
$\frac{4 \text{ mm}}{4} = \frac{4s}{4}$
$1 \text{ mm} = s$

The length of one side of the square is 1 mm.

13. The perimeter of a square parking lot is 92 yards (yd).

$P = 4s$ Replace P with 92 yd.
$92 \text{ yd} = 4s$ Divide both sides by 4.
$\frac{92 \text{ yd}}{4} = \frac{4s}{4}$
$23 \text{ yd} = s$

The length of one side of the parking lot is 23 yards.

15. The perimeter of a square closet is 8 feet (ft).

$P = 4s$ Replace P with 8 ft.
$8 \text{ ft} = 4s$ Divide both sides by 4.
$\frac{8 \text{ ft}}{4} = \frac{4s}{4}$
$2 \text{ ft} = s$

The length of one side of the closet is 2 ft.

17. $P = 2l + 2w$ Replace l with 8 yd and w with 6 yd.
$P = 2 \cdot 8 \text{ yd} + 2 \cdot 6 \text{ yd}$ Multiply first.
$P = 16 \text{ yd} + 12 \text{ yd}$ Add last.
$P = 28$ yd

The perimeter of the rectangle is 28 yd.

Check: $P = 8 \text{ yd} + 8 \text{ yd} + 6 \text{ yd} + 6 \text{ yd}$
$P = 28$ yd

19. $P = 2l + 2w$ Replace l with 25 cm and w with 10 cm.
$P = 2 \cdot 25 \text{ cm} + 2 \cdot 10 \text{ cm}$ Multiply first.
$P = 50 \text{ cm} + 20 \text{ cm}$ Add last.
$P = 70$ cm

The perimeter of the rectangle is 70 cm.

Check: $P = 25 \text{ cm} + 25 \text{ cm} + 10 \text{ cm} + 10 \text{ cm}$
$P = 70$ cm

21.

20 ft

16 ft 16 ft

20 ft

$P = 2l + 2w$	Replace l with 20 ft and w with 16 ft.
$P = 2 \cdot 20\text{ ft} + 2 \cdot 16\text{ ft}$	Multiply first.
$P = 40\text{ ft} + 32\text{ ft}$	Add last.
$P = 72\text{ ft}$	

The perimeter of the rectangular living room is 72 ft.

23.

5 in.

8 in. 8 in.

5 in.

$P = 2l + 2w$	Replace l with 8 in. and w with 5 in.
$P = 2 \cdot 8\text{ in.} + 2 \cdot 5\text{ in.}$	Multiply first.
$P = 16\text{ in.} + 10\text{ in.}$	Add last.
$P = 26\text{ in.}$	

The perimeter of the rectangular piece of paper is 26 in.

25.

$$\begin{aligned} P &= 2l + 2w && \text{Replace } P \text{ with 30 cm and } w \text{ with 6 cm.} \\ 30\text{ cm} &= 2l + 2 \cdot 6\text{ cm} && \text{Multiply on the right.} \\ 30\text{ cm} &= 2l + 12\text{ cm} && \text{Add } {}^{-}12\text{ cm to both sides.} \\ {}^{-}12\text{ cm} & \quad\;\; {}^{-}12\text{ cm} && \\ \hline 18\text{ cm} &= 2l + 0\text{ cm} && \\ \frac{18\text{ cm}}{2} &= \frac{2l}{2} && \text{Divide both sides by 2.} \\ 9\text{ cm} &= l && \end{aligned}$$

The length is 9 cm.

Check:

6 cm

9 cm 9 cm

6 cm

$P = 9\text{ cm} + 9\text{ cm} + 6\text{ cm} + 6\text{ cm}$

$P = 30\text{ cm}$

30 cm matches the original perimeter, so 9 cm is the correct length.

27.

$$\begin{aligned} P &= 2l + 2w && \text{Replace } P \text{ with 10 mi and } l \text{ with 4 mi.} \\ 10\text{ mi} &= 2 \cdot 4\text{ mi} + 2w && \text{Multiply on the right.} \\ 10\text{ mi} &= 8\text{ mi} + 2w && \text{Add } {}^{-}8 \text{ to both sides.} \\ {}^{-}8\text{ mi} & \quad\;\; {}^{-}8\text{ mi} && \\ \hline 2\text{ mi} &= 0\text{ mi} + 2w && \\ \frac{2\text{ mi}}{2} &= \frac{2w}{2} && \text{Divide both sides by 2.} \\ 1\text{ mi} &= w && \end{aligned}$$

The width is 1 mi.

Check:

4 mi

1 mi 1 mi

4 mi

$P = 4\text{ mi} + 4\text{ mi} + 1\text{ mi} + 1\text{ mi}$

$P = 10\text{ mi}$

10 mi matches the original perimeter, so 1 mi is the correct width.

29.

$$\begin{aligned} P &= 2l + 2w && \text{Replace } l \text{ with 6 ft and } P \text{ with 16 ft.} \\ 16\text{ ft} &= 2 \cdot 6\text{ ft} + 2w && \text{Multiply on the right.} \\ 16\text{ ft} &= 12\text{ ft} + 2w && \text{Add } {}^{-}12 \text{ to both sides.} \\ {}^{-}12\text{ ft} & \quad\;\; {}^{-}12\text{ ft} && \\ \hline 4\text{ ft} &= 0\text{ ft} + 2w && \\ \frac{4\text{ ft}}{2} &= \frac{2w}{2} && \text{Divide both sides by 2.} \\ 2\text{ ft} &= w && \end{aligned}$$

Check:

6 ft

2 ft 2 ft

6 ft

$P = 6\text{ ft} + 6\text{ ft} + 2\text{ ft} + 2\text{ ft}$

$P = 16\text{ ft}$

The width of the rectangle is 2 ft.

31.

$$\begin{aligned} P &= 2l + 2w && \text{Replace } w \text{ with 1 m and } P \text{ with 6 m.} \\ 6\text{ m} &= 2l + 2 \cdot 1\text{ m} && \text{Multiply on the right.} \\ 6\text{ m} &= 2l + 2\text{ m} && \text{Add } {}^{-}2\text{ m to both sides.} \\ {}^{-}2\text{ m} & \quad\;\; {}^{-}2\text{ m} && \\ \hline 4\text{ m} &= 2l + 0\text{ m} && \\ \frac{4\text{ m}}{2} &= \frac{2}{2}l && \text{Divide both sides by 2.} \\ 2\text{ m} &= l && \end{aligned}$$

Check:

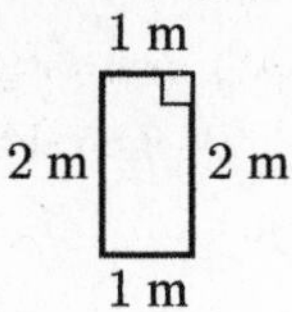

$P = 2\text{ m} + 2\text{ m} + 1\text{ m} + 1\text{ m}$
$P = 6\text{ m}$

The length of the rectangle is 2 m.

33. Add all four sides to find the perimeter.

$P = 58\text{ m} + 58\text{ m} + 46\text{ m} + 46\text{ m}$
$P = 208\text{ m}$

The perimeter is 208 m.

35. Add all four sides to find the perimeter.

$P = 100\text{ ft} + 60\text{ ft} + 100\text{ ft} + 60\text{ ft}$
$P = 320\text{ ft}$

The perimeter is 320 ft.

37. Add all three sides to find the perimeter of the triangle.

$P = 12\text{ mm} + 26\text{ mm} + 16\text{ mm}$
$P = 54\text{ mm}$

The perimeter is 54 mm.

39. Add all six sides to find the perimeter.

$P = 4\text{ ft} + 12\text{ ft} + 12\text{ ft} + 3\text{ ft} + 8\text{ ft} + 9\text{ ft}$
$P = 48\text{ ft}$

The perimeter is 48 ft.

41. Add all six sides to find the perimeter.

$P = 13\text{ in.} + 8\text{ in.} + 18\text{ in.} + 13\text{ in.} + 18\text{ in.} + 8\text{ in.}$
$P = 78\text{ in.}$

The perimeter is 78 in.

43. Add all five sides to find the perimeter.

$P = 34\text{ m} + 22\text{ m} + 20\text{ m} + 22\text{ m} + 27\text{ m}$
$P = 125\text{ m}$

The perimeter is 125 m.

45. The perimeter is 115 cm.

$P = 10\text{ cm} + 10\text{ cm} + 30\text{ cm} + 25\text{ cm} + ?$
$P = 75\text{ cm} + ?$

Because the perimeter is 115 cm, replace P with 115 cm.

$115\text{ cm} = 75\text{ cm} + ?$

$115\text{ cm} = 75\text{ cm} + ?$ Add $^{-}75$ cm to both sides.

$$\begin{array}{rcl} 115\text{ cm} & = & 75\text{ cm} + ? \\ \underline{^{-}75\text{ cm}} & & \underline{^{-}75\text{ cm}} \\ 40\text{ cm} & = & 0\text{ cm} + ? \\ 40\text{ cm} & = & ? \end{array}$$

The length of the unknown side is 40 cm.

47. The perimeter is 78 in.

$78\text{ in.} = 15\text{ in.} + 6\text{ in.} + 6\text{ in.} + 6\text{ in.} + 6\text{ in.} + 6\text{ in.} + 9\text{ in.} + ? + 6\text{ in.} + 6\text{ in.}$

$$\begin{array}{rcl} 78\text{ in.} & = & 66\text{ in.} + ? \\ \underline{^{-}66\text{ in.}} & & \underline{^{-}66\text{ in.}} \\ 12\text{ in.} & = & ? \end{array}$$

The length of the unknown side is 12 in.

49. **(a)** Sketches will vary.

(b) Formula for perimeter of an equilateral triangle is $P = 3s$, where s is the length of one side.

(c) The formula will not work for other kinds of triangles because the sides will have different lengths.

51. Use $d = rt$ with $r = 70$ miles per hour.

(a) In $t = 2$ hours, you will travel $d = 70 \cdot 2 = 140$ miles.

(b) In $t = 5$ hours, you will travel $d = 70 \cdot 5 = 350$ miles.

(c) In $t = 8$ hours, you will travel $d = 70 \cdot 8 = 560$ miles.

52. Use $d = rt$ with $r = 35$ miles per hour.

(a) In $t = 2$ hours, you will travel $d = 35 \cdot 2 = 70$ miles.

(b) In $t = 5$ hours, you will travel $d = 35 \cdot 5 = 175$ miles.

(c) In $t = 8$ hours, you will travel $d = 35 \cdot 8 = 280$ miles.

(d) The rate is half of 70 miles per hour, so in each case the distance traveled will be half as far. Divide each result in Exercise 51 by 2.

53. Use $rt = d$ with $d = 3000$ miles and r in miles per hour. In each case, divide by r.

(a) $60t = 3000$

$$\frac{60t}{60} = \frac{3000}{60}$$

$t = 50$ hours

(b) $50t = 3000$

$\frac{50t}{50} = \frac{3000}{50}$

$t = 60$ hours

(c) $20t = 3000$

$\frac{20t}{20} = \frac{3000}{20}$

$t = 150$ hours

54. Use $rt = d$. In each case, divide by t.

(a) $r \cdot 11 = 671$

$\frac{11r}{11} = \frac{671}{11}$

$r = 61$ miles per hour

(b) $r \cdot 27 = 1539$

$\frac{27r}{27} = \frac{1539}{27}$

$r = 57$ miles per hour

(c) $r \cdot 16 = 1040$

$\frac{16r}{16} = \frac{1040}{16}$

$r = 65$ miles per hour

3.2 Problem Solving: Area

3.2 Section Exercises

1. The figure is a rectangle.

$A = l \cdot w$ Replace l with 11 ft and w with 7 ft.

$A = 11 \text{ ft} \cdot 7 \text{ ft}$

$A = 77 \text{ ft}^2$

The area is 77 ft^2, or 77 square feet.

3. The figure is a square.

$A = s^2$

$A = s \cdot s$ Remember, s^2 means $s \cdot s$.

$A = 10 \text{ m} \cdot 10 \text{ m}$ Replace s with 10 m.

$A = 100 \text{ m}^2$

The area is 100 m^2, or 100 square meters.

5. The figure is a parallelogram.

$A = bh$

Turn your book sideways to identify that the height is 25 mm and the base is 31 mm. Replace b with 31 mm and h with 25 mm.

$A = 31 \text{ mm} \cdot 25 \text{ mm}$

$A = 775 \text{ mm}^2$

The area is 775 mm^2, or 775 square millimeters.

7. The figure is a square.

$A = s^2$

$A = s \cdot s$ Remember, s^2 means $s \cdot s$.

$A = 6 \text{ in.} \cdot 6 \text{ in.}$ Replace s with 6 in.

$A = 36 \text{ in.}^2$

The area is 36 in.^2, or 36 square inches.

9.

15 cm

7 cm

$A = l \cdot w$ Replace l with 15 cm and w with 7 cm.

$A = 15 \text{ cm} \cdot 7 \text{ cm}$

$A = 105 \text{ cm}^2$

The area of the calculator is 105 cm^2.

11.

Area of a parallelogram.

$A = b \cdot h$ Replace b with 8 ft and h with 9 ft.

$A = 8 \text{ ft} \cdot 9 \text{ ft}$

$A = 72 \text{ ft}^2$

The area of the parallelogram is 72 ft^2.

13.

25 mi

25 mi

$A = s^2$ Area formula for a square

$A = s \cdot s$ Replace s with 25 mi.

$A = 25 \text{ mi} \cdot 25 \text{ mi}$

$A = 625 \text{ mi}^2$

The fire burned an area of 625 mi^2.

15.

1 m

1 m

$A = s^2$ Area of a square

$A = s \cdot s$ Replace s with 1 m.

$A = 1 \text{ m} \cdot 1 \text{ m}$

$A = 1 \text{ m}^2$

The area is 1 m^2.

17. $A = l \cdot w$ Replace A with 18 ft^2, and w with 3 ft.

$18 \text{ ft}^2 = l \cdot 3 \text{ ft}$ Divide both sides by 3 ft.

$$\frac{18 \text{ ft} \cdot \cancel{\text{ft}}}{3 \cancel{\text{ft}}} = \frac{l \cdot 3 \text{ ft}}{3 \text{ ft}}$$

$6 \text{ ft} = l$

The length of the desk is 6 ft.

6 ft

3 ft 3 ft

6 ft

Check: $A = l \cdot w$

$A = 6 \text{ ft} \cdot 3 \text{ ft}$

$A = 18 \text{ ft}^2$

19. $A = l \cdot w$ Replace A with 7200 yd^2, and l with 90 yd.

$7200 \text{ yd}^2 = 90 \text{ yd} \cdot w$ Divide both sides by 90 yd.

$$\frac{7200 \text{ yd} \cdot \cancel{\text{yd}}}{90 \cancel{\text{yd}}} = \frac{90 \text{ yd} \cdot w}{90 \text{ yd}}$$

$80 \text{ yd} = w$

The width of the parking lot is 80 yd.

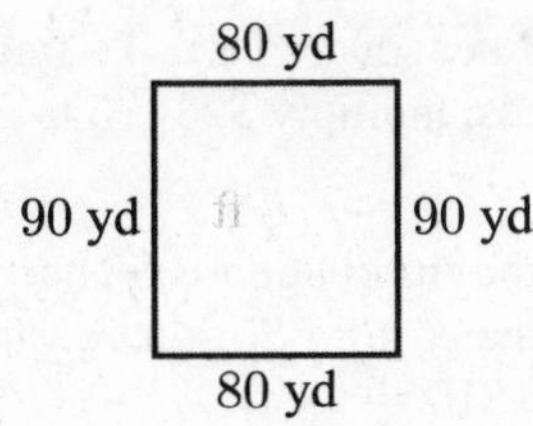

Check: $A = l \cdot w$

$A = 90 \text{ yd} \cdot 80 \text{ yd}$

$A = 7200 \text{ yd}^2$

21. $A = l \cdot w$ Replace A with 154 in.^{2}, and w with 11 in.

$154 \text{ in.}^2 = l \cdot 11 \text{ in.}$ Divide both sides by 11 in.

$$\frac{154 \text{ in.} \cdot \cancel{\text{in.}}}{11 \cancel{\text{in.}}} = \frac{l \cdot 11 \text{ in.}}{11 \text{ in.}}$$

$14 \text{ in.} = l$

The length of the photo is 14 in.

14 in.

11 in. Photo $A = 154 \text{ in.}^2$ 11 in.

14 in.

Check: $A = l \cdot w$

$A = 14 \text{ in.} \cdot 11 \text{ in.}$

$A = 154 \text{ in.}^2$

23. $A = s^2$ Area formula for a square

$A = s \cdot s$ Replace A with 36 m^2.

$36 \text{ m}^2 = s \cdot s$ By inspection, what number times itself is 36?

$36 \text{ m}^2 = 6 \text{ m} \cdot 6 \text{ m}$ $6 \cdot 6$ is 36, so $6 \text{ m} \cdot 6 \text{ m}$ is 36 m^2.

The length of one side of the floor is 6 m.

25. $A = s^2$ Area formula for a square

$A = s \cdot s$ Replace A with 4 ft^2.

$4 \text{ ft}^2 = s \cdot s$ By inspection, what number times itself is 4?

$4 \text{ ft}^2 = 2 \text{ ft} \cdot 2 \text{ ft}$ $2 \cdot 2$ is 4, so $2 \text{ ft} \cdot 2 \text{ ft}$ is 4 ft^2.

The length of one side of the sign is 2 ft.

27. Use the area formula for a parallelogram. Replace A with 500 cm^2 and b with 25 cm.

$A = b \cdot h$

$500 \text{ cm}^2 = 25 \text{ cm} \cdot h$ Divide both sides by 25 cm.

$$\frac{500 \text{ cm} \cdot \cancel{\text{cm}}}{25 \cancel{\text{cm}}} = \frac{25 \text{ cm} \cdot h}{25 \text{ cm}}$$

$20 \text{ cm} = h$

The height is 20 cm.

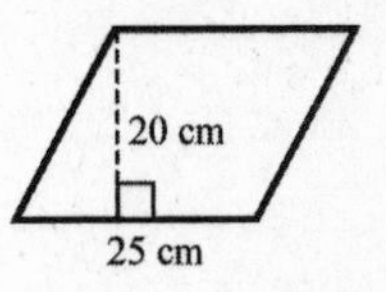

Check: $A = b \cdot h$

$A = 25 \text{ cm} \cdot 20 \text{ cm}$

$A = 500 \text{ cm}^2$

29. Use the area formula for a parallelogram. Replace A with 221 in.^{2} and h with 13 in.

$$A = b \cdot h$$

$$221 \text{ in.}^2 = b \cdot 13 \text{ in.}$$ Divide both sides by 13 in.

$$\frac{221 \text{ in.} \cdot \cancel{\text{in.}}}{13 \cancel{\text{in.}}} = \frac{b \cdot 13 \text{ in.}}{13 \text{ in.}}$$

$$17 \text{ in.} = b$$

The base is 17 in.

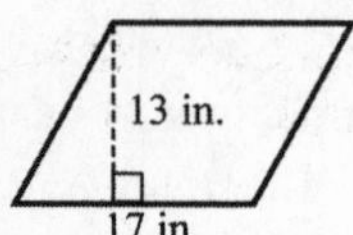

Check: $A = b \cdot h$
$A = 17 \text{ in.} \cdot 13 \text{ in.}$
$A = 221 \text{ in.}^2$

31. Use the area formula for a parallelogram. Replace A with 9 m^2 and b with 9 m.

$$A = b \cdot h$$

$$9 \text{ m}^2 = 9 \text{ m} \cdot h$$ Divide both sides by 9 m.

$$\frac{9 \text{ m} \cdot \cancel{\text{m}}}{9 \cancel{\text{m}}} = \frac{9 \text{ m} \cdot h}{9 \text{ m}}$$

$$1 \text{ m} = h$$

The height is 1 m.

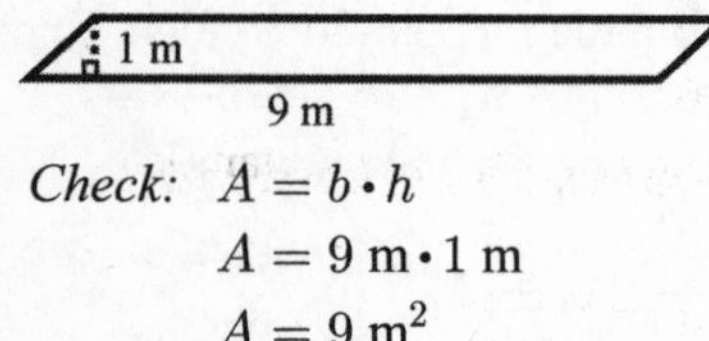

Check: $A = b \cdot h$
$A = 9 \text{ m} \cdot 1 \text{ m}$
$A = 9 \text{ m}^2$

33. The height of the parallelogram is not part of the perimeter. Also, square units are used for area, not perimeter.

$P = 25 \text{ cm} + 25 \text{ cm} + 25 \text{ cm} + 25 \text{ cm}$
$P = 100 \text{ cm}$

35. The mat is a square.

12 m
12 m 12 m
12 m

$P = 4 \cdot s$	$A = s \cdot s$
$P = 4 \cdot 12 \text{ m}$	$A = 12 \text{ m} \cdot 12 \text{ m}$
$P = 48 \text{ m}$	$A = 144 \text{ m}^2$

37. 5 m
4 m

Tyra is decorating the top edges of her walls, so find the perimeter of her ceiling.

$P = 2 \cdot l + 2 \cdot w$ Perimeter of a rectangle.

$P = 2 \cdot 5 \text{ m} + 2 \cdot 4 \text{ m}$ Replace l with 5 m and w with 4 m.

$P = 18 \text{ m}$

The strip costs \$6 per meter. To find the cost of 18 meters, multiply \$6 • 18 to get \$108.

To have the top edges of her walls decorated, Tyra will spend \$108.

39. 5 yd
5 yd

Mr. and Mrs. Gomez are *covering* the bedroom floor, so find the area.

$A = s^2$ Area of a square
$A = s \cdot s$
$A = 5 \text{ yd} \cdot 5 \text{ yd}$
$A = 25 \text{ yd}^2$

The carpet cost is \$23 per square yard. To find the cost for 25 square yards, multiply \$23 • 25 to get \$575.

The cost of padding and installation is \$6 per square yard. To find the cost for 25 square yards, multiply \$6 • 25 to get \$150.

To have the bedroom carpeted, Mr. and Mrs. Gomez will spend \$575 + \$150, or \$725 total.

41. The football field is a rectangle with length 100 yards and area 5300 yd^2.

$$A = l \cdot w$$ Area formula for a rectangle.

$$5300 \text{ yd}^2 = 100 \text{ yd} \cdot w$$ Replace A with 5300 yd^2 and l with 100 yd.

$$\frac{5300 \text{ yd} \cdot \cancel{\text{yd}}}{100 \cancel{\text{yd}}} = \frac{100 \text{ yd} \cdot w}{100 \text{ yd}}$$ Divide both sides by 100 yd.

$$53 \text{ yd} = w$$

The width of the field is 53 yards.

43. The Panoramic photo is a rectangle with base 14 in. and height 4 in.

$P = 2 \cdot l + 2 \cdot w$ Formula for perimeter of rectangle

$P = 2 \cdot 14 \text{ in.} + 2 \cdot 4 \text{ in.}$ Substitute 14 in. for l and 4 in. for w.

$P = 28 \text{ in.} + 8 \text{ in.}$

$P = 36 \text{ in.}$

$A = lw$ Formula for area of rectangle

$A = 14 \text{ in.} \cdot 4 \text{ in.}$ Replace l with 14 in. and w with 4 in.

$A = 56 \text{ in.}^2$

Using the above formulas for the other rectangular photos, the 4 in. × 6 in. has perimeter

$P = 2 \cdot 6 \text{ in.} + 2 \cdot 4 \text{ in.}$

$P = 20 \text{ in.}$

and area

$A = 6 \text{ in.} \cdot 4 \text{ in.}$

$A = 24 \text{ in.}^2$

Likewise, the 4 in. × 7 in. photo has perimeter

$P = 2 \cdot 7 \text{ in.} + 2 \cdot 4 \text{ in.}$

$P = 22 \text{ in.}$

and area

$A = 7 \text{ in.} \cdot 4 \text{ in.}$

$A = 28 \text{ in.}^2$

45. Since perimeter $P = 2 \cdot l + 2 \cdot w$, if $l = 1$ and $P = 12$, then $2 \cdot w = 12 - 2 \cdot 1 = 10$ and $w = 5$. Similarly, if $l = 2$, $w = 4$, and if $l = 3$, $w = 3$. So there are only three possibilities for whole number plots with perimeter 12 ft.

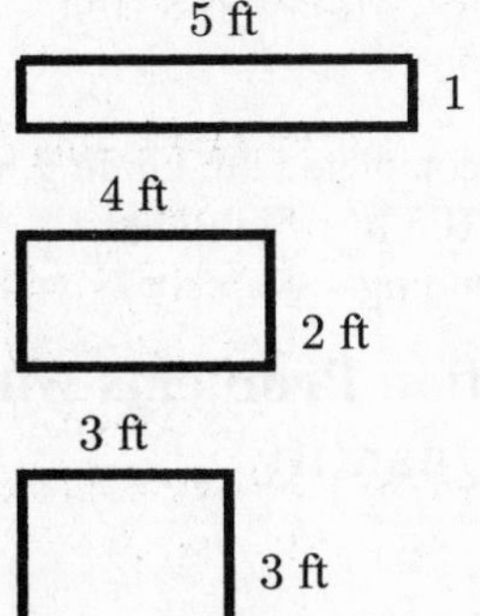

46. **(a)** Use the area formula for a rectangle, $A = l \cdot w$.

l	w	$A = l \cdot w$
1 ft	5 ft	5 ft²
2 ft	4 ft	8 ft²
3 ft	3 ft	9 ft²

(b) The table in part (a) shows that the 3 ft by 3 ft plot has the greatest area.

47. As in Exercise 45, we let l equal a whole number and find w.

l	$2w = 16 - 2 \cdot l$	w
1	14	7
2	12	6
3	10	5
4	8	4

So there are only four possibilities for whole number plots with perimeter 16 ft.

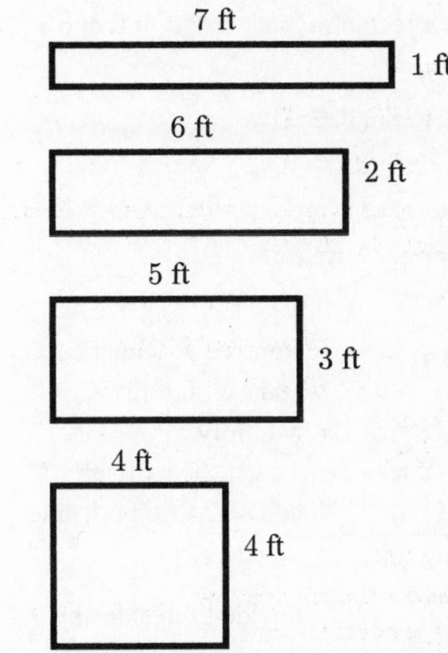

48. **(a)** Use the area formula for a rectangle, $A = l \cdot w$.

l	w	$A = l \cdot w$
1 ft	7 ft	7 ft²
2 ft	6 ft	12 ft²
3 ft	5 ft	15 ft²
4 ft	4 ft	16 ft²

(b) Based on Exercises 46 (a) and 48 (a), square plots have the greatest area.

Summary Exercises on Perimeter and Area

1. The figure is a rectangle with base 3 m and height 13 m.

$P = 13 \text{ m} + 13 \text{ m} + 3 \text{ m} + 3 \text{ m}$

$P = 32 \text{ m}$

$A = b \cdot h$

$A = 3 \text{ m} \cdot 13 \text{ m}$

$A = 39 \text{ m}^2$

3. The figure is a parallelogram with base 7 yd and height 8 yd.

$$P = 10 \text{ yd} + 10 \text{ yd} + 7 \text{ yd} + 7 \text{ yd}$$
$$P = 34 \text{ yd}$$

$$A = b \cdot h$$
$$A = 7 \text{ yd} \cdot 8 \text{ yd}$$
$$A = 56 \text{ yd}^2$$

5. The figure is a square.

$$P = 4s = 4 \cdot 9 \text{ in.} = 36 \text{ in.}$$

$$A = s^2 = s \cdot s = 9 \text{ in.} \cdot 9 \text{ in.} = 81 \text{ in.}^2$$

7. The figure is a rectangle with length 9 ft and width 4 ft.

$$P = 9 \text{ ft} + 9 \text{ ft} + 4 \text{ ft} + 4 \text{ ft}$$
$$P = 26 \text{ ft}$$

$$A = b \cdot h$$
$$A = 9 \text{ ft} \cdot 4 \text{ ft}$$
$$A = 36 \text{ ft}^2$$

9.

$$\begin{array}{rcll} P &=& 2l + 2w & \text{Replace } P \text{ with 14 ft and } l \text{ with 6 ft.} \\ 14 \text{ ft} &=& 2 \cdot 6 \text{ ft} + 2w & \text{Multiply.} \\ 14 \text{ ft} &=& 12 \text{ ft} + 2w & \text{To get } 2w \text{ by itself,} \\ {}^-12 \text{ ft} && {}^-12 \text{ ft} & \text{add } {}^-12 \text{ ft to both sides.} \\ \hline 2 \text{ ft} &=& 0 + 2w & \\ \dfrac{2 \text{ ft}}{2} &=& \dfrac{2w}{2} & \text{Divide both sides by 2.} \\ 1 \text{ ft} &=& w & \end{array}$$

The width is 1 ft.

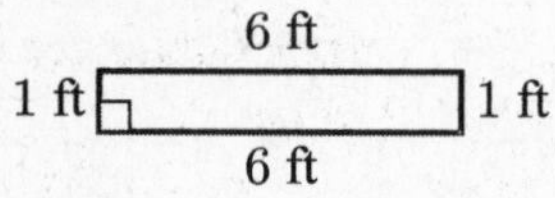

Check: $P = 6 \text{ ft} + 6 \text{ ft} + 1 \text{ ft} + 1 \text{ ft} = 14 \text{ ft}$

11.

$$\begin{array}{rcll} A &=& s^2 & \text{Replace } A \text{ with 81 in.}^2. \\ 81 \text{ in.}^2 &=& s^2 & \\ 81 \text{ in.}^2 &=& s \cdot s & \\ 81 \text{ in.}^2 &=& 9 \text{ in.} \cdot 9 \text{ in.} & \end{array}$$

The length of one side of the clock face is 9 in.

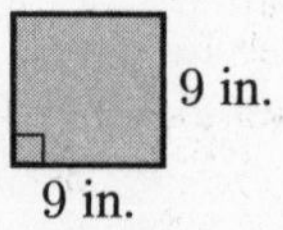

Check:
$$A = s \cdot s$$
$$A = 9 \text{ in.} \cdot 9 \text{ in.}$$
$$A = 81 \text{ in.}^2$$

13.

$$\begin{array}{rcll} P &=& 4s & \text{Perimeter formula for a square} \\ 64 \text{ cm} &=& 4s & \text{Replace } P \text{ with 64 cm.} \\ \dfrac{64 \text{ cm}}{4} &=& \dfrac{4s}{4} & \text{Divide both sides by 4.} \\ 16 \text{ cm} &=& s & \end{array}$$

The length of one side is 16 cm.

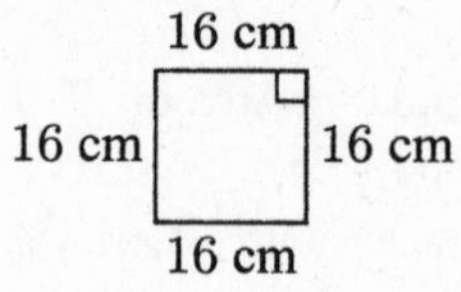

Check:
$P = 16 \text{ cm} + 16 \text{ cm} + 16 \text{ cm} + 16 \text{ cm} = 64 \text{ cm}$

15. The figure is a triangle.

$$P = 168 \text{ m} + 168 \text{ m} + 168 \text{ m}$$
$$P = 504 \text{ m}$$

504 meters of fencing are needed.

17.

$$\begin{array}{rcll} P &=& 2l + 2w & \text{Replace } w \text{ with 12 in. and } P \text{ with 42 in.} \\ 42 \text{ in.} &=& 2l + 2 \cdot 12 \text{ in.} & \text{Multiply on the right.} \\ 42 \text{ in.} &=& 2l + 24 \text{ in.} & \text{Add } {}^-24 \text{ in. to both sides.} \\ {}^-24 \text{ in.} && {}^-24 \text{ in.} & \\ \hline 18 \text{ in.} &=& 2l + 0 \text{ in.} & \\ \dfrac{18 \text{ in.}}{2} &=& \dfrac{2l}{2} & \text{Divide both sides by 2.} \\ 9 \text{ in.} &=& l & \end{array}$$

The height of the screen is 9 inches.

19. Each flag is rectangular in shape.

$$P = 3 \text{ ft} + 5 \text{ ft} + 3 \text{ ft} + 5 \text{ ft}$$
$$P = 16 \text{ ft}$$

$$A = l \cdot w$$
$$A = 5 \text{ ft} \cdot 3 \text{ ft}$$
$$A = 15 \text{ ft}^2$$

To make seven flags, complete with binding, we will need $7(15 \text{ ft}^2) = 105 \text{ ft}^2$ of fabric and $7(16 \text{ ft}) = 112$ ft of binding.

3.3 Solving Application Problems with One Unknown Quantity

3.3 Section Exercises

1. 14 plus a number

$$14 + x \quad \text{or} \quad x + 14$$

3. $^-5$ added to a number

$$^-5 + x \quad \text{or} \quad x + {}^-5$$

5. 20 minus a number

$$20 - x$$

7. 9 less than a number

$$x - 9$$

9. Subtract 4 from a number

$$x - 4$$

11. $^{-}6$ times a number

$$^{-}6x$$

13. Double a number

$$2x$$

15. A number divided by 2

$$\frac{x}{2} \text{ or } \frac{1}{2}x$$

17. Twice a number added to 8

$$8 + 2x \text{ or } 2x + 8$$

19. 10 fewer than seven times a number

$$7x - 10$$

21. The sum of twice a number and the number

$$2x + x \text{ or } x + 2x$$

23. Let n represent the unknown number.

four times a number	decreased by 2	result is	26.
↓	↓	↓	↓
$4n$	-2	$=$	26

$$\begin{aligned} 4n - 2 &= 26 \\ 4n - 2 + 2 &= 26 + 2 \\ 4n &= 28 \\ \frac{4n}{4} &= \frac{28}{4} \\ n &= 7 \end{aligned}$$

The number is 7.

Check: Four times 7 [= 28] is decreased by 2 [= 26]. *True*

25. Let n represent the unknown number.

Twice a number	added to	the number	is	$^{-}15$.
↓	↓	↓	↓	↓
$2n$	$+$	n	$=$	$^{-}15$

$$\begin{aligned} 2n + n &= {}^{-}15 \\ 3n &= {}^{-}15 \\ \frac{3n}{3} &= \frac{^{-}15}{3} \\ n &= {}^{-}5 \end{aligned}$$

The number is $^{-}5$.

Check: Twice $^{-}5$ [= $^{-}10$] is added to $^{-}5$ [= $^{-}15$]. *True*

27. Let n represent the unknown number.

Product of a number and 5	increased by	12	the result is	7 times the number.
↓	↓	↓	↓	↓
$5n$	$+$	12	$=$	$7n$

$$\begin{aligned} 5n + 12 &= 7n \\ 5n + 12 - 5n &= 7n - 5n \\ 12 &= 2n \\ \frac{12}{2} &= \frac{2n}{2} \\ 6 &= n \end{aligned}$$

The number is 6.

Check: The product of 6 and 5 [= 30] is increased by 12 [= 42] is seven times 6 [= 42]. *True*

29. Let n represent the unknown number.

30	subtract	3 times a number	is	2	plus	the number
↓	↓	↓	↓	↓	↓	↓
30	$-$	$3n$	$=$	2	$+$	n

$$\begin{aligned} 30 - 3n &= 2 + n \\ 30 - 3n + 3n &= 2 + n + 3n \\ 30 &= 2 + 4n \\ 30 - 2 &= 2 + 4n - 2 \\ 28 &= 4n \\ \frac{28}{4} &= \frac{4n}{4} \\ 7 &= n \end{aligned}$$

The number is 7.

Check: Three times 7 [= 21] is subtracted from 30 [= 9] is 2 plus 7 [= 9]. *True*

31. *Step 1*
The problem is about Ricardo's weight.

Unknown: his original weight

Known: He gained 15 pounds, lost 28, and then regained 5 pounds to weigh 177.

Step 2(a)
Let w represent Ricardo's original weight.

Step 3

weight at beginning	+	pounds gained	−	pounds lost	+	pounds regained	=	final weight
↓		↓		↓		↓		↓
w	$+$	15	$-$	28	$+$	5	$=$	177

Step 4

$$\begin{aligned} w + 15 - 28 + 5 &= 177 \quad \text{Combine like terms.} \\ w - 8 &= 177 \\ w &= 185 \end{aligned}$$

Step 5
He weighed 185 pounds originally.

Step 6
$185 + 15 - 28 + 5 = 177$

The correct answer is 185 pounds.

33. *Step 1*
The problem is about the number of cookies in the cookie jar.

Unknown: number of cookies the children ate

Known: 18 cookies at the start, three dozen cookies put in the jar, and 49 cookies at the end

Step 2(a)
Let c represent the number of cookies the children ate.

Step 3

cookies at the start	–	cookies eaten	+	cookies added	=	ended up with 49 cookies
↓		↓		↓		↓
18	–	c	+	36	=	49

Step 4

$$18 - c + 36 = 49$$ Write the understood $1c$. Change subtraction to adding the opposite.

$$\begin{aligned} 54 + (^-1c) &= 49 \\ ^-54 \quad & \quad ^-54 \\ \hline 0 + (^-1c) &= {^-5} \\ 0 + (^-1c) &= {^-5} \\ \frac{^-1c}{^-1} &= \frac{^-5}{^-1} \\ c &= 5 \end{aligned}$$

Step 5
The children ate 5 cookies.

Step 6
There were 18 cookies at the start. The children ate 5 cookies, so that left $18 - 5 = 13$ cookies in the jar.

Three dozen cookies were added, so there were $13 + 36 = 49$ cookies.

The correct answer is 5 cookies.

35. *Step 1*
The problem is about boxes of pens.

Unknown: the number of pens in each box

Known: The bookstore started with 6 boxes of pens. The bookstore sold 32 pens, then 35 pens, and 5 pens were left.

Step 2(a)
Let p represent the number of pens in each box.

Step 3

total number of boxes	•	number of pens in each box	–	32	–	35	=	5
↓		↓		↓		↓		↓
6	•	p	–	32	–	35	=	5

Step 4

$$\begin{aligned} 6p - 32 - 35 &= 5 \\ 6p - 67 &= 5 \\ ^+67 \quad & \quad ^+67 \\ \hline 6p + 0 &= 72 \\ \frac{6p}{6} &= \frac{72}{6} \\ p &= 12 \end{aligned}$$

Step 5
There were 12 pens in each box.

Step 6
The bookstore ordered 6 boxes of red pens each containing 12 pens. $6(12) = 72$ pens
32 were sold: $72 - 32 = 40$
35 were sold: $40 - 35 = 5$
5 pens were left on the shelf.

The correct answer is 12 pens per box.

37. *Step 1*
The problem is about the bank account of a music club.

Unknown: the dues paid by each member

Known: 14 total members, earned \$340, spent \$575, and account is overdrawn by \$25

Step 2(a)
Let d represent the dues paid by each member.

Step 3

total dues paid by 14 members	+	earned \$340	–	spent \$575	=	ended up with an overdrawn bank account
↓		↓		↓		↓
$14 \cdot d$	+	340	–	575	=	$^-25$

Step 4

$$\begin{aligned} 14d + 340 - 575 &= {^-25} \\ 14d + 340 + (^-575) &= {^-25} \\ 14d + (^-235) &= {^-25} \\ ^+235 \quad & \quad ^+235 \\ \hline 14d + 0 &= 210 \\ \frac{14d}{14} &= \frac{210}{14} \\ d &= 15 \end{aligned}$$

Step 5
Each member paid \$15 in dues.

Step 6
14 members paid dues of \$15 each:
14(\$15) = \$210 in the bank account.

Earned \$340: \$210 + \$340 = \$550 in the account.

Spent \$575: \$550 – \$575 = $^-$\$25 in the account.

The account was overdrawn by \$25.

The correct answer is \$15 dues.

39. *Step 1*
The problem is about Tamu's age.

Unknown: how old Tamu is right now

Known: Tamu's age equals the number you get after multiplying his age by 4 and then subtracting 75.

Step 2(a)
Let a represent Tamu's age.

Step 3

Tamu's age	=	his age multiplied by 4	subtract	75
↓		↓	↓	↓
a	=	$4a$	–	75

Step 4

$$\begin{aligned} a &= 4a - 75 \\ {}^-a & \quad {}^-a \\ \hline 0 &= 3a - 75 \\ {}^+75 & \quad {}^+75 \\ \hline 75 &= 3a \\ \frac{75}{3} &= \frac{3a}{3} \\ 25 &= a \end{aligned}$$

Step 5
Tamu is 25 years old.

Step 6
75 subtracted from 4 times 25 :
4(25) – 75 = 100 – 75 = 25

The correct solution is 25 years old.

41. *Step 1*
The problem is about spending money for clothes.

Unknown: amount spent on clothes by Brenda

Known: Consuelo spent \$3 less than twice the amount that Brenda spent, Consuelo spent \$81

Step 2(a)
Let m represent the amount of money that Brenda spent.

Step 3

Consuelo spent	=	\$3 less than twice the amount that Brenda spent
↓		↓
81	=	$2m - 3$

Step 4

$$\begin{aligned} 81 &= 2m - 3 \\ 81 &= 2m + (^-3) \\ {}^+3 & \quad\quad {}^+3 \\ \hline \frac{84}{2} &= \frac{2m}{2} \\ 42 &= m \end{aligned}$$

Step 5
Brenda spent \$42 on clothes.

Step 6
Consuelo spent \$3 less than twice the amount that Brenda spent, or
2 • \$42 + ($^-$\$3) = \$84 – (\$3) = \$81.

The correct answer is \$42.

43. *Step 1*
The problem concerns bags of candy.

Unknown: the number of pieces of candy in each bag

Known: There were originally 5 bags. She gave 3 pieces of candy each to 48 children. Afterwards 1 bag remained.

Step 2(a)
Let b represent the number of pieces in each bag.

Step 3

total amount of candy at the beginning	subtract	total amount of candy given out	results in	amount of candy in one bag
↓	↓	↓	↓	↓
$5 \cdot b$	–	$3 \cdot 48$	=	b

Step 4

$$\begin{aligned} 5b - 3 \cdot 48 &= b \\ 5b - 144 &= b \\ +144 & \quad +144 \\ \hline 5b + 0 &= b + 144 \\ {}^-b & \quad {}^-b \\ \hline 4b &= 144 \\ \frac{4b}{4} &= \frac{144}{4} \\ b &= 36 \end{aligned}$$

Step 5
There were 36 pieces of candy in each bag.

Step 6
5 bags of 36 pieces of candy: $5 \cdot 36 = 180$
3 pieces were handed out to each of 48 children:
$3 \cdot 48 = 144$
What remained: $180 - 144 = 36$
Equals the amount in one bag: 36

The correct answer is 36.

45. *Step 1*
The problem is about the recommended daily iron intake.

Unknown: the recommended daily iron intake for an infant

Known: The recommended daily iron intake for an adult female is 3 mg more than twice that for an infant. The amount for an adult female is 15 mg.

Step 2(a)
Let d represent the daily amount for an infant.

Step 3

the amount for an adult female	is	3 mg	more than	twice	the amount for an infant
↓	↓	↓	↓	↓	↓
15	=	3	+	2 •	d

Step 4

$$\begin{aligned} 15 &= 3 + 2d \\ {}^-3 &\quad {}^-3 \\ \hline 12 &= 0 + 2d \\ \frac{12}{2} &= \frac{2d}{2} \\ 6 &= d \end{aligned}$$

Step 5
The daily recommended iron intake for a newborn infant is 6 mg.

Step 6
3 more than twice 6: $3 + 2 \cdot 6 = 15$
Equals the amount for an adult female: 15

The correct answer is 6 mg.

3.4 Solving Application Problems with Two Unknown Quantities

3.4 Section Exercises

1. *Step 1*
The problem is about someone's age.

Unknowns: how old I am; how old my sister is

Known: My sister is 9 years older than I am. The sum of our ages is 51.

Step 2(b)
Let a represent how old I am. Since my sister is 9 years older, let $a + 9$ represent her age.

Step 3

My age	+	sister's age	=	sum of ages
↓		↓		↓
a	+	$a + 9$	=	51

Step 4

$$\begin{aligned} a + a + 9 &= 51 \\ 2a + 9 &= 51 \\ {}^-9 &\quad {}^-9 \\ \hline 2a + 0 &= 42 \\ \frac{2a}{2} &= \frac{42}{2} \\ a &= 21 \end{aligned}$$

Step 5
I am 21 years old, and my sister is $21 + 9 = 30$ years old.

Step 6
30 is 9 more than 21 and the sum of our ages is $21 + 30 = 51$.

3. *Step 1*
The problem concerns a couple's earnings.

Unknowns: how much each earned

Known: Lien earned \$1500 more than her husband. Together they earned \$37,500.

Step 2(b)
Let m represent the husband's earnings. Then Lien earned $m + 1500$.

Step 3

Husband's earnings	+	Lien's earnings	=	amount earned together
↓		↓		↓
m	+	$m + 1500$	=	37,500

Step 4

$$\begin{aligned} m + m + 1500 &= 37{,}500 \\ 2m + 1500 &= 37{,}500 \\ {}^-1500 &\quad {}^-1500 \\ \hline 2m + 0 &= 36{,}000 \\ \frac{2m}{2} &= \frac{36{,}000}{2} \\ m &= 18{,}000 \\ m + 1500 &= 18{,}000 + 1500 \\ &= 19{,}500 \end{aligned}$$

Step 5
Lien earned \$19,500 and her husband earned \$18,000.

Step 6
\$19,500 is \$1500 more than \$18,000 and the sum of \$18,000 and \$19,500 is \$37,500.

5. *Step 1*
The problem is about the price of a computer and the price of a printer.

Unknowns: the computer's price and the printer's price

Known: The computer's price is five times the printer's price. The total paid is \$1320.

Step 2(b)
Let m represent the price of the printer. The price of the computer can be represented by $5 \cdot m$, or $5m$.

Step 3
The total price for both is \$1320.

$$\begin{array}{ccccc} \text{printer price} & + & \text{computer price} & = & \text{total price} \\ \downarrow & & \downarrow & & \downarrow \\ m & + & 5m & = & 1320 \end{array}$$

Step 4

$$\begin{aligned} m + 5m &= 1320 \\ \frac{6m}{6} &= \frac{1320}{6} \\ m &= 220 \end{aligned}$$

Step 5
The printer's price is \$220. The computer's price is $5 \cdot \$220 = \$1100$.

Step 6
Five times \$220 is \$1100 and the sum of \$220 and \$1100 is \$1320.

7. *Step 1*
The problem concerns cutting a board into two pieces.

Unknowns: the length of each piece of board

Known: The board had a total length of 78 cm. One piece is 10 cm longer than the other.

Step 2(b)
Let x represent the shorter length. Then the longer length is $x + 10$ cm.

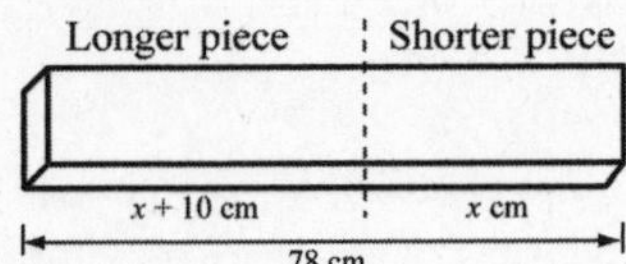

Step 3
The sum of the lengths is 78 cm, so $x + (x + 10) = 78$.

Step 4

$$\begin{aligned} x + x + 10 &= 78 \\ 2x + 10 &= 78 \\ {}^{-}10 & \quad {}^{-}10 \\ \hline 2x + 0 &= 68 \\ \frac{2x}{2} &= \frac{68}{2} \\ x &= 34 \end{aligned}$$

Step 5
The shorter length is $x = 34$ cm, so the longer length is $34 + 10 = 44$ cm.

Step 6
44 cm is 10 cm longer than 34 cm, and the sum of 34 cm and 44 cm is 78 cm.

9. *Step 1*
The problem is about the lengths of two pieces of wire.

Unknowns: the length of each piece

Known: One piece is 7 ft shorter than the other. The wire was 31 ft long before it was cut into two pieces.

Step 2(b)
Let x represent the length of the long piece. Then $x - 7$ represents the piece that is 7 ft shorter.

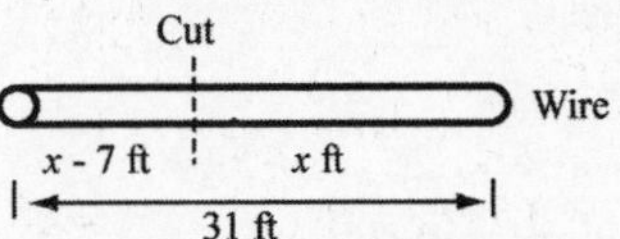

Step 3
The lengths of the two pieces together should be 31, so $x + (x - 7) = 31$.

Step 4

$$\begin{aligned} x + x - 7 &= 31 \\ 2x + ({}^{-}7) &= 31 \\ 7 & \quad 7 \\ \hline 2x + 0 &= 38 \\ 2x &= 38 \\ x &= 19 \end{aligned}$$

Step 5
The long piece is 19 ft. The short piece is $19 - 7 = 12$ ft.

Step 6
12 ft is 7 ft shorter than 19 ft and the sum of 19 ft and 12 ft is 31 ft.

11. *Step 1*
The problem is about the number of Senators and Representatives in Congress.

Unknowns: The number of Senators and the number of Representatives

Known: The number of Representatives is 65 less than 5 times the number of Senators. The total number of Representatives and Senators is 535.

Step 2(b)
Let s represent the number of Senators. Then the number of Representatives is $5s - 65$.

Step 3
Since the total number in both houses of Congress equals 535, we have $s + 5s - 65 = 535$.

Step 4

$$\begin{aligned} s + 5s - 65 &= 535 \\ 6s - 65 &= 535 \\ +65 \quad & \quad +65 \\ \hline 6s + 0 &= 600 \\ \frac{6s}{6} &= \frac{600}{6} \\ s &= 100 \end{aligned}$$

Step 5
The number of Senators is 100 and the number of Representatives is
$5s - 65 = 5(100) - 65 = 435$.

Step 6
Since $5(100) - 65 = 435$ and
$100 + 435 = 535$, this answer is correct.

13. *Step 1*
The problem is about cutting a fence into parts.

Unknown: The length of each part

Known: Two parts are of equal length. The third part is 25 m longer than each of the other parts. The fence is 706 m long.

Step 2(b)
Let x represent the length of one of the two equal parts. Then the third part is $x + 25$ m long.

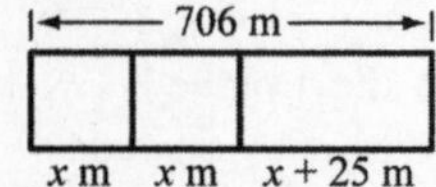

Step 3
Since the sum of the parts is 706 m,
$x + x + (x + 25) = 706$.

Step 4

$$\begin{aligned} x + x + x + 25 &= 706 \\ 3x + 25 &= 706 \\ {}^{-}25 \quad & \quad {}^{-}25 \\ \hline 3x + 0 &= 681 \\ \frac{3x}{3} &= \frac{681}{3} \\ x &= 227 \end{aligned}$$

Step 5
The two equal parts are each 227 m long, so the longer part is $x + 25 = 227 + 25 = 252$ m in length.

Step 6
Since $252 - 227 = 25$ and
$227 + 227 + 252 = 706$, this answer is correct.

15. *Step 1*
The problem is about the length l of a rectangle.

Unknown: Rectangle's length

Known: Rectangle's width and perimeter

Step 2(a)
Let x represent the length.

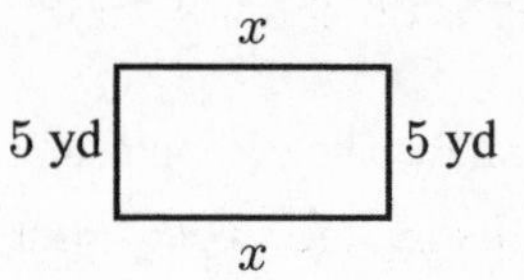

Step 3
$P = 2l + 2w$ Perimeter = 48 yd
$48 = 2 \cdot x + 2 \cdot 5$

Step 4

$$\begin{aligned} 48 &= 2x + 10 \\ {}^{-}10 \quad & \quad {}^{-}10 \\ \hline 38 &= 2x + 0 \\ \frac{38}{2} &= \frac{2x}{2} \\ 19 &= x \end{aligned}$$

Step 5
The rectangle's length is 19 yd.

Step 6
$P = 2l + 2w$
$P = 2 \cdot 19 + 2 \cdot 5$
$P = 38 + 10$
$P = 48$

17. *Step 1*
The problem is about the dimensions of a rectangular dog pen.

Unknowns: The length and width of the dog pen

Known: The perimeter is 36 ft. The length is twice the width.

Step 2(b)
Let w represent the width. Then the length equals $2w$.

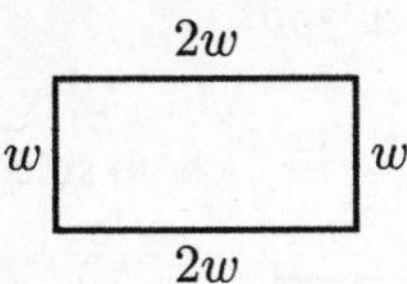

Step 3
$P = 2l + 2w$ Perimeter = 36 ft
$36 = 2 \cdot 2w + 2w$

Step 4
$36 = 4w + 2w$
$36 = 6w$
$\frac{36}{6} = \frac{6w}{6}$
$6 = w$

Step 5
The width is 6 ft, so the length is $2w = 2 \cdot 6 = 12$ ft.

Step 6
Since 12 is twice 6 and $2 \cdot 12 + 2 \cdot 6 = 24 + 12 = 36$, this answer is correct.

19. *Step 1*
The problem is about finding the length and the width of a jewelry box.

Unknowns: The length and width of the box

Known: The length is 3 in. more than twice the width. The perimeter is 36 in.

Step 2(b)
Let w represent the width. Then, since the length is 3 in. more than twice the width, use $2w + 3$ to represent the length.

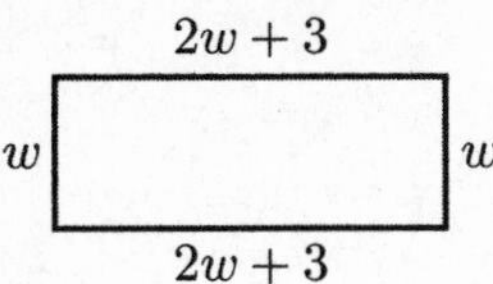

Step 3
$$2l + 2w = P$$
$$2 \cdot (2w + 3) + 2 \cdot w = 36$$

Step 4
$$\begin{aligned} 2(2w + 3) + 2w &= 36 \\ 4w + 6 + 2w &= 36 \\ 4w + 2w + 6 &= 36 \\ 6w + 6 &= 36 \\ {}^{-}6 & \quad {}^{-}6 \\ \frac{6w}{6} &= \frac{30}{6} \\ w &= 5 \end{aligned}$$

Step 5
The width is 5 in. The length is $2(5) + 3 = 13$ in.

Step 6
The length of 13 in. is 3 in. longer than twice the width of 5 in.

$P = 2l + 2w$
$P = 2 \cdot 13 + 2 \cdot 5$
$P = 26 + 10$
$P = 36$ in.

21. *Step 1*
The problem is about a framed photo.

Unknowns: The outside perimeter and total area of the photo and frame.

Known: The photo inside the frame is a rectangle with length 10 in. and width 8 in. The frame is 2 in. wide.

Step 2(b)
Let P represent the outside perimeter and A represent the total area of the photo and frame. Note that $P = 2l + 2w$, where $l = 10 + 2 \cdot 2$ and $w = 8 + 2 \cdot 2$ are the length and the width of the frame, respectively.

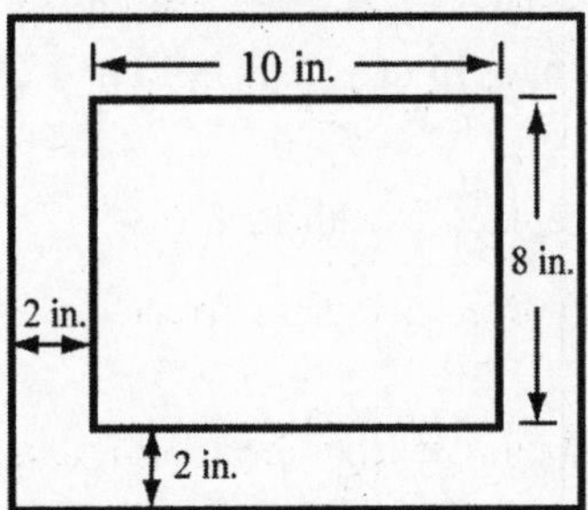

Step 3
$P = 2(10 + 2 \cdot 2) + 2(8 + 2 \cdot 2)$ and
$A = lw$

Step 4
$$\begin{aligned} P &= 2(10 + 4) + 2(8 + 4) \\ &= 28 + 24 \\ &= 52 \text{ in.} \end{aligned}$$

$$\begin{aligned} A &= lw \\ &= (10 + 4)(8 + 4) \\ &= 14 \cdot 12 \\ &= 168 \text{ in.}^2 \end{aligned}$$

Step 5
The outside perimeter is 52 in. and the total area of the photo and frame is 168 in.^{2}.

Step 6
Since the dimensions of the photo are 8 in. and 10 in. and the frame is 2 in. wide, the dimensions of the frame are 12 in. and 14 in. Therefore, the outside perimeter is $12 + 14 + 12 + 14 = 52$ in. The total area of the photo and frame is $12 \cdot 14 = 168$ in.^{2}.

Chapter 3 Review Exercises

1. The figure is a square.

$P = 4s$
$P = 4 \cdot 28$ cm
$P = 112$ cm

2. The figure is a rectangle.

$P = 2 \cdot l + 2 \cdot w$
$P = 2 \cdot 8$ mi $+ 2 \cdot 3$ mi
$P = 16$ mi $+ 6$ mi
$P = 22$ mi

3. The figure is a parallelogram.

$P =$ sum of all four sides
$P = 7$ yd $+ 14$ yd $+ 7$ yd $+ 14$ yd
$P = 42$ yd

4. $P =$ sum of all sides
$P = 26$ m $+ 44$ m $+ 14$ m $+ 20$ m $+ 13$ m $+ 24$ m
$P = 141$ m

5. $P = 4s$ Replace P with 12 ft.
$12 \text{ ft} = 4s$
$\frac{12 \text{ ft}}{4} = \frac{4s}{4}$
$3 \text{ ft} = s$

One side of the table is 3 ft.

6. $P = 2l + 2w$ Replace P with 128 yd and w with 31 yd.

$128 \text{ yd} = 2l + 2 \cdot 31 \text{ yd}$

$128 \text{ yd} = 2l + 62 \text{ yd}$ To get $2l$ by itself, add $^-62$ yd to both sides.

$\underline{^-62 \text{ yd} \qquad\qquad ^-62 \text{ yd}}$
$66 \text{ yd} = 2l + 0$

$\frac{66 \text{ yd}}{2} = \frac{2l}{2}$ To get l by itself, divide both sides by 2.

$33 \text{ yd} = l$

The length of the playground is 33 yd.

7. $P = 2l + 2w$ Replace P with 72 in. and l with 21 in.

$72 \text{ in.} = 2 \cdot 21 \text{ in.} + 2w$

$72 \text{ in.} = 42 \text{ in.} + 2w$ Add $^-42$ in. to both sides.
$\underline{^-42 \text{ in.} \qquad ^-42 \text{ in.}}$
$30 \text{ in.} = 0 + 2w$

$\frac{30 \text{ in.}}{2} = \frac{2w}{2}$ Divide both sides by 2.
$15 \text{ in.} = w$

The width of the painting is 15 in.

8.

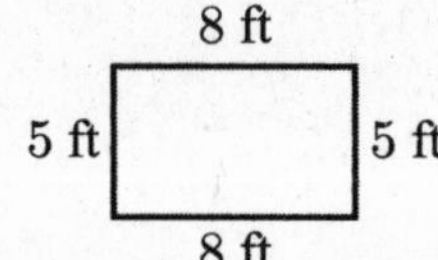

$A = l \cdot w$ The tablecloth is a rectangle.
$A = 8 \text{ ft} \cdot 5 \text{ ft}$
$A = 40 \text{ ft}^2$

The area of the tablecloth is 40 ft^2.

9.

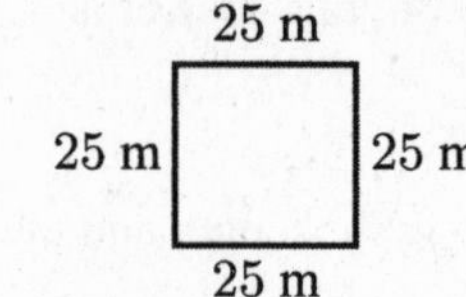

$A = s^2$
$A = s \cdot s$
$A = 25 \text{ m} \cdot 25 \text{ m}$
$A = 625 \text{ m}^2$

The area of the dance floor is 625 m^2.

10.

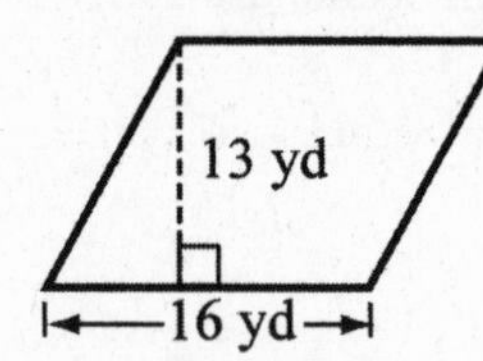

$A = b \cdot h$
$A = 16 \text{ yd} \cdot 13 \text{ yd}$
$A = 208 \text{ yd}^2$

The area of the lot is 208 yd^2.

11. 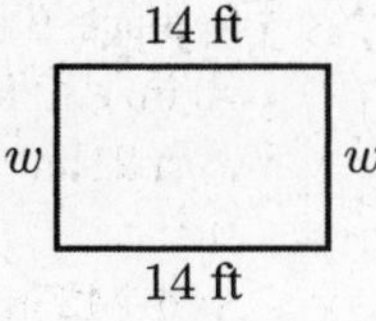

$$A = l \cdot w$$
$$126 \text{ ft}^2 = 14 \text{ ft} \cdot w$$
$$\frac{126 \text{ ft} \cdot \cancel{\text{ft}}}{14 \cancel{\text{ft}}} = \frac{14 \text{ ft} \cdot w}{14 \text{ ft}}$$
$$9 \text{ ft} = w$$

The width of the patio is 9 ft.

12. 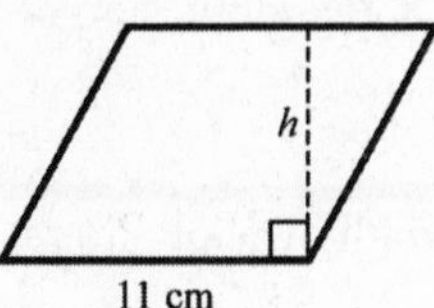

$$A = b \cdot h$$ Replace A with 88 cm^2 and b with 11 cm.

$$88 \text{ cm}^2 = 11 \text{ cm} \cdot h$$
$$\frac{88 \text{ cm} \cdot \cancel{\text{cm}}}{11 \cancel{\text{cm}}} = \frac{11 \text{ cm} \cdot h}{11 \text{ cm}}$$
$$8 \text{ cm} = h$$

The height is 8 cm.

13. 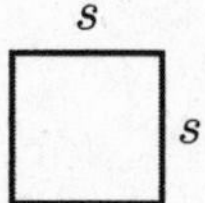

$$A = s^2$$ Replace A with 100 mi^2 and solve by inspection.

$$100 \text{ mi}^2 = s^2$$
$$100 \text{ mi}^2 = s \cdot s$$
$$100 \text{ mi}^2 = 10 \text{ mi} \cdot 10 \text{ mi}$$

The length of one side of the piece of land is 10 mi.

14. A number subtracted from 57

$$57 - x$$

(*Note:* $x - 57$ would be the translation of "57 subtracted from a number.")

15. The sum of 15 and twice a number

$$15 + 2x \text{ or } 2x + 15$$

16. The product of $^-9$ and a number

$$^-9x$$

17. The sum of four times a number and 6 is $^-30$.

$$4n + 6 = {}^-30$$

$$\begin{aligned} 4n + 6 &= {}^-30 \\ {}^-6 &\quad {}^-6 \\ 4n + 0 &= {}^-36 \\ \frac{4n}{4} &= \frac{{}^-36}{4} \\ n &= {}^-9 \end{aligned}$$

The number is $^-9$.

18. When twice a number is subtracted from 10 the result is 4 plus the number

$$10 - 2n = 4 + n$$

$$\begin{aligned} 10 - 2n &= 4 + n \\ +2n &\quad +2n \\ 10 + 0 &= 4 + 3n \\ 10 &= 4 + 3n \\ {}^-4 &\quad {}^-4 \\ 6 &= 0 + 3n \\ \frac{6}{3} &= \frac{3n}{3} \\ 2 &= n \end{aligned}$$

The number is 2.

19. *Step 1*
Unknown: Amount in Grace's account before writing her rent check

Known: \$600 check, \$750 deposit, \$75 deposit, \$309 ending balance

Step 2(a)
Let m represent the amount of money originally in the account.

Step 3
$m - 600 + 750 + 75 = 309$

Step 4
$$\begin{aligned} m + ({}^-600) + 750 + 75 &= 309 \\ m + 225 &= 309 \\ {}^-225 &\quad {}^-225 \\ m + 0 &= 84 \\ m &= 84 \end{aligned}$$

Step 5
\$84 was originally in Grace's account.

Step 6
$\$84 - \$600 + \$750 + \$75$
$= {}^{-}\$516 + \$750 + \$75$
$= \$234 + \$75$
$= \$309$ (ending balance checks)

20. *Step 1*
Unknown: Number of candles in each box

Known: There were 4 boxes, one candle on each of 25 tables, 23 candles left.

Step 2(a)
Let c represent the number of candles per box.

Step 3

$$\underbrace{\text{Total number of candles in 4 boxes}}_{4 \cdot c} - \underbrace{\text{number of candles used}}_{25} = \underbrace{\text{number of candles left}}_{23}$$

Step 4

$$\begin{aligned} 4c - 25 &= 23 \\ +25 &\quad +25 \\ 4c + 0 &= 48 \\ \frac{4c}{4} &= \frac{48}{4} \\ c &= 12 \end{aligned}$$

Step 5
There were 12 candles in each box.

Step 6
4 boxes with 12 candles per box = 48 candles.

25 candles used: 48 − 25 = 23 left (matches)

21. *Step 1*
Unknowns: How much Reggie earned and how much Donald earned

Known: \$1000 prize, Donald should get \$300 more than Reggie

Step 2(b)
There are two unknowns. You know the least about how much Reggie received in prize money. Let p represent the prize money earned by Reggie. Then $p + 300$ represents the prize money earned by Donald.

Step 3

$$\underbrace{\text{Reggie's prize money}}_{p} + \underbrace{\text{Donald's prize money}}_{p+300} = \underbrace{\text{Total prize money}}_{1000}$$

Step 4

$$\begin{aligned} p + p + 300 &= 1000 \\ 2p + 300 &= 1000 \\ {}^{-}300 &\quad {}^{-}300 \\ 2p + 0 &= 700 \\ \frac{2p}{2} &= \frac{700}{2} \\ p &= 350 \end{aligned}$$

Step 5
Reggie earns \$350 and Donald earns \$350 + \$300 = \$650.

Step 6
\$650 is \$300 more than \$350 and the sum of \$650 and \$350 is \$1000.

22. *Step 1*
Unknowns: length and width of photograph

Known: Perimeter is 84 cm, the photo is twice as long as it is wide

Step 2(b)
There are two unknowns. You know the least about the width. Let w represent the width. Then $2w$ represents the length.

Step 3
$2l + 2w = P$

Step 4

$$\begin{aligned} 2l + 2w &= 84 \text{ cm} \qquad P = 84 \text{ cm} \\ 2 \cdot 2w + 2w &= 84 \text{ cm} \qquad \text{length } l = 2w \\ 4w + 2w &= 84 \text{ cm} \\ 6w &= 84 \text{ cm} \\ \frac{6w}{6} &= \frac{84}{6} \text{ cm} \\ w &= 14 \text{ cm} \end{aligned}$$

Step 5
The width is 14 cm and the length is 28 cm.

Step 6

$$\begin{aligned} 2 \cdot 14 \text{ cm} + 2 \cdot 28 \text{ cm} &= P \\ 28 \text{ cm} + 56 \text{ cm} &= P \\ 84 \text{ cm} &= P \quad \text{(matches)} \end{aligned}$$

23. **[3.1] (a)** Kit #2 has 36 feet of fencing to make a square dog pen.

$P = 4s$ Replace P with 36 feet.

$$\begin{aligned} 36 \text{ ft} &= 4s \\ \frac{36 \text{ ft}}{4} &= \frac{4s}{4} \\ 9 \text{ ft} &= s \end{aligned}$$

The length of one side of the dog pen is 9 ft.

[3.2] (b) The area of the square dog pen

$A = s^2$

$A = s \cdot s$

$A = 9\text{ ft} \cdot 9\text{ ft}$

$A = 81\text{ ft}^2$

The area of the dog pen is 81 ft^2.

24. **[3.2]** Rectangles will vary. To create your rectangles, choose a length and width that add up to 10, which is one-half of the perimeter of 20. Two possibilities are:

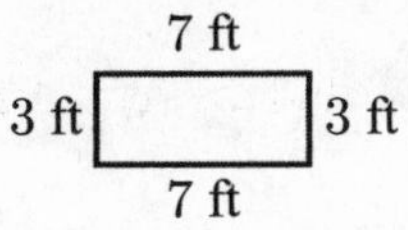

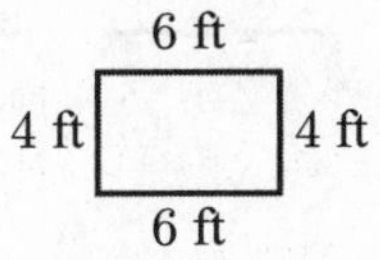

$A = l \cdot w$

$A = 7\text{ ft} \cdot 3\text{ ft}$

$A = 21\text{ ft}^2$

$A = l \cdot w$

$A = 6\text{ ft} \cdot 4\text{ ft}$

$A = 24\text{ ft}^2$

25. **[3.3]** *Step 1*
Unknown: Amount of fencing Timotha put around her garden

Known: Kit #1 contains 20 ft, Kit #2 contains 36 ft, 41 ft of fencing available for dog pen

Step 2(a)
Let f represent the number of feet of fencing for the garden.

Step 3

Fencing in Kit #2		Fencing around garden		Fencing in Kit #1		Fencing for dog pen
↓		↓		↓		↓
36	−	f	+	20	=	41

Step 4

$$\begin{aligned} 36 - f + 20 &= 41 \\ 56 - f &= 41 \\ {}^{-}56 \quad & \quad {}^{-}56 \\ \hline {}^{-}f &= {}^{-}15 \\ \frac{{}^{-}1f}{{}^{-}1} &= \frac{{}^{-}15}{{}^{-}1} \\ f &= 15 \end{aligned}$$

Step 5
15 ft of fencing went around the garden.

Step 6
15 ft + 41 ft = 56 ft matches the total fencing from Kits #1 and #2 [20 + 36].

26. **[3.4]** *Step 1*
Unknowns: The width and the length of the dog pen

Known: 36 ft of fencing was used, the length was 2 ft more than the width

Step 2(b)
There are two unknowns. You know the least about the width. Let w represent the width. Then $w + 2$ represents the length.

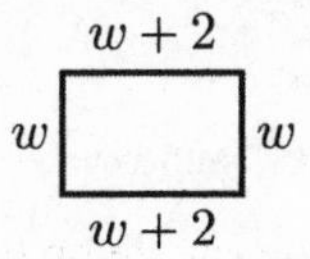

Step 3

$2l + 2w = P$

$2 \cdot (w + 2) + 2w = 36$

Step 4

$$\begin{aligned} 2(w + 2) + 2w &= 36 \\ 2w + 4 + 2w &= 36 \\ 4w + 4 &= 36 \\ {}^{-}4 \quad & \quad {}^{-}4 \\ \hline 4w &= 32 \\ \frac{4w}{4} &= \frac{32}{4} \\ w &= 8 \end{aligned}$$

Step 5
The width is 8 ft and the length is 8 + 2 = 10 ft.

Step 6
10 ft is 2 ft longer than 8 ft.

$P = 2 \cdot 10\text{ ft} + 2 \cdot 8\text{ ft}$

$P = 20\text{ ft} + 16\text{ ft}$

$P = 36\text{ ft}$ (matches)

Chapter 3 Test

1. $P = 59\text{ m} + 72\text{ m} + 59\text{ m} + 72\text{ m}$
$P = 262\text{ m}$

2. $P = 8\text{ in.} + 4\text{ in.} + 4\text{ in.} + 4\text{ in.} + 8\text{ in.} + 4\text{ in.} + 4\text{ in.} + 4\text{ in.}$
$P = 40\text{ in.}$

3. $P = 4s$
$P = 4 \cdot 3\text{ miles}$
$P = 12\text{ miles}$

4. $P = 2l + 2w$
$P = 2 \cdot 4\text{ ft} + 2 \cdot 2\text{ ft}$
$P = 8\text{ ft} + 4\text{ ft}$
$P = 12\text{ ft}$

5. $P = 45\text{ cm} + 50\text{ cm} + 15\text{ cm}$
$P = 110\text{ cm}$

6. $A = l \cdot w$
$A = 27\text{ mm} \cdot 18\text{ mm}$
$A = 486\text{ mm}^2$

7. *Note:* The base is perpendicular to the height.

$$A = b \cdot h$$
$$A = 14 \text{ cm} \cdot 10 \text{ cm}$$
$$A = 140 \text{ cm}^2$$

8.
$$A = l \cdot w$$
$$A = 68 \text{ mi} \cdot 55 \text{ mi}$$
$$A = 3740 \text{ mi}^2$$

9.
$$A = s^2$$
$$A = s \cdot s$$
$$A = 6 \text{ m} \cdot 6 \text{ m}$$
$$A = 36 \text{ m}^2$$

10.
$$P = 4s$$
$$12 \text{ ft} = 4s$$
$$\frac{12 \text{ ft}}{4} = \frac{4s}{4}$$
$$3 \text{ ft} = s$$

The length of one side of the table is 3 ft.

11.
$$P = 2l + 2w$$
$$34 \text{ ft} = 2l + 2(6 \text{ ft})$$
$$34 \text{ ft} = 2l + 12 \text{ ft}$$
$$\underline{^-12 \text{ ft}} \qquad \underline{^-12 \text{ ft}}$$
$$22 \text{ ft} = 2l + 0$$
$$\frac{22 \text{ ft}}{2} = \frac{2l}{2}$$
$$11 \text{ ft} = l$$

The length of the plot is 11 ft.

12.
$$A = bh$$
$$65 \text{ in.}^2 = 13 \text{ in.} \cdot h$$
$$\frac{65 \text{ in.} \cdot \cancel{\text{in.}}}{13 \cancel{\text{in.}}} = \frac{13 \text{ in.} \cdot h}{13 \text{ in.}}$$
$$5 \text{ in.} = h$$

The parallelogram's height is 5 in.

13.
$$A = l \cdot w$$
$$12 \text{ cm}^2 = 4 \text{ cm} \cdot w$$
$$\frac{12 \text{ cm} \cdot \cancel{\text{cm}}}{4 \cancel{\text{cm}}} = \frac{4 \text{ cm} \cdot w}{4 \text{ cm}}$$
$$3 \text{ cm} = w$$

The width of the postage stamp is 3 cm.

14.
$$A = s^2$$
$$A = s \cdot s$$

$16 \text{ ft}^2 = s \cdot s$ Replace A with 16 ft² and solve by inspection.

$$16 \text{ ft}^2 = 4 \text{ ft} \cdot 4 \text{ ft}$$

Each side of the bulletin board is 4 feet.

15. Linear units like ft are used to measure length, width, height, and perimeter. Area is measured in square units like ft² (squares that measure 1 ft on each side).

16. If 40 is added to four times a number the result is zero.

$$40 + 4n = 0$$

$$40 + 4n = 0$$
$$\underline{^-40} \qquad \underline{^-40}$$
$$0 + 4n = {}^-40$$
$$\frac{4n}{4} = \frac{^-40}{4}$$
$$n = {}^-10$$

The number is $^-10$.

17. When 7 times a number is decreased by 23 the result is the number plus 7.

$$7n - 23 = n + 7$$

$$7n - 23 = n + 7$$
$$\underline{^-n} \qquad \underline{^-n}$$
$$6n - 23 = 7$$
$$\underline{+23} \qquad \underline{+23}$$
$$6n + 0 = 30$$
$$6n = 30$$
$$\frac{6n}{6} = \frac{30}{6}$$
$$n = 5$$

The number is 5.

18. *Step 1*

Unknown: Amount of money used by son

Known: \$43 originally, some spent by son on groceries, \$16 put into wallet, and \$44 at the end

Step 2(a)

Let m represent the money used by Josephine's son.

Step 3

$$43 - m + 16 = 44$$

Step 4

$$43 - m + 16 = 44$$
$$59 - m = 44$$
$$\underline{^-59} \qquad \underline{^-59}$$
$$0 - m = {}^-15$$
$$\frac{^-1m}{^-1} = \frac{^-15}{^-1}$$
$$m = 15$$

Step 5
Her son spent \$15.

Step 6
\$43 − \$15 + \$16 = \$44 (matches)

19. *Step 1*
Unknown: Daughter's age

Known: Ray is 39 years old and his age is 4 years more than five times his daughter's age.

Step 2(a)
Let d represent the daughter's age.

Step 3

$$\underbrace{\text{Ray's age}}_{39} \quad \overset{\text{is}}{\downarrow} \quad \underbrace{\text{4 years more than five times his daughter's age.}}_{4+5d}$$

$$39 = 4 + 5d$$

Step 4

$$\begin{aligned} 39 &= 4 + 5d \\ {}^{-}4 & \quad {}^{-}4 \\ \hline 35 &= 0 + 5d \\ \frac{35}{5} &= \frac{5d}{5} \\ 7 &= d \end{aligned}$$

Step 5
The daughter is 7 years old.

Step 6
Four years more than five times 7 years is $4 + 5 \cdot 7 = 4 + 35 = 39$ yrs (matches).

20. *Step 1*
Unknowns: The length of each piece

Known: Total board length is 118 cm, one piece is 4 cm longer than the other

Step 2(b)
There are two unknowns. Let p represent the shorter piece. Then $p + 4$ represents the longer piece.

Step 3

$$\underbrace{\text{long piece}}_{p+4} + \underbrace{\text{short piece}}_{p} = \underbrace{\text{total length}}_{118}$$

Step 4

$$\begin{aligned} p + 4 + p &= 118 \\ 2p + 4 &= 118 \\ {}^{-}4 & \quad {}^{-}4 \\ \hline 2p + 0 &= 114 \\ \frac{2p}{2} &= \frac{114}{2} \\ p &= 57 \end{aligned}$$

Step 5
One piece is 57 cm and the other is $57 + 4 = 61$ cm.

Step 6
61 cm is 4 cm longer than 57 cm and the sum of 61 cm and 57 cm is 118 cm.

21. *Step 1*
Unknowns: Length and width

Known: Perimeter is 420 ft, length is four times as long as the width

Step 2(b)
Let w represent the width.
Then $4w$ represents the length.

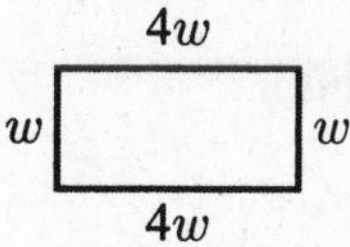

Step 3

$$P = 2l + 2w$$
$$420 \text{ ft} = 2 \cdot 4w + 2 \cdot w$$

Step 4

$$\begin{aligned} 420 &= 2 \cdot 4w + 2 \cdot w \\ 420 &= 8w + 2w \\ 420 &= 10w \\ \frac{420}{10} &= \frac{10w}{10} \\ 42 &= w \end{aligned}$$

Step 5
The width is 42 ft and the length is $4(42) = 168$ ft.

Step 6
168 ft is four times 42 ft.

$$P = 2 \cdot 168 \text{ ft} + 2 \cdot 42 \text{ ft}$$
$$P = 336 \text{ ft} + 84 \text{ ft}$$
$$P = 420 \text{ ft} \qquad \text{(matches)}$$

22. *Step 1*
Unknowns: How many hours each person worked

Known: 19 hours worked by both, Tim worked 3 hours less than Marcella

Step 2(b)
Let h represent the hours worked by Marcella.
Then $h - 3$ represents the hours worked by Tim.

Step 3

$$\underbrace{\text{Tim's hours}}_{h-3} + \underbrace{\text{Marcella's hours}}_{h} = \underbrace{\text{Total hours}}_{19}$$

Step 4

$$\begin{aligned} h - 3 + h &= 19 \\ 2h - 3 &= 19 \\ +3 \quad & \quad +3 \\ \hline 2h + 0 &= 22 \\ \frac{2h}{2} &= \frac{22}{2} \\ h &= 11 \end{aligned}$$

Step 5
Marcella worked 11 hours and Tim worked $11 - 3 = 8$ hours.

Step 6
8 hours is 3 hours less than 11 hours and the sum of 8 and 11 is 19.

Cumulative Review Exercises (Chapters 1–3)

1. 4,000,206,300 in words: four billion, two hundred six thousand, three hundred

2. Seventy million, five thousand, four hundred eighty-nine written using digits is 70,005,489.

3. $^-7$ lies to the *left* of $^-1$ on the number line. Write $^-7 < {}^-1$.

0 lies to the *right* of $^-5$ on the number line. Write $0 > {}^-5$.

4. **(a)** $1(97) = 97$; multiplication property of 1

(b) $^-10 + 0 = {}^-10$; addition property of 0

(c) $(3 \cdot {}^-7) \cdot 6 = 3 \cdot ({}^-7 \cdot 6)$; associative property of multiplication

5. **(a)** Underline the tens place. 37$\underline{9}$5
The next digit is 5 or more. Add one to 9. Write 0 and regroup one to the hundreds place. Change 5 to 0. **3800**

(b) Underline the ten-thousands place. 4$\underline{9}$3,662
The next digit is 4 or less. Leave 9 as 9 and change 3, 6, 6, and 2 to zeros. **490,000**

6.
$$\begin{aligned} &{}^-12 - 12 \\ &= {}^-12 + ({}^-12) \\ &= {}^-24 \end{aligned}$$

7. $^-3({}^-9) = 27$

8.
$$\begin{aligned} &|7| - |{}^-10| \\ &= 7 - 10 \\ &= 7 + ({}^-10) \\ &= {}^-3 \end{aligned}$$

9.
$$\begin{aligned} &{}^-40 \div 2 \cdot 5 && \text{Divide.} \\ &= {}^-20 \cdot 5 && \text{Multiply.} \\ &= {}^-100 \end{aligned}$$

10.
$$\begin{aligned} &3 - 8 + 10 \\ &= 3 + ({}^-8) + 10 \\ &= {}^-5 + 10 \\ &= 5 \end{aligned}$$

11. $\frac{0}{{}^-6} = 0$
Zero divided by any nonzero number is 0.

12.
$$\begin{aligned} &{}^-8 + 5(2 - 3) && \text{Parentheses} \\ &= {}^-8 + 5({}^-1) && \text{Multiply.} \\ &= {}^-8 + {}^-5 && \text{Add.} \\ &= {}^-13 \end{aligned}$$

13.
$$\begin{aligned} &({}^-3)^2 + 4^2 && \text{Exponents} \\ &= {}^-3 \cdot {}^-3 + 4 \cdot 4 && \text{Multiply.} \\ &= 9 + 16 && \text{Add.} \\ &= 25 \end{aligned}$$

14.
$$\begin{aligned} &4 - 3({}^-6 \div 3) + 7(0 - 6) \\ &= 4 - 3 \cdot ({}^-2) + 7 \cdot ({}^-6) \\ &= 4 - ({}^-6) + ({}^-42) \\ &= 4 + ({}^+6) + ({}^-42) \\ &= 10 + ({}^-42) \\ &= {}^-32 \end{aligned}$$

15.
$$\frac{4 - 2^3 + 5^2 - 3({}^-2)}{{}^-1(3) - 6({}^-2) - 9}$$

Numerator:

$$\begin{aligned} &4 - 2^3 + 5^2 - 3({}^-2) \\ &= 4 - (2 \cdot 2 \cdot 2) + (5 \cdot 5) - 3 \cdot ({}^-2) \\ &= 4 - 8 + 25 - ({}^-6) \\ &= 4 + ({}^-8) + 25 + ({}^+6) \\ &= {}^-4 + 25 + 6 \\ &= 21 + 6 \\ &= 27 \end{aligned}$$

Denominator:

$$\begin{aligned} &{}^-1(3) - 6({}^-2) - 9 \\ &= {}^-3 - ({}^-12) - 9 \\ &= {}^-3 + ({}^+12) + ({}^-9) \\ &= 9 + ({}^-9) \\ &= 0 \end{aligned}$$

Last step is division: $\frac{27}{0}$ is undefined.

16. $10w^2xy^4$ can be written as

$$10 \cdot w \cdot w \cdot x \cdot y \cdot y \cdot y \cdot y.$$

17. $^{-}6cd^3$

$= {}^{-}6 \cdot c \cdot d \cdot d \cdot d$ Replace c with 5 and d with $^{-}2$.

$= {}^{-}6 \cdot 5 \cdot {}^{-}2 \cdot {}^{-}2 \cdot {}^{-}2$

$= {}^{-}30 \cdot {}^{-}2 \cdot {}^{-}2 \cdot {}^{-}2$

$= 60 \cdot {}^{-}2 \cdot {}^{-}2$

$= {}^{-}120 \cdot {}^{-}2$

$= 240$

18. $^{-}4k + k + 5k$

$= {}^{-}4k + 1k + 5k$

$= {}^{-}3k + 5k$

$= 2k$

19. $m^2 + 2m + 2m^2 = \underline{1m^2} + \underline{2m^2} + 2m = 3m^2 + 2m$

20. $xy^3 - xy^3 = 0$

Anything minus itself equals zero.

21. $5({}^{-}4a) = (5 \cdot {}^{-}4)a = {}^{-}20a$

22. $^{-}8 + x + 5 - 2x^2 - x$

$= {}^{-}8 + 1x + 5 + ({}^{-}2x^2) + ({}^{-}1x)$

$= {}^{-}8 + 5 + 1x + ({}^{-}1x) + ({}^{-}2x^2)$

$= {}^{-}3 + 0x + ({}^{-}2x^2)$

$= {}^{-}3 + ({}^{-}2x^2)$ or $^{-}2x^2 - 3$

23. $^{-}3(4n + 3) + 10$

$= {}^{-}3 \cdot 4n + {}^{-}3 \cdot 3 + 10$

$= {}^{-}12n + ({}^{-}9) + 10$

$= {}^{-}12n + 1$

24.

$$\begin{aligned} 6 - 20 &= 2x - 9x \\ 6 + ({}^{-}20) &= 2x + ({}^{-}9x) \\ {}^{-}14 &= {}^{-}7x \\ \frac{{}^{-}14}{{}^{-}7} &= \frac{{}^{-}7x}{{}^{-}7} \\ 2 &= x \end{aligned}$$

Check:

$$\begin{aligned} 6 - 20 &= 2(2) - 9(2) \\ 6 - 20 &= 4 - 18 \\ 6 + ({}^{-}20) &= 4 + ({}^{-}18) \\ {}^{-}14 &= {}^{-}14 \quad \text{Balances} \end{aligned}$$

The solution is 2.

25.

$$\begin{aligned} {}^{-}5y &= y + 6 \\ {}^{-}5y &= 1y + 6 \\ \underline{{}^{-}1y} & \quad \underline{{}^{-}1y} \\ {}^{-}6y &= 0 + 6 \\ \frac{{}^{-}6y}{{}^{-}6} &= \frac{6}{{}^{-}6} \\ y &= {}^{-}1 \end{aligned}$$

Check: $^{-}5({}^{-}1) = {}^{-}1 + 6$

$5 = 5$ Balances

The solution is $^{-}1$.

26.

$$\begin{aligned} 3b - 9 &= 19 - 4b \\ \underline{+4b} & \quad \underline{+4b} \\ 7b - 9 &= 19 + 0 \\ 7b - 9 &= 19 \\ \underline{+9} & \quad \underline{+9} \\ 7b + 0 &= 28 \\ \frac{7b}{7} &= \frac{28}{7} \\ b &= 4 \end{aligned}$$

The solution is 4.

27.

$$\begin{aligned} {}^{-}16 - h + 2 &= h - 10 \\ {}^{-}16 + ({}^{-}h) + 2 &= h - 10 \\ {}^{-}14 + ({}^{-}h) &= h - 10 \\ \underline{+h} & \quad \underline{+h} \\ {}^{-}14 + 0 &= 2h - 10 \\ {}^{-}14 &= 2h - 10 \\ \underline{+10} & \quad \underline{+10} \\ {}^{-}4 &= 2h + 0 \\ \frac{{}^{-}4}{2} &= \frac{2h}{2} \\ {}^{-}2 &= h \end{aligned}$$

The solution is $^{-}2$.

28.

$$\begin{aligned} {}^{-}5(2x + 4) &= 3x - 20 \\ {}^{-}10x + ({}^{-}20) &= 3x - 20 \\ \underline{+20} & \quad \underline{+20} \\ {}^{-}10x + 0 &= 3x + 0 \\ {}^{-}10x &= 3x \\ \underline{+10x} & \quad \underline{+10x} \\ 0 &= 13x \\ \frac{0}{13} &= \frac{13x}{13} \\ 0 &= x \end{aligned}$$

The solution is 0.

29.

$$\begin{aligned} 6 + 4(a + 8) &= {}^{-}8a + 5 + a \\ 6 + 4a + 32 &= {}^{-}8a + 5 + 1a \\ 38 + 4a &= {}^{-}7a + 5 \\ \underline{+7a} & \quad \underline{+7a} \\ 38 + 11a &= 0 + 5 \\ 38 + 11a &= 5 \\ \underline{{}^{-}38} & \quad \underline{{}^{-}38} \\ 0 + 11a &= {}^{-}33 \\ \frac{11a}{11} &= \frac{{}^{-}33}{11} \\ a &= {}^{-}3 \end{aligned}$$

The solution is $^{-}3$.

30. $P = 12 \text{ in.} + 18 \text{ in.} + 12 \text{ in.} + 18 \text{ in.}$
$P = 60 \text{ in.}$

$A = b \cdot h$
$A = 18 \text{ in.} \cdot 11 \text{ in.}$
$A = 198 \text{ in.}^2$

31. $P = 4s$ $\quad A = s^2$
$P = 4 \cdot 15 \text{ m}$ $\quad A = s \cdot s$
$P = 60 \text{ m}$ $\quad A = 15 \text{ m} \cdot 15 \text{ m}$
$A = 225 \text{ m}^2$

32. $P = 2l + 2w$ $\quad A = l \cdot w$
$P = 2 \cdot 8 \text{ ft} + 2 \cdot 4 \text{ ft}$ $\quad A = 8 \text{ ft} \cdot 4 \text{ ft}$
$P = 16 \text{ ft} + 8 \text{ ft}$ $\quad A = 32 \text{ ft}^2$
$P = 24 \text{ ft}$

33. $\underbrace{\text{}^-50 \text{ is added to five times a number}}_{5n + {}^-50}$ $\underbrace{\text{the result is}}_{=}$ $\underbrace{\text{zero}}_{0}$

$$\begin{aligned} 5n + {}^-50 &= 0 \\ +50 \quad & \quad +50 \\ 0 + 5n &= 50 \\ \frac{5n}{5} &= \frac{50}{5} \\ n &= 10 \end{aligned}$$

The number is 10.

34. $\underbrace{\text{Three times a number is subtracted from 10}}_{10 - 3n}$ $\underbrace{\text{the result is}}_{=}$ $\underbrace{\text{two times the number.}}_{2n}$

$$\begin{aligned} 10 - 3n &= 2n \\ +3n \quad & \quad +3n \\ 10 + 0 &= 5n \\ \frac{10}{5} &= \frac{5n}{5} \\ 2 &= n \end{aligned}$$

The number is 2.

35. *Step 1*
Unknown: Number of people originally in line

Known: 3 left, 6 got in line, 2 left, 5 still in line

Step 2(a)
Let p represent the number of people in line originally.

Step 3
$p - 3 + 6 - 2 = 5$

Step 4
$$\begin{aligned} p + ({}^-3) + 6 + ({}^-2) &= 5 \\ p + 1 &= 5 \\ {}^-1 \quad & \quad {}^-1 \\ p + 0 &= 4 \\ p &= 4 \end{aligned}$$

Step 5
There were 4 people in line originally.

Step 6
4 in line
3 left $\quad 4 - 3 = 1$
6 got in line $\quad 1 + 6 = 7$
2 left $\quad 7 - 2 = 5$
5 left $\quad$ 5 left (matches)

36. *Step 1*
Unknown: Amount paid by each player

Known: 12 players, \$2200 total expenses, account overdrawn by \$40

Step 2(a)
Let m be the amount paid by each player.

Step 3
$\underbrace{\text{amount collected}}_{12 \cdot m}$ $-$ $\underbrace{\text{total expenses}}_{2200}$ $=$ $\underbrace{\text{amount in the team bank account}}_{{}^-40}$

Step 4
$$\begin{aligned} 12m - 2200 &= {}^-40 \\ +2200 &= +2200 \\ 12m + 0 &= 2160 \\ \frac{12m}{12} &= \frac{2160}{12} \\ m &= 180 \end{aligned}$$

Step 5
Each player paid \$180.

Step 6
$12 \cdot \$180 = \$2160$ collected
$\$2160 - \$2200 = {}^-\$40$ (matches)

37. *Step 1*
Unknowns: Number of students in small group and in large group

Known: 192 students, one group was three times the size of the other group

Step 2(b)
There are 2 unknowns. You know the least about the small group. Let g be the size of the small group. Then $3g$ is the size of the large group.

Step 3
$\underbrace{\text{small group}}_{g}$ $+$ $\underbrace{\text{large group}}_{3g}$ $=$ $\underbrace{\text{total number of students}}_{192}$

Step 4
$g + 3g = 192$
$4g = 192$
$g = 48$

Step 5
There were 48 students in the small group and $3(48) = 144$ students in the large group.

Step 6
144 is 3 times 48 and the sum of 48 and 144 is 192.

38. *Step 1*
Unknowns: length and width

Known: Perimeter is 92 ft, length is 14 ft longer than the width

Step 2(b)
You know the least about the width.
Let w represent the width.
Then $w + 14$ represents the length.

Step 3
$P = 2l + 2w$
$92 = 2 \cdot (w + 14) + 2 \cdot w$

Step 4

$$\begin{aligned} 92 &= 2(w+14)+2w \\ 92 &= 2w+28+2w \\ 92 &= 4w+28 \\ {}^{-}28 & \quad\; {}^{-}28 \\ \frac{64}{4} &= \frac{4w}{4} \\ 16 &= w \end{aligned}$$

Step 5
The width is 16 ft and the length is $16 + 14 = 30$ ft.

Step 6
30 ft is 14 ft longer than 16 ft.

$P = 2 \cdot 30\text{ ft} + 2 \cdot 16\text{ ft}$
$P = 60\text{ ft} + 32\text{ ft}$
$P = 92\text{ ft}$ (matches)

CHAPTER 4 RATIONAL NUMBERS: POSITIVE AND NEGATIVE FRACTIONS

4.1 Introduction to Signed Fractions

4.1 Section Exercises

1. The figure has 8 equal parts. The 5 shaded parts are represented by the fraction $\frac{5}{8}$; the unshaded parts by $\frac{3}{8}$.

3. The figure has 3 equal parts. The 2 shaded parts are represented by the fraction $\frac{2}{3}$; the unshaded part by $\frac{1}{3}$.

5. An area equal to 3 of the $\frac{1}{2}$ parts is shaded: $\frac{3}{2}$

   An area equal to 1 of the $\frac{1}{2}$ parts is unshaded: $\frac{1}{2}$

7. An area equal to 11 of the $\frac{1}{6}$ parts is shaded: $\frac{11}{6}$

   An area equal to 1 of the $\frac{1}{6}$ parts is unshaded: $\frac{1}{6}$

9. Two of the 11 coins are dimes: $\frac{2}{11}$

   Three of the 11 coins are pennies: $\frac{3}{11}$

   Four of the 11 coins are nickels: $\frac{4}{11}$

11. 8 students out of the "whole" class of 25 are deaf, so $\frac{8}{25}$ of the students are deaf. $25 - 8 = 17$, so $\frac{17}{25}$ of the students are *not* deaf.

13. $71 - 58 = 13$

    $\frac{13}{71}$ of the computers are *not* laptops.

    $\frac{58}{71}$ of the computers are laptops.

15. 6 of 20 women would like flowers delivered at work: $\frac{6}{20}$

17. $13 + 6 = 19$ of 20 women would like flowers delivered either at home or at work: $\frac{19}{20}$

19. $\frac{3}{4}$ $\begin{array}{l}\leftarrow 3 \text{ is the Numerator} \\ \leftarrow 4 \text{ is the Denominator}\end{array}$

    The denominator is 4, so there are 4 parts in the whole.

21. $\frac{12}{7}$ $\begin{array}{l}\leftarrow 12 \text{ is the Numerator} \\ \leftarrow 7 \text{ is the Denominator}\end{array}$

    The denominator is 7, so there are 7 parts in the whole.

23. Proper fractions have a numerator that is *smaller* than the denominator.

    $\frac{1}{3}, \frac{5}{8}, \frac{7}{16}$

    Improper fractions have a numerator that is *equal to or greater* than the denominator.

    $\frac{8}{5}, \frac{6}{6}, \frac{12}{2}$

25. Proper fractions: $\frac{3}{4}, \frac{9}{11}, \frac{7}{15}$

    Improper fractions: $\frac{3}{2}, \frac{5}{5}, \frac{19}{18}$

27. Fractions will vary. The denominator shows the number of equal parts in the whole and the numerator shows how many of the parts are being considered. The fraction bar separates the numerator from the denominator. Show your drawing to your instructor.

29. To graph $\frac{1}{4}$ and $-\frac{1}{4}$ on the number line, divide the space between 0 and 1 into 4 equal parts. Start at zero and count 1 part to the right. This spot represents $\frac{1}{4}$. Now divide the space between -1 and 0 into 4 equal parts. Start at zero and count 1 part to the left. This spot represents $-\frac{1}{4}$.

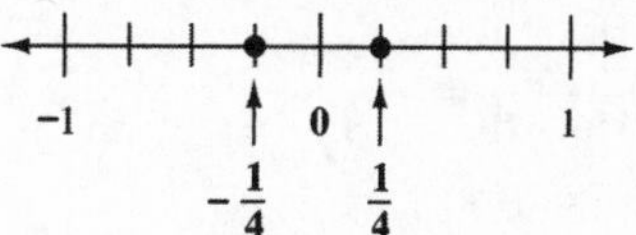

31. Graph $-\frac{3}{5}$ and $\frac{3}{5}$ on the number line.

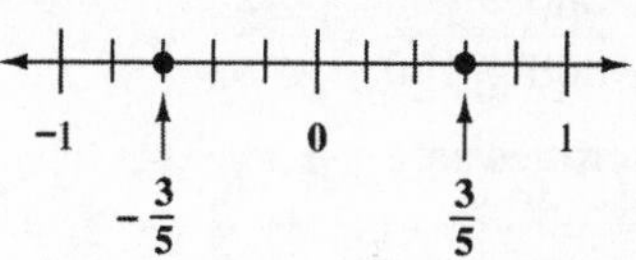

33. Graph $\frac{7}{8}$ and $-\frac{7}{8}$ on the number line.

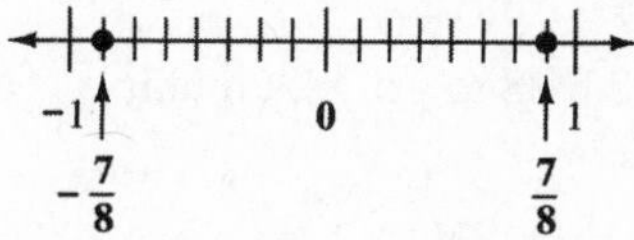

35. "lost" $\frac{3}{4}$ pound: $-\frac{3}{4}$ pound

37. $\frac{1}{2}$ quart "below": $-\frac{1}{2}$ quart

39. $\frac{3}{10}$ mile long: $+\frac{3}{10}$ mile or $\frac{3}{10}$ mile

41. $\left|-\frac{2}{5}\right| = \frac{2}{5}$ because the distance from 0 to $-\frac{2}{5}$ on the number line is $\frac{2}{5}$.

43. $\left|\frac{9}{10}\right| = \frac{9}{10}$ because the distance from 0 to $\frac{9}{10}$ on the number line is $\frac{9}{10}$.

45. $\left|-\frac{13}{6}\right| = \frac{13}{6}$ because the distance from 0 to $-\frac{13}{6}$ on the number line is $\frac{13}{6}$.

47. (a) $\frac{1}{2} = \frac{1 \cdot 12}{2 \cdot 12} = \frac{12}{24}$

    (b) $\frac{1}{3} = \frac{1 \cdot 8}{3 \cdot 8} = \frac{8}{24}$

    (c) $\frac{2}{3} = \frac{2 \cdot 8}{3 \cdot 8} = \frac{16}{24}$

(d) $\frac{1}{4} = \frac{1 \cdot 6}{4 \cdot 6} = \frac{6}{24}$

(e) $\frac{3}{4} = \frac{3 \cdot 6}{4 \cdot 6} = \frac{18}{24}$

(f) $\frac{1}{6} = \frac{1 \cdot 4}{6 \cdot 4} = \frac{4}{24}$

(g) $\frac{5}{6} = \frac{5 \cdot 4}{6 \cdot 4} = \frac{20}{24}$

(h) $\frac{1}{8} = \frac{1 \cdot 3}{8 \cdot 3} = \frac{3}{24}$

(i) $\frac{3}{8} = \frac{3 \cdot 3}{8 \cdot 3} = \frac{9}{24}$

(j) $\frac{5}{8} = \frac{5 \cdot 3}{8 \cdot 3} = \frac{15}{24}$

49. **(a)** $-\frac{2}{6} = -\frac{2 \div 2}{6 \div 2} = -\frac{1}{3}$

(b) $-\frac{4}{6} = -\frac{4 \div 2}{6 \div 2} = -\frac{2}{3}$

(c) $-\frac{12}{18} = -\frac{12 \div 6}{18 \div 6} = -\frac{2}{3}$

(d) $-\frac{6}{18} = -\frac{6 \div 6}{18 \div 6} = -\frac{1}{3}$

(e) $-\frac{200}{300} = -\frac{200 \div 100}{300 \div 100} = -\frac{2}{3}$

(f) Some possibilities are: $-\frac{4}{12} = -\frac{1}{3}$; $-\frac{8}{24} = -\frac{1}{3}$; $-\frac{20}{30} = -\frac{2}{3}$; $-\frac{24}{36} = -\frac{2}{3}$.

51. **(a)** $\frac{3}{8} = \frac{3 \cdot 489}{8 \cdot 489} = \frac{1467}{3912}$

(b) Divide 3912 by 8 to get 489; multiply 3 by 489 to get 1467.

52. **(a)** $\frac{7}{9} = \frac{7 \cdot 608}{9 \cdot 608} = \frac{4256}{5472}$

(b) Divide 5472 by 9 to get 608; multiply 7 by 608 to get 4256.

53. **(a)** $-\frac{697}{3485} = -\frac{697 \div 697}{3485 \div 697} = -\frac{1}{5}$

(b) Divide 3485 by 2, by 3, and by 5 to see that dividing by 5 gives 697. Or divide 3485 by 697 to get 5.

54. **(a)** $-\frac{817}{4902} = -\frac{817 \div 817}{4902 \div 817} = -\frac{1}{6}$

(b) Divide 4902 by 4, by 6, and by 8 to see that dividing by 6 gives 817. Or divide 4902 by 817 to get 6.

55. Since $2028 \div 12 = 169$, divide the denominator by 169 also.

$\frac{1183 \div 169}{2028 \div 169} = \frac{7}{12}$ Replace ? with 7.

56. Since $6105 \div 11 = 555$, divide the numerator by 555 also.

$\frac{2775}{6105} = \frac{2775 \div 555}{6105 \div 555} = \frac{5}{11}$ Replace ? with 5.

57. Since $1157 \div 13 = 89$, divide the denominator by 89 also.

$\frac{1157}{1335} = \frac{1157 \div 89}{1335 \div 89} = \frac{13}{15}$ Replace ? with 15.

58. Since $891 \div 9 = 99$, divide the denominator by 99 also.

$\frac{891}{1584} = \frac{891 \div 99}{1584 \div 99} = \frac{9}{16}$ Replace ? with 16.

59. Multiply or divide the numerator and denominator by the same nonzero number. Some possibilities:

$\frac{2}{3} = \frac{2 \cdot 4}{3 \cdot 4} = \frac{8}{12}$ and $\frac{10}{16} = \frac{10 \div 2}{16 \div 2} = \frac{5}{8}$

61. You cannot do it if you want the numerator to be a whole number, because 5 does not divide into 18 evenly. You could use multiples of 5 as the denominator, such as 10, 15, 20, etc.

63. $\frac{10}{1} = 10 \div 1 = 10$

65. $-\frac{16}{16} = -16 \div 16 = -1$

67. $-\frac{18}{3} = -18 \div 3 = -6$

69. $\frac{24}{8} = 24 \div 8 = 3$

71. $\frac{12}{12} = 12 \div 12 = 1$

73. $\frac{14}{7} = 14 \div 7 = 2$

75. $-\frac{5}{1} = -5 \div 1 = -5$

77. $-\frac{45}{9} = -45 \div 9 = -5$

79. $\frac{150}{150} = 150 \div 150 = 1$

81. $-\frac{32}{4} = -32 \div 4 = -8$

83.

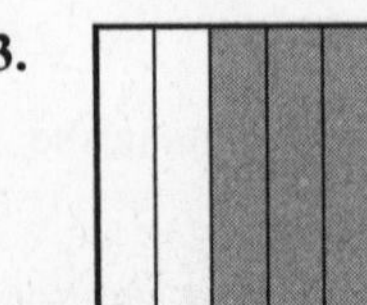

$\frac{3}{5}$ is shaded; $\frac{2}{5}$ is unshaded.

85. 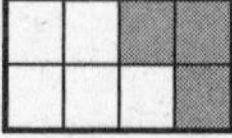

$\frac{3}{8}$ is shaded; $\frac{5}{8}$ is unshaded.

87. 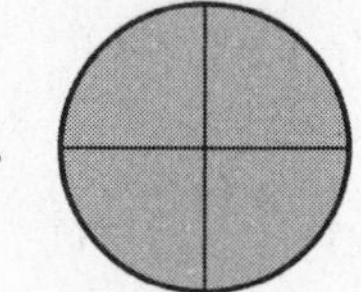 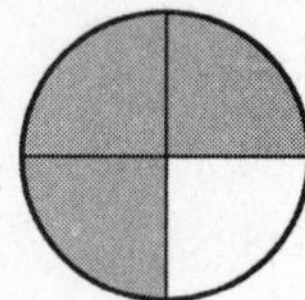

$\frac{7}{4}$ is shaded; $\frac{1}{4}$ is unshaded.

89. 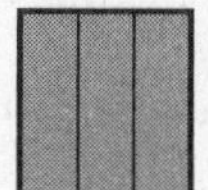 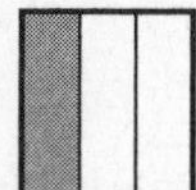

$\frac{4}{3}$ is shaded; $\frac{2}{3}$ is unshaded.

91. 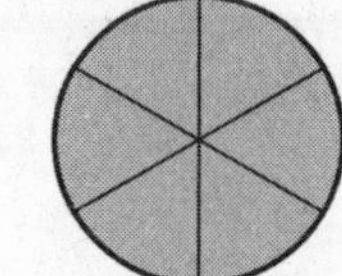

$\frac{6}{6}$ is shaded; $\frac{0}{6}$ is unshaded.

93. 

$\frac{10}{5}$ is shaded; $\frac{0}{5}$ is unshaded.

95. Drawings will vary.

97. One possibility is shown.

99. One possibility is shown. (!!)!!!, . . . ???

4.2 Writing Fractions in Lowest Terms

4.2 Section Exercises

1. **(a)** $-\frac{3}{10}$ is in lowest terms since the numerator and denominator have no common factor other than 1.

   (b) $\frac{10}{15}$ is not in lowest terms. A common factor of 10 and 15 is 5.

   (c) $\frac{9}{16}$ is in lowest terms.

   (d) $-\frac{4}{21}$ is in lowest terms.

   (e) $\frac{6}{9}$ is not in lowest terms. A common factor of 6 and 9 is 3.

   (f) $-\frac{7}{28}$ is not in lowest terms. A common factor of 7 and 28 is 7.

3. **(a)** $\frac{10}{15} = \frac{10 \div 5}{15 \div 5} = \frac{2}{3}$

   (b) $\frac{6}{9} = \frac{6 \div 3}{9 \div 3} = \frac{2}{3}$

   (c) $-\frac{7}{28} = -\frac{7 \div 7}{28 \div 7} = -\frac{1}{4}$

   (d) $-\frac{25}{50} = -\frac{25 \div 25}{50 \div 25} = -\frac{1}{2}$

   (e) $\frac{16}{18} = \frac{16 \div 2}{18 \div 2} = \frac{8}{9}$

   (f) $-\frac{8}{20} = -\frac{8 \div 4}{20 \div 4} = -\frac{2}{5}$

5. Prime: 2, 5, 11
   Composite: 9 since 9 is divisible by 3
   8 since 8 is divisible by 2 and 4
   10 since 10 is divisible by 2 and 5
   21 since 21 is divisible by 3 and 7

   1 is neither prime nor composite.

7. $6 \div 2 = 3$ Divide 6 by 2.
   $3 \div 3 = 1$ 3 is not divisible by 2; use 3. Quotient is 1.

   The prime factorization of 6 is $2 \cdot 3$.

9. $20 \div 2 = 10$ Divide 20 by 2.
   $10 \div 2 = 5$ Divide 10 by 2.
   $5 \div 5 = 1$ 5 is not divisible by 2 or 3; use 5. Quotient is 1.

   The prime factorization of 20 is $2 \cdot 2 \cdot 5$.

11. 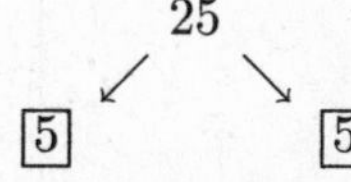

   The prime factorization of 25 is $5 \cdot 5$.

13. 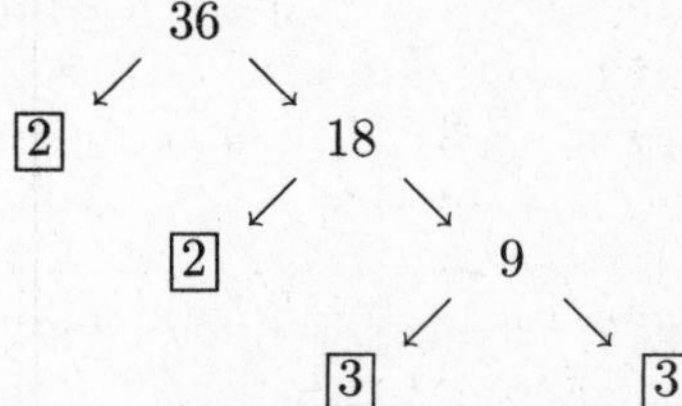

   The prime factorization of 36 is $2 \cdot 2 \cdot 3 \cdot 3$.

15. **(a)** 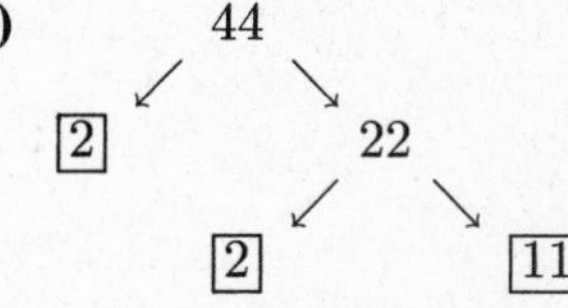

   The prime factorization of 44 is $2 \cdot 2 \cdot 11$.

   (b) 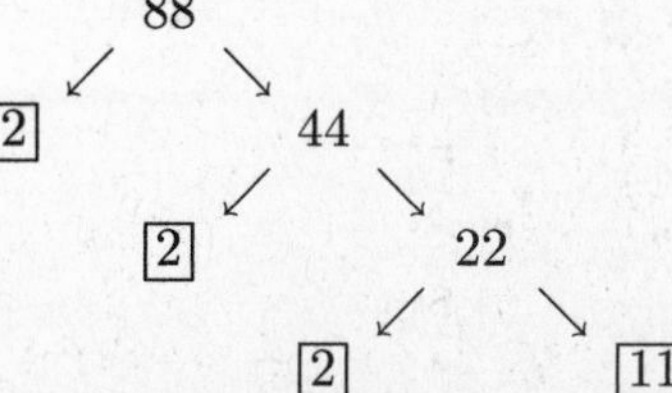

   The prime factorization of 88 is $2 \cdot 2 \cdot 2 \cdot 11$.

17. (a)

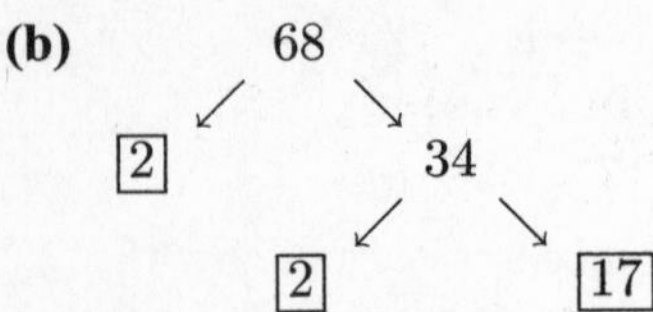

The prime factorization of 75 is $3 \cdot 5 \cdot 5$.

(b)

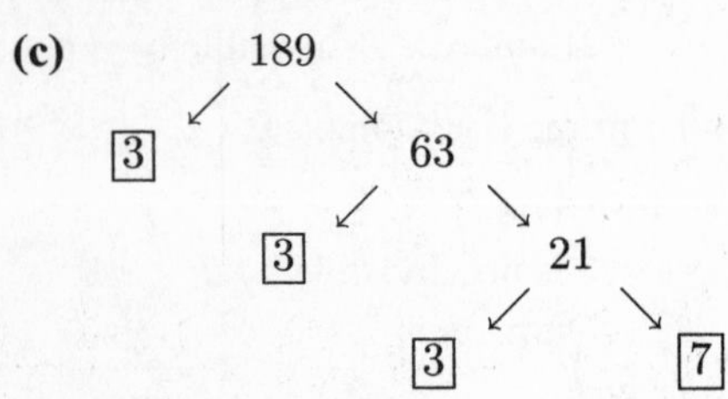

The prime factorization of 68 is $2 \cdot 2 \cdot 17$.

(c)

189 → [3], 63
63 → [3], 21
21 → [3], [7]

The prime factorization of 189 is $3 \cdot 3 \cdot 3 \cdot 7$.

19. $\dfrac{8}{16} = \dfrac{\overset{1}{\cancel{2}} \cdot \overset{1}{\cancel{2}} \cdot \overset{1}{\cancel{2}}}{\underset{1}{\cancel{2}} \cdot \underset{1}{\cancel{2}} \cdot \underset{1}{\cancel{2}} \cdot 2} = \dfrac{1}{2}$

21. $\dfrac{32}{48} = \dfrac{\overset{1}{\cancel{2}} \cdot \overset{1}{\cancel{2}} \cdot \overset{1}{\cancel{2}} \cdot \overset{1}{\cancel{2}} \cdot 2}{\underset{1}{\cancel{2}} \cdot \underset{1}{\cancel{2}} \cdot \underset{1}{\cancel{2}} \cdot \underset{1}{\cancel{2}} \cdot 3} = \dfrac{2}{3}$

23. $\dfrac{14}{21} = \dfrac{2 \cdot \overset{1}{\cancel{7}}}{3 \cdot \underset{1}{\cancel{7}}} = \dfrac{2}{3}$

25. $\dfrac{36}{42} = \dfrac{\overset{1}{\cancel{2}} \cdot 2 \cdot \overset{1}{\cancel{3}} \cdot 3}{\underset{1}{\cancel{2}} \cdot \underset{1}{\cancel{3}} \cdot 7} = \dfrac{6}{7}$

27. $\dfrac{50}{63} = \dfrac{2 \cdot 5 \cdot 5}{3 \cdot 3 \cdot 7}$

The fraction is already in lowest terms.

29. $\dfrac{27}{45} = \dfrac{\overset{1}{\cancel{3}} \cdot \overset{1}{\cancel{3}} \cdot 3}{\underset{1}{\cancel{3}} \cdot \underset{1}{\cancel{3}} \cdot 5} = \dfrac{3}{5}$

31. $\dfrac{12}{18} = \dfrac{\overset{1}{\cancel{2}} \cdot 2 \cdot \overset{1}{\cancel{3}}}{\underset{1}{\cancel{2}} \cdot \underset{1}{\cancel{3}} \cdot 3} = \dfrac{2}{3}$

33. $\dfrac{35}{40} = \dfrac{\overset{1}{\cancel{5}} \cdot 7}{2 \cdot 2 \cdot 2 \cdot \underset{1}{\cancel{5}}} = \dfrac{7}{8}$

35. $\dfrac{90}{180} = \dfrac{\overset{1}{\cancel{2}} \cdot \overset{1}{\cancel{3}} \cdot \overset{1}{\cancel{3}} \cdot \overset{1}{\cancel{5}}}{\underset{1}{\cancel{2}} \cdot 2 \cdot \underset{1}{\cancel{3}} \cdot \underset{1}{\cancel{3}} \cdot \underset{1}{\cancel{5}}} = \dfrac{1}{2}$

37. $\dfrac{210}{315} = \dfrac{2 \cdot \overset{1}{\cancel{3}} \cdot \overset{1}{\cancel{5}} \cdot \overset{1}{\cancel{7}}}{3 \cdot \underset{1}{\cancel{3}} \cdot \underset{1}{\cancel{5}} \cdot \underset{1}{\cancel{7}}} = \dfrac{2}{3}$

39. $\dfrac{429}{495} = \dfrac{\overset{1}{\cancel{3}} \cdot \overset{1}{\cancel{11}} \cdot 13}{\underset{1}{\cancel{3}} \cdot 3 \cdot 5 \cdot \underset{1}{\cancel{11}}} = \dfrac{13}{15}$

41. 60 minutes in an hour ⇒ 60 parts in the whole

(a) $\dfrac{15}{60} = \dfrac{\overset{1}{\cancel{3}} \cdot \overset{1}{\cancel{5}}}{2 \cdot 2 \cdot \underset{1}{\cancel{3}} \cdot \underset{1}{\cancel{5}}} = \dfrac{1}{4}$ of an hour

(b) $\dfrac{30}{60} = \dfrac{\overset{1}{\cancel{2}} \cdot \overset{1}{\cancel{3}} \cdot \overset{1}{\cancel{5}}}{2 \cdot \underset{1}{\cancel{2}} \cdot \underset{1}{\cancel{3}} \cdot \underset{1}{\cancel{5}}} = \dfrac{1}{2}$ of an hour

(c) $\dfrac{6}{60} = \dfrac{\overset{1}{\cancel{2}} \cdot \overset{1}{\cancel{3}}}{\underset{1}{\cancel{2}} \cdot 2 \cdot \underset{1}{\cancel{3}} \cdot 5} = \dfrac{1}{10}$ of an hour

(d) $\dfrac{60}{60} = 1$ hour

43. (a) $\dfrac{500}{1500} = \dfrac{5 \cdot \overset{1}{\cancel{100}}}{15 \cdot \underset{1}{\cancel{100}}} = \dfrac{5}{15} = \dfrac{\overset{1}{\cancel{5}}}{3 \cdot \underset{1}{\cancel{5}}} = \dfrac{1}{3}$

(b) $\dfrac{300}{1500} = \dfrac{3 \cdot \overset{1}{\cancel{100}}}{15 \cdot \underset{1}{\cancel{100}}} = \dfrac{3}{15} = \dfrac{\overset{1}{\cancel{3}}}{\underset{1}{\cancel{3}} \cdot 5} = \dfrac{1}{5}$

(c) $1500 − $500 − $300 = $700

$\dfrac{700}{1500} = \dfrac{7 \cdot \overset{1}{\cancel{100}}}{15 \cdot \underset{1}{\cancel{100}}} = \dfrac{7}{15}$

45. (a) $\dfrac{10}{48} = \dfrac{\overset{1}{\cancel{2}} \cdot 5}{\underset{1}{\cancel{2}} \cdot 2 \cdot 2 \cdot 2 \cdot 3} = \dfrac{5}{24}$

(b) $\dfrac{32}{48} = \dfrac{\overset{1}{\cancel{2}} \cdot \overset{1}{\cancel{2}} \cdot \overset{1}{\cancel{2}} \cdot \overset{1}{\cancel{2}} \cdot 2}{\underset{1}{\cancel{2}} \cdot \underset{1}{\cancel{2}} \cdot \underset{1}{\cancel{2}} \cdot \underset{1}{\cancel{2}} \cdot 3} = \dfrac{2}{3}$

(c) $\dfrac{6}{48} = \dfrac{\overset{1}{\cancel{2}} \cdot \overset{1}{\cancel{3}}}{\underset{1}{\cancel{2}} \cdot 2 \cdot 2 \cdot 2 \cdot \underset{1}{\cancel{3}}} = \dfrac{1}{8}$

47. (a) The result of dividing 3 by 3 is 1, so 1 should be written above and below all the slashes. The numerator is $1 \cdot 1$, so the correct answer is $\frac{1}{4}$.

$\dfrac{9}{36} = \dfrac{\overset{1}{\cancel{3}} \cdot \overset{1}{\cancel{3}}}{2 \cdot 2 \cdot \underset{1}{\cancel{3}} \cdot \underset{1}{\cancel{3}}} = \dfrac{1}{4}$

(b) You must divide numerator and denominator by the *same* number. The fraction is already in lowest terms because 9 and 16 have no common factor besides 1.

49. $\frac{16c}{40} = \frac{\overset{1}{\cancel{2}}\cdot\overset{1}{\cancel{2}}\cdot\overset{1}{\cancel{2}}\cdot 2\cdot c}{\underset{1}{\cancel{2}}\cdot\underset{1}{\cancel{2}}\cdot\underset{1}{\cancel{2}}\cdot 5} = \frac{2c}{5}$

51. $\frac{20x}{35x} = \frac{2\cdot 2\cdot\overset{1}{\cancel{5}}\cdot\overset{1}{\cancel{x}}}{\underset{1}{\cancel{5}}\cdot 7\cdot\underset{1}{\cancel{x}}} = \frac{4}{7}$

53. $\frac{18r^2}{15rs} = \frac{2\cdot\overset{1}{\cancel{3}}\cdot 3\cdot\overset{1}{\cancel{r}}\cdot r}{\underset{1}{\cancel{3}}\cdot 5\cdot\underset{1}{\cancel{r}}\cdot s} = \frac{6r}{5s}$

55. $\frac{6m}{42mn^2} = \frac{\overset{1}{\cancel{2}}\cdot\overset{1}{\cancel{3}}\cdot\overset{1}{\cancel{m}}}{\underset{1}{\cancel{2}}\cdot\underset{1}{\cancel{3}}\cdot 7\cdot\underset{1}{\cancel{m}}\cdot n\cdot n} = \frac{1}{7n^2}$

57. $\frac{9x^2}{16y^2} = \frac{3\cdot 3\cdot x\cdot x}{2\cdot 2\cdot 2\cdot 2\cdot y\cdot y}$

There are no common factors, so the fraction is already in lowest terms.

59. $\frac{7xz}{9xyz} = \frac{7\cdot\overset{1}{\cancel{x}}\cdot\overset{1}{\cancel{z}}}{3\cdot 3\cdot\underset{1}{\cancel{x}}\cdot y\cdot\underset{1}{\cancel{z}}} = \frac{7}{9y}$

61. $\frac{21k^3}{6k^2} = \frac{\overset{1}{\cancel{3}}\cdot 7\cdot\overset{1}{\cancel{k}}\cdot\overset{1}{\cancel{k}}\cdot k}{2\cdot\underset{1}{\cancel{3}}\cdot\underset{1}{\cancel{k}}\cdot\underset{1}{\cancel{k}}} = \frac{7k}{2}$

63. $\frac{13a^2bc^3}{39a^2bc^3} = \frac{\overset{1}{\cancel{13}}\cdot\overset{1}{\cancel{a}}\cdot\overset{1}{\cancel{a}}\cdot\overset{1}{\cancel{b}}\cdot\overset{1}{\cancel{c}}\cdot\overset{1}{\cancel{c}}\cdot\overset{1}{\cancel{c}}}{3\cdot\underset{1}{\cancel{13}}\cdot\underset{1}{\cancel{a}}\cdot\underset{1}{\cancel{a}}\cdot\underset{1}{\cancel{b}}\cdot\underset{1}{\cancel{c}}\cdot\underset{1}{\cancel{c}}\cdot\underset{1}{\cancel{c}}} = \frac{1}{3}$

65. $\frac{14c^2d}{14cd^2} = \frac{\overset{1}{\cancel{2}}\cdot\overset{1}{\cancel{7}}\cdot\overset{1}{\cancel{c}}\cdot c\cdot\overset{1}{\cancel{d}}}{\underset{1}{\cancel{2}}\cdot\underset{1}{\cancel{7}}\cdot\underset{1}{\cancel{c}}\cdot\underset{1}{\cancel{d}}\cdot d} = \frac{c}{d}$

67. $\frac{210ab^3c}{35b^2c^2} = \frac{2\cdot 3\cdot\overset{1}{\cancel{5}}\cdot\overset{1}{\cancel{7}}\cdot a\cdot\overset{1}{\cancel{b}}\cdot\overset{1}{\cancel{b}}\cdot b\cdot\overset{1}{\cancel{c}}}{\underset{1}{\cancel{5}}\cdot\underset{1}{\cancel{7}}\cdot\underset{1}{\cancel{b}}\cdot\underset{1}{\cancel{b}}\cdot\underset{1}{\cancel{c}}\cdot c} = \frac{6ab}{c}$

69. $\frac{25m^3rt^2}{36n^2s^3w^2} = \frac{5\cdot 5\cdot m\cdot m\cdot m\cdot r\cdot t\cdot t}{2\cdot 2\cdot 3\cdot 3\cdot n\cdot n\cdot s\cdot s\cdot s\cdot w\cdot w}$

There are no common factors, so the fraction is already in lowest terms.

71. $\frac{33e^2fg^3}{11efg} = \frac{3\cdot\overset{1}{\cancel{11}}\cdot\overset{1}{\cancel{e}}\cdot e\cdot\overset{1}{\cancel{f}}\cdot\overset{1}{\cancel{g}}\cdot g\cdot g}{\underset{1}{\cancel{11}}\cdot\underset{1}{\cancel{e}}\cdot\underset{1}{\cancel{f}}\cdot\underset{1}{\cancel{g}}} = 3eg^2$

4.3 Multiplying and Dividing Signed Fractions

4.3 Section Exercises

1. $-\frac{3}{8}\cdot\frac{1}{2} = -\frac{3\cdot 1}{8\cdot 2} = -\frac{3}{16}$

3. $\left(-\frac{3}{8}\right)\left(-\frac{12}{5}\right) = \frac{3\cdot 3\cdot\overset{1}{\cancel{4}}}{2\cdot\underset{1}{\cancel{4}}\cdot 5} = \frac{9}{10}$

5. $\frac{21}{30}\left(\frac{5}{7}\right) = \frac{\overset{1}{\cancel{3}}\cdot\overset{1}{\cancel{7}}\cdot\overset{1}{\cancel{5}}}{2\cdot\underset{1}{\cancel{3}}\cdot\underset{1}{\cancel{5}}\cdot\underset{1}{\cancel{7}}} = \frac{1}{2}$

7. $10\left(-\frac{3}{5}\right) = \frac{10}{1}\left(-\frac{3}{5}\right) = -\frac{2\cdot\overset{1}{\cancel{5}}\cdot 3}{1\cdot\underset{1}{\cancel{5}}} = -\frac{6}{1} = -6$

9. $\frac{4}{9}$ of $81 = \frac{4}{9}\cdot\frac{81}{1} = \frac{4\cdot\overset{1}{\cancel{9}}\cdot 9}{\underset{1}{\cancel{9}}\cdot 1} = \frac{36}{1} = 36$

11. $\left(\frac{3x}{4}\right)\left(\frac{5}{xy}\right) = \frac{3\cdot\overset{1}{\cancel{x}}\cdot 5}{4\cdot\underset{1}{\cancel{x}}\cdot y} = \frac{15}{4y}$

13. $\frac{1}{6}\div\frac{1}{3} = \frac{1}{6}\cdot\frac{3}{1} = \frac{1\cdot\overset{1}{\cancel{3}}}{2\cdot\underset{1}{\cancel{3}}\cdot 1} = \frac{1}{2}$

15. $-\frac{3}{4}\div\left(-\frac{5}{8}\right) = -\frac{3}{4}\cdot\left(-\frac{8}{5}\right) = \frac{3\cdot 2\cdot\overset{1}{\cancel{4}}}{\underset{1}{\cancel{4}}\cdot 5} = \frac{6}{5}$

17. $6\div\left(-\frac{2}{3}\right) = \frac{6}{1}\cdot\left(-\frac{3}{2}\right) = -\frac{\overset{1}{\cancel{2}}\cdot 3\cdot 3}{1\cdot\underset{1}{\cancel{2}}}$

$= -\frac{9}{1} = -9$

19. $-\frac{2}{3}\div 4 = -\frac{2}{3}\div\frac{4}{1} = -\frac{2}{3}\cdot\frac{1}{4} = -\frac{\overset{1}{\cancel{2}}\cdot 1}{3\cdot\underset{1}{\cancel{2}}\cdot 2} = -\frac{1}{6}$

21. $\frac{11c}{5d}\div 3c = \frac{11c}{5d}\div\frac{3c}{1} = \frac{11c}{5d}\cdot\frac{1}{3c}$

$= \frac{11\cdot\overset{1}{\cancel{c}}\cdot 1}{5\cdot d\cdot 3\cdot\underset{1}{\cancel{c}}} = \frac{11}{15d}$

23. $\frac{ab^2}{c}\div\frac{ab}{c} = \frac{ab^2}{c}\cdot\frac{c}{ab} = \frac{\overset{1}{\cancel{a}}\cdot\overset{1}{\cancel{b}}\cdot b\cdot\overset{1}{\cancel{c}}}{\underset{1}{\cancel{c}}\cdot\underset{1}{\cancel{a}}\cdot\underset{1}{\cancel{b}}} = \frac{b}{1} = b$

25. **(a)** Forgot to write 1s in numerator and denominator when dividing out common factors. Answer is $\frac{1}{6}$.

(b) Used reciprocal of $\frac{2}{3}$ in multiplication, but the reciprocal is used only in division. Correct answer is $8\cdot\frac{2}{3} = \frac{8\cdot 2}{3} = \frac{16}{3}$.

27. **(a)** Forgot to use reciprocal of $\frac{4}{1}$; correct answer is

$$\frac{2}{3}\cdot\frac{1}{4} = \frac{\overset{1}{\cancel{2}}\cdot 1}{3\cdot 2\cdot\underset{1}{\cancel{2}}} = \frac{1}{6}.$$

(b) Used reciprocal of $\frac{5}{6}$ instead of reciprocal of $\frac{10}{9}$; correct answer is

$$\frac{5}{6}\cdot\frac{9}{10}=\frac{\overset{1}{\cancel{5}}\cdot\overset{1}{\cancel{3}}\cdot 3}{2\cdot\underset{1}{\cancel{3}}\cdot 2\cdot\underset{1}{\cancel{5}}}=\frac{3}{4}.$$

29. Rewrite division as multiplication. Leave the first number (dividend) the same. Change the second number (divisor) to its reciprocal by "flipping" it. Then multiply.

31. $\frac{4}{5}\div 3=\frac{4}{5}\div\frac{3}{1}=\frac{4}{5}\cdot\frac{1}{3}=\frac{4\cdot 1}{5\cdot 3}=\frac{4}{15}$

33. $-\frac{3}{8}\left(\frac{3}{4}\right)=-\frac{3\cdot 3}{8\cdot 4}=-\frac{9}{32}$

35. $\frac{3}{5}\text{ of }35=\frac{3}{5}\cdot 35=\frac{3}{5}\cdot\frac{35}{1}=\frac{3\cdot\overset{1}{\cancel{5}}\cdot 7}{\underset{1}{\cancel{5}}\cdot 1}=\frac{21}{1}=21$

37. $-9\div\left(-\frac{3}{5}\right)=-\frac{9}{1}\cdot\left(-\frac{5}{3}\right)=\frac{\overset{1}{\cancel{3}}\cdot 3\cdot 5}{1\cdot\underset{1}{\cancel{3}}}$

$$=\frac{15}{1}=15$$

39. $\frac{12}{7}\div 0$ is undefined.

41. $\left(\frac{11}{2}\right)\left(-\frac{5}{6}\right)=-\frac{11\cdot 5}{2\cdot 6}=-\frac{55}{12}$

43. $\frac{4}{7}\text{ of }14b=\frac{4}{7}\cdot 14b=\frac{4}{7}\cdot\frac{14b}{1}=\frac{4\cdot 2\cdot\overset{1}{\cancel{7}}\cdot b}{\underset{1}{\cancel{7}}\cdot 1}$

$$=\frac{8b}{1}=8b$$

45. $\frac{12}{5}\div 4d=\frac{12}{5}\div\frac{4d}{1}=\frac{12}{5}\cdot\frac{1}{4d}$

$$=\frac{3\cdot\overset{1}{\cancel{4}}\cdot 1}{5\cdot\underset{1}{\cancel{4}}\cdot d}=\frac{3}{5d}$$

47. $\frac{x^2}{y}\div\frac{w}{2y}=\frac{x^2}{y}\cdot\frac{2y}{w}=\frac{x^2\cdot 2\cdot\overset{1}{\cancel{y}}}{\underset{1}{\cancel{y}}\cdot w}=\frac{2x^2}{w}$

49. The top of a table is a rectangle.

$A=l\cdot w$

$A=\frac{4}{5}\text{ yd}\cdot\frac{3}{8}\text{ yd}$

$A=\frac{\overset{1}{\cancel{4}}\cdot 3}{5\cdot 2\cdot\underset{1}{\cancel{4}}}\text{ yd}^2$

$A=\frac{3}{10}\text{ yd}^2$ or $\frac{3}{10}$ square yard

51. Splitting 10 ounces into equal size parts indicates division. Each part will contain $\frac{1}{8}$ ounce.

$10\div\frac{1}{8}=\frac{10}{1}\cdot\frac{8}{1}=\frac{80}{1}=80$

80 eyedrop dispensers can be filled.

53. Todd must earn $\frac{3}{4}$ of the cost:

$\frac{3}{4}\cdot\$12{,}400=\frac{3}{4}\cdot\frac{12{,}400}{1}=\frac{3\cdot\overset{1}{\cancel{4}}\cdot 3100}{\underset{1}{\cancel{4}}\cdot 1}$

$$=\frac{9300}{1}=\$9300$$

Todd must borrow the rest:

$$\$12{,}400-\$9300=\$3100$$

55. The total number of cords divided by the number of cords per trip will give us the number of trips.

$6\div\frac{2}{3}=\frac{6}{1}\div\frac{2}{3}=\frac{6}{1}\cdot\frac{3}{2}=\frac{\overset{1}{\cancel{2}}\cdot 3\cdot 3}{1\cdot\underset{1}{\cancel{2}}}=\frac{9}{1}=9$

9 trips are needed to deliver 6 cords.

57. $210\cdot\frac{1}{3}=\frac{210}{1}\cdot\frac{1}{3}=\frac{70\cdot\overset{1}{\cancel{3}}\cdot 1}{1\cdot\underset{1}{\cancel{3}}}=70$

70 infield players are in the Hall of Fame.

59. The weight of the adult divided by the weight of the hatchling is

$400\div\frac{1}{8}=\frac{400}{1}\cdot\frac{8}{1}=\frac{400\cdot 8}{1\cdot 1}=3200.$

The adult weighs 3200 times the hatchling.

61. **(a)** Add the monthly incomes.

$\$4575+\$4312+\cdots+\$6458=\$58{,}000$

(b) Rent is $\frac{1}{5}$ of the income.

$\frac{1}{5}\cdot 58{,}000=\frac{1\cdot\overset{1}{\cancel{5}}\cdot 11{,}600}{\underset{1}{\cancel{5}}}=11{,}600$

The family's rent is \$11,600.

63. The circle graph shows that $\frac{5}{16}$ of their total income is spent on food and $\frac{1}{8}$ is spent on clothing.

From Exercise 61(a), their total income is \$58,000.

Food Expense:

$\frac{5}{16}\cdot 58{,}000=\frac{5}{16}\cdot\frac{58{,}000}{1}=\frac{5\cdot\overset{1}{\cancel{16}}\cdot 3625}{\underset{1}{\cancel{16}}\cdot 1}=18{,}125$

\$18,125 is spent on food.

Clothing Expense:

$$\frac{1}{8}\cdot 58{,}000 = \frac{1}{8}\cdot\frac{58{,}000}{1} = \frac{1\cdot \cancel{8}^{1}\cdot 7250}{\cancel{8}_{1}\cdot 1} = 7250$$

\$7250 is spent on clothing.

\$18,125 + \$7250 = \$25,375 was spent on food and clothing.

65. $150\cdot\frac{1}{10} = \frac{150}{1}\cdot\frac{1}{10} = \frac{15\cdot\cancel{10}^{1}\cdot 1}{1\cdot\cancel{10}_{1}} = 15$

15 million U.S. pets are birds.

67. Dogs: $150\cdot\frac{2}{5} = \frac{150}{1}\cdot\frac{2}{5} = \frac{\cancel{5}^{1}\cdot 30\cdot 2}{1\cdot\cancel{5}_{1}} = 60$

Cats: $150\cdot\frac{23}{50} = \frac{150\cdot 23}{1\cdot 50} = \frac{3\cdot\cancel{50}^{1}\cdot 23}{\cancel{50}_{1}} = 69$

60 + 69 = 129 million U.S. pets are dogs and cats.

4.4 Adding and Subtracting Signed Fractions

4.4 Section Exercises

1. **(a)** The LCD of $\frac{2}{5}$ and $\frac{1}{2}$ is 10.

Since the larger denominator, 5, is not divisible by the smaller denominator, 2, check multiples of 5: 5, 10, 15, 20, etc. 10 is the smallest one divisible by 2.

(b) The LCD for $\frac{5}{6}$ and $\frac{5}{18}$ is 18, since the larger denominator, 18, is divisible by the smaller denominator, 6.

(c) The LCD of $\frac{3}{10}$ and $\frac{1}{4}$ is 20.

Since the larger denominator, 10, is not divisible by the smaller denominator, 4, check multiples of 10: 10, 20, 30, etc. 20 is the smallest one divisible by 4.

(d) The LCD of $\frac{1}{3}$ and $\frac{6}{7}$ is 21.

Since the larger denominator, 7, is not divisible by the smaller denominator, 3, check multiples of 7: 7, 14, 21, etc. 21 is the smallest one divisible by 3.

3. **(a)** $\frac{5}{6}$ and $\frac{7}{16}$

$$\left.\begin{array}{l}6 = 2\cdot 3\\ 16 = 2\cdot 2\cdot 2\cdot 2\end{array}\right\}\text{LCD} = 2\cdot 2\cdot 2\cdot 2\cdot 3 = 48$$

(b) $\frac{11}{15}$ and $\frac{11}{12}$

$$\left.\begin{array}{l}15 = 3\cdot 5\\ 12 = 2\cdot 2\cdot 3\end{array}\right\}\text{LCD} = 2\cdot 2\cdot 3\cdot 5 = 60$$

(c) $\frac{1}{24}$ and $\frac{10}{21}$

$$\left.\begin{array}{l}24 = 2\cdot 2\cdot 2\cdot 3\\ 21 = 3\cdot 7\end{array}\right\}\text{LCD} = 2\cdot 2\cdot 2\cdot 3\cdot 7 = 168$$

5. $\frac{2}{3} - \frac{1}{6}$

Step 1
The LCD is 6.

Step 2
$\frac{2}{3} = \frac{2\cdot 2}{3\cdot 2} = \frac{4}{6}$, $\frac{1}{6}$ has the LCD.

Step 3
$\frac{2}{3} - \frac{1}{6} = \frac{4}{6} - \frac{1}{6} = \frac{4-1}{6} = \frac{3}{6}$

Step 4
$\frac{3}{6} = \frac{\cancel{3}^{1}}{2\cdot\cancel{3}_{1}} = \frac{1}{2}$

7. $\frac{3}{4} + \frac{1}{8}$

Step 1
The LCD is 8.

Step 2
$\frac{3}{4} = \frac{3\cdot 2}{4\cdot 2} = \frac{6}{8}$, $\frac{1}{8}$ has the LCD.

Step 3
$\frac{3}{4} + \frac{1}{8} = \frac{6}{8} + \frac{1}{8} = \frac{6+1}{8} = \frac{7}{8}$

Step 4
$\frac{7}{8}$ is already in lowest terms.

9. $\frac{3}{8} - \frac{3}{5}$

Step 1
The LCD is $8\cdot 5 = 40$.

Step 2
$\frac{3}{8} = \frac{3\cdot 5}{8\cdot 5} = \frac{15}{40}$, $\frac{3}{5} = \frac{3\cdot 8}{5\cdot 8} = \frac{24}{40}$

Step 3
$\frac{3}{8} - \frac{3}{5} = \frac{15}{40} - \frac{24}{40} = \frac{15-24}{40} = -\frac{9}{40}$

Step 4
$-\frac{9}{40}$ is already in lowest terms.

11. $-\frac{7}{20} - \frac{5}{20}$

Step 1
The LCD is 20.

Step 2
Each fraction has the LCD.

Step 3
$-\frac{7}{20} - \frac{5}{20} = \frac{-7-5}{20} = -\frac{12}{20}$

Step 4

$$-\frac{12}{20} = -\frac{\overset{1}{\cancel{4}}\cdot 3}{\underset{1}{\cancel{4}}\cdot 5} = -\frac{3}{5}$$

13. $-\frac{3}{10} + 1 = -\frac{3}{10} + \frac{1}{1}$

Step 1
The LCD is 10.

Step 2
$-\frac{3}{10}$ has the LCD, $\frac{1}{1} = \frac{1\cdot 10}{1\cdot 10} = \frac{10}{10}$

Step 3
$-\frac{3}{10} + \frac{1}{1} = -\frac{3}{10} + \frac{10}{10} = \frac{-3+10}{10} = \frac{7}{10}$

Step 4
$\frac{7}{10}$ is already in lowest terms.

15. $2 - \frac{6}{7} = \frac{2}{1} - \frac{6}{7}$

Step 1
The LCD is 7.

Step 2
$2 = \frac{2\cdot 7}{7} = \frac{14}{7}$, $\frac{6}{7}$ has the LCD.

Step 3
$2 - \frac{6}{7} = \frac{14}{7} - \frac{6}{7} = \frac{14-6}{7} = \frac{8}{7}$

Step 4
$\frac{8}{7}$ is already in lowest terms.

17. $-\frac{1}{2} + \frac{3}{24}$

Step 1
The LCD is 24, the larger denominator.

Step 2
$-\frac{1}{2} = -\frac{1\cdot 12}{2\cdot 12} = -\frac{12}{24}$, $\frac{3}{24}$ has the LCD.

Step 3
$$-\frac{1}{2} + \frac{3}{24} = -\frac{12}{24} + \frac{3}{24} = \frac{-12+3}{24}$$
$$= \frac{-9}{24} \text{ or } -\frac{9}{24}$$

Step 4

$$-\frac{9}{24} = -\frac{\overset{1}{\cancel{3}}\cdot 3}{\underset{1}{\cancel{3}}\cdot 8} = -\frac{3}{8}$$

19. $\frac{1}{5} + \frac{c}{3}$

Step 1
The LCD is $5\cdot 3 = 15$.

Step 2
$\frac{1}{5} = \frac{1\cdot 3}{5\cdot 3} = \frac{3}{15}$, $\frac{c}{3} = \frac{c\cdot 5}{3\cdot 5} = \frac{5c}{15}$

Step 3
$\frac{1}{5} + \frac{c}{3} = \frac{3}{15} + \frac{5c}{15} = \frac{3+5c}{15}$

Step 4
$\frac{3+5c}{15}$ is already in lowest terms.

21. $\frac{5}{m} - \frac{1}{2}$

Step 1
The LCD is $2\cdot m$ or $2m$.

Step 2
$\frac{5}{m} = \frac{5\cdot 2}{m\cdot 2} = \frac{10}{2m}$, $\frac{1}{2} = \frac{1\cdot m}{2\cdot m} = \frac{1m}{2m}$

Step 3
$$\frac{5}{m} - \frac{1}{2} = \frac{10}{2m} - \frac{1m}{2m}$$
$$= \frac{10-1m}{2m} \text{ or } \frac{10-m}{2m}$$

Step 4
$\frac{10-m}{2m}$ is already in lowest terms.

23. $\frac{3}{b^2} + \frac{5}{b^2} = \frac{3+5}{b^2} = \frac{8}{b^2}$, which is in lowest terms.

25. $\frac{c}{7} + \frac{3}{b}$

Step 1
The LCD is $7\cdot b$ or $7b$.

Step 2
$\frac{c}{7} = \frac{c\cdot b}{7\cdot b} = \frac{bc}{7b}$, $\frac{3}{b} = \frac{3\cdot 7}{b\cdot 7} = \frac{21}{7b}$

Step 3
$\frac{c}{7} + \frac{3}{b} = \frac{bc}{7b} + \frac{21}{7b} = \frac{bc+21}{7b}$

Step 4
$\frac{bc+21}{7b}$ is already in lowest terms.

27. $-\frac{4}{c^2} - \frac{d}{c}$

Step 1
The LCD is c^2.

Step 2
$\frac{4}{c^2}$ has the LCD, $\frac{d}{c} = \frac{d\cdot c}{c\cdot c} = \frac{cd}{c^2}$.

Step 3
$-\frac{4}{c^2} - \frac{d}{c} = -\frac{4}{c^2} - \frac{cd}{c^2} = \frac{-4-cd}{c^2}$

Step 4
$\frac{-4-cd}{c^2}$ is already in lowest terms.

29. $-\frac{11}{42}-\frac{11}{70}$

Step 1
Use prime factorization to find the LCD.

$$\left.\begin{array}{l}42=2\cdot3\cdot7\\70=2\cdot5\cdot7\end{array}\right\}\text{LCD}=2\cdot3\cdot5\cdot7=210$$

Step 2
$$-\frac{11}{42}=-\frac{11\cdot5}{42\cdot5}=-\frac{55}{210},\ \frac{11}{70}=\frac{11\cdot3}{70\cdot3}=\frac{33}{210}$$

Step 3
$$-\frac{11}{42}-\frac{11}{70}=-\frac{55}{210}-\frac{33}{210}=\frac{-55-33}{210}$$
$$=\frac{-55+(-33)}{210}$$
$$=\frac{-88}{210}\text{ or }-\frac{88}{210}$$

Step 4
$$-\frac{88}{210}=-\frac{2\cdot44}{2\cdot105}=-\frac{44}{105}$$

31. You cannot add or subtract until all the fractional pieces are the same size.

33. **(a)** You cannot add fractions with unlike denominators; use 20 as the LCD.

$$\frac{3}{4}+\frac{2}{5}=\frac{15}{20}+\frac{8}{20}=\frac{23}{20}$$

(b) When rewriting fractions with 18 as the denominator, you must multiply denominator and numerator by the same number.

$$\frac{5}{6}-\frac{4}{9}=\frac{15}{18}-\frac{8}{18}=\frac{7}{18}$$

35. **(a)** $-\frac{2}{3}+\frac{3}{4}=-\frac{8}{12}+\frac{9}{12}=\frac{-8+9}{12}=\frac{1}{12}$

$$\frac{3}{4}+\left(-\frac{2}{3}\right)=\frac{9}{12}+\left(-\frac{8}{12}\right)=\frac{9+(-8)}{12}=\frac{1}{12}$$

Both sums are $\frac{1}{12}$ because addition is commutative.

(b) $\frac{5}{6}-\frac{1}{2}=\frac{5}{6}-\frac{3}{6}=\frac{5-3}{6}=\frac{2}{6}=\frac{\overset{1}{\cancel{2}}}{\underset{1}{\cancel{2}}\cdot3}=\frac{1}{3}$

$$\frac{1}{2}-\frac{5}{6}=\frac{3}{6}-\frac{5}{6}=\frac{3-5}{6}=\frac{-2}{6}=-\frac{2}{6}$$
$$=\frac{-\overset{1}{\cancel{2}}}{\underset{1}{\cancel{2}}\cdot3}=-\frac{1}{3}$$

The answers are different because subtraction is not commutative.

(c) $-\frac{2}{3}\cdot\frac{9}{10}=-\frac{\overset{1}{\cancel{2}}\cdot\overset{1}{\cancel{3}}\cdot3}{\underset{1}{\cancel{3}}\cdot\underset{1}{\cancel{2}}\cdot5}=-\frac{3}{5}$

$$\frac{9}{10}\cdot\left(-\frac{2}{3}\right)=-\frac{\overset{1}{\cancel{3}}\cdot3\cdot\overset{1}{\cancel{2}}}{\underset{1}{\cancel{2}}\cdot5\cdot\underset{1}{\cancel{3}}}=-\frac{3}{5}$$

Multiplication is commutative.

(d) $\frac{2}{5}\div\frac{1}{15}=\frac{2}{5}\cdot\frac{15}{1}=\frac{2\cdot3\cdot\overset{1}{\cancel{5}}}{\underset{1}{\cancel{5}}\cdot1}=6$

$$\frac{1}{15}\div\frac{2}{5}=\frac{1}{15}\cdot\frac{5}{2}=\frac{1\cdot\overset{1}{\cancel{5}}}{3\cdot\underset{1}{\cancel{5}}\cdot2}=\frac{1}{6}$$

Division is *not* commutative.

36. **(a)** $-\frac{7}{12}+\frac{7}{12}=\frac{-7+7}{12}=\frac{0}{12}=0$

$$\frac{3}{5}+\left(-\frac{3}{5}\right)=\frac{3+(-3)}{5}=\frac{0}{5}=0$$

The sum of a number and its opposite is 0.

(b) $-\frac{13}{16}\div\left(-\frac{13}{16}\right)=-\frac{\overset{1}{\cancel{13}}}{\underset{1}{\cancel{16}}}\cdot\left(-\frac{\overset{1}{\cancel{16}}}{\underset{1}{\cancel{13}}}\right)=1$

$$\frac{1}{8}\div\frac{1}{8}=\frac{1}{\cancel{8}}\cdot\frac{\overset{1}{\cancel{8}}}{1}=1$$

When a nonzero number is divided by itself, the quotient is 1.

(c) $\frac{5}{6}\cdot1=\frac{5}{6}$

$$1\left(-\frac{17}{20}\right)=-\frac{17}{20}$$

Multiplying by 1 leaves a number unchanged.

(d) $\left(-\frac{4}{5}\right)\left(-\frac{5}{4}\right)=\left(-\frac{\overset{1}{\cancel{4}}}{\underset{1}{\cancel{5}}}\right)\left(-\frac{\overset{1}{\cancel{5}}}{\underset{1}{\cancel{4}}}\right)=1$

$$7\cdot\frac{1}{7}=\frac{\overset{1}{\cancel{7}}}{1}\cdot\frac{1}{\underset{1}{\cancel{7}}}=1$$

A number times its reciprocal is 1.

37. $\frac{1}{5}+\frac{1}{3}+\frac{1}{4}=\frac{12}{60}+\frac{20}{60}+\frac{15}{60}=\frac{12+20+15}{60}=\frac{47}{60}$

The total length is $\frac{47}{60}$ in.

39. $\frac{1}{3}+\frac{3}{8}+\frac{1}{4}=\frac{8}{24}+\frac{9}{24}+\frac{6}{24}=\frac{8+9+6}{24}=\frac{23}{24}$

$\frac{23}{24}$ cubic yard of material was ordered.

41. Fraction of an acre planted:

$$\frac{5}{12}+\frac{11}{12}=\frac{16}{12}$$

Lost $\frac{7}{12}$ acre due to fire: $-\frac{7}{12}$

Remaining seedlings:

$$\frac{16}{12}+\left(-\frac{7}{12}\right)=\frac{16+(-7)}{12}=\frac{9}{12}=\frac{3}{4}$$

$\frac{3}{4}$ acre of seedlings remained.

43. $\frac{2}{5}+\frac{3}{50}=\frac{20}{50}+\frac{3}{50}=\frac{20+3}{50}=\frac{23}{50}$

$\frac{23}{50}$ of workers are self-taught or learned from friends or family.

45. $\frac{2}{5}-\frac{6}{25}=\frac{10}{25}-\frac{6}{25}=\frac{10-6}{25}=\frac{4}{25}$

$\frac{4}{25}$ of workers is the difference.

47. $\frac{1}{8}+\frac{1}{6}=\frac{3}{24}+\frac{4}{24}=\frac{3+4}{24}=\frac{7}{24}$

$$\frac{7}{24}\cdot 24=\frac{7}{24}\cdot\frac{24}{1}=\frac{7\cdot \cancel{24}}{\cancel{24}}=7$$

$\frac{7}{24}$ of the day (or 7 hours) was spent in class and study.

49. $\frac{7}{24}-\frac{1}{6}=\frac{7}{24}-\frac{4}{24}=\frac{3}{24}=\frac{\overset{1}{\cancel{3}}}{\underset{1}{\cancel{3}}\cdot 8}=\frac{1}{8}$

$\frac{1}{8}$ of the day more was spent sleeping than studying.

51. rightmost size minus leftmost size:

$$\frac{1}{2}-\frac{3}{16}=\frac{8}{16}-\frac{3}{16}=\frac{8-3}{16}=\frac{5}{16}$$

The rightmost driver fits a nut that is $\frac{5}{16}$ inch larger than the nut for the leftmost driver.

53. $\frac{3}{8}-\frac{11}{32}=\frac{12}{32}-\frac{11}{32}=\frac{12-11}{32}=\frac{1}{32}$

The nut size for the middle driver is $\frac{1}{32}$ inch less than the nut size for the blue-handled driver.

55. If the total perimeter is $\frac{7}{8}$ mile, use subtraction to find the length of the fourth side.

$$\begin{aligned}\text{Length}&=\frac{7}{8}-\frac{1}{4}-\frac{1}{6}-\frac{3}{8}\\&=\frac{7}{8}+\left(-\frac{1}{4}\right)+\left(-\frac{1}{6}\right)+\left(-\frac{3}{8}\right)\\&=\frac{21}{24}+\left(-\frac{6}{24}\right)+\left(-\frac{4}{24}\right)+\left(-\frac{9}{24}\right)\\&=\frac{21+(-6)+(-4)+(-9)}{24}\\&=\frac{2}{24}=\frac{1}{12}\text{ mile}\end{aligned}$$

The fourth side is $\frac{1}{12}$ mile long.

4.5 Problem Solving: Mixed Numbers and Estimating

4.5 Section Exercises

1. Graph $2\frac{1}{3}$ and $-2\frac{1}{3}$.

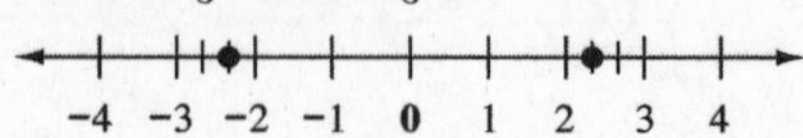

3. Graph $\frac{3}{2}=1\frac{1}{2}$ and $-\frac{3}{2}=-1\frac{1}{2}$.

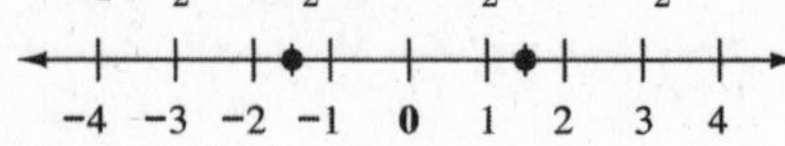

5. $4\frac{1}{2}$

Step 1: $2\cdot 4=8;\ 8+1=9$

Step 2: $4\frac{1}{2}=\frac{9}{2}$

7. $-1\frac{3}{5}$

Step 1: $5\cdot 1=5;\ 5+3=8$

Step 2: $-1\frac{3}{5}=-\frac{8}{5}$

9. $2\frac{3}{8}$

Step 1: $8\cdot 2=16;\ 16+3=19$

Step 2: $2\frac{3}{8}=\frac{19}{8}$

11. $-5\frac{7}{10}$

Step 1: $10\cdot 5=50;\ 50+7=57$

Step 2: $-5\frac{7}{10}=-\frac{57}{10}$

13. $10\frac{11}{15}$

Step 1: $15\cdot 10=150;\ 150+11=161$

Step 2: $10\frac{11}{15}=\frac{161}{15}$

15. Divide 13 by 3.

$$\begin{array}{r}4\\3\overline{)1\,3}\\\underline{1\,2}\\1\end{array}\quad\text{so}\quad\frac{13}{3}=4\frac{1}{3}$$

17. Divide 10 by 4.

$$\begin{array}{r}2\\4\overline{)1\,0}\\\underline{8}\\2\end{array}\quad\text{so}\quad-\frac{10}{4}=-2\frac{2}{4}=-2\frac{1}{2}$$

19. Divide 22 by 6.

$$\begin{array}{r}3\\6\overline{)2\,2}\\\underline{1\,8}\\4\end{array}\quad\text{so}\quad\frac{22}{6}=3\frac{4}{6}=3\frac{2}{3}$$

21. Divide 51 by 9.

$$9\overline{)51}\quad 5,\; 45,\; 6 \quad \text{so} \quad -\frac{51}{9} = -5\frac{6}{9} = -5\frac{2}{3}$$

23. **(a)** $10\frac{1}{6}$ rounds to 10 since the numerator, 1, is less than half of the denominator, 6.

(b) $1\frac{2}{3}$ rounds up to 2 since the numerator, 2, is more than half of the denominator, 3.

(c) $7\frac{7}{16}$ rounds to 7 since the numerator, 7, is less than half of the denominator, 16.

(d) $4\frac{1}{2}$ rounds up to 5. 1 is half of 2.

(e) $3\frac{5}{9}$ rounds up to 4 since the numerator, 5, is more than half of the denominator, 9.

(f) $9\frac{2}{5}$ rounds to 9 since the numerator, 2, is less than half of the denominator, 5.

25. $2\frac{1}{4}$ rounds to 2. $3\frac{1}{2}$ rounds to 4.

Estimate: $2 \cdot 4 = 8$

Exact: $2\frac{1}{4} \cdot 3\frac{1}{2} = \frac{9}{4} \cdot \frac{7}{2} = \frac{63}{8} = 7\frac{7}{8}$

27. $3\frac{1}{4}$ rounds to 3. $2\frac{5}{8}$ rounds to 3.

Estimate: $3 \div 3 = 1$

Exact: $3\frac{1}{4} \div 2\frac{5}{8} = \frac{13}{4} \div \frac{21}{8}$

$$= \frac{13}{4} \cdot \frac{8}{21} = \frac{13 \cdot 2 \cdot \cancel{4}^{1}}{\cancel{4}_{1} \cdot 21} = \frac{26}{21} = 1\frac{5}{21}$$

29. $3\frac{2}{3}$ rounds to 4. $1\frac{5}{6}$ rounds to 2.

Estimate: $4 + 2 = 6$

Exact: $3\frac{2}{3} + 1\frac{5}{6} = \frac{11}{3} + \frac{11}{6} = \frac{22}{6} + \frac{11}{6}$

$$= \frac{22 + 11}{6} = \frac{33}{6}$$

$$= 5\frac{3}{6} = 5\frac{1}{2}$$

31. $4\frac{1}{4}$ rounds to 4. $\frac{7}{12}$ rounds to 1.

Estimate: $4 - 1 = 3$

Exact: $4\frac{1}{4} - \frac{7}{12} = \frac{17}{4} - \frac{7}{12} = \frac{51}{12} - \frac{7}{12}$

$$= \frac{51 - 7}{12} = \frac{44}{12} = \frac{11}{3} = 3\frac{2}{3}$$

33. $5\frac{2}{3}$ rounds to 6.

Estimate: $6 \div 6 = 1$

Exact: $5\frac{2}{3} \div 6 = \frac{17}{3} \div \frac{6}{1} = \frac{17}{3} \cdot \frac{1}{6} = \frac{17}{18}$

35. $1\frac{4}{5}$ rounds to 2.

Estimate: $8 - 2 = 6$

Exact: $8 - 1\frac{4}{5} = \frac{8}{1} - \frac{9}{5} = \frac{40}{5} - \frac{9}{5}$

$$= \frac{40 - 9}{5} = \frac{31}{5} = 6\frac{1}{5}$$

37. The figure is a square.

$$\text{Perimeter} = 4s = 4 \cdot 1\frac{3}{4} = \frac{4}{1} \cdot \frac{7}{4} = \frac{\cancel{4}^{1} \cdot 7}{1 \cdot \cancel{4}_{1}}$$

$$= \frac{7}{1} = 7 \text{ inches}$$

$$\text{Area} = s \cdot s = 1\frac{3}{4} \cdot 1\frac{3}{4} = \frac{7}{4} \cdot \frac{7}{4} = \frac{49}{16}$$

$$= 3\frac{1}{16} \text{ square inches}$$

39. The figure is a rectangle.

$$P = 2l + 2w = 2 \cdot 6\frac{1}{2} + 2 \cdot 3\frac{1}{4}$$

$$= \frac{2}{1} \cdot \frac{13}{2} + \frac{2}{1} \cdot \frac{13}{4} = \frac{26}{2} + \frac{13}{2} = \frac{39}{2} = 19\frac{1}{2}$$

The perimeter is $19\frac{1}{2}$ yd.

$$A = lw = 6\frac{1}{2} \cdot 3\frac{1}{4} = \frac{13}{2} \cdot \frac{13}{4} = \frac{169}{8} = 21\frac{1}{8}$$

The area is $21\frac{1}{8}$ yd^2.

41. $12\frac{1}{2}$ ft rounds to 13 ft.

$8\frac{2}{3}$ ft rounds to 9 ft.

"In all" implies addition.

Estimate: $13 + 9 = 22$ ft

Exact: $12\frac{1}{2} + 8\frac{2}{3} = \frac{25}{2} + \frac{26}{3} = \frac{75}{6} + \frac{52}{6}$

$$= \frac{127}{6} = 21\frac{1}{6} \text{ ft of trim}$$

43. $1\frac{3}{4}$ ounces/gallon rounds to 2 ounces/gallon.

$5\frac{1}{2}$ gallons rounds to 6 gallons.

Estimate: $2 \cdot 6 = 12$ ounces

Exact: $1\frac{3}{4} \cdot 5\frac{1}{2} = \frac{7}{4} \cdot \frac{11}{2}$

$$= \frac{77}{8} = 9\frac{5}{8} \text{ ounces}$$

45. $1\frac{7}{10}$ miles rounds to 2 miles.

Amount left to be picked up implies subtraction.

Estimate: $4 - 2 = 2$ miles

Exact: $$4 - 1\frac{7}{10} = \frac{4}{1} - \frac{17}{10} = \frac{40}{10} - \frac{17}{10} = \frac{23}{10} = 2\frac{3}{10}$$

$2\frac{3}{10}$ miles of highway remain to be picked up by the Boy Scout troop.

47. $3\frac{3}{4}$ yd rounds to 4 yd.

Estimate: $4 \cdot 5 = 20$ yd

Exact: $$3\frac{3}{4} \cdot 5 = \frac{15}{4} \cdot \frac{5}{1} = \frac{75}{4} = 18\frac{3}{4} \text{ yd}$$

49. $9\frac{5}{8}$ cubic yards rounds to 10 cubic yards.

$1\frac{1}{2}$ cubic yards rounds to 2 cubic yards.

3 cubic yards rounds to 3 cubic yards.

Amount remaining in the truck implies subtraction.

Estimate: $10 - 2 - 3 = 5$ cubic yards

Exact: $$9\frac{5}{8} - 1\frac{1}{2} - 3 = \frac{77}{8} - \frac{3}{2} - \frac{3}{1} = \frac{77}{8} - \frac{12}{8} - \frac{24}{8} = \frac{77 - 12 - 24}{8} = \frac{41}{8} = 5\frac{1}{8}$$

$5\frac{1}{8}$ cubic yards of peat moss remain in the truck.

51. $18\frac{3}{4}$ hours rounds to 19 hours.

Estimate: $19 \div 5 = \frac{19}{5} = 3\frac{4}{5}$, which is about 4 hours.

Exact: $$18\frac{3}{4} \div 5 = \frac{75}{4} \div 5 = \frac{75}{4} \div \frac{5}{1} = \frac{75}{4} \cdot \frac{1}{5} = \frac{\overset{1}{\cancel{5}} \cdot 15 \cdot 1}{4 \cdot \underset{1}{\cancel{5}}} = \frac{15}{4} = 3\frac{3}{4} \text{ hours}$$

She worked $3\frac{3}{4}$ hours each day.

53. $6\frac{1}{4}$ in. rounds to 6 in.

$1\frac{7}{8}$ in. rounds to 2 in.

$29\frac{1}{2}$ in. rounds to 30 in.

Estimate: $30 - 6 - 2 = 22$ in.

Exact: $$29\frac{1}{2} - 6\frac{1}{4} - 1\frac{7}{8} = \frac{59}{2} - \frac{25}{4} - \frac{15}{8} = \frac{236}{8} - \frac{50}{8} - \frac{15}{8} = \frac{236 - 50 - 15}{8} = \frac{171}{8} = 21\frac{3}{8} \text{ in.}$$

The length of the arrow shaft is $21\frac{3}{8}$ in.

55. $23\frac{3}{4}$ in. rounds to 24 in.

$34\frac{1}{2}$ in. rounds to 35 in.

Estimate: $24 + 35 + 24 + 35 = 118$ in.

Exact: $$23\frac{3}{4} + 34\frac{1}{2} + 23\frac{3}{4} + 34\frac{1}{2} = \frac{95}{4} + \frac{69}{2} + \frac{95}{4} + \frac{69}{2} = \frac{95}{4} + \frac{138}{4} + \frac{95}{4} + \frac{138}{4} = \frac{466}{4} = 116\frac{1}{2} \text{ in.}$$

The length of lead stripping needed is $116\frac{1}{2}$ in.

57. $10\frac{3}{8}$ pounds rounds to 10 pounds.

25,730 pounds of steel are available and each anchor requires about 10 pounds. Using estimated numbers it is easier to see that you need *division.*

Estimate: $25{,}730 \div 10 = 2573$ anchors

Exact: $$25{,}730 \div 10\frac{3}{8} = \frac{25{,}730}{1} \div \frac{83}{8} = \frac{25{,}730}{1} \cdot \frac{8}{83} = \frac{310 \cdot \overset{1}{\cancel{83}} \cdot 8}{1 \cdot \underset{1}{\cancel{83}}} = 2480 \text{ anchors}$$

59. Round the finished lengths, $21\frac{7}{8}$, $22\frac{5}{8}$, and $23\frac{1}{2}$, to 22, 23, and 24. Round the $\frac{3}{4}$ in. seam allowance to 1 inch, so the rounded lengths (including the seam allowance) are 23, 24, and 25 in.

Estimate: $(4 \cdot 23) + (5 \cdot 24) + (3 \cdot 25) = 287$ in.

Exact:

$$4\left(21\frac{7}{8}+\frac{3}{4}\right)+5\left(22\frac{5}{8}+\frac{3}{4}\right)+3\left(23\frac{1}{2}+\frac{3}{4}\right)$$
$$=4\left(\frac{175}{8}+\frac{6}{8}\right)+5\left(\frac{181}{8}+\frac{6}{8}\right)+3\left(\frac{47}{2}+\frac{3}{4}\right)$$
$$=\frac{4}{1}\cdot\frac{181}{8}+\frac{5}{1}\cdot\frac{187}{8}+\frac{3}{1}\cdot\frac{97}{4}$$
$$=\frac{724}{8}+\frac{935}{8}+\frac{291}{4}$$
$$=\frac{724}{8}+\frac{935}{8}+\frac{582}{8}$$
$$=\frac{2241}{8}=280\frac{1}{8}\text{ in.}$$

Summary Exercises on Fractions

1. (a) 3 of 8 equally sized portions are shaded: $\frac{3}{8}$

   5 of 8 equally sized portions are unshaded: $\frac{5}{8}$

   (b) 4 of 5 equally sized portions are shaded: $\frac{4}{5}$

   1 of 5 equally sized portions are unshaded: $\frac{1}{5}$

3. (a) $30 \div 5 = 6$

   $$-\frac{4}{5}=-\frac{4\cdot 6}{5\cdot 6}=-\frac{24}{30}$$

   (b) $14 \div 7 = 2$

   $$\frac{2}{7}=\frac{2\cdot 2}{7\cdot 2}=\frac{4}{14}$$

5. (a)

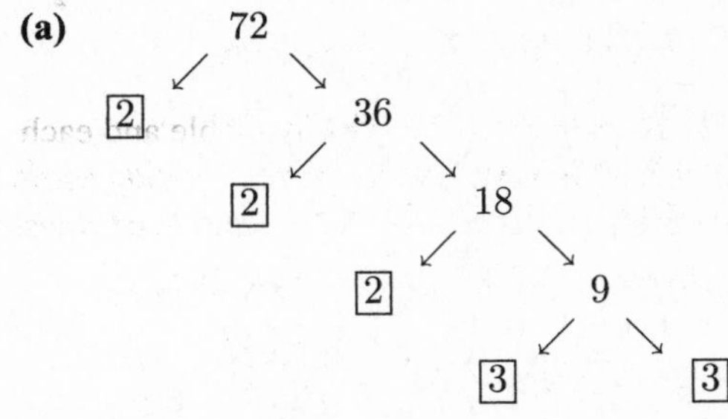

   $72 = 2\cdot 2\cdot 2\cdot 3\cdot 3$

   (b)

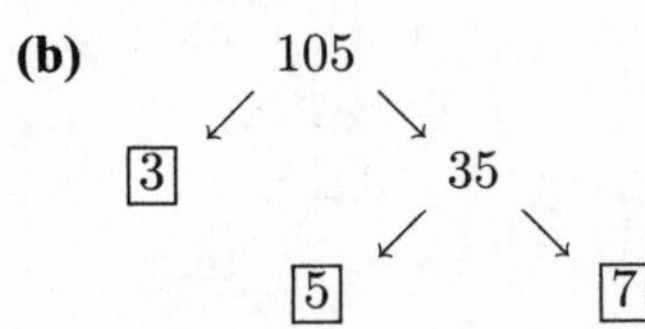

   $105 = 3\cdot 5\cdot 7$

7. $$\left(-\frac{3}{4}\right)\left(-\frac{2}{3}\right)=\frac{\overset{1}{\cancel{3}}\cdot\overset{1}{\cancel{2}}}{\underset{1}{\cancel{2}}\cdot 2\cdot\underset{1}{\cancel{3}}}=\frac{1}{2}$$

9. $$\frac{7}{16}+\frac{5}{8}=\frac{7}{16}+\frac{10}{16}=\frac{7+10}{16}=\frac{17}{16}$$

11. $$\frac{2}{3}-\frac{4}{5}=\frac{2\cdot 5}{3\cdot 5}-\frac{4\cdot 3}{5\cdot 3}=\frac{10}{15}-\frac{12}{15}=\frac{10-12}{15}=-\frac{2}{15}$$

13. $$-21\div\left(-\frac{3}{8}\right)=\frac{21}{1}\cdot\left(\frac{8}{3}\right)=\frac{\overset{1}{\cancel{3}}\cdot 7\cdot 8}{\underset{1}{\cancel{3}}}=56$$

15. $$-\frac{35}{45}\div\frac{10}{15}=-\frac{35}{45}\cdot\frac{15}{10}=-\frac{\overset{1}{\cancel{5}}\cdot 7\cdot\overset{1}{\cancel{3}}\cdot\overset{1}{\cancel{5}}}{\underset{1}{\cancel{5}}\cdot 3\cdot\underset{1}{\cancel{3}}\cdot 2\cdot\underset{1}{\cancel{5}}}=-\frac{7}{6}$$

17. $$\frac{7}{12}+\frac{5}{6}+\frac{2}{3}=\frac{7}{12}+\frac{10}{12}+\frac{8}{12}=\frac{7+10+8}{12}=\frac{25}{12}$$

19. $4\frac{3}{4}$ rounds to 5. $2\frac{5}{6}$ rounds to 3.

    Estimate: $5+3=8$

    Exact: $4\frac{3}{4}+2\frac{5}{6}=\frac{19}{4}+\frac{17}{6}=\frac{57}{12}+\frac{34}{12}$
    $$=\frac{91}{12}=7\frac{7}{12}$$

21. $2\frac{7}{10}$ rounds to 3.

    Estimate: $6-3=3$

    Exact: $6-2\frac{7}{10}=\frac{60}{10}-\frac{27}{10}=\frac{60-27}{10}$
    $$=\frac{33}{10}=3\frac{3}{10}$$

23. $4\frac{2}{3}$ rounds to 5. $1\frac{1}{6}$ rounds to 1.

    Estimate: $5\div 1=5$

    Exact: $4\frac{2}{3}\div 1\frac{1}{6}=\frac{14}{3}\div\frac{7}{6}=\frac{14}{3}\cdot\frac{6}{7}$
    $$=\frac{2\cdot\overset{1}{\cancel{7}}\cdot 2\cdot\overset{1}{\cancel{3}}}{\underset{1}{\cancel{3}}\cdot\underset{1}{\cancel{7}}}=\frac{4}{1}=4$$

25. (a) $\frac{1}{4}+\frac{11}{16}+1\frac{1}{8}=\frac{4}{16}+\frac{11}{16}+\frac{18}{16}=\frac{33}{16}=2\frac{1}{16}$

    The total length is $2\frac{1}{16}$ in.

    (b) $2\frac{1}{16}-1\frac{3}{4}=\frac{33}{16}-\frac{7}{4}=\frac{33}{16}-\frac{28}{16}$
    $$=\frac{33-28}{16}=\frac{5}{16}$$

    $\frac{5}{16}$ in. will stick out the back of the board.

27. $$9\div\frac{3}{4}=\frac{9}{1}\cdot\frac{4}{3}=\frac{3\cdot\overset{1}{\cancel{3}}\cdot 4}{1\cdot\underset{1}{\cancel{3}}}=12$$

    You can make 12 batches of cookies.

29. Not sure: $\frac{3}{20}\cdot 1500 = \frac{3\cdot 75\cdot \overset{1}{\cancel{20}}}{\underset{1}{\cancel{20}}} = 225$ adults

Real: $\frac{9}{20}\cdot 1500 = \frac{9\cdot 75\cdot \overset{1}{\cancel{20}}}{\underset{1}{\cancel{20}}} = 675$ adults

Imaginary: $\frac{2}{5}\cdot 1500 = \frac{2\cdot 300\cdot \overset{1}{\cancel{5}}}{\underset{1}{\cancel{5}}} = 600$ adults

31. $15\frac{1}{3} \div \frac{2}{3} = \frac{46}{3}\cdot\frac{3}{2} = \frac{\overset{1}{\cancel{2}}\cdot 23\cdot \overset{1}{\cancel{3}}}{\underset{1}{\cancel{3}}\cdot\underset{1}{\cancel{2}}} = 23$

23 bottles can be filled.

4.6 Exponents, Order of Operations, and Complex Fractions

4.6 Section Exercises

1. $\left(-\frac{3}{4}\right)^2 = \left(-\frac{3}{4}\right)\left(-\frac{3}{4}\right) = \frac{3\cdot 3}{4\cdot 4} = \frac{9}{16}$

3. $\left(\frac{2}{5}\right)^3 = \frac{2}{5}\cdot\frac{2}{5}\cdot\frac{2}{5}$

$= \frac{4}{25}\cdot\frac{2}{5}$

$= \frac{8}{125}$

5. $\left(-\frac{1}{3}\right)^3 = \left(-\frac{1}{3}\right)\left(-\frac{1}{3}\right)\left(-\frac{1}{3}\right)$

$= \frac{1}{9}\left(-\frac{1}{3}\right)$

$= -\frac{1}{27}$

7. $\left(\frac{1}{2}\right)^5 = \frac{1}{2}\cdot\frac{1}{2}\cdot\frac{1}{2}\cdot\frac{1}{2}\cdot\frac{1}{2}$

$= \frac{1}{4}\cdot\frac{1}{2}\cdot\frac{1}{2}\cdot\frac{1}{2}$

$= \frac{1}{8}\cdot\frac{1}{2}\cdot\frac{1}{2}$

$= \frac{1}{16}\cdot\frac{1}{2}$

$= \frac{1}{32}$

9. $\left(\frac{7}{10}\right)^2 = \frac{7}{10}\cdot\frac{7}{10} = \frac{49}{100}$

11. $\left(-\frac{6}{5}\right)^2 = \left(-\frac{6}{5}\right)\left(-\frac{6}{5}\right)$

$= \frac{36}{25}$ or $1\frac{11}{25}$

13. $\frac{15}{16}\left(\frac{4}{5}\right)^3$ Exponent first

$= \frac{15}{16}\left(\frac{4}{5}\cdot\frac{4}{5}\cdot\frac{4}{5}\right)$

$= \frac{15}{16}\left(\frac{64}{125}\right)$ Multiply.

$= \frac{3\cdot\overset{1}{\cancel{5}}\cdot\overset{1}{\cancel{16}}\cdot 4}{\underset{1}{\cancel{16}}\cdot\underset{1}{\cancel{5}}\cdot 25}$ Reduce.

$= \frac{12}{25}$

15. $\left(\frac{1}{3}\right)^4\left(\frac{9}{10}\right)^2 = \left(\frac{1}{3}\cdot\frac{1}{3}\cdot\frac{1}{3}\cdot\frac{1}{3}\right)\left(\frac{9}{10}\cdot\frac{9}{10}\right)$

$= \frac{1\cdot 1\cdot 1\cdot 1\cdot \overset{1}{\cancel{3}}\cdot\overset{1}{\cancel{3}}\cdot\overset{1}{\cancel{3}}\cdot\overset{1}{\cancel{3}}}{\underset{1}{\cancel{3}}\cdot\underset{1}{\cancel{3}}\cdot\underset{1}{\cancel{3}}\cdot\underset{1}{\cancel{3}}\cdot 10\cdot 10}$

$= \frac{1}{100}$

17. $\left(-\frac{3}{2}\right)^3\left(-\frac{2}{3}\right)^2$

$= \left(-\frac{3}{2}\right)\left(-\frac{3}{2}\right)\left(-\frac{3}{2}\right)\left(-\frac{2}{3}\right)\left(-\frac{2}{3}\right)$

$= -\frac{\overset{1}{\cancel{3}}\cdot\overset{1}{\cancel{3}}\cdot 3\cdot\overset{1}{\cancel{2}}\cdot\overset{1}{\cancel{2}}}{\underset{1}{\cancel{2}}\cdot\underset{1}{\cancel{2}}\cdot 2\cdot\underset{1}{\cancel{3}}\cdot\underset{1}{\cancel{3}}}$

$= -\frac{3}{2}$ or $-1\frac{1}{2}$

19. **(a)** $\left(-\frac{1}{2}\right)^2 = \left(-\frac{1}{2}\right)\left(-\frac{1}{2}\right)$

$= \frac{1}{4}$

$\left(-\frac{1}{2}\right)^3 = \left(-\frac{1}{2}\right)^2\left(-\frac{1}{2}\right)$

$= \left(\frac{1}{4}\right)\left(-\frac{1}{2}\right) = -\frac{1}{8}$

$\left(-\frac{1}{2}\right)^4 = \left(-\frac{1}{2}\right)^3\left(-\frac{1}{2}\right)$

$= \left(-\frac{1}{8}\right)\left(-\frac{1}{2}\right) = \frac{1}{16}$

$\left(-\frac{1}{2}\right)^5 = \left(-\frac{1}{2}\right)^4\left(-\frac{1}{2}\right)$

$= \left(\frac{1}{16}\right)\left(-\frac{1}{2}\right) = -\frac{1}{32}$

$\left(-\frac{1}{2}\right)^6 = \left(-\frac{1}{2}\right)^5\left(-\frac{1}{2}\right)$

$= \left(-\frac{1}{32}\right)\left(-\frac{1}{2}\right) = \frac{1}{64}$

$$\left(-\frac{1}{2}\right)^7 = \left(-\frac{1}{2}\right)^6\left(-\frac{1}{2}\right) = \left(\frac{1}{64}\right)\left(-\frac{1}{2}\right) = -\frac{1}{128}$$

$$\left(-\frac{1}{2}\right)^8 = \left(-\frac{1}{2}\right)^7\left(-\frac{1}{2}\right) = \left(-\frac{1}{128}\right)\left(-\frac{1}{2}\right) = \frac{1}{256}$$

$$\left(-\frac{1}{2}\right)^9 = \left(-\frac{1}{2}\right)^8\left(-\frac{1}{2}\right) = \left(\frac{1}{256}\right)\left(-\frac{1}{2}\right) = -\frac{1}{512}$$

(b) When a negative number is raised to an even power, the answer is positive. When a negative number is raised to an odd power, the answer is negative.

20. **(a)** Ask yourself, "What number, times itself, is 4?" This is the numerator. Then ask, "What number, times itself, is 9?" This is the denominator. The number under the ketchup is either $\frac{2}{3}$ or $-\frac{2}{3}$.

(b) The number under the ketchup is $-\frac{1}{3}$ because $\left(-\frac{1}{3}\right)\left(-\frac{1}{3}\right)\left(-\frac{1}{3}\right) = -\frac{1}{27}$.

(c) Either $\frac{1}{2}$ or $-\frac{1}{2}$.

(d) No real number works, because both $\left(\frac{3}{4}\right)^2$ and $\left(-\frac{3}{4}\right)^2$ give a *positive* result.

(e) Either $\frac{1}{3}$ or $-\frac{1}{3}$ inside one set of parentheses and $\frac{1}{2}$ or $-\frac{1}{2}$ inside the other.

21. $\frac{1}{5} - 6\left(\frac{7}{10}\right)$ Rewrite 6 as $\frac{6}{1}$.

$= \frac{1}{5} - \frac{6}{1}\left(\frac{7}{10}\right)$ Multiply.

$= \frac{1}{5} - \frac{42}{10}$ Subtract. LCD is 10.

$= \frac{2}{10} - \frac{42}{10}$

$= \frac{2-42}{10}$

$= \frac{-40}{10}$ Reduce.

$= -4$

23. $\left(\frac{4}{3} \div \frac{8}{3}\right) + \left(-\frac{3}{4}\cdot\frac{1}{4}\right)$

$= \left(\frac{4}{3}\cdot\frac{3}{8}\right) + \left(-\frac{3}{4}\cdot\frac{1}{4}\right)$ Change division.

$= \left(\frac{\overset{1}{\cancel{4}}\cdot\overset{1}{\cancel{3}}}{\underset{1}{\cancel{3}}\cdot 2\cdot\underset{1}{\cancel{4}}}\right) + \left(-\frac{3}{16}\right)$ Parentheses first

$= \frac{1}{2} - \frac{3}{16}$ Add. LCD is 16.

$= \frac{8}{16} - \frac{3}{16}$

$= \frac{8-3}{16} = \frac{5}{16}$

25. $-\frac{3}{10} \div \frac{3}{5}\left(-\frac{2}{3}\right)$ Change division.

$= -\frac{3}{10}\cdot\frac{5}{3}\left(-\frac{2}{3}\right)$

$= -\frac{\overset{1}{\cancel{3}}\cdot\overset{1}{\cancel{5}}}{2\cdot\underset{1}{\cancel{5}}\cdot\underset{1}{\cancel{3}}}\left(-\frac{2}{3}\right)$

$= -\frac{1}{2}\left(-\frac{2}{3}\right)$

$= \frac{1\cdot\overset{1}{\cancel{2}}}{\underset{1}{\cancel{2}}\cdot 3}$

$= \frac{1}{3}$

27. $\frac{8}{3}\left(\frac{1}{4} - \frac{1}{2}\right)^2$ LCD is 4.

$= \frac{8}{3}\left(\frac{1}{4} - \frac{2}{4}\right)^2$ Subtract.

$= \frac{8}{3}\left(-\frac{1}{4}\right)^2$ Exponent

$= \frac{8}{3}\left(-\frac{1}{4}\right)\left(-\frac{1}{4}\right)$

$= \frac{8}{3}\left(\frac{1}{16}\right)$

$= \frac{\overset{1}{\cancel{8}}\cdot 1}{3\cdot 2\cdot\underset{1}{\cancel{8}}}$

$= \frac{1}{6}$

29. $-\frac{3}{8}+\frac{2}{3}\left(-\frac{2}{3}+\frac{1}{6}\right)$ LCD inside parentheses is 6.

$=-\frac{3}{8}+\frac{2}{3}\left(-\frac{4}{6}+\frac{1}{6}\right)$ Add.

$=-\frac{3}{8}+\frac{2}{3}\left(\frac{-4+1}{6}\right)$

$=-\frac{3}{8}+\frac{2}{3}\left(\frac{-3}{6}\right)$ Reduce.

$=-\frac{3}{8}+\frac{2}{3}\left(-\frac{1}{2}\right)$ Multiply.

$=-\frac{3}{8}+\left(-\frac{1}{3}\right)$ Add. LCD is 24.

$=-\frac{9}{24}+\left(-\frac{8}{24}\right)$

$=\frac{-9+(-8)}{24}$

$=\frac{-17}{24}$ or $-\frac{17}{24}$

31. $2\left(\frac{1}{3}\right)^3-\frac{2}{9}$ Exponent

$=2\left(\frac{1}{3}\right)\left(\frac{1}{3}\right)\left(\frac{1}{3}\right)-\frac{2}{9}$

$=2\left(\frac{1}{9}\right)\left(\frac{1}{3}\right)-\frac{2}{9}$

$=2\left(\frac{1}{27}\right)-\frac{2}{9}$

$=\frac{2}{27}-\frac{2}{9}$ Subtract. LCD is 27.

$=\frac{2}{27}-\frac{6}{27}$

$=\frac{2-6}{27}=-\frac{4}{27}$

33. $\left(-\frac{2}{3}\right)^3\left(\frac{1}{8}-\frac{1}{2}\right)-\frac{2}{3}\left(\frac{1}{8}\right)$ Parentheses first

$=\left(-\frac{2}{3}\right)^3\left(\frac{1}{8}-\frac{4}{8}\right)-\frac{2}{3}\left(\frac{1}{8}\right)$

$=\left(-\frac{2}{3}\right)^3\left(-\frac{3}{8}\right)-\frac{2}{3}\left(\frac{1}{8}\right)$ Exponent

$=\left(-\frac{2}{3}\right)\left(-\frac{2}{3}\right)\left(-\frac{2}{3}\right)\left(-\frac{3}{8}\right)-\frac{2}{3}\left(\frac{1}{8}\right)$

$=\left(-\frac{8}{27}\right)\left(-\frac{3}{8}\right)-\frac{2}{3}\left(\frac{1}{8}\right)$ Multiply.

$=\frac{\overset{1}{\cancel{8}}\cdot\overset{1}{\cancel{3}}}{\underset{1}{\cancel{3}}\cdot 9\cdot\underset{1}{\cancel{8}}}-\frac{\overset{1}{\cancel{2}}\cdot 1}{3\cdot\underset{1}{\cancel{2}}\cdot 4}$

$=\frac{1}{9}-\frac{1}{12}$ Subtract. LCD is 36.

$=\frac{4}{36}-\frac{3}{36}$

$=\frac{4-3}{36}$

$=\frac{1}{36}$

35. $A=s^2=\left(\frac{3}{8}\right)^2$

$=\left(\frac{3}{8}\right)\left(\frac{3}{8}\right)$

$=\frac{3\cdot 3}{8\cdot 8}$

$=\frac{9}{64}$ in.^{2}

37. $P=2l+2w$

$=2\left(\frac{7}{10}\right)+2\left(\frac{1}{4}\right)$ Rewrite as $\frac{2}{1}$.

$=\frac{2}{1}\left(\frac{7}{10}\right)+\frac{2}{1}\left(\frac{1}{4}\right)$ Multiply.

$=\frac{\overset{1}{\cancel{2}}\cdot 7}{1\cdot\underset{1}{\cancel{2}}\cdot 5}+\frac{\overset{1}{\cancel{2}}\cdot 1}{1\cdot\underset{1}{\cancel{2}}\cdot 2}$ Reduce.

$=\frac{7}{5}+\frac{1}{2}$ Add. LCD is 10.

$=\frac{14}{10}+\frac{5}{10}$

$=\frac{14+5}{10}$

$=\frac{19}{10}$ mi or $1\frac{9}{10}$ miles

39. $\frac{-\frac{7}{9}}{-\frac{7}{36}}=-\frac{7}{9}\div\left(-\frac{7}{36}\right)$

$=-\frac{7}{9}\cdot\left(-\frac{36}{7}\right)$

$=\frac{\overset{1}{\cancel{7}}\cdot\overset{1}{\cancel{9}}\cdot 4}{\underset{1}{\cancel{9}}\cdot\underset{1}{\cancel{7}}}$

$=\frac{4}{1}=4$

41. $\frac{-15}{\frac{6}{5}}=-15\div\frac{6}{5}=-\frac{15}{1}\div\frac{6}{5}$

$=-\frac{15}{1}\cdot\frac{5}{6}$

$=-\frac{\overset{1}{\cancel{3}}\cdot 5\cdot 5}{1\cdot 2\cdot\underset{1}{\cancel{3}}}$

$=-\frac{25}{2}$ or $-12\frac{1}{2}$

43. $\dfrac{\frac{4}{7}}{8} = \dfrac{4}{7} \div 8$

$= \dfrac{4}{7} \div \dfrac{8}{1}$

$= \dfrac{4}{7} \cdot \dfrac{1}{8}$

$= \dfrac{\overset{1}{\cancel{4}} \cdot 1}{7 \cdot \underset{1}{\cancel{4}} \cdot 2}$

$= \dfrac{1}{14}$

45. $\dfrac{-\frac{2}{3}}{-2\frac{2}{5}} = \dfrac{-\frac{2}{3}}{-\frac{12}{5}}$ Write as division.

$= \dfrac{2}{3} \div \dfrac{12}{5}$ Change division.

$= \dfrac{2}{3} \cdot \dfrac{5}{12}$

$= \dfrac{\overset{1}{\cancel{2}} \cdot 5}{3 \cdot \underset{1}{\cancel{2}} \cdot 6}$

$= \dfrac{5}{18}$

47. $\dfrac{-4\frac{1}{2}}{\left(\frac{3}{4}\right)^2} = \dfrac{-4\frac{1}{2}}{\frac{9}{16}}$ Exponent first

$= \dfrac{-\frac{9}{2}}{\frac{9}{16}}$ Write as division.

$= -\dfrac{9}{2} \div \dfrac{9}{16}$ Change division.

$= -\dfrac{9}{2} \cdot \dfrac{16}{9}$

$= -\dfrac{\overset{1}{\cancel{9}} \cdot \overset{1}{\cancel{2}} \cdot 8}{\underset{1}{\cancel{2}} \cdot \underset{1}{\cancel{9}}}$

$= -\dfrac{8}{1} = -8$

49. $\dfrac{\left(\frac{2}{5}\right)^2}{\left(-\frac{4}{3}\right)^2}$ Exponents first

$= \dfrac{\frac{4}{25}}{\frac{16}{9}}$ Write as division.

$= \dfrac{4}{25} \div \dfrac{16}{9}$ Change division.

$= \dfrac{4}{25} \cdot \dfrac{9}{16}$

$= \dfrac{\overset{1}{\cancel{4}} \cdot 9}{25 \cdot \underset{1}{\cancel{4}} \cdot 4} = \dfrac{9}{100}$

4.7 Problem Solving: Equations Containing Fractions

4.7 Section Exercises

1. $\dfrac{1}{3}a = 10$ Multiply both sides by $\frac{3}{1}$, the reciprocal of $\frac{1}{3}$.

$\dfrac{\overset{1}{\cancel{3}}}{1}\left(\dfrac{1}{\underset{1}{\cancel{3}}}a\right) = \dfrac{3}{1}(10)$

$a = \dfrac{3}{1} \cdot \dfrac{10}{1}$

$a = 30$

Check: $\dfrac{1}{3}(30) = 10$

$\dfrac{30}{3} = 10$

$10 = 10$ Balances

The solution is 30.

3. $-20 = \dfrac{5}{6}b$ Multiply both sides by $\frac{6}{5}$, the reciprocal of $\frac{5}{6}$.

$\dfrac{6}{5}(-20) = \dfrac{\overset{1}{\cancel{6}}}{\underset{1}{\cancel{5}}}\left(\dfrac{\overset{1}{\cancel{5}}}{\underset{1}{\cancel{6}}}b\right)$

$-\dfrac{6 \cdot 4 \cdot \overset{1}{\cancel{5}}}{\underset{1}{\cancel{5}}} = b$

$-24 = b$

Check: $-20 = \dfrac{5}{6}(-24)$

$-20 = -20$ Balances

The solution is -24.

5. $-\dfrac{7}{2}c = -21$ Multiply both sides by $-\frac{2}{7}$, the reciprocal of $-\frac{7}{2}$.

$-\dfrac{\overset{1}{\cancel{2}}}{\underset{1}{\cancel{7}}}\left(-\dfrac{\overset{1}{\cancel{7}}}{\underset{1}{\cancel{2}}}c\right) = -\dfrac{2}{\underset{1}{\cancel{7}}}\left(-\dfrac{\overset{3}{\cancel{21}}}{1}\right)$

$c = \dfrac{6}{1}$

$c = 6$

Check: $-\dfrac{7}{2}(6) = -21$

$-\dfrac{42}{2} = -21$

$-21 = -21$ Balances

The solution is 6.

7. $\frac{9}{16} = \frac{3}{4}m$ Multiply both sides by $\frac{4}{3}$, the reciprocal of $\frac{3}{4}$.

$$\frac{\overset{1}{\cancel{4}}}{\underset{1}{\cancel{3}}} \cdot \frac{\overset{3}{\cancel{9}}}{\underset{4}{\cancel{16}}} = \frac{\overset{1}{\cancel{4}}}{\underset{1}{\cancel{3}}}\left(\frac{\overset{1}{\cancel{3}}}{\underset{1}{\cancel{4}}}m\right)$$

$$\frac{3}{4} = m$$

Check: $\frac{9}{16} = \frac{3}{4}\left(\frac{3}{4}\right)$

$\frac{9}{16} = \frac{9}{16}$ Balances

The solution is $\frac{3}{4}$.

9. $\frac{3}{10} = -\frac{1}{4}d$ Multiply both sides by $-\frac{4}{1}$, the reciprocal of $-\frac{1}{4}$.

$$-\frac{4}{1}\left(\frac{3}{10}\right) = -\frac{\overset{1}{\cancel{4}}}{1}\left(\frac{-1}{\underset{1}{\cancel{4}}}d\right)$$

$$-\frac{12}{10} = d$$

$$-\frac{6}{5} = d$$

Check: $\frac{3}{10} = -\frac{1}{4}\left(-\frac{6}{5}\right)$

$$\frac{3}{10} = \frac{1 \cdot \overset{1}{\cancel{2}} \cdot 3}{\underset{1}{\cancel{2}} \cdot 2 \cdot 5}$$

$\frac{3}{10} = \frac{3}{10}$ Balances

The solution is $-\frac{6}{5}$.

11. $\frac{1}{6}n + 7 = 9$ Add -7 to both sides.

$\underline{\quad -7 \quad} \quad \underline{-7}$

$\frac{1}{6}n = 2$ Multiply both sides by $\frac{6}{1}$, the reciprocal of $\frac{1}{6}$.

$$\frac{\overset{1}{\cancel{6}}}{1} \cdot \frac{1}{\underset{1}{\cancel{6}}}n = \frac{6}{1} \cdot 2$$

$$n = 12$$

Check: $\frac{1}{6}(12) + 7 = 9$

$2 + 7 = 9$

$9 = 9$ Balances

The solution is 12.

13. $-10 = \frac{5}{3}r + 5$ Add -5 to both sides.

$\underline{\quad -5 \quad} \quad \underline{\quad -5 \quad}$

$-15 = \frac{5}{3}r + 0$ Multiply both sides by $\frac{3}{5}$.

$$\frac{3}{5}\left(-\frac{15}{1}\right) = \frac{\overset{1}{\cancel{3}}}{\underset{1}{\cancel{5}}}\left(\frac{\overset{1}{\cancel{5}}}{\underset{1}{\cancel{3}}}r\right)$$

$$-\frac{3 \cdot 3 \cdot \overset{1}{\cancel{5}}}{\underset{1}{\cancel{5}}} = r$$

$$-9 = r$$

Check: $-10 = \frac{5}{3}(-9) + 5$

$-10 = -\frac{45}{3} + 5$

$-10 = -15 + 5$

$-10 = -10$ Balances

The solution is -9.

15. $\frac{3}{8}x - 9 = 0$ Add 9 to both sides.

$\underline{\quad 9 \quad} \quad \underline{9}$

$\frac{3}{8}x = 9$ Multiply both sides by $\frac{8}{3}$.

$$\frac{\overset{1}{\cancel{8}}}{\underset{1}{\cancel{3}}} \cdot \frac{\overset{1}{\cancel{3}}}{\underset{1}{\cancel{8}}}x = \frac{8}{\underset{1}{\cancel{3}}} \cdot \frac{\overset{3}{\cancel{9}}}{1}$$

$$x = 24$$

Check: $\frac{3}{8}(24) - 9 = 0$

$9 - 9 = 0$

$0 = 0$ Balances

The solution is 24.

17. $7 - 2 = \frac{1}{5}y - 4$ Add the opposite.

$7 + (-2) = \frac{1}{5}y + (-4)$ Combine like terms.

$5 = \frac{1}{5}y + (-4)$ Add 4 to both sides.

$\underline{\quad 4 \quad} \quad \underline{\quad 4 \quad}$

$9 = \frac{1}{5}y + 0$ Multiply both sides by 5.

$$\frac{5}{1}(9) = \frac{\overset{1}{\cancel{5}}}{1}\left(\frac{1}{\underset{1}{\cancel{5}}}y\right)$$

$$45 = y$$

The solution is 45.

19.
$$4 + \frac{2}{3}n = -10 + 2 \quad \text{Combine like terms.}$$
$$4 + \frac{2}{3}n = -8 \quad \text{Add } -4 \text{ to both sides.}$$
$$\underline{-4 \qquad\quad -4}$$
$$\frac{2}{3}n = -12 \quad \text{Multiply both sides by } \frac{3}{2}.$$
$$\frac{\overset{1}{\cancel{3}}}{\underset{1}{\cancel{2}}}\left(\frac{\overset{1}{\cancel{2}}}{\underset{1}{\cancel{3}}}n\right) = \frac{3}{2}(-12)$$
$$n = -\frac{36}{2}$$
$$n = -18$$

The solution is -18.

21.
$$3x + \frac{1}{2} = \frac{3}{4}$$
$$\underline{-\frac{1}{2} \qquad -\frac{1}{2}}$$
$$3x + 0 = \frac{3}{4} + \left(-\frac{1}{2}\right)$$
$$3x = \frac{3}{4} + \left(-\frac{2}{4}\right)$$
$$3x = \frac{3 + (-2)}{4}$$
$$3x = \frac{1}{4}$$
$$\frac{1}{\underset{1}{\cancel{3}}}(\overset{1}{\cancel{3}}x) = \frac{1}{3}\left(\frac{1}{4}\right)$$
$$x = \frac{1}{12}$$

The solution is $\frac{1}{12}$.

23.
$$\frac{3}{10} = -4b - \frac{1}{5}$$
$$\underline{\frac{1}{5} \qquad\qquad \frac{1}{5}}$$
$$\frac{3}{10} + \frac{1}{5} = -4b$$
$$\frac{3}{10} + \frac{2}{10} = -4b$$
$$\frac{5}{10} = -4b$$
$$-\frac{1}{4} \cdot \frac{5}{10} = -\frac{1}{4}(-4b)$$
$$-\frac{5}{40} = b$$
$$-\frac{\overset{1}{\cancel{5}}}{\underset{1}{\cancel{5}} \cdot 8} = b$$
$$-\frac{1}{8} = b$$

The solution is $-\frac{1}{8}$.

25. **(a)**
$$\frac{1}{6}x + 1 = -2 \quad \text{Let } x = 18.$$
$$\frac{1}{6}\left(\frac{18}{1}\right) + 1 = -2$$
$$\frac{18}{6} + 1 = -2$$
$$3 + 1 = -2$$
$$4 \neq -2$$

No, it does not balance. $x = 18$ is not the solution, so we'll solve the equation.

$$\frac{1}{6}x + 1 = -2$$
$$\underline{-1 \qquad\quad -1}$$
$$\frac{1}{6}x + 0 = -3$$
$$\frac{1}{6}x = -3$$
$$\frac{\overset{1}{\cancel{6}}}{1}\left(\frac{1}{\underset{1}{\cancel{6}}}x\right) = \frac{6}{1}(-3)$$
$$x = -18$$

The correct solution is -18.

(b)
$$-\frac{3}{2} = \frac{9}{4}k \qquad \text{Let } k = -\frac{2}{3}.$$
$$-\frac{3}{2} = \frac{9}{4}\left(-\frac{2}{3}\right)$$
$$-\frac{3}{2} = -\frac{18}{12}$$
$$-\frac{3}{2} = -\frac{6 \cdot 3}{6 \cdot 2}$$
$$-\frac{3}{2} = -\frac{3}{2} \qquad \text{Balances}$$

Yes, $k = -\frac{2}{3}$ is the correct solution because the equation balances.

27. Some possibilities are:
$\frac{1}{2}x = 4$; $-\frac{1}{4}a = -2$; $\frac{3}{4}b = 6$.

29. Let a be the man's age.
$$100 + \frac{a}{2} = \text{Systolic Blood Pressure}$$
$$100 + \frac{a}{2} = 109$$
$$\underline{-100 \qquad\quad -100}$$
$$0 + \frac{a}{2} = 9$$
$$\frac{a}{2} = 9$$
$$\frac{1}{2}a = 9$$
$$\frac{\overset{1}{\cancel{2}}}{1}\left(\frac{1}{\underset{1}{\cancel{2}}}a\right) = \frac{2}{1}(9)$$
$$a = 18$$

The man is 18 years old.

31. Let a be the woman's age.

$$100 + \frac{a}{2} = \text{Systolic Blood Pressure}$$
$$100 + \frac{a}{2} = 122$$
$$\underline{-100} \quad \underline{-100}$$
$$0 + \frac{a}{2} = 22$$
$$\frac{\overset{1}{\cancel{2}}}{1}\left(\frac{1}{\underset{1}{\cancel{2}}}a\right) = 2(22)$$
$$a = 44$$

The woman is 44 years old.

33. Let p be the penny size.

$$\frac{p}{4} + \frac{1}{2} = 3$$
$$\underline{-\frac{1}{2}} \quad \underline{-\frac{1}{2}}$$
$$\frac{p}{4} = 3 - \frac{1}{2}$$
$$\frac{p}{4} = \frac{6}{2} - \frac{1}{2}$$
$$\frac{p}{4} = \frac{5}{2}$$
$$\overset{1}{\cancel{4}} \cdot \frac{p}{\underset{1}{\cancel{4}}} = \overset{2}{\cancel{4}} \cdot \frac{5}{\underset{1}{\cancel{2}}}$$
$$p = 10$$

The penny size is 10.

35. Let p be the penny size.

$$\frac{p}{4} + \frac{1}{2} = \frac{5}{2}$$
$$\underline{-\frac{1}{2}} \quad \underline{-\frac{1}{2}}$$
$$\frac{p}{4} = 2$$
$$\overset{1}{\cancel{4}} \cdot \frac{p}{\underset{1}{\cancel{4}}} = 4 \cdot 2$$
$$p = 8$$

The penny size is 8.

37. Let h be the man's height.

$$\frac{11}{2}h - 220 = \text{Recommended weight}$$
$$\frac{11}{2}h - 220 = 209$$
$$\underline{\quad 220} \quad \underline{220 \quad}$$
$$\frac{11}{2}h + 0 = 429$$
$$\frac{\overset{1}{\cancel{2}}}{\underset{1}{\cancel{11}}}\left(\frac{\overset{1}{\cancel{11}}}{\underset{1}{\cancel{2}}}h\right) = \frac{2}{11}(429)$$
$$h = \frac{2 \cdot \overset{1}{\cancel{11}} \cdot 39}{\underset{1}{\cancel{11}}}$$
$$h = 78$$

The man is 78 inches tall.

39. Let h be the woman's height.

$$\frac{11}{2}h - 220 = \text{Recommended weight}$$
$$\frac{11}{2}h - 220 = 132$$
$$\underline{\quad 220} \quad \underline{220 \quad}$$
$$\frac{11}{2}h = 352$$
$$\frac{\overset{1}{\cancel{2}}}{\underset{1}{\cancel{11}}}\left(\frac{\overset{1}{\cancel{11}}}{\underset{1}{\cancel{2}}}h\right) = \frac{2}{11}(352)$$
$$h = \frac{2 \cdot \overset{1}{\cancel{11}} \cdot 32}{\underset{1}{\cancel{11}}}$$
$$h = 64$$

The woman is 64 inches tall.

4.8 Geometry Applications: Area and Volume

4.8 Section Exercises

1. $P = 58\text{ m} + 72\text{ m} + 72\text{ m} = 202\text{ m}$

$$A = \frac{1}{2}bh$$
$$A = \frac{1}{2}(58\text{ m})(66\text{ m})$$
$$A = \frac{1}{2} \cdot \frac{58\text{ m}}{1} \cdot \frac{66\text{ m}}{1}$$
$$A = \frac{1 \cdot \overset{1}{\cancel{2}} \cdot 29\text{ m} \cdot 66\text{ m}}{\underset{1}{\cancel{2}} \cdot 1 \cdot 1}$$
$$A = 1914\text{ m}^2$$

3. $P = 2\frac{1}{4}\text{ ft} + 1\frac{1}{2}\text{ ft} + 1\frac{1}{4}\text{ ft}$

$$P = \frac{9}{4}\text{ ft} + \frac{3}{2}\text{ ft} + \frac{5}{4}\text{ ft}$$
$$P = \frac{9}{4}\text{ ft} + \frac{6}{4}\text{ ft} + \frac{5}{4}\text{ ft}$$
$$P = \frac{9+6+5}{4}\text{ ft}$$
$$P = \frac{20}{4}\text{ ft} = 5\text{ ft}$$

$$A = \frac{1}{2}bh$$
$$A = \frac{1}{2}\left(2\frac{1}{4}\text{ ft}\right)\left(\frac{3}{4}\text{ ft}\right)$$
$$A = \frac{1}{2}\left(\frac{9}{4}\text{ ft}\right)\left(\frac{3}{4}\text{ ft}\right)$$
$$A = \frac{1 \cdot 9 \cdot 3}{2 \cdot 4 \cdot 4}\text{ ft}^2 = \frac{27}{32}\text{ ft}^2$$

5. $P = 9\text{ yd} + 7\text{ yd} + 10\frac{1}{4}\text{ yd}$

$$P = 16\text{ yd} + 10\frac{1}{4}\text{ yd}$$
$$P = 26\frac{1}{4}\text{ yd}$$

$$A = \frac{1}{2}bh$$
$$A = \frac{1}{2}\left(10\frac{1}{4}\text{ yd}\right)(6\text{ yd})$$
$$A = \frac{1}{2}\left(\frac{41}{4}\text{ yd}\right)\left(\frac{6}{1}\text{ yd}\right)$$
$$A = \frac{1 \cdot 41 \cdot \overset{1}{\cancel{2}} \cdot 3}{\underset{1}{\cancel{2}} \cdot 4 \cdot 1}\text{ yd}^2$$
$$A = \frac{123}{4}\text{ yd}^2 \text{ or } 30\frac{3}{4}\text{ yd}^2$$

7. $P = 12\frac{3}{5}\text{ yd} + 7\frac{2}{3}\text{ yd} + 10\text{ yd}$

$$P = \frac{63}{5}\text{ yd} + \frac{23}{3}\text{ yd} + 10\text{ yd}$$
$$P = \frac{189}{15}\text{ yd} + \frac{115}{15}\text{ yd} + \frac{150}{15}\text{ yd}$$
$$P = \frac{454}{15}\text{ yd} \text{ or } 30\frac{4}{15}\text{ yd}$$

$$A = \frac{1}{2}bh$$
$$A = \frac{1}{2}(10\text{ yd})\left(7\frac{2}{3}\text{ yd}\right)$$
$$A = (5\text{ yd})\left(7\frac{2}{3}\text{ yd}\right)$$
$$A = \left(\frac{5}{1}\text{ yd}\right)\left(\frac{23}{3}\text{ yd}\right)$$
$$A = \frac{115}{3}\text{ yd}^2 \text{ or } 38\frac{1}{3}\text{ yd}^2$$

9. The entire figure is shaded.

Area of square $= s^2$

$$A = (12\text{ m})^2$$
$$A = 12\text{ m} \cdot 12\text{ m}$$
$$A = 144\text{ m}^2$$

Area of triangle $= \frac{1}{2}bh$

$$A = \frac{1}{2} \cdot 12\text{ m} \cdot 9\text{ m}$$
$$A = \frac{1}{2} \cdot \frac{12\text{ m}}{1} \cdot \frac{9\text{ m}}{1}$$
$$A = \frac{1 \cdot \overset{1}{\cancel{2}} \cdot 6\text{ m} \cdot 9\text{ m}}{\underset{1}{\cancel{2}} \cdot 1 \cdot 1}$$
$$A = 54\text{ m}^2$$

Total area $= 144\text{ m}^2 + 54\text{ m}^2 = 198\text{ m}^2$

11. Area of rectangle – Area of triangle = Area of shaded portion

$$lw - \frac{1}{2}bh = (37\text{ m})(52\text{ m}) - \frac{1}{2}(52\text{ m})(8\text{ m})$$
$$= 1924\text{ m}^2 - 208\text{ m}^2$$
$$= 1716\text{ m}^2$$

The shaded area is 1716 m^2.

13. **Perimeter** is the distance around the outside edges of a flat shape and is measured in linear units.

Area is the space inside a flat shape and is measured in square units.

Volume is the space inside a solid shape and is measured in cubic units.

15. $A = \frac{1}{2}bh$

$$A = \frac{1}{2}\left(3\frac{1}{2}\text{ ft}\right)\left(4\frac{1}{2}\text{ ft}\right)$$
$$A = \frac{1}{2}\left(\frac{7}{2}\text{ ft}\right)\left(\frac{9}{2}\text{ ft}\right)$$
$$A = \frac{63}{8}\text{ ft}^2 \text{ or } 7\frac{7}{8}\text{ ft}^2$$

$7\frac{7}{8}\text{ ft}^2$ of material is needed.

17. Amount of curbing to go "around" the space implies perimeter.

$$P = 33\text{ m} + 55\text{ m} + 44\text{ m}$$
$$P = 132\text{ m}$$

132 m of curbing will be needed.

Amount of sod to "cover" the space implies area.

The figure is a triangle.

continued

$A = \frac{1}{2}bh$

$A = \frac{1}{2} \cdot 44 \text{ m} \cdot 33 \text{ m}$

$A = \frac{1}{2} \cdot \frac{44 \text{ m}}{1} \cdot \frac{33 \text{ m}}{1}$

$A = \frac{1 \cdot \overset{1}{\cancel{2}} \cdot 22 \text{ m} \cdot 33 \text{ m}}{\underset{1}{\cancel{2}} \cdot 1 \cdot 1}$

$A = 726 \text{ m}^2$

726 m² of sod will be needed.

19. **(a)** The frontage of the lot is

$$100 \text{ yd} + 75 \text{ yd} = 175 \text{ yd}.$$

(b) Draw a vertical line to break the region into a triangle and a rectangle. For the triangular region, the base is 75 yd and the height is 100 yd.

$A = \frac{1}{2} \cdot b \cdot h$
$= \frac{1}{2} \cdot 75 \text{ yd} \cdot 100 \text{ yd}$
$= 3750 \text{ yd}^2$

Rectangular area:

$A = l \cdot w$
$= 100 \text{ yd} \cdot 50 \text{ yd}$
$= 5000 \text{ yd}^2$

Total area $= 3750 \text{ yd}^2 + 5000 \text{ yd}^2 = 8750 \text{ yd}^2$

21. The figure is a rectangular solid.

$V = lwh$
$V = (12 \text{ cm})(4 \text{ cm})(11 \text{ cm})$
$V = 528 \text{ cm}^3$

23. The figure is a rectangular solid or cube.

$V = lwh$

$V = 2\frac{1}{2} \text{ in.} \cdot 2\frac{1}{2} \text{ in.} \cdot 2\frac{1}{2} \text{ in.}$

$V = \frac{5}{2} \text{ in.} \cdot \frac{5}{2} \text{ in.} \cdot \frac{5}{2} \text{ in.}$

$V = \frac{125}{8} \text{ in.}^3 \text{ or } 15\frac{5}{8} \text{ in.}^3$

25. The figure is a pyramid.

First find the area of the rectangular base.

$B = l \cdot w$
$= 8 \text{ cm} \cdot 15 \text{ cm}$
$= 120 \text{ cm}^2$

Now find the volume of the pyramid.

$V = \frac{B \cdot h}{3}$
$= \frac{120 \text{ cm}^2 \cdot 20 \text{ cm}}{3}$
$= 800 \text{ cm}^3$

27. Use the formula for the volume of a pyramid.

First find the area of the square base.

$B = s \cdot s$
$= 8 \text{ ft} \cdot 8 \text{ ft}$
$= 64 \text{ ft}^2$

Now find the volume of the pyramid.

$V = \frac{B \cdot h}{3}$
$= \frac{64 \text{ ft}^2 \cdot 5 \text{ ft}}{3}$
$= \frac{320}{3} \text{ ft}^3 \text{ or } 106\frac{2}{3} \text{ ft}^3$

29. The figure is a rectangular box.

$V = lwh$
$= 8 \text{ in.} \cdot 3 \text{ in.} \cdot \frac{3}{4} \text{ in.}$
$= \frac{8 \text{ in.}}{1} \cdot \frac{3 \text{ in.}}{1} \cdot \frac{3 \text{ in.}}{4}$
$= \frac{2 \cdot \overset{1}{\cancel{4}} \text{ in.} \cdot 3 \text{ in.} \cdot 3 \text{ in.}}{1 \cdot 1 \cdot \underset{1}{\cancel{4}}}$
$= 18 \text{ in.}^3$

31. Use the formula for the volume of a pyramid.

First find the area of the square base.

$B = s \cdot s$
$= 145 \text{ m} \cdot 145 \text{ m}$
$= 21{,}025 \text{ m}^2$

Now find the volume of the pyramid.

$V = \frac{B \cdot h}{3}$
$= \frac{21{,}025 \text{ m}^2 \cdot 93 \text{ m}}{3}$
$= 651{,}775 \text{ m}^3$

The volume of the ancient stone pyramid is 651,775 m³.

33. The volume of the base is

$V = l \cdot w \cdot h$
$= 11 \text{ cm} \cdot 9 \text{ cm} \cdot 3 \text{ cm}$
$= 297 \text{ cm}^3$

The volume of the column is

$V = l \cdot w \cdot h$
$= 9 \text{ cm} \cdot 2 \text{ cm} \cdot 12 \text{ cm}$
$= 216 \text{ cm}^3$

Thus, the volume of shape is

$$297 \text{ cm}^3 + 216 \text{ cm}^3 = 513 \text{ cm}^3.$$

35. The correct answers are $A = 135$ ft^2 (square feet, not squaring 135) and $P = 5\frac{1}{2}$ in. (perimeter is in linear units, not square units).

Chapter 4 Review Exercises

1. 2 of the 5 figures are squares: $\frac{2}{5}$

1 of the 5 figures is a circle: $\frac{1}{5}$

2. Shaded: $\frac{3}{10}$ Unshaded: $\frac{7}{10}$

3. Graph $-\frac{1}{2}$ and $1\frac{1}{2}$ on the number line.

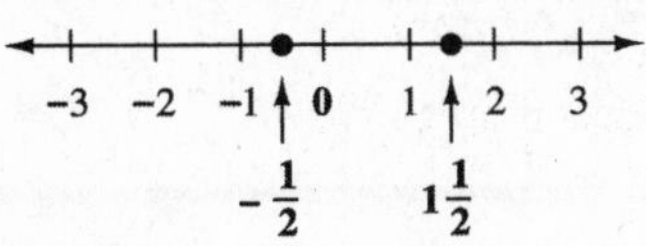

4. **(a)** $-\frac{20}{5} = -\frac{20 \div 5}{5 \div 5} = -\frac{4}{1} = -4$

(b) $\frac{8}{1} = 8 \div 1 = 8$

(c) $-\frac{3}{3} = -3 \div 3 = -1$

5. $\frac{28}{32} = \frac{\cancel{2} \cdot \cancel{2} \cdot 7}{\cancel{2} \cdot \cancel{2} \cdot 2 \cdot 2 \cdot 2} = \frac{7}{8}$

6. $\frac{54}{90} = \frac{\cancel{2} \cdot \cancel{3} \cdot \cancel{3} \cdot 3}{\cancel{2} \cdot \cancel{3} \cdot \cancel{3} \cdot 5} = \frac{3}{5}$

7. $\frac{16}{25} = \frac{2 \cdot 2 \cdot 2 \cdot 2}{5 \cdot 5} = \frac{16}{25}$

The fraction is already in lowest terms.

8. $\frac{15x^2}{40x} = \frac{3 \cdot \cancel{5} \cdot \cancel{x} \cdot x}{2 \cdot 2 \cdot 2 \cdot \cancel{5} \cdot \cancel{x}} = \frac{3x}{8}$

9. $\frac{7a^3}{35a^3b} = \frac{\cancel{7} \cdot \cancel{a} \cdot \cancel{a} \cdot \cancel{a}}{5 \cdot \cancel{7} \cdot \cancel{a} \cdot \cancel{a} \cdot \cancel{a} \cdot b} = \frac{1}{5b}$

10. $\frac{12mn^2}{21m^3n} = \frac{2 \cdot 2 \cdot \cancel{3} \cdot \cancel{m} \cdot \cancel{n} \cdot n}{\cancel{3} \cdot 7 \cdot \cancel{m} \cdot m \cdot m \cdot \cancel{n}} = \frac{4n}{7m^2}$

11. $-\frac{3}{8} \div (-6) = -\frac{3}{8} \div \frac{-6}{1} = -\frac{3}{8} \cdot \frac{1}{-6}$

$= -\frac{3}{8}\left(-\frac{1}{6}\right)$

$= \frac{\cancel{3} \cdot 1}{8 \cdot 2 \cdot \cancel{3}} = \frac{1}{16}$

12. $\frac{2}{5}$ of (-30)

$\frac{2}{5} \cdot (-30) = \frac{2}{5} \cdot -\frac{30}{1} = -\frac{2 \cdot 30}{5 \cdot 1}$

$= -\frac{2 \cdot \cancel{5} \cdot 6}{\cancel{5} \cdot 1} = -\frac{12}{1} = -12$

13. $\frac{4}{9}\left(\frac{2}{3}\right) = \frac{4 \cdot 2}{9 \cdot 3} = \frac{8}{27}$

14. $\left(\frac{7}{3x^3}\right)\left(\frac{x^2}{14}\right) = \frac{\cancel{7} \cdot \cancel{x} \cdot \cancel{x}}{3 \cdot \cancel{x} \cdot \cancel{x} \cdot x \cdot 2 \cdot \cancel{7}} = \frac{1}{6x}$

15. $\frac{ab}{5} \div \frac{b}{10a} = \frac{ab}{5} \cdot \frac{10a}{b} = \frac{a \cdot \cancel{b} \cdot 2 \cdot \cancel{5} \cdot a}{\cancel{5} \cdot \cancel{b}} = 2a^2$

16. $\frac{18}{7} \div 3k = \frac{18}{7} \div \frac{3k}{1}$

$= \frac{18}{7} \cdot \frac{1}{3k}$

$= \frac{6 \cdot \cancel{3} \cdot 1}{7 \cdot \cancel{3} \cdot k}$

$= \frac{6}{7k}$

17. $-\frac{5}{12} + \frac{5}{8}$

Step 1
The LCD is 24.

Step 2
$-\frac{5}{12} = -\frac{5 \cdot 2}{12 \cdot 2} = -\frac{10}{24}$
$\frac{5}{8} = \frac{5 \cdot 3}{8 \cdot 3} = \frac{15}{24}$

Step 3
$-\frac{5}{12} + \frac{5}{8} = -\frac{10}{24} + \frac{15}{24}$
$= \frac{-10 + 15}{24} = \frac{5}{24}$

Step 4
$\frac{5}{24}$ is in lowest terms.

18. $\frac{2}{3} - \frac{4}{5}$

Step 1
The LCD is 15.

Step 2
$\frac{2}{3} = \frac{2 \cdot 5}{3 \cdot 5} = \frac{10}{15}$
$\frac{4}{5} = \frac{4 \cdot 3}{5 \cdot 3} = \frac{12}{15}$

Step 3

$$\frac{2}{3}-\frac{4}{5}=\frac{10}{15}-\frac{12}{15}=\frac{10-12}{15}=-\frac{2}{15}$$

Step 4

$-\frac{2}{15}$ is in lowest terms.

19. $4-\frac{5}{6}=\frac{4}{1}-\frac{5}{6}$

Step 1

The LCD is 6.

Step 2

$$\frac{4}{1}=\frac{4\cdot 6}{1\cdot 6}=\frac{24}{6}$$

Step 3

$$\frac{4}{1}-\frac{5}{6}=\frac{24}{6}-\frac{5}{6}=\frac{24-5}{6}=\frac{19}{6}\text{ or }3\frac{1}{6}$$

Step 4

$\frac{19}{6}$ is in lowest terms.

20. $\frac{7}{9}+\frac{13}{18}$

Step 1

The LCD is 18.

Step 2

$$\frac{7}{9}=\frac{7\cdot 2}{9\cdot 2}=\frac{14}{18}$$

Step 3

$$\frac{7}{9}+\frac{13}{18}=\frac{14}{18}+\frac{13}{18}=\frac{14+13}{18}=\frac{27}{18}$$

Step 4

$$\frac{27}{18}=\frac{3\cdot\overset{1}{\cancel{9}}}{2\cdot\underset{1}{\cancel{9}}}=\frac{3}{2}\text{ or }1\frac{1}{2}$$

21. $\frac{n}{5}+\frac{3}{4}$

Step 1

The LCD is 20.

Step 2

$$\frac{n}{5}=\frac{n\cdot 4}{5\cdot 4}=\frac{4n}{20},\ \frac{3}{4}=\frac{3\cdot 5}{4\cdot 5}=\frac{15}{20}$$

Step 3

$$\frac{n}{5}+\frac{3}{4}=\frac{4n}{20}+\frac{15}{20}=\frac{4n+15}{20}$$

Step 4

$\frac{4n+15}{20}$ is in lowest terms.

22. $\frac{3}{10}-\frac{7}{y}$

Step 1

The LCD is $10\cdot y$ or $10y$.

Step 2

$$\frac{3}{10}=\frac{3\cdot y}{10\cdot y}=\frac{3y}{10y},\ \frac{7}{y}=\frac{7\cdot 10}{y\cdot 10}=\frac{70}{10y}$$

Step 3

$$\frac{3}{10}-\frac{7}{y}=\frac{3y}{10y}-\frac{70}{10y}=\frac{3y-70}{10y}$$

Step 4

$\frac{3y-70}{10y}$ is in lowest terms.

23. $2\frac{1}{4}$ rounds to 2. $1\frac{5}{8}$ rounds to 2.

Estimate: $2\div 2=1$

Exact: $2\frac{1}{4}\div 1\frac{5}{8}=\frac{9}{4}\div\frac{13}{8}=\frac{9}{4}\cdot\frac{8}{13}$

$$=\frac{9\cdot 2\cdot\overset{1}{\cancel{4}}}{\underset{1}{\cancel{4}}\cdot 13}=\frac{18}{13}=1\frac{5}{13}$$

24. $7\frac{1}{3}$ rounds to 7. $4\frac{5}{6}$ rounds to 5.

Estimate: $7-5=2$

Exact: $7\frac{1}{3}-4\frac{5}{6}=\frac{22}{3}-\frac{29}{6}=\frac{44}{6}-\frac{29}{6}$

$$=\frac{15}{6}=\frac{\overset{1}{\cancel{3}}\cdot 5}{\underset{1}{\cancel{3}}\cdot 2}=\frac{5}{2}=2\frac{1}{2}$$

25. $1\frac{3}{4}$ rounds to 2. $2\frac{3}{10}$ rounds to 2.

Estimate: $2+2=4$

Exact: $1\frac{3}{4}+2\frac{3}{10}=\frac{7}{4}+\frac{23}{10}=\frac{35}{20}+\frac{46}{20}$

$$=\frac{81}{20}=4\frac{1}{20}$$

26.

$$\left(-\frac{3}{4}\right)^3=-\frac{3}{4}\left(-\frac{3}{4}\right)\left(-\frac{3}{4}\right)=\frac{9}{16}\left(-\frac{3}{4}\right)=-\frac{27}{64}$$

27.

$$\left(\frac{2}{3}\right)^2\left(-\frac{1}{2}\right)^4=\frac{2}{3}\cdot\frac{2}{3}\left(-\frac{1}{2}\right)\left(-\frac{1}{2}\right)\left(-\frac{1}{2}\right)\left(-\frac{1}{2}\right)=\frac{\overset{1}{\cancel{2}}\cdot\overset{1}{\cancel{2}}}{3\cdot 3\cdot\underset{1}{\cancel{2}}\cdot\underset{1}{\cancel{2}}\cdot 2\cdot 2}=\frac{1}{36}$$

28. $\frac{2}{5}+\frac{3}{10}(-4)=\frac{2}{5}+\frac{3}{10}\cdot\left(\frac{-4}{1}\right)$
$=\frac{2}{5}+\left(-\frac{12}{10}\right)$
$=\frac{2}{5}+\left(-\frac{6}{5}\right)=\frac{2+(-6)}{5}$
$=\frac{-4}{5}$ or $-\frac{4}{5}$

29. $-\frac{5}{8}\div\left(-\frac{1}{2}\right)\left(\frac{14}{15}\right)$
$=-\frac{5}{8}\div\left(-\frac{1}{2}\right)\cdot\left(\frac{14}{15}\right)$ Change division to multiplication.
$=-\frac{5}{8}\left(-\frac{2}{1}\right)\left(\frac{14}{15}\right)$ Multiply.
$=\frac{\overset{1}{\cancel{5}}\cdot\overset{1}{\cancel{2}}\cdot\overset{1}{\cancel{2}}\cdot 7}{\underset{1}{\cancel{2}}\cdot\underset{1}{\cancel{2}}\cdot 2\cdot 3\cdot\underset{1}{\cancel{5}}}$
$=\frac{7}{6}=1\frac{1}{6}$

30. $\frac{\frac{5}{8}}{\frac{1}{16}}=\frac{5}{8}\div\frac{1}{16}=\frac{5}{8}\cdot\frac{16}{1}$
$=\frac{5\cdot 2\cdot\overset{1}{\cancel{8}}}{\underset{1}{\cancel{8}}\cdot 1}$
$=\frac{10}{1}=10$

31. $\frac{\frac{8}{9}}{-6}=\frac{8}{9}\div(-6)=\frac{8}{9}\div\left(-\frac{6}{1}\right)$
$=\frac{8}{9}\left(-\frac{1}{6}\right)$
$=-\frac{\overset{1}{\cancel{2}}\cdot 4\cdot 1}{9\cdot\underset{1}{\cancel{2}}\cdot 3}=-\frac{4}{27}$

32. $-12=-\frac{3}{5}w$
$-\frac{5}{3}\left(-\frac{12}{1}\right)=-\frac{5}{3}\left(-\frac{3}{5}w\right)$
$-\frac{5}{\underset{1}{\cancel{3}}}\left(-\frac{\overset{1}{\cancel{3}}\cdot 4}{1}\right)=-\frac{\overset{1}{\cancel{5}}}{\underset{1}{\cancel{3}}}\left(-\frac{\overset{1}{\cancel{3}}}{\underset{1}{\cancel{5}}}\right)w$
$20=w$

The solution is 20.

33. $18+\frac{6}{5}r=0$ Add -18 to both sides.
$\underline{-18\qquad\quad -18}$ Multiply both sides by $\frac{5}{6}$.
$0+\frac{6}{5}r=-18$

$\frac{5}{6}\cdot\left(\frac{6}{5}r\right)=\frac{5}{6}(-18)$
$r=-15$

The solution is -15.

34. $3x-\frac{2}{3}=\frac{5}{6}$
$\underline{\quad+\frac{2}{3}\qquad+\frac{2}{3}}$
$3x+0=\frac{5}{6}+\frac{2}{3}$ $\qquad\frac{5}{6}+\frac{2}{3}=\frac{5}{6}+\frac{4}{6}=\frac{9}{6}$
$3x=\frac{9}{6}$ $\qquad\frac{9}{6}=\frac{\overset{1}{\cancel{3}}\cdot 3}{\underset{1}{\cancel{3}}\cdot 2}=\frac{3}{2}$
$\frac{1}{3}(3x)=\frac{1}{3}\left(\frac{3}{2}\right)$
$x=\frac{1}{2}$

The solution is $\frac{1}{2}$.

35. $A=\frac{1}{2}bh$
$A=\frac{1}{2}\cdot 8\text{ ft}\cdot 3\frac{1}{2}\text{ ft}$
$A=\frac{1}{2}\cdot\frac{8}{1}\text{ ft}\cdot\frac{7}{2}\text{ ft}$
$A=\frac{1\cdot\overset{1}{\cancel{2}}\cdot\overset{1}{\cancel{2}}\cdot 2\cdot 7}{\underset{1}{\cancel{2}}\cdot 1\cdot\underset{1}{\cancel{2}}}\text{ ft}^2$
$A=14\text{ ft}^2$

36. The figure is a rectangular solid.
$V=l\cdot w\cdot h$
$V=4\text{ in.}\cdot 3\frac{1}{4}\text{ in.}\cdot 2\frac{1}{2}\text{ in.}$
$V=\frac{4}{1}\text{ in.}\cdot\frac{13}{4}\text{ in.}\cdot\frac{5}{2}\text{ in.}$
$V=\frac{\overset{1}{\cancel{4}}\cdot 13\cdot 5}{1\cdot\underset{1}{\cancel{4}}\cdot 2}\text{ in.}^3$
$V=\frac{65}{2}\text{ in.}^3$ or $32\frac{1}{2}\text{ in.}^3$

37. The figure is a pyramid.
The base is a rectangle, so the area of the base is:
$B=l\cdot w$
$B=8\text{ m}\cdot 5\text{ m}$
$B=40\text{ m}^2$

$V=\frac{1}{3}B\cdot h$
$V=\frac{1}{3}\cdot\frac{40\text{ m}^2}{1}\cdot\frac{7\text{ m}}{1}$
$V=\frac{280}{3}\text{ m}^3$ or $93\frac{1}{3}\text{ m}^3$

38. **[4.3]** The $2\frac{1}{2}$ pounds of meat will be used to make 10 servings:

"$2\frac{1}{2}$ pounds divided or separated into 10 servings"

$$2\frac{1}{2} \div 10 = \frac{5}{2} \div \frac{10}{1} = \frac{5}{2} \cdot \frac{1}{10} = \frac{\overset{1}{\cancel{5}} \cdot 1}{2 \cdot 2 \cdot \underset{1}{\cancel{5}}} = \frac{1}{4}$$

$\frac{1}{4}$ pound will be used in each serving.

For 30 servings:

$$30\left(\frac{1}{4}\right) = \frac{30}{1} \cdot \frac{1}{4} = \frac{\overset{1}{\cancel{2}} \cdot 15 \cdot 1}{1 \cdot \underset{1}{\cancel{2}} \cdot 2} = \frac{15}{2} = 7\frac{1}{2}$$

You would need $7\frac{1}{2}$ pounds of meat.

39. **[4.4]** To find out how much longer she worked on Monday than on Friday, use subtraction.

$$4\frac{1}{2} - 3\frac{2}{3} = \frac{9}{2} - \frac{11}{3} \quad \text{LCD is 6.}$$
$$= \frac{27}{6} - \frac{22}{6}$$
$$= \frac{5}{6} \text{ hour}$$

She worked $\frac{5}{6}$ hour longer on Monday than on Friday.

To find how many hours she worked in all, use addition.

$$4\frac{1}{2} + 2\frac{3}{4} + 3\frac{2}{3}$$
$$= \frac{9}{2} + \frac{11}{4} + \frac{11}{3} \quad \text{LCD is 12.}$$
$$= \frac{54}{12} + \frac{33}{12} + \frac{44}{12}$$
$$= \frac{54 + 33 + 44}{12} = \frac{131}{12} = 10\frac{11}{12} \text{ hours}$$

She worked a total of $10\frac{11}{12}$ hours.

40. **[4.2]** 60 total children is the "whole."

$\frac{1}{5}$ of the "whole" are preschoolers:

$$\frac{1}{5} \cdot 60 = \frac{1}{5} \cdot \frac{60}{1} = \frac{60}{5} = 12 \text{ preschoolers}$$

$\frac{2}{3}$ of the "whole" are toddlers:

$$\frac{2}{3} \cdot 60 = \frac{2}{3} \cdot \frac{60}{1} = \frac{120}{3} = 40 \text{ toddlers}$$

"The rest" are infants; use subtraction:

$$60 - 12 - 40 = 8 \text{ infants}$$

41. **[4.4]** $P = 2l + 2w$

$$P = \frac{2}{1} \cdot \frac{3}{4} \text{ mi} + \frac{2}{1} \cdot \frac{3}{10} \text{ mi}$$
$$P = \frac{3}{2} \text{ mi} + \frac{3}{5} \text{ mi}$$
$$P = \frac{15}{10} \text{ mi} + \frac{6}{10} \text{ mi}$$
$$P = \frac{21}{10} \text{ miles or } 2\frac{1}{10} \text{ miles}$$

$$A = l \cdot w$$
$$A = \frac{3}{4} \text{ mi} \cdot \frac{3}{10} \text{ mi}$$
$$A = \frac{9}{40} \text{ mi}^2$$

Chapter 4 Test

1. Shaded: $\frac{5}{6}$ Unshaded: $\frac{1}{6}$

2. Graph $-\frac{2}{3}$ and $2\frac{1}{3}$ on the number line.

$-\frac{2}{3}$ $2\frac{1}{3}$

-3 -2 -1 0 1 2 3

3. $$\frac{21}{84} = \frac{\overset{1}{\cancel{3}} \cdot \overset{1}{\cancel{7}}}{2 \cdot 2 \cdot \underset{1}{\cancel{3}} \cdot \underset{1}{\cancel{7}}} = \frac{1}{4}$$

4. $\frac{25}{54} = \frac{5 \cdot 5}{2 \cdot 3 \cdot 3 \cdot 3}$ is already in lowest terms.

5. $$\frac{6a^2b}{9b^2} = \frac{2 \cdot \overset{1}{\cancel{3}} \cdot a \cdot a \cdot \overset{1}{\cancel{b}}}{\underset{1}{\cancel{3}} \cdot 3 \cdot \underset{1}{\cancel{b}} \cdot b} = \frac{2a^2}{3b}$$

6. $$\frac{1}{6} + \frac{7}{10} \quad \text{LCD is 30}$$
$$= \frac{5}{30} + \frac{21}{30} = \frac{5 + 21}{30} = \frac{26}{30} = \frac{13}{15}$$

7. $$-\frac{3}{4} \div \frac{3}{8} = -\frac{3}{4} \cdot \frac{8}{3} = -\frac{\overset{1}{\cancel{3}} \cdot 2 \cdot \overset{1}{\cancel{4}}}{\underset{1}{\cancel{4}} \cdot \underset{1}{\cancel{3}}}$$
$$= -\frac{2}{1} = -2$$

8. $$\frac{5}{8} - \frac{4}{5} = \frac{25}{40} - \frac{32}{40} = \frac{25 - 32}{40}$$
$$= \frac{-7}{40} \text{ or } -\frac{7}{40}$$

9. $$(-20)\left(-\frac{7}{10}\right) = -\frac{20}{1}\left(-\frac{7}{10}\right) = \frac{2 \cdot \overset{1}{\cancel{10}} \cdot 7}{1 \cdot \underset{1}{\cancel{10}}}$$
$$= \frac{14}{1} = 14$$

10. $\frac{\frac{4}{9}}{-6} = \frac{4}{9} \div (-6) = \frac{4}{9} \div \left(-\frac{6}{1}\right)$

$= \frac{4}{9}\left(-\frac{1}{6}\right) = -\frac{\overset{1}{\cancel{2}} \cdot 2 \cdot 1}{9 \cdot \underset{1}{\cancel{2}} \cdot 3}$

$= -\frac{2}{27}$

11. $4 - \frac{7}{8} = \frac{4}{1} - \frac{7}{8} = \frac{32}{8} - \frac{7}{8} = \frac{32-7}{8}$

$= \frac{25}{8}$ or $3\frac{1}{8}$

12. $-\frac{2}{9} + \frac{2}{3} = -\frac{2}{9} + \frac{6}{9} = \frac{-2+6}{9} = \frac{4}{9}$

13. $\frac{21}{24}\left(\frac{9}{14}\right) = \frac{\overset{1}{\cancel{3}} \cdot \overset{1}{\cancel{7}} \cdot 3 \cdot 3}{\underset{1}{\cancel{3}} \cdot 8 \cdot 2 \cdot \underset{1}{\cancel{7}}} = \frac{9}{16}$

14. $\frac{12x}{7y} \div 3x = \frac{12x}{7y} \div \frac{3x}{1} = \frac{12x}{7y} \cdot \frac{1}{3x}$

$= \frac{\overset{1}{\cancel{3}} \cdot 4 \cdot \overset{1}{\cancel{x}} \cdot 1}{7 \cdot y \cdot \underset{1}{\cancel{3}} \cdot \underset{1}{\cancel{x}}} = \frac{4}{7y}$

15. $\frac{6}{n} - \frac{1}{4} = \frac{24}{4n} - \frac{n}{4n} = \frac{24-n}{4n}$

16. $\frac{2}{3} + \frac{a}{5} = \frac{10}{15} + \frac{3a}{15} = \frac{10+3a}{15}$

17. $\left(\frac{5}{9b^2}\right)\left(\frac{b}{10}\right) = \frac{\overset{1}{\cancel{5}} \cdot \overset{1}{\cancel{b}}}{3 \cdot 3 \cdot \underset{1}{\cancel{b}} \cdot b \cdot 2 \cdot \underset{1}{\cancel{5}}} = \frac{1}{18b}$

18. $\left(-\frac{1}{2}\right)^3\left(\frac{2}{3}\right)^2 = -\frac{1}{2}\left(-\frac{1}{\underset{1}{\cancel{2}}}\right)\left(-\frac{1}{\underset{1}{\cancel{2}}}\right) \cdot \frac{\overset{1}{\cancel{2}}}{3} \cdot \frac{\overset{1}{\cancel{2}}}{3}$

$= -\frac{1}{18}$

19. $\frac{1}{6} + 4\left(\frac{2}{5} - \frac{7}{10}\right)$ Parentheses

$= \frac{1}{6} + 4\left(\frac{4}{10} - \frac{7}{10}\right)$

$= \frac{1}{6} + 4\left(\frac{-3}{10}\right)$

$= \frac{1}{6} + \frac{4}{1} \cdot \frac{-3}{10}$ $\quad \frac{4}{1} \cdot \frac{-3}{10} = \frac{\overset{1}{\cancel{2}} \cdot 2 \cdot (-3)}{1 \cdot \underset{1}{\cancel{2}} \cdot 5} = \frac{-6}{5}$

$= \frac{1}{6} + \frac{-6}{5}$

$= \frac{5}{30} + \frac{-36}{30}$

$= \frac{5 + (-36)}{30}$

$= -\frac{31}{30}$ or $-1\frac{1}{30}$

20. $4\frac{4}{5}$ rounds to 5. $1\frac{1}{8}$ rounds to 1.

Estimate: $5 \div 1 = 5$

Exact: $4\frac{4}{5} \div 1\frac{1}{8} = \frac{24}{5} \div \frac{9}{8} = \frac{24}{5} \cdot \frac{8}{9}$

$= \frac{\overset{1}{\cancel{3}} \cdot 8 \cdot 8}{5 \cdot \underset{1}{\cancel{3}} \cdot 3}$

$= \frac{64}{15}$ or $4\frac{4}{15}$

21. $3\frac{2}{5}$ rounds to 3. $1\frac{9}{10}$ rounds to 2.

Estimate: $3 - 2 = 1$

Exact: $3\frac{2}{5} - 1\frac{9}{10} = \frac{17}{5} - \frac{19}{10} = \frac{34}{10} - \frac{19}{10}$

$= \frac{15}{10} = \frac{3}{2}$ or $1\frac{1}{2}$

22.

$7 = \frac{1}{5}d$

$\frac{5}{1}(7) = \frac{5}{1}\left(\frac{1}{5}d\right)$

$35 = d$

The solution is 35.

23.

$-\frac{3}{10}t = \frac{9}{14}$

$-\frac{10}{3}\left(-\frac{3}{10}t\right) = -\frac{10}{3}\left(\frac{9}{14}\right)$

$t = -\frac{\overset{1}{\cancel{2}} \cdot 5 \cdot \overset{1}{\cancel{3}} \cdot 3}{\underset{1}{\cancel{3}} \cdot \underset{1}{\cancel{2}} \cdot 7}$

$t = -\frac{15}{7}$ or $-2\frac{1}{7}$

The solution is $-\frac{15}{7}$.

24.

$0 = \frac{1}{4}b - 2$ Add 2 to both sides.

$\underline{+2 \qquad\quad +2}$

$2 = \frac{1}{4}b + 0$

$\frac{4}{1}(2) = \frac{4}{1}\left(\frac{1}{4}b\right)$

$8 = b$

The solution is 8.

25.

$\frac{4}{3}x + 7 = -13$ Add -7 to both sides.

$\underline{\qquad -7 \qquad -7}$

$\frac{4}{3}x + 0 = -20$

$\frac{3}{4}\left(\frac{4}{3}x\right) = \frac{3}{4}(-20)$

$x = -15$

The solution is -15.

26. $A = \frac{1}{2}bh$

$A = \frac{1}{2} \cdot 13 \text{ m} \cdot 8 \text{ m}$

$A = \frac{1}{2} \cdot \frac{13 \text{ m}}{1} \cdot \frac{8 \text{ m}}{1}$

$A = \frac{13 \cdot \overset{1}{\cancel{2}} \cdot 4}{\underset{1}{\cancel{2}}} \text{ m}^2$

$A = 52 \text{ m}^2$

27. $A = \frac{1}{2}bh$

$A = \frac{1}{2} \cdot 9 \text{ yd} \cdot 13 \text{ yd}$

$A = \frac{1}{2} \cdot \frac{9 \text{ yd}}{1} \cdot \frac{13 \text{ yd}}{1}$

$A = \frac{117}{2} \text{ or } 58\frac{1}{2} \text{ yd}^2$

28. The figure is a rectangular solid.

$V = l \cdot w \cdot h$

$V = 30 \text{ m} \cdot 18 \text{ m} \cdot 12 \text{ m}$

$V = 6480 \text{ m}^3$

29. The figure is a pyramid.

The base is a rectangle.
The area of the base is:

$B = l \cdot w$

$B = 4 \text{ yd} \cdot 3 \text{ yd}$

$B = 12 \text{ yd}^2$

$V = \frac{1}{3}Bh$

$V = \frac{1}{3} \cdot 12 \text{ yd}^2 \cdot 4 \text{ yd}$

$V = \frac{48}{3} \text{ yd}^3$

$V = 16 \text{ yd}^3$

30. To find her total training hours, use addition.

$4\frac{5}{6} + 6\frac{2}{3} + 3\frac{1}{4}$

$= \frac{29}{6} + \frac{20}{3} + \frac{13}{4}$ LCD is 12

$= \frac{58}{12} + \frac{80}{12} + \frac{39}{12}$

$= \frac{177}{12} = \frac{59}{4} \text{ or } 14\frac{3}{4}$

She spent $14\frac{3}{4}$ hours training.

Use subtraction to find how many more hours she trained on Tuesday than on Monday.

$6\frac{2}{3} - 4\frac{5}{6}$

$= \frac{20}{3} - \frac{29}{6}$ LCD is 6

$= \frac{40}{6} - \frac{29}{6}$

$= \frac{11}{6} \text{ or } 1\frac{5}{6}$

She spent $1\frac{5}{6}$ hours more on Tuesday than on Monday.

31. $8\frac{3}{4}$ ounces can be synthesized in how many days if $2\frac{1}{2}$ ounces can be synthesized in a day?

Estimation will help. 9 ounces is needed. 3 ounces can be synthesized per day.

$9 \div 3 = 3$ days

Exact: $8\frac{3}{4} \div 2\frac{1}{2}$

$= \frac{35}{4} \div \frac{5}{2}$

$= \frac{35}{4} \cdot \frac{2}{5}$

$= \frac{\overset{1}{\cancel{5}} \cdot 7 \cdot \overset{1}{\cancel{2}}}{\underset{1}{\cancel{2}} \cdot 2 \cdot \underset{1}{\cancel{5}}}$

$= \frac{7}{2} \text{ days or } 3\frac{1}{2} \text{ days}$

32. $\frac{7}{8}$ of the 8448 students work:

$\frac{7}{8} \cdot 8448 = \frac{7 \cdot \overset{1}{\cancel{8}} \cdot 1056}{\underset{1}{\cancel{8}}} = 7392$

7392 students work.

Cumulative Review Exercises (Chapters 1–4)

1. In words, 505,008,238 is five hundred five million, eight thousand, two hundred thirty-eight.

2. Thirty-five billion, six hundred million, nine hundred sixteen can be written as 35,600,000,916.

3. **(a)** 60,719

Underline the hundreds place: 60,$\underline{7}$19
The next digit is 4 or less. Leave the 7, change 1 and 9 to 0. **60,700**

(b) 99,505

Underline the thousands place: 9$\underline{9}$,505
The next digit is 5 or more, so add one to 9.
Write 0 and regroup 1 to the ten-thousands place.
Write 0 and regroup 1 to the hundred-thousands place. Change 5 and 5 to 0. **100,000**

(c) 3206

Underline the tens place: 32<u>0</u>6
The next digit is 5 or more, so add one to 0.
Change 6 to 0. **3210**

4. **(a)** $-3 \cdot 6 = 6 \cdot (-3)$

Commutative property of multiplication

(b) $(7+18)+2 = 7+(18+2)$

Associative property of addition

(c) $5(-10+7) = 5 \cdot (-10) + 5 \cdot 7$

Distributive property

5. $9(-6) = -54$

6. $-10-10$
$= -10+(-10)$
$= -20$

7. $\frac{-14}{0}$ is undefined.

8. $6+3(2-7)^2$ Parentheses first
$= 6+3[2+(-7)]^2$
$= 6+3(-5)^2$ Exponents next
$= 6+3 \cdot 25$ Multiply.
$= 6+75$ Add.
$= 81$

9. $|-8|+|2|$
$= 8+2$
$= 10$

10. $-8+24 \div 2$ Division first
$= -8+12$
$= 4$

11. $(-4)^2 - 2^5$ Exponents
$= -4(-4) - 2 \cdot 2 \cdot 2 \cdot 2 \cdot 2$
$= 16-32$
$= -16$

12. $\dfrac{-45 \div (-5)(3)}{-5-4(0-8)}$

Numerator:

$-45 \div (-5)(3) = 9(3) = 27$

Denominator:

$-5-4(0-8)$
$= -5-4(-8)$
$= -5-(-32)$
$= -5+32$
$= 27$

Last step is division: $\frac{27}{27} = 1$

13. $-3 \div \frac{3}{8} = -\frac{3}{1} \div \frac{3}{8} = -\frac{3}{1} \cdot \frac{8}{3} = -\frac{\overset{1}{\cancel{3}} \cdot 8}{1 \cdot \underset{1}{\cancel{3}}}$
$= -\frac{8}{1} = -8$

14. $-\frac{5}{6}(-42) = -\frac{5}{6}\left(-\frac{42}{1}\right) = \frac{5 \cdot \overset{1}{\cancel{6}} \cdot 7}{\underset{1}{\cancel{6}} \cdot 1}$
$= \frac{35}{1} = 35$

15. $\left(\frac{5a^2}{12}\right)\left(\frac{18}{a}\right) = \frac{5 \cdot a \cdot \overset{1}{\cancel{a}} \cdot 3 \cdot \overset{1}{\cancel{6}}}{2 \cdot \underset{1}{\cancel{6}} \cdot \underset{1}{\cancel{a}}} = \frac{15a}{2}$

16. $\frac{7}{x^2} \div \frac{7y^2}{3x} = \frac{7}{x^2} \cdot \frac{3x}{7y^2} = \frac{\overset{1}{\cancel{7}} \cdot 3 \cdot \overset{1}{\cancel{x}}}{x \cdot \underset{1}{\cancel{x}} \cdot \underset{1}{\cancel{7}} \cdot y^2}$
$= \frac{3}{xy^2}$

17. $\frac{3}{10} - \frac{5}{6} = \frac{9}{30} - \frac{25}{30} = \frac{9-25}{30}$
$= \frac{-16}{30} = \frac{-16 \div 2}{30 \div 2}$
$= \frac{-8}{15}$ or $-\frac{8}{15}$

18. $-\frac{3}{8} + \frac{11}{16} = -\frac{6}{16} + \frac{11}{16}$
$= \frac{-6+11}{16} = \frac{5}{16}$

19. $\frac{2}{3} - \frac{b}{7} = \frac{14}{21} - \frac{3b}{21} = \frac{14-3b}{21}$

20. $\frac{8}{5} + \frac{3}{n} = \frac{8n}{5n} + \frac{15}{5n} = \frac{8n+15}{5n}$

21. $3\frac{1}{4} \div 2\frac{1}{4} = \frac{13}{4} \div \frac{9}{4} = \frac{13}{4} \cdot \frac{4}{9}$
$= \frac{13 \cdot \overset{1}{\cancel{4}}}{\underset{1}{\cancel{4}} \cdot 9} = \frac{13}{9}$ or $1\frac{4}{9}$

22. $2\frac{2}{5} - 1\frac{3}{4} = \frac{12}{5} - \frac{7}{4} = \frac{48}{20} - \frac{35}{20}$
$= \frac{48-35}{20} = \frac{13}{20}$

23. $\left(\frac{1}{2}\right)^3 (-2)^3$
$= \frac{1}{2} \cdot \frac{1}{2} \cdot \frac{1}{2}(-2)(-2)(-2)$
$= \frac{1}{8}(-8)$
$= \frac{1}{8}\left(\frac{-8}{1}\right)$
$= \frac{-8}{8} = -1$

24. $\dfrac{\frac{7}{12}}{-\frac{14}{15}} = \dfrac{7}{12} \div \left(-\dfrac{14}{15}\right) = \dfrac{7}{12} \cdot \left(-\dfrac{15}{14}\right)$

$= -\dfrac{\overset{1}{\cancel{7}} \cdot \overset{1}{\cancel{3}} \cdot 5}{\underset{1}{\cancel{3}} \cdot 4 \cdot 2 \cdot \underset{1}{\cancel{7}}}$

$= -\dfrac{5}{8}$

25. $-5a + 2a = 12 - 15$

$-5a + 2a = 12 + (-15)$

$-3a = -3$ Divide both sides by -3.

$\dfrac{-3a}{-3} = \dfrac{-3}{-3}$

$a = 1$

The solution is 1.

26.
$$\begin{aligned} y &= -7y - 40 \\ 1y &= -7y + (-40) \\ +7y &\quad +7y \\ 8y &= 0 + (-40) \\ \frac{8y}{8} &= \frac{-40}{8} \\ y &= -5 \end{aligned}$$

The solution is -5.

27.
$$\begin{aligned} -10 &= 6 + \frac{4}{9}k \\ -6 &\quad -6 \\ -16 &= 0 + \frac{4}{9}k \\ \frac{9}{4}(-16) &= \frac{9}{4}\left(\frac{4}{9}k\right) \\ -36 &= k \end{aligned}$$

The solution is -36.

28.
$$\begin{aligned} 20 + 5x &= -2x - 8 \\ +2x &\quad +2x \\ 20 + 7x &= 0 - 8 \\ 20 + 7x &= -8 \\ -20 &\quad -20 \\ 0 + 7x &= -28 \\ \frac{7x}{7} &= \frac{-28}{7} \\ x &= -4 \end{aligned}$$

The solution is -4.

29.
$$\begin{aligned} 3(-2m + 5) &= -4m + 9 + 1m \\ -6m + 15 &= -3m + 9 \\ +6m &\quad +6m \\ 0 + 15 &= 3m + 9 \\ 15 &= 3m + 9 \\ -9 &\quad -9 \\ 6 &= 3m + 0 \\ \frac{6}{3} &= \frac{3m}{3} \\ 2 &= m \end{aligned}$$

The solution is 2.

30. The figure is a square.

$P = 4s$

$P = 4 \cdot 4\frac{1}{2}$ in.

$P = \frac{4}{1} \cdot \frac{9}{2}$ in.

$P = 18$ in.

$A = s^2$

$A = s \cdot s$

$A = 4\frac{1}{2}$ in. $\cdot 4\frac{1}{2}$ in.

$A = \frac{9}{2}$ in. $\cdot \frac{9}{2}$ in.

$A = \frac{81}{4}$ in.^{2} or $20\frac{1}{4}$ in.^{2}

31. The figure is a triangle.

$P = 6\frac{1}{4}$ yd $+ 6\frac{1}{4}$ yd $+ 7\frac{1}{2}$ yd

$P = \frac{25}{4}$ yd $+ \frac{25}{4}$ yd $+ \frac{15}{2}$ yd

$P = \frac{25}{4}$ yd $+ \frac{25}{4}$ yd $+ \frac{30}{4}$ yd

$P = \frac{25 + 25 + 30}{4}$ yd

$P = \frac{80}{4}$ yd $= 20$ yd

$A = \frac{1}{2}bh$

$A = \frac{1}{2} \cdot 7\frac{1}{2}$ yd $\cdot 5$ yd

$A = \frac{1}{2} \cdot \frac{15}{2}$ yd $\cdot \frac{5}{1}$ yd

$A = \frac{75}{4}$ yd^2 or $18\frac{3}{4}$ yd^2

32. The figure is a parallelogram.

$P = 15$ mm $+ 24$ mm $+ 15$ mm $+ 24$ mm

$P = 78$ mm

$A = bh$

$A = 24$ mm $\cdot 12$ mm

$A = 288$ mm^2

33. *Step 1*
Unknown: Weight of each bag

Known: Each bag weighs the same, 16 pounds of seed used, 25 pounds of seed used, 79 pounds of seed left

Step 2
Let b be the original weight of each bag.

Step 3

Original weight of 3 bags	–	pounds of seed used	=	pounds of seed remaining
$3b$	–	$(16+25)$	=	79

Step 4

$$3b-(16+25)=79$$
$$3b-41=79$$
$$+41 \quad +41$$
$$3b+0=120$$
$$\frac{3b}{3}=\frac{120}{3}$$
$$b=40$$

Step 5
Each bag originally weighed 40 pounds.

Step 6
First bag: $40-16=24$ pounds
Second bag: $40-25=15$ pounds
Third bag: 40 pounds
The sum is $24+15+40=79$ pounds, which is the amount of seed left.

34. *Step 1*
Unknowns: Length and width

Known: Perimeter is 102 yd, lot is twice as long as it is wide.

Step 2
Let w represent the width.
Then $2w$ represents the length.

Step 3
$P=2l+2w$

Step 4

$$102\text{ yd}=2(2w)+2(w)$$
$$102\text{ yd}=4w+2w$$
$$102\text{ yd}=6w$$
$$\frac{102\text{ yd}}{6}=\frac{6w}{6}$$
$$17\text{ yd}=w$$

Step 5
The width is 17 yd and the length is $2(17)=34$ yd.

Step 6
34 yd is twice as long as 17 yd.

$$P=2l+2w$$
$$P=2\cdot 34\text{ yd}+2\cdot 17\text{ yd}$$
$$P=68\text{ yd}+34\text{ yd}$$
$$P=102\text{ yd}$$

CHAPTER 5 RATIONAL NUMBERS: POSITIVE AND NEGATIVE DECIMALS

5.1 Reading and Writing Decimal Numbers

5.1 Section Exercises

1. **(a)** The figure has 10 equal parts; 1 part is shaded.

$\frac{1}{10}$; 0.1; one tenth

(b) The figure has 10 equal parts; 3 parts are shaded.

$\frac{3}{10}$; 0.3; three tenths

(c) The figure has 10 equal parts; 9 parts are shaded.

$\frac{9}{10}$; 0.9; nine tenths

3.

tens	ones		tenths		
7	0	.	4	8	9

5.

			hundredths	thousandths	ten-thousandths
0	.	2	5	1	8

7.

			tenths		thousandths	ten-thousandths	
9	3	.	0	1	4	7	2

9.

hundreds	tens			tenths		
3	1	4	.	6	5	8

11.

hundreds		ones			hundredths		
1	4	9	.	0	8	3	2

13. 0 ones, 5 hundredths, 1 ten, 4 hundreds, 2 tenths:

410.25

15. 3 thousandths, 4 hundredths, 6 ones, 2 ten-thousandths, 5 tenths:
6.5432

17. 4 hundredths, 4 hundreds, 0 tens, 0 tenths, 5 thousandths, 5 thousands, 6 ones:
5406.045

19. $0.7 = \frac{7}{10}$

21. $13.4 = 13\frac{4}{10} = 13\frac{4 \div 2}{10 \div 2} = 13\frac{2}{5}$

23. $0.35 = \frac{35}{100} = \frac{35 \div 5}{100 \div 5} = \frac{7}{20}$

25. $0.66 = \frac{66}{100} = \frac{66 \div 2}{100 \div 2} = \frac{33}{50}$

27. $10.17 = 10\frac{17}{100}$

29. $0.06 = \frac{6}{100} = \frac{6 \div 2}{100 \div 2} = \frac{3}{50}$

31. $0.205 = \frac{205}{1000} = \frac{205 \div 5}{1000 \div 5} = \frac{41}{200}$

33. $5.002 = 5\frac{2}{1000} = 5\frac{2 \div 2}{1000 \div 2} = 5\frac{1}{500}$

35. $0.686 = \frac{686}{1000} = \frac{686 \div 2}{1000 \div 2} = \frac{343}{500}$

37. 0.5 is five tenths.

39. 0.78 is seventy-eight hundredths.

41. 0.105 is one hundred five thousandths.

43. 12.04 is twelve and four hundredths.

45. 1.075 is one and seventy-five thousandths.

47. Six and seven tenths:

$6\frac{7}{10} = 6.7$

49. Thirty-two hundredths:

$\frac{32}{100} = 0.32$

51. Four hundred twenty and eight thousandths:

$420\frac{8}{1000} = 420.008$

53. Seven hundred three ten-thousandths:

$\frac{703}{10,000} = 0.0703$

55. Seventy-five and thirty thousandths:

$75\frac{30}{1000} = 75.030$

57. Anne should not say "and" because that denotes a decimal point.

59. 8-pound test line has a diameter of 0.010 inch. 0.010 inch is read ten thousandths of an inch.

$$0.010 = \frac{10}{1000} = \frac{10 \div 10}{1000 \div 10} = \frac{1}{100} \text{ inch}$$

61. $\frac{13}{1000}$ inch $= 0.013$ inch

A diameter of 0.013 inch corresponds to a test strength of 12 pounds.

63. Six tenths is 0.6, so the correct part number is 3-C.

65. One and six thousandths is 1.006, which is part number 4-A.

67. The size of part number 4-E is 1.602 centimeters, which in words is one and six hundred two thousandths centimeters.

69. Use the whole number place value chart on text page 2 to find the names after hundred-thousands. Then change "*s*" to "*ths*" in those names. So millions becomes million*ths* when used on the right side of the decimal point, and so on for ten-millionths, hundred-millionths, and billionths.

70. First place to the left of the decimal point is ones. If the first place to the right were one*ths*, it would mean a fraction with a denominator of 1, which would equal 1 or more. Anything that is 1 or more is to the *left* of the decimal point.

71. The eighth place value to the right of the decimal point is hundred-millionths, so 0.72436955 is seventy-two million four hundred thirty-six thousand nine hundred fifty-five hundred-millionths.

72. The ninth place value to the right of the decimal point is billionths, so 0.000678554 is six hundred seventy-eight thousand five hundred fifty-four billionths.

73. 8006.500001 is eight thousand six and five hundred thousand one millionths.

74. 20,060.000505 is twenty thousand, sixty and five hundred five millionths.

75. Three hundred two thousand forty ten-millionths:

0.0302040

76. Nine billion, eight hundred seventy-six million, five hundred forty-three thousand, two hundred ten and one hundred million two hundred thousand three hundred billionths:

9,876,543,210.100200300

5.2 Rounding Decimal Numbers

5.2 Section Exercises

1. 16.8974 to the nearest tenth
Draw a cut-off line after the tenths place:
16.8|974
The first digit cut is 9, which is 5 or more, so round up the tenths place.

$$\begin{array}{r} 16.8 \\ +0.1 \\ \hline 16.9 \end{array}$$

Answer: ≈ 16.9

3. 0.95647 to the nearest thousandth
Draw a cut-off line after the thousandths place:
0.956|47
The first digit cut is 4, which is 4 or less. The part you keep stays the same.

Answer: ≈ 0.956

5. 0.799 to the nearest hundredth
Draw a cut-off line after the hundredths place:
0.79|9
The first digit cut is 9, which is 5 or more, so round up the hundredths place.

$$\begin{array}{r} 0.79 \\ +0.01 \\ \hline 0.80 \end{array}$$

Answer: ≈ 0.80

7. 3.66062 to the nearest thousandth
Draw a cut-off line after the thousandths place:
3.660|62
The first digit cut is 6, which is 5 or more, so round up the thousandths place.

$$\begin{array}{r} 3.660 \\ +0.001 \\ \hline 3.661 \end{array}$$

Answer: ≈ 3.661

9. 793.988 to the nearest tenth
Draw a cut-off line after the tenths place:
793.9|88
The first digit cut is 8, which is 5 or more, so round up the tenths place.

$$\begin{array}{r} 793.9 \\ +0.1 \\ \hline 794.0 \end{array}$$

Answer: ≈ 794.0

11. 0.09804 to the nearest ten-thousandth
Draw a cut-off line after the ten-thousandths place: 0.0980|4
The first digit cut is 4, which is 4 or less. The part you keep stays the same.

Answer: ≈ 0.0980

13. 48.512 to the nearest one
Draw a cut-off line after the ones place: 48.|512
The first digit cut is 5, which is 5 or more, so round up the ones place.

$$\begin{array}{r} 48 \\ +1 \\ \hline 49 \end{array}$$

Answer: ≈ 49

15. 9.0906 to the nearest hundredth
Draw a cut-off line after the hundredths place: 9.09|06
The first digit cut is 0, which is 4 or less. The part you keep stays the same.

Answer: ≈ 9.09

17. 82.000151 to the nearest ten-thousandth
Draw a cut-off line after the ten-thousandths place: 82.0001|51
The first digit cut is 5, which is 5 or more, so round up the ten-thousandths place.

$$\begin{array}{r} 82.0001 \\ +0.0001 \\ \hline 82.0002 \end{array}$$

Answer: ≈ 82.0002

19. Round \$0.81666 to the nearest cent.
Draw a cut-off line: \$0.81|666
The first digit cut is 6, which is 5 or more, so round up.

$$\begin{array}{r} \$0.81 \\ +0.01 \\ \hline \$0.82 \end{array}$$

Nardos pays \$0.82.

21. Round \$1.2225 to the nearest cent.
Draw a cut-off line: \$1.22|25
The first digit cut is 2, which is 4 or less. The part you keep stays the same.

Nardos pays \$1.22.

23. Round \$0.4983 to the nearest cent.
Draw a cut-off line: \$0.49|83
The first digit cut is 8, which is 5 or more, so round up.

$$\begin{array}{r} \$0.49 \\ +0.01 \\ \hline \$0.50 \end{array}$$

Nardos pays \$0.50.

25. Round \$48,649.60 to the nearest dollar.
Draw a cut-off line: \$48,649.|60
The first digit cut is 6, which is 5 or more, so round up.

$$\begin{array}{r} \$48{,}649 \\ +\quad 1 \\ \hline \$48{,}650 \end{array}$$

Income from job: $\approx$ \$48,650

27. Round \$310.08 to the nearest dollar.
Draw a cut-off line: \$310.|08
The first digit cut is 0, which is 4 or less. The part you keep stays the same.

Union dues: $\approx$ \$310

29. Round \$848.91 to the nearest dollar.
Draw a cut-off line: \$848.|91
The first digit cut is 9, which is 5 or more, so round up.

$$\begin{array}{r} \$848 \\ +\ 1 \\ \hline \$849 \end{array}$$

Donations to charity: $\approx$ \$849

31. \$499.98 to the nearest dollar
Draw a cut-off line: \$499.|98
The first digit cut is 9, which is 5 or more, so round up.

$$\begin{array}{r} \$499 \\ +\ 1 \\ \hline \$500 \end{array}$$

Answer: $\approx$ \$500

33. \$0.996 to the nearest cent
Draw a cut-off line: \$0.99|6
The first digit cut is 6, which is 5 or more, so round up.

$$\begin{array}{r} \$0.99 \\ +0.01 \\ \hline \$1.00 \end{array}$$

Answer: $\approx$ \$1.00

35. \$999.73 to the nearest dollar
Draw a cut-off line: \$999.|73
The first digit cut is 7, which is 5 or more, so round up.

$$\begin{array}{r} \$999 \\ +\quad 1 \\ \hline \$1000 \end{array}$$

Answer: $\approx$ \$1000

37. **(a)** Round 322.15 to the nearest whole number.
Draw a cut-off line: 322.|15
The first digit cut is 1, which is 4 or less. The part you keep stays the same.

The record speed for a motorcycle is about 322 miles per hour.

(b) Round 106.9 to the nearest whole number.
Draw a cut-off line: 106.|9
The first digit cut is 9, which is 5 or more, so round up.

$$\begin{array}{r} 106 \\ +1 \\ \hline 107 \end{array}$$

The fastest roller-coaster speed record is about 107 miles per hour.

39. **(a)** Round 185.981 to the nearest tenth.
Draw a cut-off line: 185.9|81
The first digit cut is 8, which is 5 or more, so round up.

$$\begin{array}{r} 185.9 \\ +\ 0.1 \\ \hline 186.0 \end{array}$$

The Indianapolis 500 fastest average winning speed is about 186.0 miles per hour.

(b) Round 763.04 to the nearest tenth.
Draw a cut-off line: 763.0|4
The first digit cut is 4, which is 4 or less. The part you keep stays the same.

The land speed record is about 763.0 miles per hour.

41. If you round \$0.499 to the nearest dollar, it will round to \$0 (zero dollars) because \$0.499 is closer to \$0 than to \$1.

42. Round \$0.499 to the nearest cent to get \$0.50.
Guideline: Round amounts less than \$1.00 to the nearest cent instead of the nearest dollar.

43. If you round \$0.0015 to the nearest cent, it will round to \$0.00 (zero cents) because \$0.0015 is closer to \$0.00 than to \$0.01.

44. Both \$0.5968 and \$0.6014 round to \$0.60.
Rounding to nearest thousandth (tenth of a cent) would allow you to identify \$0.597 as less than \$0.601.

5.3 Adding and Subtracting Signed Decimal Numbers

5.3 Section Exercises

1.
$$\begin{array}{r} {}^{1\,2}\\ 5.69 \\ 0.24 \\ +\,11.79 \\ \hline 17.72 \end{array}$$
Line up decimal points.

3.
$$\begin{array}{r} 0.38 \\ 7.00 \\ +\,4.60 \\ \hline 11.98 \end{array}$$
Line up decimal points and write in zeros.

5.
$$\begin{array}{r} {}^{2\,1\,1\,1}\\ 14.230 \\ 8.000 \\ 74.630 \\ 18.715 \\ +\,0.286 \\ \hline 115.861 \end{array}$$
Line up decimal points and write in zeros.

7.
$$\begin{array}{r} {}^{2\,2\,1\,1}\\ 27.650 \\ 18.714 \\ 9.749 \\ +\,3.210 \\ \hline 59.323 \end{array}$$
Line up decimal points and write in zeros.

9. He did not line up the decimal points; 6 should be written as 6.00.

$$\begin{array}{r} 0.72 \\ 6.00 \\ +\,39.50 \\ \hline 46.22 \end{array}$$

11. $0.3000 = \dfrac{3000}{10{,}000} = \dfrac{3000 \div 1000}{10{,}000 \div 1000} = \dfrac{3}{10} = 0.3$

13. 90.5 – 0.8

$$\begin{array}{r} 90.5 \\ -\ \ 0.8 \\ \hline 89.7 \end{array}$$
Line up decimal points. *Subtract as usual.*

15. 0.4 less 0.291

$$\begin{array}{r} 0.400 \\ -\,0.291 \\ \hline 0.109 \end{array}$$
Line up decimal points and write in zeros. *Subtract as usual.*

17. 6 minus 5.09

$$\begin{array}{rl} 6.00 & \textit{Line up decimal points} \\ -\ 5.09 & \textit{and write in zeros.} \\ \hline 0.91 & \textit{Subtract as usual.} \end{array}$$

19. Subtract 8.339 from 15

$$\begin{array}{rl} 15.000 & \textit{Line up decimal points} \\ -\ 8.339 & \textit{and write in zeros.} \\ \hline 6.661 & \textit{Subtract as usual.} \end{array}$$

21. "Subtract 7.45 from 15.32" requires 15.32 to be on top.

$$\begin{array}{rl} 15.32 & \textit{Line up decimal points.} \\ -\ 7.45 & \\ \hline 7.87 & \textit{Subtract as usual.} \end{array}$$

23. **(a)** Humerus 14.35 in., radius 10.4 in.

$$\begin{array}{rl} 14.35 & \textit{Line up decimal points} \\ +\ 10.40 & \textit{and write in zero.} \\ \hline 24.75 & \end{array}$$

The combined length is 24.75 inches.

(b)
$$\begin{array}{r} 14.35 \\ -\ 10.40 \\ \hline 3.95 \end{array}$$

The difference is 3.95 inches.

25. **(a)** Humerus 14.35 in., ulna 11.1 in., femur 19.88 in., tibia 16.94 in.

$$\begin{array}{rl} 14.35 & \textit{Line up decimal points} \\ 11.10 & \textit{and write in zero.} \\ 19.88 & \\ +\ 16.94 & \\ \hline 62.27 & \end{array}$$

The sum of the lengths is 62.27 inches.

(b) 8th rib 9.06 in., 7th rib 9.45 in.

$$\begin{array}{r} 9.45 \\ -\ 9.06 \\ \hline 0.39 \end{array}$$

The 8th rib is 0.39 in. shorter than the 7th rib.

27. $24.008 + (-0.995)$

24.008 has the larger absolute value, so the sum will be positive. Subtract the absolute values.

$$\begin{array}{r} 24.008 \\ -\ 0.995 \\ \hline 23.013 \end{array}$$

29. $-6.05 + (-39.7)$

The addends are the same sign, so the sum will be negative.

$$\begin{array}{r} 6.05 \\ +\ 39.70 \\ \hline 45.75 \end{array}$$

$-6.05 + (-39.7) = -45.75$

31. $0.9 - 7.59 = 0.9 + (-7.59)$

-7.59 has the larger absolute value, so the sum will be negative. Subtract the absolute values.

$$\begin{array}{r} 7.59 \\ -\ 0.90 \\ \hline 6.69 \end{array}$$

$0.9 - 7.59 = -6.69$

33. $-2 - 4.99 = -2 + (-4.99)$

The addends are the same sign, so the sum will be negative.

$$\begin{array}{r} 2.00 \\ +\ 4.99 \\ \hline 6.99 \end{array}$$

$-2 - 4.99 = -6.99$

35. $-5.009 + 0.73$

-5.009 has the larger absolute value, so the sum will be negative. Subtract the absolute values.

$$\begin{array}{r} 5.009 \\ -\ 0.730 \\ \hline 4.279 \end{array}$$

$-5.009 + 0.73 = -4.279$

37. $-1.7035 - (5 - 6.7)$

Work inside parentheses first.

$5 - 6.7 = 5 + (-6.7) = -1.7$

Now the problem becomes:

$$\begin{aligned} &= -1.7035 - (-1.7) \\ &= -1.7035 + 1.7 \\ &= -0.0035 \end{aligned}$$

39. $8000 - (8002.63 - 8)$

Work inside parentheses first.

$8002.63 - 8 = 8002.63 + (-8) = 7994.63$

Now the problem becomes:

$$\begin{aligned} &8000 - (7994.63) \\ &= 8000 + (-7994.63) \\ &= 5.37 \end{aligned}$$

41. We can estimate $18 - 11.725$ as $20 - 10 = 10$, so 6.275 is the most reasonable answer.

43. We can estimate $-6.25 + 0.7$ as $-7 + 1 = -6$, so -5.8 is the most reasonable answer.

45. We can estimate $-42.671 - 194.9$ as $-40 + (-200) = -240$, so -237.571 is the most reasonable answer.

47. We can estimate $8.4 - (-50.83)$ as $8 + (+50) = 58$, so 59.23 is the most reasonable answer.

49. Canada: 15.4 million; South Korea: 19 million

Estimate:	*Exact:*
19	19.0
− 15	− 15.4
4	3.6

There are 3.6 million fewer Internet users in Canada than there are in South Korea.

51. Add the number of users from all the countries.

Estimate:	*Exact:*
100	134.6
30	33.9
20	22.5
20	19.0
20	15.4
+ 9	+ 9.0
199	234.4

There are 234.4 million Internet users in all the countries listed in the table.

53. Add the 3 heights.

Estimate:	*Exact:*
2	1.83
2	2.16
+ 2	+ 2.10
6 meters	6.09 meters

The NBA stars' combined height is 0.31 meter less than the rhino's height of 6.4 meters.

55. Subtract \$9.12 from \$20.

Estimate:	*Exact:*
\$20	\$20.00
− 9	− 9.12
\$11	\$10.88

He received \$10.88 in change.

57. Subtract the price of the regular fishing line, \$4.84, from the price of the fluorescent fishing line, \$5.14.

Estimate:	*Exact:*
\$5	\$5.14
− 5	− 4.84
\$0	\$0.30

The fluorescent fishing line costs \$0.30 more than the regular fishing line.

59. Add the price of the four items and the sales tax.

Estimate:	*Exact:*
\$19	\$18.84 *spinning reel*
2	2.07 *tin split shot*
2	2.07 *tin split shot*
10	9.96 *tackle box*
+ 2	+ 2.31 *sales tax*
\$35	\$35.25

The total cost of the four items and the sales tax is \$35.25.

61. Add the monthly expenses.

\$994.00
190.78
205.00
39.95
19.95
40.80
57.32
186.81
97.75
+ 107.00
\$1939.36

Olivia's total expenses are \$1939.36 per month.

63. Subtract to find the difference.

\$190.78
− 186.81
\$3.97

The difference in the amounts spent for groceries and for the car payment is \$3.97.

65. Subtract the two inside measurements from the total length.

3.00 *Total length*
− 0.91 *Leftmost measurement*
2.09
− 0.70 *Middle measurement*
1.39

$b = 1.39$ cm.

67. Add the given lengths and subtract the sum from the total length.

$$\begin{array}{r} 2.981 \\ +\,2.981 \\ \hline 5.962 \text{ feet} \end{array} \qquad \begin{array}{r} 29.805 \\ -\,5.962 \\ \hline 23.843 \text{ feet} \end{array}$$

$q = 23.843$ feet

5.4 Multiplying Signed Decimal Numbers

5.4 Section Exercises

1. Multiply the numbers as if they were whole numbers.

$$\begin{array}{rl} 0.042 & \leftarrow \textit{3 decimal places} \\ \times\ \ 3.2 & \leftarrow \textit{1 decimal place} \\ \hline 84 & \\ 126\ \ & \\ \hline 0.1344 & \leftarrow \textit{4 decimal places} \end{array}$$

Count 4 places. Write in the decimal point and zero. The factors have the *same* sign, so the product is *positive.*

3. $-21.5(7.4)$

$$\begin{array}{rl} 21.5 & \leftarrow \textit{1 decimal place} \\ \times\ \ 7.4 & \leftarrow \textit{1 decimal place} \\ \hline 8\ 60 & \\ 150\ 5\ \ & \\ \hline -159.10 & \leftarrow \textit{2 decimal places} \end{array}$$

The factors have *different* signs, so the product is *negative.*

5. Use a calculator.

$$(-23.4)(-0.66) = 15.444$$

The factors have the *same* sign, so the product is *positive.*

7. Use a calculator.

$$\begin{array}{r} \$51.88 \\ \times\ \ \ 665 \\ \hline \$34{,}500.20 \end{array}$$

9. $72(-0.6) = -43.2$

The factors have *different* signs, so the product is *negative.*

72 has 0 decimal places. 0.6 has 1 decimal place. } → Answer has 1 decimal place.

11. $(7.2)(0.06) = 0.432$

The factors have the *same* sign, so the product is *positive.*

7.2 has 1 decimal place. 0.06 has 2 decimal places. } → Answer has 3 decimal places.

13. $-0.72(-0.06) = 0.0432$

The factors have the *same* sign, so the product is *positive.*

0.72 has 2 decimal places. 0.06 has 2 decimal places. } → Answer has 4 decimal places.

15. $(0.0072)(0.6) = 0.00432$

The factors have the *same* sign, so the product is *positive.*

0.0072 has 4 decimal places. 0.6 has 1 decimal place. } → Answer has 5 decimal places.

17. $(0.006)(0.0052)$

$$\begin{array}{rl} 0.0052 & \leftarrow \textit{4 decimal places} \\ \times\ 0.006 & \leftarrow \textit{3 decimal places} \\ \hline 0.0000312 & \leftarrow \textit{7 decimal places} \end{array}$$

19. $(-0.003)^2$ means $(-0.003)(-0.003)$.

$$\begin{array}{rl} 0.003 & \leftarrow \textit{3 decimal places} \\ \times\ 0.003 & \leftarrow \textit{3 decimal places} \\ \hline 0.000009 & \leftarrow \textit{6 decimal places} \end{array}$$

Write in 5 zeros to get 6 decimal places. The factors have the *same* sign, so the product is *positive.*

21. $(5.96)(10) = \underline{59.6}$ $\quad (3.2)(10) = \underline{32}$
$(0.476)(10) = \underline{4.76}$ $\quad (80.35)(10) = \underline{803.5}$
$(722.6)(10) = \underline{7226}$ $\quad (0.9)(10) = \underline{9}$

Multiplying by 10, decimal point moves one place to the right; by 100, two places to the right; by 1000, three places to the right.

22. $(59.6)(0.1) = \underline{5.96}$ $\quad (3.2)(0.1) = \underline{0.32}$
$(0.476)(0.1) = \underline{0.0476}$ $\quad (80.35)(0.1) = \underline{8.035}$
$(65)(0.1) = \underline{6.5}$ $\quad (523)(0.1) = \underline{52.3}$

Multiplying by 0.1, decimal point moves one place to the left; by 0.01, two places to the left; by 0.001, three places to the left.

23. *Estimate:* *Exact:*

$$\begin{array}{r} 40 \\ \times 5 \\ \hline 200 \end{array} \qquad \begin{array}{rl} 39.6 & \leftarrow \textit{1 decimal place} \\ \times\ \ 4.8 & \leftarrow \textit{1 decimal place} \\ \hline 31\ 68 & \\ 158\ 4\ \ & \\ \hline 190.08 & \leftarrow \textit{2 decimal places} \end{array}$$

25. *Estimate:* *Exact:*

$$\begin{array}{r} 40 \\ \times 40 \\ \hline 1600 \end{array} \qquad \begin{array}{rl} 37.1 & \leftarrow \textit{1 decimal place} \\ \times\ \ 42 & \leftarrow \textit{0 decimal places} \\ \hline 74\,2 & \\ 1484\ \ & \\ \hline 1558.2 & \leftarrow \textit{1 decimal place} \end{array}$$

27. *Estimate:* *Exact:*

```
   7        6.53   ← 2 decimal places
 × 5      × 4.6    ← 1 decimal place
  35       3 918
          26 12
          30.038   ← 3 decimal places
```

29. *Estimate:* *Exact:*

```
   3        2.809  ← 3 decimal places
 × 7      × 6.85   ← 2 decimal places
  21       14045
          2 2472
         16 854
         19.24165  ← 5 decimal places
```

31. An $28.90 car payment is *unreasonable.* A reasonable answer would be $289.00.

33. A height of 60.5 inches (about 5 feet) is *reasonable.*

35. A gallon of milk for $349 is *unreasonable.* A reasonable answer would be $3.49.

37. 0.095 pounds for a baby's weight is *unreasonable.* A reasonable answer would be 9.5 pounds.

39. Multiply her pay per hour times the hours she worked.

```
  $18.73   ← 2 decimal places
 × 50.5    ← 1 decimal place
   9 365
  936 50
 $945.865  ← 3 decimal places
```

Round $945.865 to the nearest cent. LaTasha made $945.87 (rounded).

41. Multiply the cost of one meter of canvas by the number of meters needed.

```
 $4.09
 × 0.6
$2.454
```

$2.454 rounds to $2.45. Sid will spend $2.45 on the canvas.

43. Multiply the number of gallons that she pumped into her pickup truck by the price per gallon.

```
     18.583
 ×   $2.879
 $53.500457
```

Round $53.500457 to the nearest cent. Michelle paid $53.50 for the gas.

45. Multiply the cost of the home by 0.07.

```
  $289,500
 ×    0.07
$20,265.00
```

Ms. Rolack's fee was $20,265.

47. **(a) Area before 1929:**

```
      7.4218
 ×     3.125
      371090
     148436
     74218
  22 2654
  23.1931250
```

Rounding to the nearest tenth gives us 23.2 in.^{2}.

Area after 1929:

```
     6.14
   × 2.61
      614
    3 684
   12 28
  16.0254
```

Rounding to the nearest tenth gives us 16.0 in.^{2}.

(b) Subtract to find the difference.

```
   23.2
 − 16.0
    7.2
```

The difference in rounded areas is 7.2 in.^{2}.

49. **(a)** Multiply the thickness of one bill times the number of bills.

```
 0.0043
 ×  100
 0.4300
```

A pile of 100 bills would be 0.43 inch high.

(b)

```
  0.0043
 × 1000
  4.3000
```

A pile of 1000 bills would be 4.3 inches high.

51. Multiply the monthly cost of cable times 24 months (two years).

$$\begin{array}{rl} \$38.96 & \textit{basic per month} \\ \times \quad 24 & \\ \hline 155\ 84 & \\ 779\ 2\ \ & \\ \hline \$935.04 & \textit{monthly total} \\ +\ 49.00 & \textit{one-time installation} \\ \hline \$984.04 & \textit{two-year total} \end{array}$$

The total cost for basic cable is $984.04.

$$\begin{array}{rl} \$89.95 & \textit{deluxe per month} \\ \times \quad 24 & \\ \hline 359\ 80 & \\ 1799\ 0\ \ & \\ \hline \$2158.80 & \textit{monthly total} \\ +\ \ 49.00 & \textit{one-time installation} \\ \hline \$2207.80 & \textit{two-year total} \end{array}$$

The total cost for deluxe cable is $2207.80.

53. Multiply the number of sheets by the cost per sheet.

$$\begin{array}{r} 5100 \\ \times\ \$0.015 \\ \hline 25\ 500 \\ 51\ 00\ \ \\ \hline \$76.500 \end{array}$$

The library will pay $76.50 for the paper.

55. Multiply to find the cost of the rope, then multiply to find the cost of the wire. Add the results to find Barry's total purchases. Subtract the purchases from $15 (three $5 bills is $15).

Cost of rope	**Cost of wire**
16.5	$1.05
× $0.47	× 3
$7.755	$3.15

The cost of the rope rounds to $7.76.

Purchases	**Change**
$7.76 *rope*	$15.00
+ 3.15 *wire*	− 10.91
$10.91	$4.09

Barry received $4.09 in change.

57. Find the cost of the 4 long-sleeve, solid color shirts.

$$\begin{array}{r} \$18.95 \\ \times \quad 4 \\ \hline \$75.80 \end{array}$$

Then find the cost of the 2 short-sleeve, striped shirts.

$$\begin{array}{r} \$16.75 \\ \times \quad 2 \\ \hline \$33.50 \end{array}$$

Add these two amounts and the $2 per shirt charge for the XXL size.

$$\begin{array}{r} \$2 \\ \times\ 6 \\ \hline \$12 \end{array} \qquad \begin{array}{r} \$75.80 \\ 33.50 \\ +\ 12.00 \\ \hline \$121.30 \end{array}$$

The total cost is $121.30 plus shipping, or $121.30 + $7.95 = $129.25.

59. **(a)** Find the cost for 3 short-sleeved, solid-color shirts.

$$\begin{array}{r} \$14.75 \\ \times \quad 3 \\ \hline \$44.25 \end{array}$$

Based on this subtotal, shipping is $5.95. Find the cost of the 3 monograms.

$$\begin{array}{r} \$4.95 \\ \times \quad 3 \\ \hline \$14.85 \end{array}$$

Add these amounts, plus $5.00 for a gift box.

$$\begin{array}{r} \$44.25 \\ 5.95 \\ 14.85 \\ +\ 5.00 \\ \hline \$70.05 \end{array}$$

The total cost is $70.05.

(b) Subtract the cost of the shirts to find the difference.

$$\begin{array}{r} \$70.05 \\ -\ 44.25 \\ \hline \$25.80 \end{array}$$

The monograms, gift box, and shipping added $25.80 to the cost of the gift.

5.5 Dividing Signed Decimal Numbers

5.5 Section Exercises

1. **(a)** $27.3 \div (-7) = -3.9$

The numbers have different signs, so the quotient is negative. Divide as if both numbers are whole numbers.

$$\begin{array}{r l} 3.\,9 & \textit{Line up decimal points.} \\ 7\overline{)2\,7.\,3} & \\ 2\,1 & \\ \hline 6\,3 & \\ 6\,3 & \\ \hline 0 & \end{array}$$

(b) $\dfrac{4.23}{9} = 0.47$

The numbers have the same sign, so the quotient is positive. Divide as if both numbers are whole numbers.

$$\begin{array}{r l} 0.\,4\,7 & \textit{Line up decimal points.} \\ 9\overline{)4.\,2\,3} & \\ 3\,6 & \\ \hline 6\,3 & \\ 6\,3 & \\ \hline 0 & \end{array}$$

3. **(a)** $0.65 \div 4 = 0.1625$

$$\begin{array}{r l} 0.\,1\,6\,2\,5 & \\ 4\overline{)0.\,6\,5\,0\,0} & \leftarrow \textit{Write two zeros.} \\ 4 & \\ \hline 2\,5 & \\ 2\,4 & \\ \hline 1\,0 & \\ 8 & \\ \hline 2\,0 & \\ 2\,0 & \\ \hline 0 & \end{array}$$

(b) $\dfrac{138.5}{-50} = -2.77$

The numbers have different signs, so the quotient is negative. Divide as if both numbers are whole numbers.

$$\begin{array}{r l} 2.\,7\,7 & \\ 50\overline{)1\,3\,8.\,5\,0} & \leftarrow \textit{Write one zero.} \\ 1\,0\,0 & \\ \hline 3\,8\,5 & \\ 3\,5\,0 & \\ \hline 3\,5\,0 & \\ 3\,5\,0 & \\ \hline 0 & \end{array}$$

5. $-20.01 \div (-0.05) = 400.2$

The numbers have the same sign, so the quotient is positive.

$$\begin{array}{r l} 4\,0\,0.\,2 & \\ 0.05_\wedge\overline{)2\,0.\,0\,1_\wedge\,0} & \textit{Move decimal point} \\ 2\,0 & \textit{in divisor and dividend} \\ \hline 0\,0\,1\,0 & \textit{2 places; write one zero.} \\ 1\,0 & \\ \hline 0 & \end{array}$$

7. $54 \div 1.5 = 36$

The numbers have the same sign, so the quotient is positive.

$$\begin{array}{r l} 3\,6 & \\ 1.5_\wedge\overline{)5\,4.\,0_\wedge} & \textit{Move decimal point} \\ 4\,5 & \textit{in divisor and dividend} \\ \hline 9\,0 & \textit{1 place; write } 0. \\ 9\,0 & \\ \hline 0 & \end{array}$$

9. Given: $108 \div 18 = 6$

Find: $1.8\overline{)0.108}$ or $0.108 \div 1.8$

The new dividend, 0.108, has the effect of moving the decimal place in the answer *left* three places. The new divisor, 1.8, has the effect of moving the decimal place in the answer *right* one place. So the new answer has the decimal point moved to the left two places.

$\mathbf{0.108 \div 1.8 = 0.06}$

11. Given: $108 \div 18 = 6$

Find: $0.018\overline{)108}$ or $108 \div 0.018$

The new divisor, 0.018, has the effect of moving the decimal place in the answer *right* three places.

$\mathbf{108 \div 0.018 = 6000}$

13. Given: $108 \div 18 = 6$

Find: $0.18\overline{)10.8}$ or $10.8 \div 0.18$

The new dividend, 10.8, has the effect of moving the decimal place in the answer *left* one place. The new divisor, 0.18, has the effect of moving the decimal place in the answer *right* two places. So the new answer has the decimal point moved to the right one place.

$\mathbf{10.8 \div 0.18 = 60}$

15. Given: $108 \div 18 = 6$

Find: $18\overline{)0.0108}$ or $0.0108 \div 18$

The new dividend, 0.108, has the effect of moving the decimal place in the answer *left* four places. So the new answer has the decimal point moved to the left four places.

$\mathbf{0.0108 \div 18 = 0.0006}$

17. $116.38 \div 4.6 = 25.3$

```
          2 5. 3   Line up decimal points.
    4.6^)1 1 6. 3^ 8   Move decimal point 1
         9 2          place in dividend and
         2 4 3        divisor.
         2 3 0
           1 3 8
           1 3 8
               0
```

19. $\dfrac{-3.1}{-0.006}$

The numbers have the same sign, so the quotient is positive.

```
              5 1 6. 6 6 6   Line up decimal points.
    0.006^)3. 1 0 0^ 0 0 0
           3 0                Move decimal point in
             1 0              divisor and dividend 3
               6              places.
               4 0
               3 6
                 4 0          Write 0 in dividend.
                 3 6
                   4 0        Write 0 in dividend.
                   3 6
                     4 0      Write 0 in dividend.
                     3 6
                       4      Stop and round answer
                              to the nearest hundredth.
```

516.666 rounds to 516.67.

21. Enter on calculator: −240.8 [÷] 9 [=]

The numbers have different signs, so the quotient is negative.

Round −26.75555556 to −26.756.

23. $0.034\overline{)342.81}$

The numbers have the same sign, so the quotient is positive.

Enter on calculator: 342.81 [÷] 0.034 [=]

Round 10,082.64706 to 10,082.647.

25. $3.77 \div 10 = \underline{0.377}$ $\quad 9.1 \div 10 = \underline{0.91}$
$0.886 \div 10 = \underline{0.0886}$ $\quad 30.19 \div 10 = \underline{3.019}$
$406.5 \div 10 = \underline{40.65}$ $\quad 6625.7 \div 10 = \underline{662.57}$

(a) Dividing by 10, decimal point moves one place to the left; by 100, two places to the left; by 1000, three places to the left.

(b) The decimal point moved to the *right* when multiplying by 10, by 100, or by 1000; here it moves to the *left* when dividing by 10, by 100, or by 1000.

26. $40.2 \div 0.1 = \underline{402}$ $\quad 7.1 \div 0.1 = \underline{71}$
$0.339 \div 0.1 = \underline{3.39}$ $\quad 15.77 \div 0.1 = \underline{157.7}$
$46 \div 0.1 = \underline{460}$ $\quad 873 \div 0.1 = \underline{8730}$

(a) Dividing by 0.1, decimal point moves one place to the right; by 0.01, two places to the right; by 0.001, three places to the right.

(b) The decimal point moved to the *left* when multiplying by 0.1, 0.01, or 0.001; here it moves to the *right* when dividing by 0.1, 0.01, or 0.001.

27. $37.8 \div 8 = 47.25$

Estimate: $40 \div 8 = 5$

The answer 47.25 is *unreasonable.*

```
      4. 7 2 5
    8)3 7. 8 0 0
      3 2
        5 8
        5 6
          2 0
          1 6
            4 0
            4 0
              0
```

The correct answer is 4.725.

29. $54.6 \div 48.1 \approx 1.135$

Estimate: $50 \div 50 = 1$

The answer 1.135 is *reasonable.*

31. $307.02 \div 5.1 = 6.2$

Estimate: $300 \div 5 = 60$

The answer 6.2 is *unreasonable.*

```
           6 0. 2
    5.1^)3 0 7. 0^ 2
         3 0 6
             1 0
               0
             1 0 2
             1 0 2
                 0
```

The correct answer is 60.2.

33. $9.3 \div 1.25 = 0.744$

Estimate: $9 \div 1 = 9$

The answer 0.744 is *unreasonable.*

```
           7. 4 4
1.25^ | 9. 3 0^ 0 0
         8 7 5
         -----
           5 5  0
           5 0  0
           ------
             5  0 0
             5  0 0
             ------
                  0
```

The correct answer is 7.44.

35. Divide the cost by the number of pairs of tights.

```
   3. 9 9 6
6 | 2 3. 9 8 0
    1 8
    ---
      5  9
      5  4
      ----
        5 8
        5 4
        ---
          4 0
          3 6
          ---
            4
```

\$3.996 rounds to \$4.00.

One pair costs \$4.00 (rounded).

37. Divide the balance by the number of months.

```
        6 7. 0 7 9
21 | 1 4 0 8. 6 6 0
     1 2 6
     -----
       1 4 8
       1 4 7
       -----
           1  6 6
           1  4 7
           ------
              1 9 0  Write one zero.
              1 8 9
              -----
                  1
```

Aimee is paying \$67.08 (rounded) per month.

39. Divide the total cost by the number of bricks to find the cost per brick.

```
          0. 3 0
619 | 1 8 5. 7 0
        1 8 5  7
        --------
               0 0
                 0
                 -
                 0
```

One brick costs \$0.30.

41. Divide the total earnings by the number of hours.

```
      1 1. 9 2
40 | 4 7 6. 8 0
     4 0
     ---
       7 6
       4 0
       ---
       3 6  8
       3 6  0
       ------
            8 0
            8 0
            ---
              0
```

Darren earns \$11.92 per hour.

43. Divide the miles driven by the gallons of gas purchased.

$346.2 \div 16.35 \approx 21.17$

She got 21.2 miles per gallon (rounded).

45. First find the sum of lengths for Jackie Joyner-Kersee.

$$
\begin{array}{r}
7.49 \\
7.45 \\
7.40 \\
7.32 \\
+\,7.20 \\
\hline
36.86
\end{array}
$$

Then divide by the number of jumps, namely 5.

```
   7. 3 7 2
5 | 3 6. 8 6 0
    3 5
    ---
      1 8
      1 5
      ---
        3 6
        3 5
        ---
          1 0
          1 0
          ---
            0
```

7.372 rounds to 7.37.

The average length of the long jumps made by Jackie Joyner-Kersee is 7.37 meters (rounded).

47. Subtract to find the difference.

$$
\begin{array}{rl}
7.40 & \textit{fifth longest jump} \\
-\,7.32 & \textit{sixth longest jump} \\
\hline
0.08 & \textit{difference}
\end{array}
$$

The fifth longest jump was 0.08 meter longer than the sixth longest jump.

49. Add the lengths of the jumps of the top three athletes.

$$\begin{array}{r} 7.52 \\ 7.49 \\ +\,7.48 \\ \hline 22.49 \end{array}$$

The total length jumped by the top three athletes was 22.49 meters.

51. $7.2 - 5.2 + 3.5^2$ *Exponent*
$= 7.2 - 5.2 + 12.25$ *Subtract.*
$= 2 + 12.25$ *Add.*
$= 14.25$

53. $38.6 + 11.6(10.4 - 13.4)$ *Parentheses*
$= 38.6 + 11.6(-3)$ *Multiply.*
$= 38.6 + (-34.8)$ *Add.*
$= 3.8$

55. $-8.68 - 4.6(10.4) \div 6.4$ *Multiply.*
$= -8.68 - 47.84 \div 6.4$ *Divide.*
$= -8.68 - 7.475$ *Subtract.*
$= -16.155$

57. $33 - 3.2(0.68 + 9) + (-1.3)^2$ *Parentheses; Exponent*
$= 33 - 3.2(9.68) + 1.69$ *Multiply.*
$= 33 - 30.976 + 1.69$ *Subtract.*
$= 2.024 + 1.69$ *Add.*
$= 3.714$

59. Multiply the price per can by the number of cans.

$$\begin{array}{r} \$0.57 \\ \times\ 6 \\ \hline \$3.42 \end{array}$$

Subtract to find total savings.

$$\begin{array}{r} \$3.42 \\ -\,3.25 \\ \hline \$0.17 \end{array}$$

There are six cans, so divide by 6 to find the savings per can.

$$\begin{array}{r} 0.028 \\ 6\overline{)0.170} \\ 12 \\ \hline 50 \\ 48 \\ \hline 2 \end{array}$$

\$0.028 rounds to \$0.03.

You will save \$0.03 per can (rounded).

61. **(a)** $\frac{37{,}000{,}000}{24} \approx 1{,}541{,}667$ pieces each hour

(b) There are $24 \times 60 = 1440$ minutes in a day.

$\frac{37{,}000{,}000}{1440} \approx 25{,}694$ pieces each minute

(c) There are $24 \times 60 \times 60 = 86{,}400$ seconds in a day.

$\frac{37{,}000{,}000}{86{,}400} \approx 428$ pieces each second

63. Divide \$10,000 by 10¢.

$$\begin{array}{r} 100{,}000 \\ 0.10_\wedge\overline{)10{,}000.00_\wedge} \\ 10 \\ \hline 000000 \end{array}$$

The school would need to collect 100,000 box tops.

65. Divide 100,000 (the answer from Exercise 63) by 38.

$$\begin{array}{r} 2631.5 \\ 38\overline{)100{,}000.0} \\ 76 \\ \hline 240 \\ 228 \\ \hline 120 \\ 114 \\ \hline 60 \\ 38 \\ \hline 220 \\ 190 \\ \hline 30 \end{array}$$

2631.5 rounds to 2632.

The school needs to collect 2632 box tops (rounded) during each of the 38 weeks.

Summary Exercises on Decimals

1. $0.8 = \frac{8}{10} = \frac{8 \div 2}{10 \div 2} = \frac{4}{5}$

3. $0.35 = \frac{35}{100} = \frac{35 \div 5}{100 \div 5} = \frac{7}{20}$

5. 2.0003 is two and three ten-thousandths.

7. five hundredths:

$\frac{5}{100} = 0.05$

9. ten and seven tenths:

$10\frac{7}{10} = 10.7$

11. 0.95 to the nearest tenth

Draw a cut-off line: 0.9|5

The first digit cut is 5, which is 5 or more, so round up.

$$\begin{array}{r} 0.9 \\ +\,0.1 \\ \hline 1.0 \end{array}$$

Answer: ≈ 1.0

13. \$0.893 to the nearest cent

Draw a cut-off line: \$0.89|3

The first digit cut is 3, which is 4 or less. The part you keep stays the same.

Answer: $\approx$ \$0.89

15. \$99.64 to the nearest dollar

Draw a cut-off line: \$99.|64

The first digit cut is 6, which is 5 or more, so round up.

$$\begin{array}{r} \$99 \\ +\ 1 \\ \hline \$100 \end{array}$$

Answer: $\approx$ \$100

17. 50 – 0.3801

$$\begin{array}{rl} \overset{4}{\not 5}\,\overset{9}{\not 0}.\overset{9}{\not 0}\,\overset{9}{\not 0}\,\overset{9}{\not 0}\,\overset{10}{\not 0} & \textit{Write four zeros.} \\ -\ 0.3801 & \\ \hline 49.6199 & \end{array}$$

$$\text{Check:}\quad \begin{array}{r} 0.3801 \\ +\ 49.6199 \\ \hline 50.0000 \end{array}$$

19. $\dfrac{-90.18}{-6}$

Same signs, quotient is positive

$$\begin{array}{r} 15.03 \\ 6\overline{)90.18} \\ \underline{6} \\ 30 \\ \underline{30} \\ 01 \\ \underline{0} \\ 18 \\ \underline{18} \\ 0 \end{array}$$

21. $1.55 - 3.7 = 1.55 + (-3.7)$

Subtract absolute values.

$$\begin{array}{r} 3.\overset{6}{7}\,\overset{10}{\not 0} \\ -\ 1.55 \\ \hline 2.15 \end{array} \qquad \text{Check:}\quad \begin{array}{r} 1.55 \\ +\ 2.15 \\ \hline 3.70 \end{array}$$

Since -3.7 has the larger absolute value, the sum is negative.

$1.55 - 3.7 = -2.15$

23. $3.6 + 0.718 + 9 + 5.0829$

$$\begin{array}{rl} \overset{1\ 11}{3.6000} & \textit{Line up decimal points.} \\ 0.7180 & \textit{Write in zeros.} \\ 9.0000 & \\ +\ 5.0829 & \\ \hline 18.4009 & \end{array}$$

25.

$$\begin{array}{ll} -8.9 + 4^2 \div (-0.02) & \textit{Exponent} \\ = -8.9 + 16 \div (-0.02) & \textit{Divide.} \\ = -8.9 - 800 & \textit{Subtract.} \\ = -808.9 & \end{array}$$

27. $0.64 \div 16.3 \approx 0.0392638037 \approx 0.04$

29. To find the perimeter, add the lengths of all the sides.

$$\begin{array}{r} 2.000 \\ 1.000 \\ 1.700 \\ 0.860 \\ 2.095 \\ 1.180 \\ +\ 0.900 \\ \hline 9.735 \text{ meters} \end{array}$$

The perimeter is 9.735 meters.

31. Find the rental cost for 3 weeks.

$$\begin{array}{r} \$375 \\ \times\ 3 \\ \hline \$1125 \end{array}$$

Find the mileage fee.

$$\begin{array}{r} \$0.35 \\ \times\ 2650 \\ \hline \$927.50 \end{array}$$

Add the two amounts.

$$\begin{array}{r} \$1125.00 \\ +\ 927.50 \\ \hline \$2052.50 \end{array}$$

The rental cost was \$2052.50.

33. **(a)**

$$\begin{aligned} A &= s^2 \\ &= (12 \text{ ft})(12 \text{ ft}) \\ &= 144 \text{ ft}^2 \end{aligned}$$

(b) Cost per square foot:

$$\frac{\$99}{144 \text{ ft}^2} = \$0.6875 \approx \$0.69 \text{ (rounded)}$$

35. Multiply 0.004 and 80.

$$\begin{array}{r} 0.004 \\ \underline{\times\ 80} \\ 0.320 \end{array}$$

The average weight of food eaten each day by a queen bee is 0.32 ounce.

5.6 Fractions and Decimals

5.6 Section Exercises

1. $\frac{1}{2} = 0.5$

$$\begin{array}{r} 0.5 \\ 2\overline{)1.0} \\ \underline{1\ 0} \\ 0 \end{array}$$

3. $\frac{3}{4} = 0.75$

$$\begin{array}{r} 0.75 \\ 4\overline{)3.00} \\ \underline{2\ 8\ } \\ 2\ 0 \\ \underline{2\ 0} \\ 0 \end{array}$$

5. $\frac{3}{10} = 0.3$

$$\begin{array}{r} 0.3 \\ 10\overline{)3.0} \\ \underline{3\ 0} \\ 0 \end{array}$$

7. $\frac{9}{10} = 0.9$

$$\begin{array}{r} 0.9 \\ 10\overline{)9.0} \\ \underline{9\ 0} \\ 0 \end{array}$$

9. $\frac{3}{5} = 0.6$

$$\begin{array}{r} 0.6 \\ 5\overline{)3.0} \\ \underline{3\ 0} \\ 0 \end{array}$$

11. $\frac{7}{8} = 0.875$

$$\begin{array}{r} 0.875 \\ 8\overline{)7.000} \\ \underline{6\ 4\ \ \ } \\ 6\ 0\ \ \\ \underline{5\ 6\ \ } \\ 4\ 0 \\ \underline{4\ 0} \\ 0 \end{array}$$

13. $2\frac{1}{4} = \frac{9}{4}$

$$\begin{array}{r} 2.25 \\ 4\overline{)9.00} \\ \underline{8\ \ \ \ } \\ 1\ 0\ \ \\ \underline{8\ \ } \\ 2\ 0 \\ \underline{2\ 0} \\ 0 \end{array}$$

15. $14\frac{7}{10} = 14 + \frac{7}{10} = 14 + 0.7 = 14.7$

17. $3\frac{5}{8} = \frac{29}{8} = 3.625$

$$\begin{array}{r} 3.625 \\ 8\overline{)29.000} \\ \underline{2\ 4\ \ \ \ \ } \\ 5\ 0\ \ \ \\ \underline{4\ 8\ \ \ } \\ 2\ 0\ \ \\ \underline{1\ 6\ \ } \\ 4\ 0 \\ \underline{4\ 0} \\ 0 \end{array}$$

19. $6\frac{1}{3} = \frac{19}{3} \approx 6.333333333 \approx 6.333$

21. $\frac{5}{6} \approx 0.8333333 \approx 0.833$

23. $1\frac{8}{9} = \frac{17}{9} \approx 1.8888889 \approx 1.889$

25. **(a)** A proper fraction is less than 1. $\frac{5}{9}$ is a proper fraction so it cannot be equivalent to a decimal number greater than 1.

(b) $\frac{5}{9}$ means $5 \div 9$ or $9\overline{)5}$, so the correct answer is 0.556 (rounded). This makes sense because both the fraction and decimal are less than 1.

$$\begin{array}{r} 0.5555 \\ 9\overline{)5.0000} \\ \underline{4\ 5\ \ \ \ \ \ } \\ 5\ 0\ \ \ \ \\ \underline{4\ 5\ \ \ \ } \\ 5\ 0\ \ \ \\ \underline{4\ 5\ \ \ } \\ 5\ 0\ \\ \underline{4\ 5\ } \\ 5 \end{array}$$

26. **(a)** $2.035 = 2\frac{35}{1000} = 2\frac{7}{200}$, not $2\frac{7}{20}$.

(b) Adding the whole number part gives $2 + 0.35$, which is 2.35, not 2.035. To check, $2.35 = 2\frac{35}{100} = 2\frac{7}{20}$.

27. Just add the whole number part to 0.375. So $1\frac{3}{8} = 1.375$; $3\frac{3}{8} = 3.375$; $295\frac{3}{8} = 295.375$.

28. It works only when the fraction part has a one-digit numerator and a denominator of 10, or a two-digit numerator and a denominator of 100, and so on.

29. $0.4 = \frac{4}{10} = \frac{4 \div 2}{10 \div 2} = \frac{2}{5}$

31. $0.625 = \frac{625}{1000} = \frac{625 \div 125}{1000 \div 125} = \frac{5}{8}$

33. $0.35 = \frac{35}{100} = \frac{35 \div 5}{100 \div 5} = \frac{7}{20}$

35. $\frac{7}{20} = 0.35$

$$\begin{array}{r} 0.35 \\ 20\overline{)7.00} \\ \underline{6\ 0} \\ 1\ 00 \\ \underline{1\ 00} \\ 0 \end{array}$$

37. $0.04 = \frac{4}{100} = \frac{4 \div 4}{100 \div 4} = \frac{1}{25}$

39. $0.15 = \frac{15}{100} = \frac{15 \div 5}{100 \div 5} = \frac{3}{20}$

41. $\frac{1}{5} = 0.2$

$$\begin{array}{r} 0.2 \\ 5\overline{)1.0} \\ \underline{1\ 0} \\ 0 \end{array}$$

43. $0.09 = \frac{9}{100}$

45. Compare the two lengths.
average length → 20.80 *longer*
Charlene's baby → 20.08 *shorter*

$$\begin{array}{r} 20.80 \\ -\ 20.08 \\ \hline 0.72 \end{array}$$

Her baby is 0.72 inch *shorter* than the average length.

47. Write two zeros to the right of 0.5 so it has the same number of decimal places as 0.505. Then you can compare the numbers: $0.505 > 0.500$. There was too much calcium in each capsule. Subtract to find the difference.

$$\begin{array}{r} 0.505 \\ -\ 0.500 \\ \hline 0.005 \end{array}$$

There was 0.005 gram too much.

49. Write two zeros so that all the numbers have four decimal places.

$1.0100 > 1.0020$ *unacceptable*
$0.9991 > 0.9980$ and $0.9991 < 1.0020$ *acceptable*
$1.0007 > 0.9980$ and $1.0007 < 1.0020$ *acceptable*
$0.9900 < 0.9980$ *unacceptable*

The lengths of 0.9991 cm and 1.0007 cm are acceptable.

51. Compare the two amounts.

$3\frac{3}{4} \rightarrow 3.75$ *less*
$3.8 \rightarrow 3.80$ *more*

$$\begin{array}{r} 3.80 \\ -\ 3.75 \\ \hline 0.05 \end{array}$$

3.8 inches is 0.05 inch *more* than Ginny had hoped for.

53. **(a)** On the first number line, 0.3125 is to the *left* of 0.375, so use the $\boxed{<}$ symbol: $0.3125 < 0.375$

(b) Write $\frac{6}{8}$ in lowest terms as $\frac{3}{4}$. On the first number line, $\frac{3}{4}$ and 0.75 are at the *same point*, so use the $\boxed{=}$ symbol: $\frac{3}{4} = 0.75$

(c) On the second number line, $0.\overline{8}$ is to the *right* of $0.8\overline{3}$, so use the $\boxed{>}$ symbol: $0.\overline{8} > 0.8\overline{3}$

(d) On the second number line, 0.5 is to the *left* of $\frac{5}{9}$, so use the $\boxed{<}$ symbol: $0.5 < \frac{5}{9}$

55. 0.54, 0.5455, 0.5399

$0.54 = 0.5400$
$0.5455 = 0.5455$ *largest*
$0.5399 = 0.5399$ *smallest*

From smallest to largest: 0.5399, 0.54, 0.5455

57. 5.8, 5.79, 5.0079, 5.804

$5.8 = 5.8000$
$5.79 = 5.7900$
$5.0079 = 5.0079$ *smallest*
$5.804 = 5.8040$ *largest*

From smallest to largest: 5.0079, 5.79, 5.8, 5.804

59. 0.628, 0.62812, 0.609, 0.6009

$0.628 = 0.62800$
$0.62812 = 0.62812$ *largest*
$0.609 = 0.60900$
$0.6009 = 0.60090$ *smallest*

From smallest to largest:
0.6009, 0.609, 0.628, 0.62812

61. 5.8751, 4.876, 2.8902, 3.88

The numbers after the decimal places are irrelevant since the whole number parts are all different.

From smallest to largest:
2.8902, 3.88, 4.876, 5.8751

63. $0.043, 0.051, 0.006, \frac{1}{20}$

$0.043 = 0.043$
$0.051 = 0.051$ *largest*
$0.006 = 0.006$ *smallest*
$\frac{1}{20} = 0.050$

From smallest to largest: $0.006, 0.043, \frac{1}{20}, 0.051$

65. $\frac{3}{8}, \frac{2}{5}, 0.37, 0.4001$

$\frac{3}{8} = 0.3750$
$\frac{2}{5} = 0.4000$
$0.37 = 0.3700$ *smallest*
$0.4001 = 0.4001$ *largest*

From smallest to largest: $0.37, \frac{3}{8}, \frac{2}{5}, 0.4001$

67. **(a)** Find the largest of:
$0.018, 0.01, 0.008, 0.010$

$0.018 = 0.018$ *largest*
$0.01 = 0.010$
$0.008 = 0.008$ *smallest*
$0.010 = 0.010$

List from smallest to largest:

$0.008, 0.01 = 0.010, 0.018$

The red box, labeled 0.018 in. diameter, has the strongest line.

The green box, labeled 0.008 in. diameter, has the line with the least strength.

(b) Subtract 0.008 from 0.018.

$$\begin{array}{r} 0.018 \\ -\ 0.008 \\ \hline 0.010 \end{array}$$

The difference in diameter is 0.01 inch.

69. $1\frac{7}{16} = \frac{23}{16}$

$$\begin{array}{r} 1.43 \\ 16\,\overline{)\,23.00} \\ \underline{16} \\ 7\,0 \\ \underline{6\,4} \\ 60 \\ \underline{48} \\ 12 \end{array}$$

1.43 rounded to the nearest tenth is 1.4.
Length (a) is 1.4 inch (rounded).

71. $\frac{1}{4} = 0.25$

0.25 rounded to the nearest tenth is 0.3.
Length (c) is 0.3 inch (rounded).

73. $\frac{3}{8}$

$$\begin{array}{r} 0.37 \\ 8\,\overline{)\,3.00} \\ \underline{2\,4} \\ 60 \\ \underline{56} \\ 4 \end{array}$$

0.37 rounded to the nearest tenth is 0.4.
Length (e) is 0.4 inch (rounded).

5.7 Problem Solving with Statistics: Mean, Median, Mode, and Variability

5.7 Section Exercises

1. $\text{mean} = \frac{\text{sum of all values}}{\text{number of values}}$

$= \frac{4 + 9 + 6 + 4 + 7 + 10 + 9}{7}$

$= \frac{49}{7} = 7$

The mean (average) age of the infants at the child care center was 7 months.

3. $\text{mean} = \frac{\text{sum of all values}}{\text{number of values}}$

$= \frac{92 + 51 + 59 + 86 + 68 + 73 + 49 + 80}{8}$

$= \frac{558}{8} = 69.75$

The mean (average) final exam score was 69.8 (rounded).

5. $\text{mean} = \frac{\text{sum of all values}}{\text{number of values}}$

$= \frac{(\$31{,}900 + 32{,}850 + 34{,}930 + 39{,}712 + 38{,}340 + 60{,}000)}{6}$

$= \frac{\$237{,}732}{6} = \$39{,}622$

The mean (average) annual salary was $39,622.

7. $\text{mean} = \frac{\text{sum of all values}}{\text{number of values}}$

$= \frac{(\$75.52 + 36.15 + 58.24 + 21.86 + 47.68 + 106.57 + 82.72 + 52.14 + 28.60 + 72.92)}{10}$

$= \frac{\$582.40}{10} = \$58.24$

The mean (average) shoe sales amount was $58.24.

9.

Policy Amount ($)	Number of Policies Sold	Product ($)
10,000	6	60,000
20,000	24	480,000
25,000	12	300,000
30,000	8	240,000
50,000	5	250,000
100,000	3	300,000
250,000	2	500,000
Totals	60	2,130,000

$$\text{weighted mean} = \frac{\text{sum of products}}{\text{total number of policies}} = \frac{\$2{,}130{,}000}{60} = \$35{,}500$$

The mean (average) amount for the policies sold was $35,500.

11.

Quiz Score	Frequency	Product
3	4	12
5	2	10
6	5	30
8	5	40
9	2	18
Totals	18	110

$$\text{weighted mean} = \frac{\text{sum of products}}{\text{total number of quizzes}} = \frac{110}{18} = 6.\overline{1}$$

The mean (average) quiz score was 6.1 (rounded).

13.

Hours Worked	Frequency	Product
12	4	48
13	2	26
15	5	75
19	3	57
22	1	22
23	5	115
Totals	20	343

$$\text{weighted mean} = \frac{\text{sum of products}}{\text{total number of workers}} = \frac{343}{20} = 17.15$$

The mean (average) number of hours worked was 17.2 (rounded).

15.

Course	Credits	Grade	Credits • Grade
Biology	4	B (= 3)	4 • 3 = 12
Biology Lab	2	A (= 4)	2 • 4 = 8
Mathematics	5	C (= 2)	5 • 2 = 10
Health	1	F (= 0)	1 • 0 = 0
Psychology	3	B (= 3)	3 • 3 = 9
Totals	15		39

$$\text{GPA} = \frac{\text{sum of Credits} \cdot \text{Grade}}{\text{total number of credits}} = \frac{39}{15} = 2.60$$

17. **(a)** In Exercise 15, replace 1 • 0 with 1 • 3 to get $\text{GPA} = \frac{42}{15} = 2.80$.

(b) In Exercise 15, replace 5 • 2 with 5 • 3 to get $\text{GPA} = \frac{44}{15} = 2.9\overline{3} = 2.93$ (rounded).

(c) Making both of those changes gives us $\text{GPA} = \frac{47}{15} = 3.1\overline{3} = 3.13$ (rounded).

19. Arrange the numbers in numerical order from smallest to largest (they already are).

$$9, 12, 14, 15, 23, 24, 28$$

The list has 7 numbers. The middle number is the 4th number, so the median is 15 messages.

21. Arrange the numbers in numerical order from smallest to largest.

$$328, 420, 483, 549, 592, 715$$

The list has 6 numbers. The middle numbers are the 3rd and 4th numbers, so the median is

$$\frac{483 + 549}{2} = 516 \text{ students.}$$

23. Arrange the numbers in numerical order from smallest to largest.

$$34, 40, 40, 47, 48, 49, 51, 56, 95, 96$$

The list has 10 numbers. The middle numbers are the 5th and 6th numbers, so the median is

$$\frac{48 + 49}{2} = 48.5 \text{ pounds of shrimp.}$$

25.

$$\text{mean} = \frac{\text{sum of all values}}{\text{number of values}} = \frac{7650 + 6450 + 1100 + 5225 + 1550 + 2875}{6} = \frac{24{,}850}{6} = 4141.\overline{6}$$

The mean distance flown without refueling was 4142 miles (rounded).

27. **(a)** 1100, 1550, 2875, 5225, 6450, 7650

There are 6 distances. The middle numbers are the 3rd and 4th numbers, so the median is

$$\frac{2875 + 5225}{2} = 4050 \text{ miles.}$$

(b) List the values from smallest to largest. There are an even number of values, so find the average of the middle two values:
$(2875 + 5225) \div 2 = 4050$

29. $3, \underline{8}, 5, 1, 7, 6, \underline{8}, 4, 5, \underline{8}$

The number 8 occurs three times, which is more often than any other number. Therefore, 8 samples is the mode.

31. $\underline{74}, \underline{68}, \underline{68}, \underline{68}, 75, 75, \underline{74}, \underline{74}, 70, 77$

Because both 68 and 74 years occur three times, which is more often than any other values, each is a mode. This list is *bimodal.*

33. $5, 9, 17, 3, 2, 8, 19, 1, 4, 20, 10, 6$

No number occurs more than once. This list has *no mode.*

35. **(a)** Barrow's

$$\text{mean} = \frac{-2 - 11 - 13 - 18 - 15 - 2}{6} = \frac{-61}{6} \approx -10°\text{F}$$

Fairbank's

$$\text{mean} = \frac{3 - 7 - 10 - 4 + 11 + 31}{6} = \frac{24}{6} = 4°\text{F}$$

(b) Find the difference: $4 - (-10) = 14$

Fairbank's mean is 14 degrees warmer than Barrow's mean.

37. Barrow's range is $-2 - (-18) = 16$ degrees; Fairbanks' range is $31 - (-10) = 41$ degrees; Fairbanks' temperatures have greater variability.

39. **(a)** $\text{mean} = \dfrac{74 + 69 + 67 + 69 + 72 + 75}{6} = \dfrac{426}{6} = 71°\text{F}$

Arrange the temperatures in order:

$67, 69, 69, 72, 74, 75$

$$\text{median} = \frac{69 + 72}{2} = 70.5 \approx 71°\text{F}$$

(b) They are nearly identical because there is so little variation in the temperatures.

41. **(a) Student P:**

$$\text{mean} = \frac{92 + 80 + 61 + 49 + 82 + 53}{6} = \frac{417}{6} = 69.5$$

Arrange the data in order: $49, 53, 61, 80, 82, 92$

$$\text{median} = \frac{61 + 80}{2} = \frac{141}{2} = 70.5$$

$\text{range} = 92 - 49 = 43$

Student Q:

$$\text{mean} = \frac{70 + 76 + 77 + 60 + 67 + 72}{6} = \frac{422}{6} \approx 70.3$$

Arrange the data in order: $60, 67, 70, 72, 76, 77$

$$\text{median} = \frac{70 + 72}{2} = 71$$

$\text{range} = 77 - 60 = 17$

(b) Since $43 > 17$, Student P has the greater variability in his or her scores.

(c) Answers will vary.

42. **(a) Golfer G:**

$$\text{mean} = \frac{84 + 87 + 83 + 89 + 88}{5} = \frac{431}{5} = 86.2$$

Arrange the data in order: $83, 84, 87, 88, 89$

$\text{median} = 87$

$\text{range} = 89 - 83 = 6$

Golfer H:

$$\text{mean} = \frac{94 + 88 + 76 + 89 + 85}{5} = \frac{432}{5} = 86.4$$

Arrange the data in order: $76, 85, 88, 89, 94$

$\text{median} = 88$

$\text{range} = 94 - 76 = 18$

(b) Since $6 < 18$, Golfer G has less variability in his or her scores.

(c) Answers will vary.

43. **(a)** A's range $= 74 - 62 = 12$;
B's range $= 26 - 18 = 8$;
C's range $= 69 - 25 = 44$;

(b) Building B's ages have the least variability.

(c) Answers will vary.

44. **(a)** P's range = \$7200 – 4900 = \$2300;
Q's range = \$6000 – 5500 = \$500;
R's range = \$6400 – 5200 = \$1200;
Company P's salaries have the greatest variability.

(b) Answers will vary.

5.8 Geometry Applications: Pythagorean Theorem and Square Roots

5.8 Section Exercises

1. $\sqrt{16} = 4$ because $4 \cdot 4 = 16$.

3. $\sqrt{64} = 8$ because $8 \cdot 8 = 64$.

5. $\sqrt{11}$
Calculator shows 3.31662479; round to 3.317.

7. $\sqrt{5}$
Calculator shows 2.236067977; rounds to 2.236.

9. $\sqrt{73}$
Calculator shows 8.544003745; rounds to 8.544.

11. $\sqrt{101}$
Calculator shows 10.04987562; rounds to 10.050.

13. On a calculator, $\sqrt{361} = 19$.

15. $\sqrt{1000}$
Calculator shows 31.6227766; rounds to 31.623.

17. 30 is about halfway between 25 and 36, so $\sqrt{30}$ should be about halfway between 5 and 6, or about 5.5. Using a calculator, $\sqrt{30} \approx 5.477$. Similarly, $\sqrt{26}$ should be a little more than $\sqrt{25}$; by calculator, $\sqrt{26} \approx 5.099$. And $\sqrt{35}$ should be a little less than $\sqrt{36}$; by calculator, $\sqrt{35} \approx 5.916$.

19. legs: 15 ft and 36 ft

$$\begin{aligned} \text{hypotenuse} &= \sqrt{(\text{leg})^2 + (\text{leg})^2} \\ &= \sqrt{(15)^2 + (36)^2} \\ &= \sqrt{225 + 1296} \\ &= \sqrt{1521} \\ &= 39 \text{ ft} \end{aligned}$$

21. legs: 8 in. and 15 in.

$$\begin{aligned} \text{hypotenuse} &= \sqrt{(\text{leg})^2 + (\text{leg})^2} \\ &= \sqrt{(8)^2 + (15)^2} \\ &= \sqrt{64 + 225} \\ &= \sqrt{289} \\ &= 17 \text{ in.} \end{aligned}$$

23. hypotenuse: 20 mm, leg: 16 mm

$$\begin{aligned} \text{leg} &= \sqrt{(\text{hypotenuse})^2 - (\text{leg})^2} \\ &= \sqrt{(20)^2 - (16)^2} \\ &= \sqrt{400 - 256} \\ &= \sqrt{144} \\ &= 12 \text{ mm} \end{aligned}$$

25. legs: 8 in. and 3 in.

$$\begin{aligned} \text{hypotenuse} &= \sqrt{(\text{leg})^2 + (\text{leg})^2} \\ &= \sqrt{(8)^2 + (3)^2} \\ &= \sqrt{64 + 9} \\ &= \sqrt{73} \\ &\approx 8.5 \text{ in.} \end{aligned}$$

27. legs: 7 yd and 4 yd

$$\begin{aligned} \text{hypotenuse} &= \sqrt{(\text{leg})^2 + (\text{leg})^2} \\ &= \sqrt{(7)^2 + (4)^2} \\ &= \sqrt{49 + 16} \\ &= \sqrt{65} \\ &\approx 8.1 \text{ yd} \end{aligned}$$

29. hypotenuse: 22 cm, leg: 17 cm

$$\begin{aligned} \text{leg} &= \sqrt{(\text{hypotenuse})^2 - (\text{leg})^2} \\ &= \sqrt{(22)^2 - (17)^2} \\ &= \sqrt{484 - 289} \\ &= \sqrt{195} \\ &\approx 14.0 \text{ cm} \end{aligned}$$

31. legs: 1.3 m and 2.5 m

$$\begin{aligned} \text{hypotenuse} &= \sqrt{(\text{leg})^2 + (\text{leg})^2} \\ &= \sqrt{(1.3)^2 + (2.5)^2} \\ &= \sqrt{1.69 + 6.25} \\ &= \sqrt{7.94} \\ &\approx 2.8 \text{ m} \end{aligned}$$

33. hypotenuse: 11.5 cm, leg: 8.2 cm

$$\begin{aligned} \text{leg} &= \sqrt{(\text{hypotenuse})^2 - (\text{leg})^2} \\ &= \sqrt{(11.5)^2 - (8.2)^2} \\ &= \sqrt{132.25 - 67.24} \\ &= \sqrt{65.01} \\ &\approx 8.1 \text{ cm} \end{aligned}$$

35. hypotenuse: 21.6 km, leg: 13.2 km

$$\begin{aligned} \text{leg} &= \sqrt{(\text{hypotenuse})^2 - (\text{leg})^2} \\ &= \sqrt{(21.6)^2 - (13.2)^2} \\ &= \sqrt{466.56 - 174.24} \\ &= \sqrt{292.32} \\ &\approx 17.1 \text{ km} \end{aligned}$$

37. legs: 4 ft and 7 ft

$$\begin{aligned} \text{hypotenuse} &= \sqrt{(\text{leg})^2 + (\text{leg})^2} \\ &= \sqrt{(4)^2 + (7)^2} \\ &= \sqrt{16 + 49} \\ &= \sqrt{65} \\ &\approx 8.1 \text{ ft} \end{aligned}$$

The length of the loading ramp is about 8.1 ft.

39. hypotenuse: 1000 m, leg: 800 m

$$\begin{aligned}\text{leg} &= \sqrt{(\text{hypotenuse})^2 - (\text{leg})^2}\\ &= \sqrt{(1000)^2 - (800)^2}\\ &= \sqrt{1{,}000{,}000 - 640{,}000}\\ &= \sqrt{360{,}000}\\ &= 600 \text{ m}\end{aligned}$$

The airplane is 600 meters above the ground.

41. legs: 4.5 ft and 3.5 ft

$$\begin{aligned}\text{hypotenuse} &= \sqrt{(\text{leg})^2 + (\text{leg})^2}\\ &= \sqrt{(4.5)^2 + (3.5)^2}\\ &= \sqrt{20.25 + 12.25}\\ &= \sqrt{32.5}\\ &\approx 5.7 \text{ ft}\end{aligned}$$

The diagonal brace is about 5.7 ft long.

43.

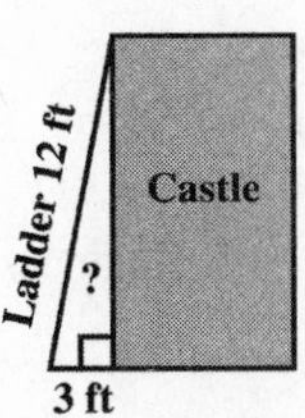

$$\begin{aligned}\text{leg} &= \sqrt{(\text{hypotenuse})^2 - (\text{leg})^2}\\ &= \sqrt{(12)^2 - (3)^2}\\ &= \sqrt{144 - 9}\\ &= \sqrt{135}\\ &\approx 11.6 \text{ ft}\end{aligned}$$

The ladder will reach about 11.6 ft high on the building.

45. The student used the formula for finding the hypotenuse but the unknown side is a leg, so $? = \sqrt{(20)^2 - (13)^2}$. Also, the final answer should be m, not m^2. Correct answer is $\sqrt{231} \approx 15.2$ m.

47. legs: 90 ft and 90 ft

$$\begin{aligned}\text{hypotenuse} &= \sqrt{(\text{leg})^2 + (\text{leg})^2}\\ &= \sqrt{(90)^2 + (90)^2}\\ &= \sqrt{8100 + 8100}\\ &= \sqrt{16{,}200}\\ &\approx 127.3 \text{ ft}\end{aligned}$$

The distance from home plate to second base is about 127.3 ft.

48. (a)

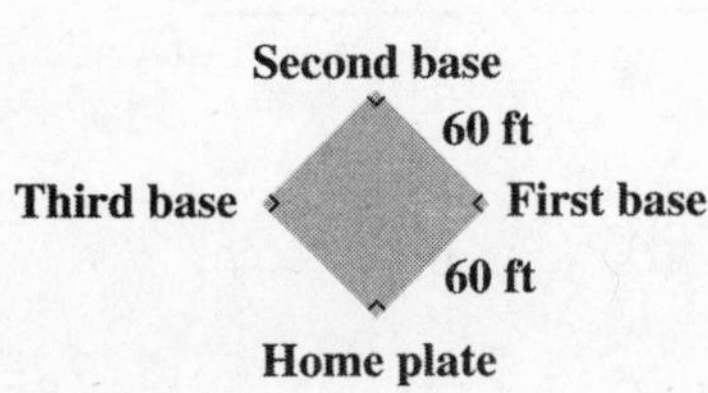

(b) legs: 60 ft and 60 ft

$$\begin{aligned}\text{hypotenuse} &= \sqrt{(\text{leg})^2 + (\text{leg})^2}\\ &= \sqrt{(60)^2 + (60)^2}\\ &= \sqrt{3600 + 3600}\\ &= \sqrt{7200}\\ &\approx 84.9 \text{ ft}\end{aligned}$$

The distance from home plate to second base is about 84.9 ft.

49. The distance from third to first is the same as the distance from home to second because the baseball diamond is a square.

50. (a) Since 80 ft < 84.9 ft, then the side length is less than 60 ft.

(b) $80^2 = 6400$

$$\frac{6400}{2} = 3200$$

$$\sqrt{3200} \approx 56.6$$

The side length of each side is about 56.6 ft.

5.9 Problem Solving: Equations Containing Decimals

5.9 Section Exercises

1. $h + 0.63 = 5.1$ To get h by itself, add -0.63 to both sides.

$$\begin{aligned}h + 0.63 - 0.63 &= 5.1 - 0.63\\ h &= 4.47\end{aligned}$$

Check: $h + 0.63 = 5.1$
$4.47 + 0.63 = 5.1$
$5.1 = 5.1$ Balances

The solution is 4.47.

3. $-20.6 + n = -22$ Add 20.6 to both sides.

$$\begin{aligned}-20.6 + n + 20.6 &= -22 + 20.6\\ n &= -1.4\end{aligned}$$

Check: $-20.6 + n = -22$
$-20.6 + (-1.4) = -22$
$-22 = -22$ Balances

The solution is -1.4.

5. $0 = b - 0.008$ Add 0.008 to both sides.

$$\begin{aligned}0 + 0.008 &= b - 0.008 + 0.008\\ 0.008 &= b\end{aligned}$$

Check: $0 = b - 0.008$
$0 = 0.008 - 0.008$
$0 = 0$ Balances

The solution is 0.008.

7. $2.03 = 7a$ Divide both sides by 7.

$$\frac{2.03}{7} = \frac{7a}{7}$$
$$0.29 = a$$

Check: $2.03 = 7a$
$2.03 = 7(0.29)$
$2.03 = 2.03$ Balances

The solution is 0.29.

9. $0.8p = -96$ Divide both sides by 0.8.

$$\frac{0.8p}{0.8} = \frac{-96}{0.8}$$
$$p = -120$$

Check: $0.8p = -96$
$0.8(-120) = -96$
$-96 = -96$ Balances

The solution is -120.

11. $-3.3t = -2.31$ Divide both sides by -3.3.

$$\frac{-3.3t}{-3.3} = \frac{-2.31}{-3.3}$$
$$t = 0.7$$

Check: $-3.3t = -2.31$
$-3.3(0.7) = -2.31$
$-2.31 = -2.31$ Balances

The solution is 0.7.

13. $7.5x + 0.15 = -6$ Add -0.15 to both sides.

$\underline{-0.15 \quad -0.15}$

$7.5x = -6.15$ Divide both sides by 7.5.

$$\frac{7.5x}{7.5} = \frac{-6.15}{7.5}$$
$$x = -0.82$$

Check: $7.5x + 0.15 = -6$
$7.5(-0.82) + 0.15 = -6$
$-6.15 + 0.15 = -6$
$-6 = -6$ Balances

The solution is -0.82.

15. $-7.38 = 2.05z - 7.38$ Add 7.38 to both sides.

$\underline{+7.38 \quad +7.38}$

$0 = 2.05z$ Divide both sides by 2.05.

$$\frac{0}{2.05} = \frac{2.05z}{2.05}$$
$$0 = z$$

Check: $-7.38 = 2.05z - 7.38$
$-7.38 = 2.05(0) - 7.38$
$-7.38 = 0 - 7.38$
$-7.38 = -7.38$ Balances

The solution is 0.

17. $3c + 10 = 6c + 8.65$ Add $-3c$ to both sides.

$\underline{-3c \quad -3c}$

$10 = 3c + 8.65$ Add -8.65 to both sides.

$\underline{-8.65 \quad -8.65}$

$1.35 = 3c$ Divide both sides by 3.

$$\frac{1.35}{3} = \frac{3c}{3}$$
$$0.45 = c$$

Check: $3c + 10 = 6c + 8.65$
$3(0.45) + 10 = 6(0.45) + 8.65$
$1.35 + 10 = 2.7 + 8.65$
$11.35 = 11.35$ Balances

The solution is 0.45.

19. $0.8w - 0.4 = -6 + w$ Add $-0.8w$ to both sides.

$\underline{-0.8w \quad -0.8w}$

$-0.4 = -6 + 0.2w$ Add 6 to both sides.

$\underline{+6 \quad +6}$

$5.6 = 0.2w$ Divide both sides by 0.2.

$$\frac{5.6}{0.2} = \frac{0.2w}{0.2}$$
$$28 = w$$

Check: $0.8w - 0.4 = -6 + w$
$0.8(28) - 0.4 = -6 + 28$
$22.4 - 0.4 = 22$
$22 = 22$ Balances

The solution is 28.

21. $-10.9 + 0.5p = 0.9p + 5.3$ Add $-0.5p$ to both sides.

$\underline{-0.5p \quad -0.5p}$

$-10.9 = 0.4p + 5.3$ Add -5.3 to both sides.

$\underline{-5.3 \quad -5.3}$

$-16.2 = 0.4p$ Divide both sides by 0.4.

$$\frac{-16.2}{0.4} = \frac{0.4p}{0.4}$$
$$-40.5 = p$$

Check:

$$\begin{aligned} -10.9 + 0.5p &= 0.9p + 5.3 \\ -10.9 + 0.5(-40.5) &= 0.9(-40.5) + 5.3 \\ -10.9 - 20.25 &= -36.45 + 5.3 \\ -31.15 &= -31.15 \quad \text{Balances} \end{aligned}$$

The solution is -40.5.

23. *Step 1*
Unknown: the adult dose

Known: child's dose is 0.3 times the adult dose; child's dose is 9 mg

Step 2
Let d be the adult dose.

Step 3

adult dose multiplied by 0.3	is	child's dose
↓	↓	↓
$0.3d$	$=$	9

Step 4

$0.3d = 9$ Divide both sides by 0.3.

$$\frac{0.3d}{0.3} = \frac{9}{0.3}$$

$$d = 30$$

Step 5
The adult dose is 30 milligrams.

Step 6
$0.3(30) = 9$ (matches)

25. *Step 1*
Unknown: number of days the saw was rented

Known: \$65.95 charge per day, \$12 sharpening fee, \$275.80 total

Step 2
Let d be the number of days.

Step 3

65.95 per day	plus	sharpening fee	equals	total bill
↓	↓	↓	↓	↓
$65.95 \cdot d$	$+$	12	$=$	275.80

Step 4

$$\begin{aligned} 65.95d + 12 &= 275.80 \\ -12 &\quad -12 \\ \hline 65.95d &= 263.80 \\ \frac{65.95d}{65.95} &= \frac{263.80}{65.95} \\ d &= 4 \end{aligned}$$

Step 5
The saw was rented for 4 days.

Step 6
Charge for 4 days: 4(\$65.95) = \$263.80
Add sharpening fee: \$263.80 + \$12 = \$275.80
This value matches the total charge.

27.

$$\begin{aligned} 0.7(220 - a) &= 140 && \text{Distributive property} \\ 154 - 0.7a &= 140 && \text{Add } -154 \text{ to both sides.} \\ -154 &\quad -154 \\ \hline -0.7a &= -14 && \text{Divide both sides by } -0.7. \\ \frac{-0.7a}{-0.7} &= \frac{-14}{-0.7} \\ a &= 20 \end{aligned}$$

The person is 20 years old.

28.

$$\begin{aligned} 0.7(220 - a) &= 126 && \text{Distributive property} \\ 154 - 0.7a &= 126 && \text{Add } -154 \text{ to both sides.} \\ -154 &\quad -154 \\ \hline -0.7a &= -28 && \text{Divide both sides by } -0.7. \\ \frac{-0.7a}{-0.7} &= \frac{-28}{-0.7} \\ a &= 40 \end{aligned}$$

The person is 40 years old.

29.

$$\begin{aligned} 0.7(220 - a) &= 134 && \text{Distributive property} \\ 154 - 0.7a &= 134 && \text{Add } -154 \text{ to both sides.} \\ -154 &\quad -154 \\ \hline -0.7a &= -20 && \text{Divide both sides by } -0.7. \\ \frac{-0.7a}{-0.7} &= \frac{-20}{-0.7} \\ a &\approx 28.57, && \text{which rounds to 29.} \end{aligned}$$

The person is about 29 years old.

30.

$$\begin{aligned} 0.7(220 - a) &= 117 && \text{Distributive property} \\ 154 - 0.7a &= 117 && \text{Add } -154 \text{ to both sides.} \\ -154 &\quad -154 \\ \hline -0.7a &= -37 && \text{Divide both sides by } -0.7. \\ \frac{-0.7a}{-0.7} &= \frac{-37}{-0.7} \\ a &\approx 52.86, && \text{which rounds to 53.} \end{aligned}$$

The person is about 53 years old.

5.10 Geometry Applications: Circles, Cylinders, and Surface Area

5.10 Section Exercises

1. The radius is 9 mm so the diameter is

$$\begin{aligned} d &= 2 \cdot r \\ &= 2 \cdot 9 \text{ mm} \\ &= 18 \text{ mm}. \end{aligned}$$

3. The diameter is 0.7 km, so the radius is

$$\frac{d}{2} = \frac{0.7 \text{ km}}{2} = 0.35 \text{ km}.$$

5. radius: 11 ft

$$\begin{aligned} C &= 2 \cdot \pi \cdot r \\ &\approx 2 \cdot 3.14 \cdot 11 \text{ ft} \\ &\approx 69.1 \text{ ft} \end{aligned}$$

$$\begin{aligned} A &= \pi \cdot r \cdot r \\ &\approx 3.14 \cdot 11 \text{ ft} \cdot 11 \text{ ft} \\ &\approx 379.9 \text{ ft}^2 \end{aligned}$$

7. diameter: 2.6 m

$$\begin{aligned} C &= \pi \cdot d \\ &\approx 3.14 \cdot 2.6 \text{ m} \\ &\approx 8.2 \text{ m} \end{aligned}$$

radius: $\dfrac{2.6 \text{ m}}{2} = 1.3 \text{ m}$

$$\begin{aligned} A &= \pi \cdot r \cdot r \\ &\approx 3.14 \cdot 1.3 \text{ m} \cdot 1.3 \text{ m} \\ &\approx 5.3 \text{ m}^2 \end{aligned}$$

9.

$$\begin{aligned} d &= 15 \text{ cm} \\ C &= \pi \cdot d \\ &\approx 3.14 \cdot 15 \text{ cm} \\ &= 47.1 \text{ cm} \\ r &= \frac{15 \text{ cm}}{2} = 7.5 \text{ cm} \\ A &= \pi \cdot r \cdot r \\ &\approx 3.14 \cdot 7.5 \text{ cm} \cdot 7.5 \text{ cm} \\ &\approx 176.6 \text{ cm}^2 \end{aligned}$$

11.

$$\begin{aligned} d &= 7\tfrac{1}{2} \text{ ft} \\ C &= \pi \cdot d \\ &\approx 3.14 \cdot 7.5 \text{ ft} \\ &\approx 23.6 \text{ ft} \\ r &= \frac{7\frac{1}{2} \text{ ft}}{2} = 3.75 \text{ ft} \\ A &= \pi \cdot r \cdot r \\ &\approx 3.14 \cdot 3.75 \text{ ft} \cdot 3.75 \text{ ft} \\ &\approx 44.2 \text{ ft}^2 \end{aligned}$$

13.

$$\begin{aligned} d &= 8.65 \text{ km} \\ C &= \pi \cdot d \\ &\approx 3.14 \cdot 8.65 \text{ km} \\ &\approx 27.2 \text{ km} \\ r &= \frac{8.65 \text{ km}}{2} = 4.325 \text{ km} \\ A &= \pi \cdot r \cdot r \\ &\approx 3.14 \cdot 4.325 \text{ km} \cdot 4.325 \text{ km} \\ &\approx 58.7 \text{ km}^2 \end{aligned}$$

15. radius: 9.8 m

Area of circle

$$\begin{aligned} A &= \pi \cdot r \cdot r \\ &\approx 3.14 \cdot 9.8 \text{ m} \cdot 9.8 \text{ m} \\ &= 301.5656 \text{ m}^2 \end{aligned}$$

Area of semicircle

$$\frac{301.5656 \text{ m}^2}{2} \approx 150.8 \text{ m}^2$$

17. Area of the circle:

$$\begin{aligned} A &= \pi \cdot r \cdot r \\ &\approx 3.14 \cdot 10 \text{ cm} \cdot 10 \text{ cm} \\ &= 314 \text{ cm}^2 \end{aligned}$$

Area of the semicircle:

$$\frac{314 \text{ cm}^2}{2} = 157 \text{ cm}^2$$

Area of the triangle:

$$\begin{aligned} A &= \frac{1}{2} \cdot b \cdot h \\ &= \frac{1}{2} \cdot 20 \text{ cm} \cdot 10 \text{ cm} \\ &= 100 \text{ cm}^2 \end{aligned}$$

The shaded area is about $157 \text{ cm}^2 - 100 \text{ cm}^2 = 57 \text{ cm}^2$.

19. Answers will vary. A sample answer follows: π is the ratio of the circumference of a circle to its diameter. If you divide the circumference of any circle by its diameter, the answer is always a little more than 3. The approximate value is 3.14, which we call π (pi). Your test question could involve finding the circumference or the area of a circle.

21.

$$\begin{aligned} A &= \pi \cdot r \cdot r \\ &\approx 3.14 \cdot 50 \text{ yd} \cdot 50 \text{ yd} \\ &= 7850 \text{ yd}^2 \end{aligned}$$

The watered area is about 7850 yd^2.

23. A point on the tire tread moves the length of the circumference in one complete turn.

$$\begin{aligned} C &= \pi \cdot d \\ &\approx 3.14 \cdot 29.10 \text{ in.} \\ &\approx 91.4 \text{ in.} \end{aligned}$$

Bonus question:

$$\frac{1 \text{ revolution}}{91.4 \text{ inches}} \cdot \frac{12 \text{ inches}}{1 \text{ foot}} \cdot \frac{5280 \text{ feet}}{1 \text{ mile}} \approx 693 \text{ revolutions/mile}$$

25.

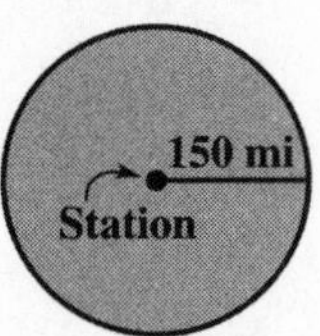

$$\begin{aligned} A &= \pi \cdot r \cdot r \\ &\approx 3.14 \cdot 150 \text{ mi} \cdot 150 \text{ mi} \\ &= 70{,}650 \text{ mi}^2 \end{aligned}$$

There are about 70,650 mi^2 in the broadcast area.

27.

Watch

Watch: $d = 1$ in., $r = \frac{1}{2}$ in.

$$\begin{aligned} C &= \pi \cdot 1 \text{ in.} \\ &\approx 3.14 \cdot 1 \text{ in.} \\ &\approx 3.1 \text{ in.} \\ A &= \pi \cdot r \cdot r \\ &\approx 3.14 \cdot \tfrac{1}{2} \text{ in.} \cdot \tfrac{1}{2} \text{ in.} \\ &\approx 0.8 \text{ in.}^2 \end{aligned}$$

Wall Clock

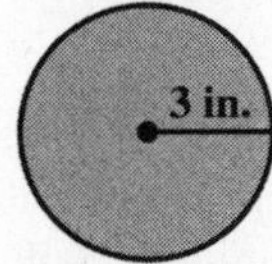

Wall clock: $r = 3$ in.

$$\begin{aligned} C &= 2 \cdot \pi \cdot 3 \text{ in.} \\ &\approx 2 \cdot 3.14 \cdot 3 \text{ in.} \\ &\approx 18.8 \text{ in.} \\ A &= \pi \cdot r \cdot r \\ &\approx 3.14 \cdot 3 \text{ in.} \cdot 3 \text{ in.} \\ &\approx 28.3 \text{ in.}^2 \end{aligned}$$

29.

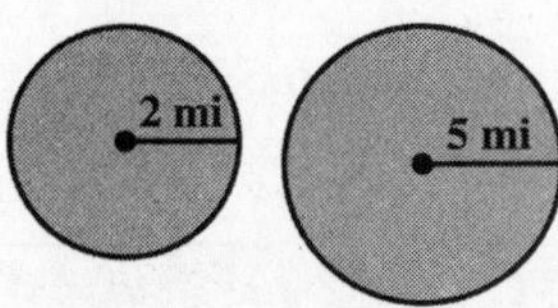

Area of larger circle:

$$\begin{aligned} A &= \pi \cdot r \cdot r \\ &\approx 3.14 \cdot 5 \text{ mi} \cdot 5 \text{ mi} \\ &= 78.5 \text{ mi}^2 \end{aligned}$$

Area of smaller circle:

$$\begin{aligned} A &= \pi \cdot r \cdot r \\ &\approx 3.14 \cdot 2 \text{ mi} \cdot 2 \text{ mi} \\ &\approx 12.6 \text{ mi}^2 \end{aligned}$$

The difference in the area covered is about

$$78.5 \text{ mi}^2 - 12.6 \text{ mi}^2 = 65.9 \text{ mi}^2.$$

31. **(a)**

$$\begin{aligned} C &= 144 \text{ cm} \\ C &= \pi \cdot d \\ 144 \text{ cm} &= \pi \cdot d \\ 144 \text{ cm} &\approx 3.14 \cdot d \\ \frac{144 \text{ cm}}{3.14} &\approx d \\ d &\approx 45.9 \text{ cm} \end{aligned}$$

The diameter is about 45.9 cm.

(b) Divide the circumference by π (144 cm $\div$ 3.14).

33. $C = 2 \cdot \pi \cdot r \approx 2 \cdot 3.14 \cdot 2.33 \text{ ft} \approx 14.6 \text{ ft}$

The rag traveled about 14.6 ft with each revolution of the wheel.

35. Find the area of the rectangle with length 29 ft and width 16 ft.

$$\begin{aligned} A &= \text{length} \cdot \text{width} \\ &= 29 \text{ ft} \cdot 16 \text{ ft} \\ &= 464 \text{ ft}^2 \end{aligned}$$

The semicircles are of equal area, and the sum of their areas equals that of a circle. The radius is 8 ft.

$$\begin{aligned} A &= \pi \cdot r \cdot r \\ &\approx 3.14 \cdot 8 \text{ ft} \cdot 8 \text{ ft} \\ &= 200.96 \text{ ft}^2 \end{aligned}$$

Total area $\approx 464 \text{ ft}^2 + 200.96 \text{ ft}^2 = 664.96 \text{ ft}^2$

Cost of the sod $= \dfrac{664.96 \text{ ft}^2}{1} \cdot \dfrac{\$1.76}{1 \text{ ft}^2} \approx \$1170.33$

37. Area of the square:

$$\begin{aligned} A &= s \cdot s \\ &= 8 \text{ ft} \cdot 8 \text{ ft} \\ &= 64 \text{ ft}^2 \end{aligned}$$

Area of the circle:

$$\begin{aligned} r &= \frac{8 \text{ ft}}{2} = 4 \text{ ft} \\ A &= \pi \cdot r \cdot r \\ &\approx 3.14 \cdot 4 \text{ ft} \cdot 4 \text{ ft} \\ &\approx 50.2 \text{ ft}^2 \end{aligned}$$

The shaded area is about
$64 \text{ ft}^2 - 50.2 \text{ ft}^2 = 13.8 \text{ ft}^2$.

39. The diameter is $7\frac{1}{2}$ in., so the radius is

$$\frac{7\frac{1}{2}\text{ in.}}{2} = 3\frac{3}{4}\text{ in. or } 3.75\text{ in.}$$

$$\begin{aligned} A &= \pi \cdot r \cdot r \\ &\approx 3.14 \cdot 3.75\text{ in.} \cdot 3.75\text{ in.} \\ &= 44.15625\text{ in.}^2 \end{aligned}$$

The area of a small pizza is about 44.2 in.^{2}.

40. The diameter is 13 in., so the radius is

$$\frac{13\text{ in.}}{2} = 6\frac{1}{2}\text{ in. or } 6.5\text{ in.}$$

$$\begin{aligned} A &= \pi \cdot r \cdot r \\ &\approx 3.14 \cdot 6.5\text{ in.} \cdot 6.5\text{ in.} \\ &= 132.665\text{ in.}^2 \end{aligned}$$

The area of a medium pizza is about 132.7 in.^{2}.

41. The diameter is 16 in., so the radius is

$$\frac{16\text{ in.}}{2} = 8\text{ in.}$$

$$\begin{aligned} A &= \pi \cdot r \cdot r \\ &\approx 3.14 \cdot 8\text{ in.} \cdot 8\text{ in.} \\ &= 200.96\text{ in.}^2 \end{aligned}$$

The area of a large pizza is about 201.0 in.^{2}.

42.

Size	Cost per square inch
Small	$\frac{\$2.80}{44.2\text{ in.}^2} \approx \$0.063$
Medium	$\frac{\$6.50}{132.7\text{ in.}^2} \approx \$0.049$
Large	$\frac{\$9.30}{201.0\text{ in.}^2} \approx \$0.046$ (*)

The best buy is the large pizza.

43.

Size	Cost per square inch
Small	$\frac{\$3.70}{44.2\text{ in.}^2} \approx \$0.084$
Medium	$\frac{\$8.95}{132.7\text{ in.}^2} \approx \$0.067$ (*)
Large	$\frac{\$14.30}{201.0\text{ in.}^2} \approx \$0.071$

The best buy is the medium pizza.

44. Cost of small pizza = \$4.35 – \$0.95 = \$3.40

Size	Cost per square inch
Small	$\frac{\$3.40}{44.2\text{ in.}^2} \approx \$0.077$ (*)
Medium	$\frac{\$10.95}{132.7\text{ in.}^2} \approx \$0.083$
Large	$\frac{\$15.65}{201.0\text{ in.}^2} \approx \$0.078$

The best buy is the small pizza.

45.

$$\begin{aligned} V &= \pi \cdot r^2 \cdot h \\ &\approx 3.14 \cdot 5\text{ ft} \cdot 5\text{ ft} \cdot 6\text{ ft} \\ &= 471\text{ ft}^3 \end{aligned}$$

$$\begin{aligned} SA &= 2\pi rh + 2\pi r^2 \\ &\approx 2 \cdot 3.14 \cdot 5\text{ ft} \cdot 6\text{ ft} + 2 \cdot 3.14 \cdot 5\text{ ft} \cdot 5\text{ ft} \\ &= 345.4\text{ ft}^2 \end{aligned}$$

47.

$$\begin{aligned} V &= lwh \\ &= 16.5\text{ m} \cdot 9.8\text{ m} \cdot 10\text{ m} \\ &= 1617\text{ m}^3 \end{aligned}$$

$$\begin{aligned} SA &= 2lw + 2lh + 2wh \\ &= 2 \cdot 16.5\text{ m} \cdot 9.8\text{ m} + 2 \cdot 16.5\text{ m} \cdot 10\text{ m} \\ &\quad + 2 \cdot 9.8\text{ m} \cdot 10\text{ m} \\ &= 849.4\text{ m}^2 \end{aligned}$$

49. $r = \frac{d}{2} = \frac{18\text{ in.}}{2} = 9\text{ in.}$

$$\begin{aligned} V &= \pi r^2 h \\ &\approx 3.14 \cdot 9\text{ in.} \cdot 9\text{ in.} \cdot 3\text{ in.} \\ &\approx 763.0\text{ in.}^3 \end{aligned}$$

$$\begin{aligned} SA &= 2\pi rh + 2\pi r^2 \\ &\approx 2 \cdot 3.14 \cdot 9\text{ in.} \cdot 3\text{ in.} + 2 \cdot 3.14 \cdot 9\text{ in.} \cdot 9\text{ in.} \\ &\approx 678.2\text{ in.}^2 \end{aligned}$$

51.

$$\begin{aligned} V &= lwh \\ &= 15\text{ mm} \cdot 10\text{ mm} \cdot 37\text{ mm} \\ &= 5550\text{ mm}^3 \end{aligned}$$

$$\begin{aligned} SA &= 2lw + 2lh + 2wh \\ &= 2 \cdot 15\text{ mm} \cdot 10\text{ mm} + 2 \cdot 15\text{ mm} \cdot 37\text{ mm} \\ &\quad + 2 \cdot 10\text{ mm} \cdot 37\text{ mm} \\ &= 2150\text{ mm}^2 \end{aligned}$$

53. Student should use radius of 3.5 cm instead of diameter of 7 cm in the formula; units for volume are cm^3, not cm^2. Correct answer is $V \approx 192.3\text{ cm}^3$.

55. Use the formula for the volume of a cylinder.

$$\text{radius} = \frac{5\text{ ft}}{2} = 2.5\text{ ft}$$

$$\begin{aligned} V &= \pi \cdot r^2 \cdot h \\ &\approx 3.14 \cdot 2.5\text{ ft} \cdot 2.5\text{ ft} \cdot 200\text{ ft} \\ &= 3925\text{ ft}^3 \end{aligned}$$

The volume of the city sewer pipe is about 3925 ft^3.

57.

$$\begin{aligned} SA &= 2lw + 2lh + 2wh \\ &= 2 \cdot 5.5\text{ in.} \cdot 2.8\text{ in.} + 2 \cdot 5.5\text{ in.} \cdot 8\text{ in.} \\ &\quad + 2 \cdot 2.8\text{ in.} \cdot 8\text{ in.} \\ &= 163.6\text{ in.}^2 \end{aligned}$$

The amount of material needed is 163.6 in.^{2}.

Chapter 5 Review Exercises

1. tenths, hundredths

2 4 3 . 0 5 9

2. ones, tenths

0 . 6 8 1 7

3. hundreds, hundredths

\$5 8 2 4 . 3 9

4. tens, tenths

8 9 6 . 5 0 3

5. tenths, ten-thousandths

2 0 . 7 3 8 6 1

6. $0.5 = \frac{5}{10} = \frac{1}{2}$

7. $0.75 = \frac{75}{100} = \frac{75 \div 25}{100 \div 25} = \frac{3}{4}$

8. $4.05 = 4\frac{5}{100} = 4\frac{5 \div 5}{100 \div 5} = 4\frac{1}{20}$

9. $0.875 = \frac{875}{1000} = \frac{875 \div 125}{1000 \div 125} = \frac{7}{8}$

10. $0.027 = \frac{27}{1000}$

11. $27.8 = 27\frac{8}{10} = 27\frac{8 \div 2}{10 \div 2} = 27\frac{4}{5}$

12. 0.8 is eight tenths.

13. 400.29 is four hundred and twenty-nine hundredths.

14. 12.007 is twelve and seven thousandths.

15. 0.0306 is three hundred six ten-thousandths.

16. eight and three tenths:

$8\frac{3}{10} = 8.3$

17. two hundred five thousandths:

$\frac{205}{1000} = 0.205$

18. seventy and sixty-six ten-thousandths:

$70\frac{66}{10{,}000} = 70.0066$

19. thirty hundredths:

$\frac{30}{100} = 0.30$

20. 275.635 to the nearest tenth:
Draw a cut-off line: 275.6|35
The first digit cut is 3, which is 4 or less. The part you keep stays the same.
Answer: ≈ 275.6

21. 72.789 to the nearest hundredth:
Draw a cut-off line: 72.78|9
The first digit cut is 9, which is 5 or more, so round up.
Answer: ≈ 72.79

22. 0.1604 to the nearest thousandth:
Draw a cut-off line: 0.160|4
The first digit cut is 4, which is 4 or less. The part you keep stays the same.
Answer: ≈ 0.160

23. 0.0905 to the nearest thousandth:
Draw a cut-off line: 0.090|5
The first digit cut is 5, which is 5 or more, so round up.
Answer: ≈ 0.091

24. 0.98 to the nearest tenth:
Draw a cut-off line: 0.9|8
The first digit cut is 8, which is 5 or more, so round up.
Answer: ≈ 1.0

25. \$15.8333 to the nearest cent:
Draw a cut-off line: \$15.83|33
The first digit cut is 3, which is 4 or less. The part you keep stays the same.
Answer: $\approx$ \$15.83

26. \$0.698 to the nearest cent:
Draw a cut-off line: \$0.69|8
The first digit cut is 8, which is 5 or more, so round up.
Answer: $\approx$ \$0.70

27. \$17,625.7906 to the nearest cent:
Draw a cut-off line: \$17,625.79|06
The first digit cut is 0, which is 4 or less. The part you keep stays the same.
Answer: $\approx$ \$17,625.79

28. \$350.48 to the nearest dollar:
Draw a cut-off line: \$350.|48
The first digit cut is 4, which is 4 or less. The part you keep stays the same.
Answer: $\approx$ \$350

29. \$129.50 to the nearest dollar:
Draw a cut-off line: \$129.|50
The first digit cut is 5, which is 5 or more, so round up.
Answer: ≈ \$130

30. \$99.61 to the nearest dollar:
Draw a cut-off line: \$99.|61
The first digit cut is 6, which is 5 or more, so round up.
Answer: ≈ \$100

31. \$29.37 to the nearest dollar:
Draw a cut-off line: \$29.|37
The first digit cut is 3, which is 4 or less. The part you keep stays the same.
Answer: ≈ \$29

32. $0.4 - 6.07$ *Add the opposite.*
$= 0.4 + (-6.07)$ *Addition of numbers with different signs.*

$$|-6.07| = 6.07, \quad |0.4| = 0.4$$

-6.07 has a larger absolute value, so the sum will be negative. Subtract absolute values.

$$\begin{array}{r} 6.07 \\ -\,0.40 \\ \hline 5.67 \end{array}$$

Answer: -5.67

33. $-20 + 19.97$ *Addition of numbers with different signs.*

$$|-20| = 20, \quad |19.97| = 19.97$$

-20 has the larger absolute value, so the sum will be negative. Subtract absolute values.

$$\begin{array}{r} 20.00 \\ -\,19.97 \\ \hline 0.03 \end{array}$$

Answer: -0.03

34. $-1.35 + 7.229$ *Addition of numbers with different signs.*

$$|-1.35| = 1.35, \quad |7.229| = 7.229$$

7.229 has the larger absolute value, so the sum will be positive. Subtract absolute values.

$$\begin{array}{r} \scriptstyle 6\ 11\,12 \\ 7.229 \\ -\,1.350 \\ \hline 5.879 \end{array}$$

Answer: 5.879

35. $0.005 + (3 - 9.44) = 0.005 + (3 + (-9.44))$
$= 0.005 + (-6.44)$
$= -6.435$

36. *Estimate:*

$$\begin{array}{rl} 80 & \text{million} \\ -\,50 & \text{million} \\ \hline 30 & \text{million} \end{array}$$

Exact:

$$\begin{array}{rl} 81.3 & \text{million} \\ -\,49.9 & \text{million} \\ \hline 31.4 & \text{million} \end{array}$$

31.4 million more people go walking than camping.

37. Add the amounts of the two checks.

Estimate:

$$\begin{array}{r} \$200 \\ +\,40 \\ \hline \$240 \end{array}$$

Exact:

$$\begin{array}{r} \$215.53 \\ +\,44.67 \\ \hline \$260.20 \end{array}$$

The total amount of the two checks was \$260.20. Now subtract from her balance of \$306.

Estimate:

$$\begin{array}{r} \$300 \\ -\,240 \\ \hline \$60 \end{array}$$

Exact:

$$\begin{array}{r} \$306.00 \\ -\,260.20 \\ \hline \$45.80 \end{array}$$

The new balance is \$45.80.

38. First total the money that Joey spent.

Estimate:

$$\begin{array}{r} \$2 \\ 5 \\ +\,20 \\ \hline \$27 \end{array}$$

Exact:

$$\begin{array}{r} \$1.59 \\ 5.33 \\ +\,18.94 \\ \hline \$25.86 \end{array}$$

Then subtract to find the change.

Estimate:

$$\begin{array}{r} \$30 \\ -\,27 \\ \hline \$3 \end{array}$$

Exact:

$$\begin{array}{r} \$30.00 \\ -\,25.86 \\ \hline \$4.14 \end{array}$$

Joey's change was \$4.14.

39. Add the kilometers that she raced each day.

Estimate:

$$\begin{array}{rl} 2 & \\ 4 & \\ +\,5 & \\ \hline 11 & \text{km} \end{array}$$

Exact:

$$\begin{array}{rl} 2.30 & \\ 4.00 & \\ +\,5.25 & \\ \hline 11.55 & \text{km} \end{array}$$

Roseanne raced 11.55 kilometers.

40. *Estimate:*

$$\begin{array}{r} 6 \\ \times\,4 \\ \hline 24 \end{array}$$

Exact:

$$\begin{array}{r} 6.138 \\ \times\,3.7 \\ \hline 4\,2966 \\ 18\,414 \\ \hline 22.7106 \end{array}$$

41. *Estimate:*

$$\begin{array}{r} 40 \\ \times\ 3 \\ \hline 120 \end{array}$$

Exact:

$$\begin{array}{r} 42.9 \\ \times\ 3.3 \\ \hline 12\ 87 \\ 128\ 7\ \ \\ \hline 141.57 \end{array}$$

42. $(-5.6)(-0.002)$

The signs are the *same*, so the product will be *positive*.

$$\begin{array}{rcl} 5.6 & \leftarrow & \text{1 decimal place} \\ \times\ 0.002 & \leftarrow & \text{3 decimal places} \\ \hline 0.0112 & \leftarrow & \text{4 decimal places} \end{array}$$

43. $(0.071)(-0.005)$

The signs are *different*, so the product will be *negative*.

$$\begin{array}{rcl} 0.071 & \leftarrow & \text{3 decimal places} \\ \times\ 0.005 & \leftarrow & \text{3 decimal places} \\ \hline -0.000355 & \leftarrow & \text{6 decimal places} \end{array}$$

44. $706.2 \div 12 = 58.85$

Estimate: $700 \div 10 = 70$

58.85 is *reasonable.*

45. $26.6 \div 2.8 = 0.95$

Estimate: $30 \div 3 = 10$

0.95 is *not reasonable.*

$$\begin{array}{r} 9.\ 5 \\ 2.8_\wedge \overline{)\,2\ 6.\ 6_\wedge\ 0} \\ 2\ 5\ 2\ \ \ \ \\ \hline 1\ 4\ \ 0 \\ 1\ 4\ \ 0 \\ \hline 0 \end{array}$$

Move decimal point 1 *place in divisor and dividend; write one zero.*

The correct answer is 9.5.

46.

$$\begin{array}{r} 1\ 4.\ 4\ 6\ 6\ 6 \\ 3\overline{)\,4\ 3.\ 4\ 0\ 0\ 0} \\ 3\ \ \ \ \ \ \ \ \ \ \ \ \ \ \ \\ \hline 1\ 3\ \ \ \ \ \ \ \ \ \ \ \ \\ 1\ 2\ \ \ \ \ \ \ \ \ \ \ \ \\ \hline 1\ 4\ \ \ \ \ \ \ \ \ \\ 1\ 2\ \ \ \ \ \ \ \ \ \\ \hline 2\ 0\ \ \ \ \ \ \\ 1\ 8\ \ \ \ \ \ \\ \hline 2\ 0\ \ \ \\ 1\ 8\ \ \ \\ \hline 2\ 0 \\ 1\ 8 \\ \hline 2 \end{array}$$

Write one zero. (each of the three times a zero is brought down)

14.4666 rounds to 14.467.

47.

$$\frac{-72}{-0.06} = -72.00 \div (-0.06)$$
$$= -7200 \div (-6) \quad \textit{Move both decimal points two places to the right.}$$
$$= 1200$$

48.

$$-0.00048 \div 0.0012$$
$$= -4.8 \div 12 \quad \textit{Move both decimal points four places to the right.}$$
$$= -0.4$$

$$\begin{array}{r} 0.\ 4 \\ 12\overline{)\,4.\ 8} \\ 4\ 8 \\ \hline 0 \end{array}$$

49. Multiply the hourly wage times the hours worked.

Pay for first 40 hours	*Pay rate after 40 hours*	*Pay for over 40 hours*
14.24	14.24	21.36
× 40	× 1.5	× 6.5
569.60	7 120	10 680
	14 24	128 16
	21.360	138.840

Add the two amounts.

$$\begin{array}{r} \$569.60 \\ 138.84 \\ \hline \$708.44 \end{array}$$

Adrienne's total earnings were \$708 (rounded).

50. Divide the cost of the book by the number of tickets in the book.

$$\begin{array}{r} 2.\ 9\ 9\ 0 \\ 12\overline{)\,3\ 5.\ 8\ 9\ 0} \\ 2\ 4\ \ \ \ \ \ \ \ \ \ \\ \hline 1\ 1\ 8\ \ \ \ \ \\ 1\ 0\ 8\ \ \ \ \ \\ \hline 1\ 0\ 9\ \ \\ 1\ 0\ 8\ \ \\ \hline 1\ 0 \end{array}$$

Round to the nearest cent.
Draw a cut-off line: \$2.99|0
The first digit cut is 0, which is 4 or less. The part you keep stays the same.

Each ticket costs \$2.99 (rounded).

51. Divide the amount of the investment by the price per share.

$$\begin{array}{r} 133.3 \\ 3.75_\wedge \overline{)500.00_\wedge 0} \\ \underline{375} \\ 1250 \\ \underline{1125} \\ 1250 \\ \underline{1125} \\ 1250 \\ \underline{1125} \\ 125 \end{array}$$

Round 133.3 down to the nearest whole number. Kenneth could buy 133 shares (rounded).

52. Multiply the price per pound by the amount to be purchased.

$$\begin{array}{r} \$0.99 \\ \times\ 3.5 \\ \hline 495 \\ 2\ 97 \\ \hline \$3.465 \end{array}$$

\$3.465 rounds to \$3.47.
Ms. Lee will pay \$3.47 (rounded).

53.

$$\begin{aligned} &3.5^2 + 8.7(-1.95) && \text{Exponent} \\ &= 12.25 + 8.7(-1.95) && \text{Multiply.} \\ &= 12.25 + (-16.965) && \text{Add.} \\ &= -4.715 \end{aligned}$$

54.

$$\begin{aligned} &11 - 3.06 \div (3.95 - 0.35) && \text{Parentheses} \\ &= 11 - 3.06 \div 3.6 && \text{Divide.} \\ &= 11 - 0.85 && \text{Subtract.} \\ &= 10.15 \end{aligned}$$

55. $3\frac{4}{5} = \frac{19}{5} = 3.8$

$$\begin{array}{r} 3.8 \\ 5\overline{)19.0} \\ \underline{15} \\ 4\ 0 \\ \underline{4\ 0} \\ 0 \end{array}$$

56. $\frac{16}{25} = 0.64$

$$\begin{array}{r} 0.64 \\ 25\overline{)16.00} \\ \underline{15\ 0} \\ 1\ 00 \\ \underline{1\ 00} \\ 0 \end{array}$$

57. $1\frac{7}{8} = \frac{15}{8} = 1.875$

$$\begin{array}{r} 1.875 \\ 8\overline{)15.000} \\ \underline{8} \\ 7\ 0 \\ \underline{6\ 4} \\ 60 \\ \underline{56} \\ 40 \\ \underline{40} \\ 0 \end{array}$$

58. $\frac{1}{9}$

$$\begin{array}{r} 0.1111 \\ 9\overline{)1.0000} \\ \underline{9} \\ 10 \\ \underline{9} \\ 10 \\ \underline{9} \\ 10 \\ \underline{9} \\ 1 \end{array}$$

0.1111 rounds to 0.111.

$\frac{1}{9} \approx 0.111$

59. 3.68, 3.806, 3.6008

3.68 = 3.6800
3.806 = 3.8060 *largest*
3.6008 = 3.6008 *smallest*

From smallest to largest:
3.6008, 3.68, 3.806

60. 0.215, 0.22, 0.209, 0.2102

0.215 = 0.2150
0.22 = 0.2200 *largest*
0.209 = 0.2090 *smallest*
0.2102 = 0.2102

From smallest to largest:
0.209, 0.2102, 0.215, 0.22

61. $0.17, \frac{3}{20}, \frac{1}{8}, 0.159$

0.17 = 0.170 *largest*

$\frac{3}{20} = 0.150$

$\frac{1}{8} = 0.125$ *smallest*

0.159 = 0.159

From smallest to largest:

$\frac{1}{8}, \frac{3}{20}, 0.159, 0.17$

62. mean

$$= \frac{(18+12+15+24+9 \\ +42+54+87+21+3)}{10}$$

$$= \frac{285}{10} = 28.5 \text{ digital cameras}$$

To find the median, arrange the data from high to low or low to high.

3, 9, 12, 15, 18, 21, 24, 42, 54, 87
↑ ↑

Since the number of items is even, the median is the average of the middle items.

$$\text{median} = \frac{18+21}{2} = 19.5 \text{ digital cameras}$$

63. mean

$$= \frac{54+28+35+43+17+37+68+75+39}{9}$$

$$= \frac{396}{9} = 44 \text{ claims}$$

To find the median, arrange the data from high to low or low to high.

17, 28, 35, 37, 39, 43, 54, 68, 75
↑

Since the number of items is odd, the median is the middle item.

median = 39 claims

64.

Dollar Value	Frequency	Product
$42	3	$126
$47	7	$329
$53	2	$106
$55	3	$165
$59	5	$295
	20	$1021

$$\text{weighted mean} = \frac{\$1021}{20} = \$51.05$$

65. (a) Arrange the data in order:

Store J: $69, $84, $107, $107, $139, $160, $160
There are two modes: $107 and $160 (bimodal);
Store K: $95, $99, $119, $119, $119, $136, $139
mode = $119

(b) Range for store J = $160 − 69 = $91
Range for store K = $139 − 95 = $44
Since $91 > $44, Store J's prices have greater variability.

66.
$$\begin{aligned}\text{hypotenuse} &= \sqrt{(\text{leg})^2 + (\text{leg})^2} \\ &= \sqrt{(15)^2 + (8)^2} \\ &= \sqrt{225+64} \\ &= \sqrt{289} \\ &= 17 \text{ in.}\end{aligned}$$

67.
$$\begin{aligned}\text{leg} &= \sqrt{(\text{hypotenuse})^2 - (\text{leg})^2} \\ &= \sqrt{(25)^2 - (24)^2} \\ &= \sqrt{625-576} \\ &= \sqrt{49} \\ &= 7 \text{ cm}\end{aligned}$$

68.
$$\begin{aligned}\text{leg} &= \sqrt{(\text{hypotenuse})^2 - (\text{leg})^2} \\ &= \sqrt{(15)^2 - (11)^2} \\ &= \sqrt{225-121} \\ &= \sqrt{104} \\ &\approx 10.2 \text{ cm}\end{aligned}$$

69.
$$\begin{aligned}\text{hypotenuse} &= \sqrt{(\text{leg})^2 + (\text{leg})^2} \\ &= \sqrt{(6)^2 + (4)^2} \\ &= \sqrt{36+16} \\ &= \sqrt{52} \\ &\approx 7.2 \text{ in.}\end{aligned}$$

70.
$$\begin{aligned}\text{hypotenuse} &= \sqrt{(\text{leg})^2 + (\text{leg})^2} \\ &= \sqrt{(2.2)^2 + (1.3)^2} \\ &= \sqrt{4.84+1.69} \\ &= \sqrt{6.53} \\ &\approx 2.6 \text{ m}\end{aligned}$$

71.
$$\begin{aligned}\text{leg} &= \sqrt{(\text{hypotenuse})^2 - (\text{leg})^2} \\ &= \sqrt{(12)^2 - (8.5)^2} \\ &= \sqrt{144-72.25} \\ &= \sqrt{71.75} \\ &\approx 8.5 \text{ km}\end{aligned}$$

72.
$$\begin{aligned}-0.1 &= b - 0.35 && \text{Add the opposite.} \\ -0.1 &= b + (-0.35) && \text{Add 0.35 to both sides.} \\ +0.35 & \quad\ \ +0.35 \\ 0.25 &= b + 0 \\ 0.25 &= b\end{aligned}$$

The solution is 0.25.

73.
$$\begin{aligned}-3.8x &= 0 && \text{Divide both sides by } -3.8. \\ \frac{-3.8x}{-3.8} &= \frac{0}{-3.8} \\ x &= 0\end{aligned}$$

The solution is 0.

74.
$$\begin{aligned} 6.8 + 0.4n &= 1.6 && \text{Add } -6.8 \text{ to both sides.} \\ -6.8 \quad & \quad -6.8 \\ \hline 0 + 0.4n &= -5.2 && \text{Divide both sides by 0.4.} \\ \frac{0.4n}{0.4} &= \frac{-5.2}{0.4} \\ n &= -13 \end{aligned}$$

The solution is -13.

75.
$$\begin{aligned} -0.375 + 1.75a &= 2a && \text{Add } -1.75a \text{ to both sides.} \\ -1.75a \quad & \quad -1.75a \\ \hline -0.375 + 0 &= 0.25a \\ -0.375 &= 0.25a && \text{Divide both sides by 0.25.} \\ \frac{-0.375}{0.25} &= \frac{0.25a}{0.25} \\ -1.5 &= a \end{aligned}$$

The solution is -1.5.

76.
$$\begin{aligned} 0.3y - 5.4 &= 2.7 + 0.8y && \text{Add } -0.3y \text{ to both sides.} \\ -0.3y \quad & \quad -0.3y \\ \hline 0 - 5.4 &= 2.7 + 0.5y \\ -5.4 &= 2.7 + 0.5y && \text{Add } -2.7 \text{ to both sides.} \\ -2.7 \quad & \quad -2.7 \\ \hline -8.1 &= 0 + 0.5y && \text{Divide both sides by 0.5.} \\ \frac{-8.1}{0.5} &= \frac{0.5y}{0.5} \\ -16.2 &= y \end{aligned}$$

The solution is -16.2.

77. $d = 2 \cdot r = 2 \cdot 68.9 \text{ m} = 137.8 \text{ m}$
The diameter of the field is 137.8 m.

78. $r = \frac{d}{2} = \frac{3 \text{ in.}}{2} = 1\frac{1}{2}$ in. or 1.5 in.

The radius of the juice can is $1\frac{1}{2}$ in. or 1.5 in.

79. radius: 1 cm, $d = 2$ cm

$$\begin{aligned} C &= \pi \cdot d \\ &\approx 3.14 \cdot 2 \text{ cm} \\ &\approx 6.3 \text{ cm} \\ A &= \pi \cdot r \cdot r \\ &\approx 3.14 \cdot 1 \text{ cm} \cdot 1 \text{ cm} \\ &\approx 3.1 \text{ cm}^2 \end{aligned}$$

80. radius: 17.4 m, $d = 34.8$ m

$$\begin{aligned} C &= \pi \cdot d \\ &\approx 3.14 \cdot 34.8 \text{ m} \\ &\approx 109.3 \text{ m} \\ A &= \pi \cdot r \cdot r \\ &\approx 3.14 \cdot 17.4 \text{ m} \cdot 17.4 \text{ m} \\ &\approx 950.7 \text{ m}^2 \end{aligned}$$

81. diameter: 12 in., radius: 6 in.

$$\begin{aligned} C &= \pi \cdot d \\ &\approx 3.14 \cdot 12 \text{ in.} \\ &\approx 37.7 \text{ in.} \\ A &= \pi \cdot r \cdot r \\ &\approx 3.14 \cdot 6 \text{ in.} \cdot 6 \text{ in.} \\ &\approx 113.0 \text{ in.}^2 \end{aligned}$$

82. The figure is a cylinder.

$$\begin{aligned} V &= \pi \cdot r^2 \cdot h \\ &\approx 3.14 \cdot 5 \text{ cm} \cdot 5 \text{ cm} \cdot 7 \text{ cm} \\ &= 549.5 \text{ cm}^3 \end{aligned}$$

$$\begin{aligned} SA &= 2\pi rh + 2\pi r^2 \\ &\approx 2 \cdot 3.14 \cdot 5 \text{ cm} \cdot 7 \text{ cm} + 2 \cdot 3.14 \cdot 5 \text{ cm} \cdot 5 \text{ cm} \\ &\approx 376.8 \text{ cm}^2 \end{aligned}$$

83. The figure is a cylinder.

$$\begin{aligned} V &= \pi \cdot r^2 \cdot h \quad (d = 24 \text{ m}; r = 12 \text{ m}) \\ &\approx 3.14 \cdot 12 \text{ m} \cdot 12 \text{ m} \cdot 4 \text{ m} \\ &\approx 1808.6 \text{ m}^3 \end{aligned}$$

$$\begin{aligned} SA &= 2\pi rh + 2\pi r^2 \\ &\approx 2 \cdot 3.14 \cdot 12 \text{ m} \cdot 4 \text{ m} + 2 \cdot 3.14 \cdot 12 \text{ m} \cdot 12 \text{ m} \\ &\approx 1205.8 \text{ m}^2 \end{aligned}$$

84. $l = 3.5$ ft, $w = 1.5$ ft, $h = 1.5$ ft

$$\begin{aligned} V &= lwh \\ &= 3.5 \text{ ft} \cdot 1.5 \text{ ft} \cdot 1.5 \text{ ft} \\ &= 7.875 \text{ ft}^3 \approx 7.9 \text{ ft}^3 \end{aligned}$$

$$\begin{aligned} SA &= 2lw + 2lh + 2wh \\ &= 2 \cdot 3.5 \text{ ft} \cdot 1.5 \text{ ft} + 2 \cdot 3.5 \text{ ft} \cdot 1.5 \text{ ft} \\ &\quad + 2 \cdot 1.5 \text{ ft} \cdot 1.5 \text{ ft} \\ &= 25.5 \text{ ft}^2 \end{aligned}$$

85. **[5.3]** $89.19 + 0.075 + 310.6 + 5$

$$\begin{array}{r} {}^{11\ 1} \\ 89.190 \\ 0.075 \\ 310.600 \\ +\ 5.000 \\ \hline 404.865 \end{array}$$

Line up decimal points.
Write in zeros.

86. **[5.4]** 72.8(−3.5)

The signs are *different*, so the product will be *negative.*

$$\begin{array}{rl} 72.8 & \leftarrow \text{1 decimal place} \\ \times\ 3.5 & \leftarrow \text{1 decimal place} \\ \hline 36\ 40 & \\ 218\ 4 & \\ \hline 254.80 & \leftarrow \text{2 decimal places} \end{array}$$

$72.8(-3.5) = -254.80$

87. **[5.5]** $1648.3 \div 0.46 \approx 3583.2609$

3583.2609 rounds to 3583.261.

88. **[5.3]**

$$\begin{array}{r} \scriptstyle 2\ 9\ \ 9\ 9\ 9\ 10 \\ \not3\ \not0\ .\not0\ \not0\ \not0\ \not0 \\ -\ 0\ .9\ 1\ 0\ 2 \\ \hline 2\ 9\ .0\ 8\ 9\ 8 \end{array}$$

89. **[5.4]** (4.38)(0.007)

$$\begin{array}{rl} 4.38 & \leftarrow \text{2 decimal places} \\ \times\ 0.007 & \leftarrow \text{3 decimal places} \\ \hline 0.03066 & \leftarrow \text{5 decimal places} \end{array}$$

90. **[5.5]**

$$\begin{array}{r} 9.\ 4 \\ 0.005_\wedge\overline{)0.047_\wedge0} \\ 4\ 5 \\ \hline 2\ 0 \\ 2\ 0 \\ \hline 0 \end{array}$$

Move decimal point 3 *places in divisor and dividend; write* 0

91. **[5.3]**

$$\begin{array}{ll} 72.105 + 8.2 - 95.37 & \text{Change subt.} \\ = 72.105 + 8.2 + (-95.37) & \text{Add.} \\ = 80.305 + (-95.37) & \text{Add.} \\ = -15.065 & \end{array}$$

92. **[5.5]** $\dfrac{-81.36}{9}$

The signs are different, so the quotient is negative.

$$\begin{array}{r} 9.04 \\ 9\overline{)81.36} \\ 81 \\ \hline 0\ 3 \\ 0 \\ \hline 36 \\ 36 \\ \hline 0 \end{array}$$

$\dfrac{-81.36}{9} = -9.04$

93. **[5.5]**

$$\begin{array}{ll} (0.6 - 1.22) + 4.8(-3.15) & \text{Change subt.} \\ = (0.6 + (-1.22)) + 4.8(-3.15) & \text{Parentheses} \\ = (-0.62) + 4.8(-3.15) & \text{Multiply.} \\ = -0.62 + (-15.12) & \text{Add.} \\ = -15.74 & \end{array}$$

94. **[5.4]** 0.455(18)

$$\begin{array}{rl} 0.455 & \leftarrow \text{3 decimal places} \\ \times\ 18 & \leftarrow \text{0 decimal places} \\ \hline 3\,640 & \\ 4\,55 & \\ \hline 8.190 & \leftarrow \text{3 decimal places} \end{array}$$

95. **[5.4]** (−1.6)(−0.58)

The signs are the same, so the product is positive.

$$\begin{array}{rl} 0.58 & \leftarrow \text{2 decimal places} \\ \times\ 1.6 & \leftarrow \text{1 decimal place} \\ \hline 348 & \\ 58 & \\ \hline 0.928 & \leftarrow \text{3 decimal places} \end{array}$$

96. **[5.5]** $0.218\overline{)7.63}$

$7.63 \div 0.218 = 35$

97. **[5.3]**

$$\begin{array}{ll} -21.059 - 20.8 & \text{Change subt.} \\ = -21.059 + (-20.8) & \text{Add.} \\ = -41.859 & \end{array}$$

98. **[5.5]**

$$\begin{array}{ll} 18.3 - 3^2 \div 0.5 & \textit{Exponent} \\ = 18.3 - 9 \div 0.5 & \textit{Divide.} \\ = 18.3 - 18 & \textit{Subtract.} \\ = 0.3 & \end{array}$$

99. **[5.5]** Men's socks are 3 pairs for $8.99. Divide the price by 3 to find the cost of one pair.

$$\begin{array}{r} 2.\ 9\ 9\ 6 \\ 3\overline{)8.\ 9\ 9\ 0} \\ 6 \\ \hline 2\ 9 \\ 2\ 7 \\ \hline 2\ 9 \\ 2\ 7 \\ \hline 2\ 0 \\ 1\ 8 \\ \hline 2 \end{array}$$

$2.996 rounds to $3.00.

One pair of men's socks costs $3.00 (rounded).

100. **[5.5]** Children's socks cost 6 pairs for $5.00. Find the cost for one pair.

$$\begin{array}{r} 0.833 \\ 6\overline{)5.000} \\ \underline{48} \\ 20 \\ \underline{18} \\ 20 \\ \underline{18} \\ 2 \end{array}$$

$0.833 rounds to $0.83.

Subtract to find how much more men's socks cost.

$$\begin{array}{r} \$3.00 \\ -\ 0.83 \\ \hline \$2.17 \end{array}$$

Men's socks cost $2.17 (rounded) more.

101. **[5.4]** A dozen pair of socks is $4 \cdot 3 = 12$ pairs of socks. So multiply 4 times $8.99.

$$\begin{array}{r} \$8.99 \\ \times\ 4 \\ \hline \$35.96 \end{array}$$

The cost for a dozen pair of men's socks is $35.96.

102. **[5.4]** Five pairs of teen jeans cost $19.95 times 5.

$$\begin{array}{r} \$19.95 \\ \times\ 5 \\ \hline \$99.75 \end{array}$$

Four pairs of women's jeans cost $24.99 times 4.

$$\begin{array}{r} \$24.99 \\ \times\ 4 \\ \hline \$99.96 \end{array}$$

Add the two amounts.

$$\begin{array}{r} \$99.75 \\ +\ 99.96 \\ \hline \$199.71 \end{array}$$

Akiko would pay $199.71.

103. **[5.3]** The highest regular price for athletic shoes is $149.50. The cheapest sale price is $71.

Subtract to find the difference.

$$\begin{array}{r} \$149.50 \\ -\ 71.00 \\ \hline \$78.50 \end{array}$$

The difference in price is $78.50.

104. **[5.9]** $4.62 = -6.6y$ Divide both sides by -6.6.

$$\frac{4.62}{-6.6} = \frac{-6.6y}{-6.6}$$
$$-0.7 = y$$

105. **[5.9]** $1.05x - 2.5 = 0.8x + 5$ Add $-0.8x$ to both sides.

$$\begin{array}{rcl} -0.8x & & -0.8x \\ \hline 0.25x - 2.5 & = & 0 + 5 \end{array}$$

$0.25x - 2.5 = 5$ Add 2.5 to both sides.

$$\begin{array}{rcl} +2.5 & & +2.5 \\ \hline 0.25x + 0 & = & 7.5 \end{array}$$

Divide both sides by 0.25.

$$\frac{0.25x}{0.25} = \frac{7.5}{0.25}$$
$$x = 30$$

106. **[5.10]** Rubber striping to go "around" the edge of the table indicates circumference of the table.

$C = \pi \cdot d$
$C = \pi \cdot 5 \text{ ft}$
$C \approx 3.14 \cdot 5 \text{ ft}$
$C \approx 15.7$ ft of rubber striping.

Since the diameter is 5, the radius is $\frac{5}{2}$ or 2.5.

$A = \pi \cdot r^2$
$A = \pi \cdot (2.5 \text{ ft})^2$
$A \approx 3.14 \cdot 2.5 \text{ ft} \cdot 2.5 \text{ ft}$
$A \approx 19.6 \text{ ft}^2$

107. **[5.7]** $\text{mean} = \dfrac{82 + 0 + 78 + 93 + 85}{5}$

$$= \frac{338}{5} = 67.6$$

Order the scores:

$$0, 78, 82, 85, 93$$

The median score for 5 scores is the third score, that is, 82.

108. **[5.8]** $\text{hypotenuse} = \sqrt{(\text{leg})^2 + (\text{leg})^2}$

$$= \sqrt{(16)^2 + (12)^2}$$
$$= \sqrt{256 + 144}$$
$$= \sqrt{400}$$
$$= 20 \text{ miles}$$

She is 20 miles from her starting point.

109. **[5.10]** A "can" is geometrically a cylinder.

$V = \pi \cdot r^2 \cdot h$ ($d = 4$ in., so $r = 2$ in.)

$V = \pi \cdot (2 \text{ in.})^2 \cdot 7 \text{ in.}$
$V \approx 3.14 \cdot 2 \text{ in.} \cdot 2 \text{ in.} \cdot 7 \text{ in.}$
$V \approx 87.9 \text{ in.}^3$

Chapter 5 Test

1. $18.4 = 18\frac{4}{10} = 18\frac{4 \div 2}{10 \div 2} = 18\frac{2}{5}$

2. $0.075 = \frac{75}{1000} = \frac{75 \div 25}{1000 \div 25} = \frac{3}{40}$

3. 60.007 is sixty and seven thousandths.

4. 0.0208 is two hundred eight ten-thousandths.

5. $7.6 + 82.0128 + 39.59$

Estimate: *Exact:*

$$\begin{array}{r} 8 \\ 80 \\ +\,40 \\ \hline 128 \end{array} \qquad \begin{array}{r} 7.6000 \\ 82.0128 \\ +\,39.5900 \\ \hline 129.2028 \end{array}$$

Line up decimals. Write in zeros.

6. $-5.79(1.2)$ The product is negative.

Estimate: *Exact:*

$$\begin{array}{r} -6 \\ \times\,1 \\ \hline -6 \end{array} \qquad \begin{array}{r} -5.79 \\ \times\,1.2 \\ \hline 1\,158 \\ 5\,79 \\ \hline -6.948 \end{array}$$

7. $-79.1 - 3.602$
$= -79.1 + (-3.602)$

Both addends have the same sign, so the sum will be negative.

Estimate: *Exact:*

$$\begin{array}{r} -80 \\ +\,-4 \\ \hline -84 \end{array} \qquad \begin{array}{r} -79.100 \\ +\,-3.602 \\ \hline -82.702 \end{array}$$

8. $-20.04 \div (-4.8)$
Same signs, quotient is positive

Estimate: *Exact:*

$$\begin{array}{r} 4 \\ 5\overline{)20} \end{array} \qquad \begin{array}{r} 4.\,1\,7\,5 \\ 4.8\overline{)2\,0.\,0_\wedge\,4\,0\,0} \\ 1\,9\,2 \\ \hline 8\,4 \\ 4\,8 \\ \hline 3\,6\,0 \\ 3\,3\,6 \\ \hline 2\,4\,0 \\ 2\,4\,0 \\ \hline 0 \end{array}$$

9. $670 - 0.996$

$$\begin{array}{r} 670.000 \\ -\,0.996 \\ \hline 669.004 \end{array}$$

Line up decimals. Write in zeros.

10.

$$\begin{array}{r} 4\,8\,0. \\ 0.15_\wedge\overline{)7\,2.\,0\,0_\wedge} \\ 6\,0 \\ \hline 1\,2\,0 \\ 1\,2\,0 \\ \hline 0\,0 \\ 0 \\ \hline 0 \end{array}$$

Move decimal point 2 places in dividend and divisor; write two zeros.

11. $(-0.006)(-0.007)$
Same signs, product is positive

$$\begin{array}{rcl} 0.006 & \leftarrow & \text{3 decimal places} \\ \times\,0.007 & \leftarrow & \text{3 decimal places} \\ \hline 0.000042 & \leftarrow & \text{6 decimal places} \end{array}$$

12. Divide to find the cost per meter.

$$\begin{array}{r} 4.\,5\,5 \\ 3.4_\wedge\overline{)\$1\,5.\,4_\wedge\,7\,0} \\ 1\,3\,6 \\ \hline 1\,8\,7 \\ 1\,7\,0 \\ \hline 1\,7\,0 \\ 1\,7\,0 \\ \hline 0 \end{array}$$

Move decimal point 1 place in dividend and divisor; write one zero.

The fabric costs \$4.55 per meter.

13. Compare the two times.

Angela: 3.500 minutes (more time)
Davida: 3.059 minutes (less time)

$$\begin{array}{r} 3.500 \\ -\,3.059 \\ \hline 0.441 \end{array}$$

Davida won by 0.441 minute.

14. Multiply the price per pound by the amount purchased.

$$\begin{array}{r} \$2.89 \\ \times\,1.85 \\ \hline 1445 \\ 2\,312 \\ 2\,89 \\ \hline \$5.3465 \end{array}$$

Round \$5.3465 to the nearest cent.
Draw a cut-off line: \$5.34|65
The first digit cut is 6, which is 5 or more, so round up.

Mr. Yamamoto paid \$5.35 for the cheese.

15.

$$\begin{aligned} -5.9 &= y + 0.25 \quad \text{Add } -0.25 \text{ to both sides.} \\ \underline{-0.25} &\quad \underline{-0.25} \\ -6.15 &= y + 0 \\ -6.15 &= y \end{aligned}$$

The solution is −6.15.

16.

$$\begin{aligned} -4.2x &= 1.47 \quad \text{Divide both sides by } -4.2. \\ \frac{-4.2x}{-4.2} &= \frac{1.47}{-4.2} \\ x &= -0.35 \end{aligned}$$

The solution is −0.35.

17.

$$\begin{aligned} 3a - 22.7 &= 10 \\ 3a + (-22.7) &= 10 \quad \text{Add 22.7 to both sides.} \\ \underline{+22.7} &\quad \underline{+22.7} \\ 3a + 0 &= 32.7 \quad \text{Divide both sides by 3.} \\ \frac{3a}{3} &= \frac{32.7}{3} \\ a &= 10.9 \end{aligned}$$

The solution is 10.9.

18.

$$\begin{aligned} -0.8n + 1.88 &= 2n - 6.1 \quad \text{Add } 0.8n \text{ to both sides.} \\ \underline{+0.8n} &\quad \underline{+0.8n} \\ 0 + 1.88 &= 2.8n - 6.1 \\ 1.88 &= 2.8n - 6.1 \quad \text{Add 6.1 to both sides.} \\ \underline{+6.1} &\quad \underline{+6.1} \\ 7.98 &= 2.8n + 0 \quad \text{Divide both sides by 2.8.} \\ \frac{7.98}{2.8} &= \frac{2.8n}{2.8} \\ 2.85 &= n \end{aligned}$$

The solution is 2.85.

19. $0.44, 0.451, \frac{9}{20}, 0.4506$

Change to ten-thousandths, if necessary, and compare.

$0.44 = 0.4400$ *smallest*

$0.451 = 0.4510$ *largest*

$\frac{9}{20} = 0.4500$

$0.4506 = 0.4506$

From smallest to largest: $0.44, \frac{9}{20}, 0.4506, 0.451$

20.

$$\begin{aligned} &6.3^2 - 5.9 + 3.4(-0.5) && \textit{Exponent} \\ &= 39.69 - 5.9 + 3.4(-0.5) && \textit{Multiply.} \\ &= 39.69 - 5.9 + (-1.7) && \textit{Add the opposite.} \\ &= 39.69 + (-5.9) + (-1.7) && \textit{Add.} \\ &= 33.79 + (-1.7) && \textit{Add.} \\ &= 32.09 \end{aligned}$$

21. mean

$$= \frac{52 + 61 + 68 + 69 + 73 + 75 + 79 + 84 + 91 + 98}{10}$$

$$= \frac{750}{10} = 75 \text{ books}$$

22. 96°, <u>104°</u>, <u>103°</u>, <u>104°</u>, <u>103°</u>, <u>104°</u>, 91°, 74°, <u>103°</u>

The mode is the value that occurs most often. The modes are 103° and 104° (bimodal).

23.

Cost	Frequency	Product
\$6	7	\$42
\$10	3	\$30
\$11	4	\$44
\$14	2	\$28
\$19	3	\$57
\$24	1	\$24
	20	\$225

$$\text{weighted mean} = \frac{\$225}{20} = \$11.25$$

24. **(a)** Arrange the prices in numerical order.

\$39.75, \$46.90, \$48, \$49.30, \$51.80, \$54.50, \$56.25, \$89

Because there is an even number of items, find the middle two numbers (4th and 5th of 8):

$$\text{median} = \frac{\$49.30 + \$51.80}{2} = \frac{\$101.10}{2} = \$50.55$$

(b) Arrange the prices in numerical order.

\$30.95, \$47.30, \$49.80, \$50.45, \$63, \$75, \$97.55

Because there is an odd number of items, find the middle number (4th of 7):

median = \$50.45

(c) Math textbook range: \$89 − \$39.75 = \$49.25

Biology textbook range:
\$97.55 − \$30.95 = \$66.60

Since \$49.25 < \$66.60, the math textbook costs have less variability.

25.

$$\begin{aligned} \text{hypotenuse} &= \sqrt{(\text{leg})^2 + (\text{leg})^2} \\ &= \sqrt{(7)^2 + (6)^2} \\ &= \sqrt{49 + 36} \\ &= \sqrt{85} \\ &\approx 9.2 \text{ cm} \end{aligned}$$

26. The unknown side is a leg of the right triangle.

$$\begin{aligned}\text{leg} &= \sqrt{(\text{hypotenuse})^2 - (\text{leg})^2}\\ &= \sqrt{(20)^2 - (11)^2}\\ &= \sqrt{400 - 121}\\ &= \sqrt{279}\\ &\approx 16.7 \text{ ft (nearest tenth)}\end{aligned}$$

27. $\text{radius} = \frac{d}{2} = \frac{25 \text{ in.}}{2} = 12.5 \text{ in. or } 12\frac{1}{2} \text{ in.}$

28. $$\begin{aligned}C &= 2 \cdot \pi \cdot r\\ &\approx 2 \cdot 3.14 \cdot 0.9 \text{ km}\\ &\approx 5.7 \text{ km}\end{aligned}$$

29. $$\begin{aligned}r &= \frac{d}{2} = \frac{16.2 \text{ cm}}{2} = 8.1 \text{ cm}\\ A &= \pi \cdot r^2\\ &\approx 3.14 \cdot 8.1 \text{ cm} \cdot 8.1 \text{ cm}\\ &\approx 206.0 \text{ cm}^2\end{aligned}$$

30. The figure is a cylinder.

$$\begin{aligned}V &= \pi \cdot r^2 \cdot h\\ &\approx 3.14 \cdot 18 \text{ ft} \cdot 18 \text{ ft} \cdot 5 \text{ ft}\\ &= 5086.8 \text{ ft}^3\end{aligned}$$

31. $$\begin{aligned}SA &= 2\pi rh + 2\pi r^2\\ &\approx 2 \cdot 3.14 \cdot 18 \text{ ft} \cdot 5 \text{ ft} + 2 \cdot 3.14 \cdot 18 \text{ ft} \cdot 18 \text{ ft}\\ &\approx 2599.9 \text{ ft}^2\end{aligned}$$

Cumulative Review Exercises (Chapters 1–5)

1. **(a)** In words, 45.0203 is

forty-five and two hundred three ten-thousandths.

(b) In words, 30,000,650,008 is

thirty billion, six hundred fifty thousand, eight.

2. **(a)** One hundred sixty million, five hundred is written as

160,000,500.

(b) Seventy-five thousandths is written as

0.075.

3. Underline the hundreds place: 46,<u>9</u>08

The next digit is 4 or less. Leave 9 and change 8 to 0.

46,900

4. Underline the hundredths place: 6.1<u>9</u>7

The next digit is 5 or more. Add 1 to 9. Write 0 and regroup 1 to the tenths place. Drop all digits to the right of the underlined place.

6.20

5. Underline the thousandths place: 0.66<u>1</u>48

The next digit is 4 or less. Leave 1 and drop all digits to the right of the underlined place.

0.661

6. Underline the hundreds place: 9<u>9</u>51

The next digit is 5 or more. Add 1 to 9. Write 0 and regroup 1 to the thousands place. Change 5 and 1 to 0.

10,000

7. $$\begin{aligned}&-5 - 8 && \text{Change subtraction.}\\ &= -5 + (-8) && \text{Add.}\\ &= -13\end{aligned}$$

8. $-0.003(0.02)$ The product is negative.

$$\begin{array}{rll} 0.003 & \leftarrow & \textit{3 decimal places}\\ \times\ 0.02 & \leftarrow & \textit{2 decimal places}\\ \hline 0.00006 & \leftarrow & \textit{5 decimal places}\end{array}$$

Answer: -0.00006

9. $\frac{-7}{0}$ is undefined.

10. $$\begin{aligned}&8 + 4(2-5)^2 && \text{Parentheses}\\ &= 8 + 4(-3)^2 && \text{Exponent}\\ &= 8 + 4(9) && \text{Multiply.}\\ &= 8 + 36 && \text{Add.}\\ &= 44\end{aligned}$$

11. $-\frac{3}{8}(-48) = -\frac{3}{8}\left(-\frac{48}{1}\right) = \frac{3 \cdot 6 \cdot \cancel{8}^{1}}{\cancel{8}_{1} \cdot 1} = \frac{18}{1} = 18$

12. $$\begin{aligned}&|4| - |-10| && \text{Absolute values}\\ &= 4 - 10 && \text{Change subtraction.}\\ &= 4 + (-10) && \text{Add.}\\ &= -6\end{aligned}$$

13. $0.721 + 55.9$

$$\begin{array}{rl} \overset{1}{\cancel{0}}.721 & \text{Line up decimals.}\\ +\ 55.900 & \text{Fill in zeros.}\\ \hline 56.621 & \end{array}$$

14. $$\begin{aligned}3\tfrac{1}{3} - 1\tfrac{5}{6} &= \frac{10}{3} - \frac{11}{6} = \frac{20}{6} - \frac{11}{6} = \frac{20-11}{6}\\ &= \frac{9}{6} = \frac{9 \div 3}{6 \div 3} = \frac{3}{2} \text{ or } 1\frac{1}{2}\end{aligned}$$

15. $\left(\frac{3}{b^2}\right)\left(\frac{b}{8}\right) = \frac{3 \cdot \cancel{b}^{1}}{\cancel{b}_{1} \cdot b \cdot 8} = \frac{3}{8b}$

16. $12 - 0.853$ Line up decimal points and fill in zeros for subtracting.

$$\begin{array}{r} 12.000 \\ -\ 0.853 \\ \hline \end{array} \qquad \begin{array}{r} {\scriptstyle 1\ 9\ 9\ 10} \\ 1\not{2}.\not{0}\not{0}\not{0} \\ -\ 0.8\,5\,3 \\ \hline 11.1\,4\,7 \end{array}$$

17. $\dfrac{3}{10} - \dfrac{3}{4} = \dfrac{6}{20} - \dfrac{15}{20} = \dfrac{6-15}{20}$

$= \dfrac{-9}{20}$ or $-\dfrac{9}{20}$

18. $\dfrac{\frac{5}{16}}{-10} = \dfrac{5}{16} \div (-10) = \dfrac{5}{16} \div \left(-\dfrac{10}{1}\right)$

$= \dfrac{5}{16} \cdot \left(-\dfrac{1}{10}\right) = -\dfrac{\overset{1}{\not{5}} \cdot 1}{16 \cdot 2 \cdot \underset{1}{\not{5}}}$

$= -\dfrac{1}{32}$

19. $-3.75 \div (-2.9)$

$= -37.5 \div (-29)$ *Move both decimal points one place to the right.*

Same signs, so the quotient is positive.

$$\begin{array}{r} 1.\ 2\ 9 \\ 29\overline{)\,3\ 7.\ 5\ 0} \\ \underline{2\ 9} \\ 8\ 5 \\ \underline{5\ 8} \\ 2\ 7\ 0 \\ \underline{2\ 6\ 1} \\ 9 \end{array}$$

$\leftarrow$ In order to round to the tenths, divide out one more place, to the hundredths.

Answer: 1.3 (rounded)

20. $\dfrac{x}{2} + \dfrac{3}{5} = \dfrac{x \cdot 5}{2 \cdot 5} + \dfrac{3 \cdot 2}{5 \cdot 2} = \dfrac{5x}{10} + \dfrac{6}{10} = \dfrac{5x+6}{10}$

21. $5 \div \left(-\dfrac{5}{8}\right) = \dfrac{5}{1} \cdot \left(-\dfrac{8}{5}\right) = -\dfrac{\overset{1}{\not{5}} \cdot 8}{1 \cdot \underset{1}{\not{5}}}$

$= -\dfrac{8}{1} = -8$

22. $2^5 - 4^3$ Exponents first

$= 32 - 64$ Change subtraction.

$= 32 + (-64)$ Add.

$= -32$

23. $\dfrac{-36 \div (-2)}{-6 - 4(0-6)}$

Numerator:

$-36 \div (-2) = \dfrac{36}{2} = \dfrac{18 \cdot \overset{1}{\not{2}}}{\underset{1}{\not{2}}} = 18$

Denominator:

$-6 - 4(0-6) = -6 - 4(-6)$

$= -6 - (-24)$

$= -6 + 24$

$= 18$

Last step is division: $\dfrac{18}{18} = 1$

24. $(0.8)^2 - 3.2 + 4(-0.8)$ Exponent

$= 0.64 - 3.2 + 4(-0.8)$ Multiply.

$= 0.64 - 3.2 + (-3.2)$ Add the opposite.

$= 0.64 + (-3.2) + (-3.2)$ Add.

$= -2.56 + (-3.2)$

$= -5.76$

25. $\dfrac{3}{4} \div \dfrac{3}{10}\left(\dfrac{1}{4} + \dfrac{2}{3}\right)$ Parentheses, LCD is 12

$= \dfrac{3}{4} \div \dfrac{3}{10}\left(\dfrac{3}{12} + \dfrac{8}{12}\right)$ Add.

$= \dfrac{3}{4} \div \dfrac{3}{10} \cdot \dfrac{11}{12}$ Change division.

$= \dfrac{3}{4} \cdot \dfrac{10}{3} \cdot \dfrac{11}{12}$ Multiply.

$= \dfrac{5}{2} \cdot \dfrac{11}{12}$ Multiply.

$= \dfrac{55}{24}$ or $2\dfrac{7}{24}$

26. To find the mean, add the ages and divide by the number of students.

$$\text{mean} = \frac{(19+23+24+19+20+29+26+35 + 20+22+26+23+25+26+20+30)}{16}$$

$= \dfrac{387}{16}$

$= 24.1875$

≈ 24.2 years (nearest tenth)

27. To find the median, first arrange the data from high to low or low to high.

19, 19, 20, 20, 20, 22, 23, 23, 24, 25, 26, 26, 26, 29, 30, 35

Since there is an even number of items (16), the median is defined as the average of the middle two numbers (8th and 9th).

$\dfrac{23+24}{2} = 23.5$

The median age is 23.5 years.

28. To find the mode, look for the items which occur most frequently.

This set has 2 modes, 20 and 26 years (each occurs 3 times). We say that the list is bimodal.

29. $3h - 4h = 16 - 12$ Add terms.

$-1h = 4$ Divide both sides by -1.

$$\frac{-1h}{-1} = \frac{4}{-1}$$
$$h = -4$$

The solution is -4.

30. $-2x = x - 15$

$-2x = 1x - 15$ Add $-1x$ to both sides.

$$\underline{-1x} \quad \underline{-1x}$$

$-3x = 0 - 15$ Divide both sides by -3.

$$\frac{-3x}{-3} = \frac{-15}{-3}$$
$$x = 5$$

The solution is 5.

31. $20 = 6r - 45.4$ Add 45.4 to both sides.

$$\underline{+45.4} \quad \underline{+45.4}$$

$65.4 = 6r + 0$ Divide both sides by 6.

$$\frac{65.4}{6} = \frac{6r}{6}$$
$$10.9 = r$$

The solution is 10.9.

32. $-3(y+4) = 7y + 8$ Distributive property

$-3y + (-12) = 7y + 8$ Add $3y$ to both sides.

$$\underline{+3y} \quad \underline{+3y}$$
$$0 + (-12) = 10y + 8$$

$-12 = 10y + 8$ Add -8 to both sides.

$$\underline{-8} \quad \underline{-8}$$

$-20 = 10y + 0$ Divide both sides by 10.

$$\frac{-20}{10} = \frac{10y}{10}$$
$$-2 = y$$

The solution is -2.

33. $-0.8 + 1.4n = 2 + 0.7n$ Add $-0.7n$ to both sides.

$$\underline{-0.7n} \quad \underline{-0.7n}$$
$$-0.8 + 0.7n = 2 + 0$$

$-0.8 + 0.7n = 2$ Add 0.8 to both sides.

$$\underline{+0.8} \quad \underline{+0.8}$$

$0 + 0.7n = 2.8$ Divide both sides by 0.7.

$$\frac{0.7n}{0.7} = \frac{2.8}{0.7}$$
$$n = 4$$

The solution is 4.

34. $P = 10 \text{ ft} + 17 \text{ ft} + 10 \text{ ft} + 17 \text{ ft}$
$P = 54 \text{ ft}$

$A = bh$
$A = 10 \text{ ft} \cdot 14 \text{ ft}$
$A = 140 \text{ ft}^2$

35. $C = \pi d$
$C \approx 3.14 \cdot 13 \text{ m}$
$C \approx 40.8 \text{ m}$ (nearest tenth)

$A = \pi \cdot r^2$ *Note:* $r = \frac{d}{2} = \frac{13}{2}$
$A = \pi \cdot r \cdot r$
$A = \pi \cdot \frac{13}{2} \text{ m} \cdot \frac{13}{2} \text{ m}$
$A \approx 3.14 \cdot 6.5 \text{ m} \cdot 6.5 \text{ m}$
$A \approx 132.7 \text{ m}^2$ (nearest tenth)

36. The figure is a right triangle, so use the Pythagorean Theorem.

$$\text{leg} = \sqrt{(\text{hypotenuse})^2 - (\text{leg})^2}$$
$$x = \sqrt{(14)^2 - (6)^2}$$
$$x = \sqrt{196 - 36}$$
$$x = \sqrt{160}$$

$x \approx 12.6 \text{ km}$ (nearest tenth)

$A = \frac{1}{2}bh$
$A \approx 0.5 \cdot 12.6 \text{ km} \cdot 6 \text{ km}$
$A \approx 37.8 \text{ km}^2$

37. The figure is a cube or rectangular solid.

$V = l \cdot w \cdot h$
$V = 1.5 \text{ in.} \cdot 1.5 \text{ in.} \cdot 1.5 \text{ in.}$
$V = 3.375 \text{ in.}^3$ or 3.4 in.^3 (rounded)

$SA = 2lw + 2lh + 2wh$
$SA = 2 \cdot 1.5 \text{ in.} \cdot 1.5 \text{ in.} + 2 \cdot 1.5 \text{ in.} \cdot 1.5 \text{ in.} + 2 \cdot 1.5 \text{ in.} \cdot 1.5 \text{ in.}$
$SA = 13.5 \text{ in.}^2$

38. \$17.96 rounds to \$20
\$0.87 rounds to \$1

Estimate: \$40 – \$20 – \$1 = \$19
Exact: \$40 – \$17.96 – \$0.87 = \$21.17

39. $2\frac{1}{3}$ yd rounds to 2 yd
$3\frac{7}{8}$ yd rounds to 4 yd

Estimate: 2 yd + 4 yd = 6 yd

$$\text{Exact: } 2\frac{1}{3}\text{ yd} + 3\frac{7}{8}\text{ yd} = \frac{7}{3}\text{ yd} + \frac{31}{8}\text{ yd} = \frac{56}{24}\text{ yd} + \frac{93}{24}\text{ yd} = \frac{149}{24}\text{ yd} = 6\frac{5}{24}\text{ yd}$$

40. 2.7 pounds rounds to 3 pounds.
\$6.18 rounds to \$6.

Estimate: \$6 ÷ 3 = \$2 per pound

Exact: \$6.18 ÷ 2.7 = $\$2.28\overline{8}$
≈ \$2.29 per pound

41. \$78,000 rounds to \$80,000
107 students rounds to 100 students.

Estimate: \$80,000 ÷ 100 = \$800

Exact: \$78,000 ÷ 107
≈ \$728.97
≈ \$729 (rounded to the nearest dollar)

42. *Step 1*
Unknowns: amount each student receives

Knowns: one student receives three times as much as the other; \$30,000 total

Step 2
Let x be the amount of the first student's money and $3x$ be the amount of the second student's money.

Step 3

first student's money		second student's money	is	scholarship total
x	+	$3 \cdot x$	=	30,000

Step 4

$$x + 3x = 30{,}000$$
$$4x = 30{,}000$$
$$\frac{4x}{4} = \frac{30{,}000}{4}$$
$$x = \$7500$$
$$3 \cdot x = 3 \cdot \$7500 = \$22{,}500$$

Step 5
One student receives \$7500 and the other student receives \$22,500.

Step 6
\$22,500 is three times \$7500 and the sum of \$7500 and \$22,500 is \$30,000.

43. *Step 1*
Unknowns: length and width of the rectangle

Knowns: length is 4 in. more than the width; perimeter is 48 in.

Step 2
Let w be the width and $w + 4$ be the length.

Step 3

$$P = 2w + 2l$$
$$48 = 2w + 2(w + 4)$$

Step 4

$$48 = 2w + 2w + 8$$
$$48 = 4w + 8$$
$$\underline{-8 \qquad -8}$$
$$40 = 4w$$
$$\frac{40}{4} = \frac{4w}{4}$$
$$10 = w, \text{ so } w + 4 = 14$$

Step 5
The width is 10 in. and the length is 14 in.

Step 6
14 in. is 4 in. longer than 10 in. and the perimeter is 2(14 in.) + 2(10 in.) = 48 in.

CHAPTER 6 RATIO, PROPORTION, AND LINE/ANGLE/ TRIANGLE RELATIONSHIPS

6.1 Ratios

6.1 Section Exercises

1. 8 days to 9 days

$$\frac{8 \text{ days}}{9 \text{ days}} = \frac{8}{9}$$

3. \$100 to \$50

$$\frac{\$100}{\$50} = \frac{100}{50} = \frac{100 \div 50}{50 \div 50} = \frac{2}{1}$$

5. 30 minutes to 90 minutes

$$\frac{30 \text{ minutes}}{90 \text{ minutes}} = \frac{30}{90} = \frac{30 \div 30}{90 \div 30} = \frac{1}{3}$$

7. 80 miles to 50 miles

$$\frac{80 \text{ miles}}{50 \text{ miles}} = \frac{80}{50} = \frac{80 \div 10}{50 \div 10} = \frac{8}{5}$$

9. 6 hours to 16 hours

$$\frac{6 \text{ hours}}{16 \text{ hours}} = \frac{6 \div 2}{16 \div 2} = \frac{3}{8}$$

11. \$4.50 to \$3.50

$$\frac{\$4.50}{\$3.50} = \frac{4.50}{3.50} = \frac{4.50 \cdot 10}{3.50 \cdot 10} = \frac{45}{35}$$
$$= \frac{45 \div 5}{35 \div 5} = \frac{9}{7}$$

13. 15 to $2\frac{1}{2}$

$$\frac{15}{2\frac{1}{2}} = \frac{\frac{15}{1}}{\frac{5}{2}} = \frac{15}{1} \div \frac{5}{2} = \frac{\overset{3}{\cancel{15}}}{1} \cdot \frac{2}{\underset{1}{\cancel{5}}} = \frac{6}{1}$$

15. $1\frac{1}{4}$ to $1\frac{1}{2}$

$$\frac{1\frac{1}{4}}{1\frac{1}{2}} = \frac{\frac{5}{4}}{\frac{3}{2}} = \frac{5}{4} \div \frac{3}{2} = \frac{5}{\underset{2}{\cancel{4}}} \cdot \frac{\overset{1}{\cancel{2}}}{3} = \frac{5}{6}$$

17. 4 feet to 30 inches

4 feet $= 4 \cdot 12$ inches $= 48$ inches

$$\frac{4 \text{ feet}}{30 \text{ inches}} = \frac{48 \text{ inches}}{30 \text{ inches}} = \frac{48}{30} = \frac{48 \div 6}{30 \div 6} = \frac{8}{5}$$

(Do not write the ratio as $1\frac{3}{5}$.)

19. 5 minutes to 1 hour

1 hour = 60 minutes

$$\frac{5 \text{ minutes}}{1 \text{ hour}} = \frac{5 \text{ minutes}}{60 \text{ minutes}} = \frac{5}{60} = \frac{5 \div 5}{60 \div 5} = \frac{1}{12}$$

21. 15 hours to 2 days

2 days $= 2 \cdot 24$ hours $= 48$ hours

$$\frac{15 \text{ hours}}{2 \text{ days}} = \frac{15 \text{ hours}}{48 \text{ hours}} = \frac{15}{48} = \frac{15 \div 3}{48 \div 3} = \frac{5}{16}$$

23. 5 gallons to 5 quarts

5 gallons $= 5 \cdot 4$ quarts $= 20$ quarts

$$\frac{5 \text{ gallons}}{5 \text{ quarts}} = \frac{20 \text{ quarts}}{5 \text{ quarts}} = \frac{20}{5} = \frac{20 \div 5}{5 \div 5} = \frac{4}{1}$$

25. Thanksgiving cards to graduation cards

$$\frac{30 \text{ million}}{60 \text{ million}} = \frac{30}{60} = \frac{30 \div 30}{60 \div 30} = \frac{1}{2}$$

27. Valentine's Day cards to Halloween cards

$$\frac{900 \text{ million}}{25 \text{ million}} = \frac{900}{25} = \frac{900 \div 25}{25 \div 25} = \frac{36}{1}$$

29. Answers will vary. One possibility is stocking cards of various types in the same ratios as those in the table.

31. *Candle in the Wind* to *White Christmas*

$$\frac{36 \text{ million}}{30 \text{ million}} = \frac{36}{30} = \frac{36 \div 6}{30 \div 6} = \frac{6}{5}$$

Candle in the Wind to *Rock Around the Clock*

$$\frac{36 \text{ million}}{17 \text{ million}} = \frac{36}{17}$$

33. To get a ratio of $\frac{3}{1}$, we can start with the smallest value in the table and see if there is a value that is three times the smallest. $3 \times 9 = 27$ and 27 million is not in the table. $3 \times 10 = 30$, so there are at least 2 pairs of songs that give a ratio of $\frac{3}{1}$: *White Christmas* to *It's Now or Never* and *White Christmas* to *I Will Always Love You.*
$3 \times 12 = 36$, so there is another pair: *Candle in the Wind* to *I Want to Hold Your Hand.*
$3 \times 17 = 51$, which is greater than the largest value in the table, so we do not need to look for any more pairs.

35. $$\frac{\text{taxes}}{\text{transportation}} = \frac{\$400}{\$200} = \frac{400 \div 200}{200 \div 200} = \frac{2}{1}$$

The ratio of taxes to transportation is $\frac{2}{1}$.

37. **(a)** Total budget $= \$225 + \$125 + \$200 + \$400 + \$300 + \$750 = \$2000$

$$\frac{\text{rent}}{\text{total budget}} = \frac{\$750}{\$2000} = \frac{750 \div 250}{2000 \div 250} = \frac{3}{8}$$

The ratio of rent to total budget is $\frac{3}{8}$.

(b) $\dfrac{\text{rent and utilities}}{\text{taxes and miscellaneous}} = \dfrac{\$750 + \$125}{\$400 + \$225} = \dfrac{\$875}{\$625} = \dfrac{875 \div 125}{625 \div 125} = \dfrac{7}{5}$

The ratio of rent and utilities to taxes and miscellaneous is $\frac{7}{5}$.

39. $\dfrac{\text{longest side}}{\text{shortest side}} = \dfrac{7 \text{ feet}}{5 \text{ feet}} = \dfrac{7}{5}$

The ratio of the length of the longest side to the length of the shortest side is $\frac{7}{5}$.

41. $\dfrac{\text{longest side}}{\text{shortest side}} = \dfrac{1.8 \text{ meters}}{0.3 \text{ meters}} = \dfrac{1.8}{0.3} = \dfrac{1.8 \cdot 10}{0.3 \cdot 10} = \dfrac{18}{3} = \dfrac{18 \div 3}{3 \div 3} = \dfrac{6}{1}$

The ratio of the length of the longest side to the length of the shortest side is $\frac{6}{1}$.

43. $\dfrac{9\frac{1}{2} \text{ inches}}{4\frac{1}{4} \text{ inches}} = \dfrac{9\frac{1}{2}}{4\frac{1}{4}} = \dfrac{\frac{19}{2}}{\frac{17}{4}} = \dfrac{19}{2} \div \dfrac{17}{4} = \dfrac{19}{\cancel{2}_1} \cdot \dfrac{\cancel{4}^2}{17} = \dfrac{38}{17}$

The ratio of the length of the longest side to the length of the shortest side is $\frac{38}{17}$.

45. The increase in price is

$$\$13.50 - \$10.80 = \$2.70.$$

$\dfrac{\$2.70}{\$10.80} = \dfrac{2.70}{10.80} = \dfrac{2.70 \cdot 10}{10.80 \cdot 10} = \dfrac{27}{108} = \dfrac{27 \div 27}{108 \div 27} = \dfrac{1}{4}$

The ratio of the increase in price to the original price is $\frac{1}{4}$.

47. $59\dfrac{1}{2} \text{ days} \div 7 = \dfrac{\cancel{119}^{17}}{2} \cdot \dfrac{1}{\cancel{7}_1} = \dfrac{17}{2} \text{ weeks}$

$\dfrac{\frac{17}{2} \text{ weeks}}{8\frac{3}{4} \text{ weeks}} = \dfrac{\frac{17}{2}}{\frac{35}{4}} = \dfrac{17}{2} \div \dfrac{35}{4} = \dfrac{17}{\cancel{2}_1} \cdot \dfrac{\cancel{4}^2}{35} = \dfrac{34}{35}$

The ratio of the first movie's filming time to the second movie's filming time is $\frac{34}{35}$.

49. $\dfrac{\text{longest side}}{\text{shortest side}} = \dfrac{2\frac{7}{12} \text{ feet}}{2\frac{7}{12} \text{ feet}} = \dfrac{1}{1}$

The ratio of the length of the longest side to the length of the shortest side is $\frac{1}{1}$.

As long as the sides all have the same length, any measurement you choose will maintain the ratio.

50. Answers will vary. Some possibilities are:

$$\frac{4}{5} = \frac{8}{10} = \frac{12}{15} = \frac{16}{20} = \frac{20}{25} = \frac{24}{30} = \frac{28}{35}.$$

51. It is not possible. Amelia would have to be older than her mother to have a ratio of 5 to 3.

52. Answers will vary, but a ratio of 3 to 1 means your income is 3 times your friend's income.

6.2 Rates

6.2 Section Exercises

1. 10 cups for 6 people

$\dfrac{10 \text{ cups} \div 2}{6 \text{ people} \div 2} = \dfrac{5 \text{ cups}}{3 \text{ people}}$

3. 15 feet in 35 seconds

$\dfrac{15 \text{ feet} \div 5}{35 \text{ seconds} \div 5} = \dfrac{3 \text{ feet}}{7 \text{ seconds}}$

5. 14 people for 28 dresses

$\dfrac{14 \text{ people} \div 14}{28 \text{ dresses} \div 14} = \dfrac{1 \text{ person}}{2 \text{ dresses}}$

7. 25 letters in 5 minutes

$\dfrac{25 \text{ letters} \div 5}{5 \text{ minutes} \div 5} = \dfrac{5 \text{ letters}}{1 \text{ minute}}$

9. \$63 for 6 visits

$\dfrac{\$63 \div 3}{6 \text{ visits} \div 3} = \dfrac{\$21}{2 \text{ visits}}$

11. 72 miles on 4 gallons

$\dfrac{72 \text{ miles} \div 4}{4 \text{ gallons} \div 4} = \dfrac{18 \text{ miles}}{1 \text{ gallon}}$

13. \$60 in 5 hours

$\dfrac{\$60 \div 5}{5 \text{ hours} \div 5} = \dfrac{\$12}{1 \text{ hour}}$

The unit rate is \$12 per hour or \$12/hour.

15. 50 eggs from 10 chickens

$\dfrac{50 \text{ eggs} \div 10}{10 \text{ chickens} \div 10} = \dfrac{5 \text{ eggs}}{1 \text{ chicken}}$

The unit rate is 5 eggs per chicken or 5 eggs/chicken.

17. 7.5 pounds for 6 people

$\dfrac{7.5 \text{ pounds} \div 6}{6 \text{ people} \div 6} = \dfrac{1.25 \text{ pounds}}{1 \text{ person}}$

The unit rate is 1.25 pounds/person.

19. \$413.20 for 4 days

$\dfrac{\$413.20 \div 4}{4 \text{ days} \div 4} = \dfrac{\$103.30}{1 \text{ day}}$

The unit rate is \$103.30/day.

21. Miles traveled:

$27{,}758.2 - 27{,}432.3 = 325.9$

Miles per gallon:

$$\frac{325.9 \text{ miles} \div 15.5}{15.5 \text{ gallons} \div 15.5} \approx 21.02 \approx 21.0$$

23. Miles traveled:

$28{,}396.7 - 28{,}058.1 = 338.6$

Miles per gallon:

$$\frac{338.6 \text{ miles} \div 16.2}{16.2 \text{ gallons} \div 16.2} \approx 20.90 \approx 20.9$$

25.

Size	Cost per Unit
2 ounces	$\frac{\$1.79}{2 \text{ ounces}} = \$0.895$
4 ounces	$\frac{\$3.09}{4 \text{ ounces}} = \$0.7725\ (*)$

The best buy is 4 ounces for \$3.09.

27.

Size	Cost per Unit
13 ounces	$\frac{\$2.80}{13 \text{ ounces}} \approx \$0.215$
15 ounces	$\frac{\$3.15}{15 \text{ ounces}} = \$0.210\ (*)$
18 ounces	$\frac{\$3.98}{18 \text{ ounces}} \approx \$0.221$

The best buy is 15 ounces for \$3.15.

29.

Size	Cost per Unit
12 ounces	$\frac{\$1.29}{12 \text{ ounces}} \approx \$0.108$
18 ounces	$\frac{\$1.79}{18 \text{ ounces}} \approx \$0.099\ (*)$
28 ounces	$\frac{\$3.39}{28 \text{ ounces}} \approx \$0.121$
40 ounces	$\frac{\$4.39}{40 \text{ ounces}} \approx \$0.110$

The best buy is 18 ounces for \$1.79.

31. Answers will vary. For example, you might choose Brand B because you like more chicken, so the cost per chicken chunk may actually be the same or less than Brand A.

33. 10.5 pounds in 6 weeks

$$\frac{10.5 \text{ pounds} \div 6}{6 \text{ weeks} \div 6} = \frac{1.75 \text{ pounds}}{1 \text{ week}}$$

Her rate of loss was 1.75 pounds/week.

35. 7 hours to earn \$85.82

$$\frac{\$85.82 \div 7}{7 \text{ hours} \div 7} = \frac{\$12.26}{1 \text{ hour}}$$

His pay rate is \$12.26/hour.

37. **(a)** Add the connection fee and five times the cost per minute.

Radiant Penny:
$\$0.39 + 5(\$0.01) = \$0.39 + \$0.05 = \$0.44$

IDT Special:
$\$0.14 + 5(\$0.022) = \$0.14 + \$0.11 = \$0.25$

Access America:
$\$0.00 + 5(\$0.047) = \$0.235$

(b) Divide the total costs by five minutes.

Radiant Penny: $\frac{\$0.44 \div 5}{5 \text{ minutes} \div 5} = \$0.088/\text{minute}$

IDT Special: $\frac{\$0.25 \div 5}{5 \text{ minutes} \div 5} = \$0.05/\text{minute}$

Access America: $\frac{\$0.235 \div 5}{5 \text{ minutes} \div 5} = \$0.047/\text{minute}$

Access America is the best buy.

39. Add the connection fee and fifteen (and the twenty) times the cost per minute.

Radiant Penny:
$\$0.39 + 15(\$0.01) = \$0.39 + \$0.15 = \$0.54$
$\$0.39 + 20(\$0.01) = \$0.39 + \$0.20 = \$0.59$

IDT Special:
$\$0.14 + 15(\$0.022) = \$0.14 + \$0.33 = \$0.47$
$\$0.14 + 20(\$0.022) = \$0.14 + \$0.44 = \$0.58$

Access America:
$\$0.00 + 15(\$0.047) = \$0.705$
$\$0.00 + 20(\$0.047) = \$0.94$

For the 15-minute call, IDT Special is the best buy at \$0.47. Alternatively, we could calculate the cost per minute.
For the 20-minute call, IDT Special is the best buy at \$0.58.

41. $\frac{44 \text{ seconds}}{400 \text{ meters}} = 0.11 \text{ second/meter}$

$$\frac{400 \text{ meters} \div 44}{44 \text{ seconds} \div 44} = 9.\overline{09} \text{ meters/second} \approx 9.1 \text{ meters/second}$$

Michael Johnson's rate was 0.11 second/meter or 9.1 meters/second (rounded).

43. One battery for \$1.79 is like getting 3 batteries for $\$1.79 \div 3 \approx \$0.597$ per battery.

An eight-pack of AAA batteries for \$4.99 is $\$4.99 \div 8 \approx \$0.624$ per battery.

The better buy is the one battery package.

45. **Brand G:** $2.39 – $0.60 coupon = $1.79

$$\frac{\$1.79}{10 \text{ ounces}} = \$0.179 \text{ per ounce}$$

Brand K: $\frac{\$3.99}{20.3 \text{ ounces}} \approx \$0.197 \text{ per ounce}$

Brand P: $3.39 – $0.50 coupon = $2.89

$$\frac{\$2.89}{16.5 \text{ ounces}} \approx \$0.175 \text{ per ounce } (*)$$

Brand P with the 50¢ coupon has the lowest cost per ounce and is the best buy.

47. The one-time activation fee is $35 (or $36) for 12 months.

$$\frac{\$35 \div 12}{12 \text{ months} \div 12} \approx \$2.92/\text{month}$$

$$(\$36/12 = \$3/\text{month})$$

The activation fee will add $2.92 (rounded) per month for Verizon, T-Mobile, and Nextel; $3 per month for Sprint.

48. Assume there are exactly 52 weeks in a year. Then there are $52 \cdot 5 = 260$ weekdays in a year.

$$\frac{260 \text{ weekdays} \div 12}{12 \text{ months} \div 12} \approx 21.7 \text{ weekdays/month}$$

For Verizon:

$$\frac{400 \text{ anytime minutes} \div 21.7}{21.7 \text{ weekdays} \div 21.7} \approx 18 \text{ minutes/weekday}$$

For T-Mobile:

$$\frac{600 \text{ anytime minutes} \div 21.7}{21.7 \text{ weekdays} \div 21.7} \approx 28 \text{ minutes/weekday}$$

For Nextel and Sprint:

$$\frac{500 \text{ anytime minutes} \div 21.7}{21.7 \text{ weekdays} \div 21.7} \approx 23 \text{ minutes/weekday}$$

49. To find the actual average cost per "anytime minute," add the monthly charge for the activation fee (from Exercise 47) to the monthly charge add then divide by the number of anytime minutes.

For Verizon:

$$\frac{\$2.92 + \$59.99}{400 \text{ minutes}} = \frac{\$62.91}{400 \text{ minutes}} \approx \$0.16/\text{minute}$$

For T-Mobile:

$$\frac{\$2.92 + \$39.99}{600 \text{ minutes}} = \frac{\$42.91}{600 \text{ minutes}} \approx \$0.07/\text{min } (*)$$

For Nextel:

$$\frac{\$2.92 + \$45.99}{500 \text{ minutes}} = \frac{\$48.91}{500 \text{ minutes}} \approx \$0.10/\text{minute}$$

For Sprint:

$$\frac{\$3 + \$55}{500 \text{ minutes}} = \frac{\$58}{500 \text{ minutes}} \approx \$0.12/\text{minute}$$

T-Mobile is the best buy at about $0.07 per "anytime minute".

50. The total cost is the sum of the activation fee, two monthly charges, and the termination fee. The number of "anytime minutes" in each case is $2 \times 100 = 200$.

For Verizon:

$$\frac{\$35 + 2(\$59.99) + \$175}{200 \text{ minutes}} = \frac{\$329.98}{200 \text{ minutes}} \approx \$1.65/\text{minute}$$

For T-Mobile:

$$\frac{\$35 + 2(\$39.99) + \$200}{200 \text{ minutes}} = \frac{\$314.98}{200 \text{ minutes}} \approx \$1.57/\text{minute}$$

For Nextel:

$$\frac{\$35 + 2(\$45.99) + \$200}{200 \text{ minutes}} = \frac{\$326.98}{200 \text{ minutes}} \approx \$1.63/\text{minute}$$

For Sprint:

$$\frac{\$36 + 2(\$55) + \$150}{200 \text{ minutes}} = \frac{\$296}{200 \text{ minutes}} = \$1.48/\text{minute}$$

6.3 Proportions

6.3 Section Exercises

1. $9 is to 12 cans as $18 is to 24 cans.

$$\frac{\$9}{12 \text{ cans}} = \frac{\$18}{24 \text{ cans}}$$

3. 200 adults is to 450 children as 4 adults is to 9 children.

$$\frac{200 \text{ adults}}{450 \text{ children}} = \frac{4 \text{ adults}}{9 \text{ children}}$$

5. 120 feet is to 150 feet as 8 feet is to 10 feet.

$\frac{120}{150} = \frac{8}{10}$ *The common units (feet) cancel.*

7. 2.2 hours is to 3.3 hours as 3.2 hours is to 4.8 hours.

$\frac{2.2}{3.3} = \frac{3.2}{4.8}$ *The common units (hours) cancel.*

9. $\frac{6}{10} = \frac{3}{5}$

$\frac{6 \div 2}{10 \div 2} = \frac{3}{5}$ and $\frac{3}{5}$

Both ratios are equivalent to $\frac{3}{5}$, so the proportion is *true*.

11. $\frac{5}{8} = \frac{25}{40}$

$\frac{5}{8}$ and $\frac{25 \div 5}{40 \div 5} = \frac{5}{8}$

Both ratios are equivalent to $\frac{5}{8}$, so the proportion is *true*.

13. $\frac{150}{200} = \frac{200}{300}$

$\frac{150 \div 50}{200 \div 50} = \frac{3}{4}$ and $\frac{200 \div 100}{300 \div 100} = \frac{2}{3}$

Because $\frac{3}{4}$ is *not* equivalent to $\frac{2}{3}$, the proportion is *false.*

15. $\frac{2}{9} = \frac{6}{27}$

Cross products: $2 \cdot 27 = 54; 9 \cdot 6 = 54$

The cross products are *equal,* so the proportion is *true.*

17. $\frac{20}{28} = \frac{12}{16}$

Cross products: $20 \cdot 16 = 320; 28 \cdot 12 = 336$

The cross products are *unequal,* so the proportion is *false.*

19. $\frac{110}{18} = \frac{160}{27}$

Cross products: $110 \cdot 27 = 2970; 18 \cdot 160 = 2880$

The cross products are *unequal,* so the proportion is *false.*

21. $\frac{3.5}{4} = \frac{7}{8}$

Cross products: $(3.5)(8) = 28; 4 \cdot 7 = 28$

The cross products are *equal,* so the proportion is *true.*

23. $\frac{18}{15} = \frac{2\frac{5}{6}}{2\frac{1}{2}}$

Cross products: $18 \cdot 2\frac{1}{2} = \frac{\overset{9}{\cancel{18}}}{1} \cdot \frac{5}{\underset{1}{\cancel{2}}} = 45$

$15 \cdot 2\frac{5}{6} = \frac{\overset{5}{\cancel{15}}}{1} \cdot \frac{17}{\underset{2}{\cancel{6}}} = \frac{85}{2} = 42\frac{1}{2}$

The cross products are *unequal,* so the proportion is *false.*

25. $\frac{6}{3\frac{2}{3}} = \frac{18}{11}$

Cross products: $6 \cdot 11 = 66;$

$3\frac{2}{3} \cdot 18 = \frac{11}{\underset{1}{\cancel{3}}} \cdot \frac{\overset{6}{\cancel{18}}}{1} = 66$

The cross products are *equal,* so the proportion is *true.*

27. (Answers may vary: proportion may be set up with "at bats" in both numerators and "hits" in both denominators. Corresponding cross products will be the same.)

Ichiro Suzuki   Sammy Sosa

$\frac{16 \text{ hits}}{50 \text{ at bats}} = \frac{128 \text{ hits}}{400 \text{ at bats}}$

Cross products: $16 \cdot 400 = 6400; 50 \cdot 128 = 6400$

The cross products are *equal,* so the proportion is *true.* Paul is correct; the two men hit equally well.

29. $\frac{1}{3} = \frac{x}{12}$

$3 \cdot x = 1 \cdot 12$ *Cross products are equivalent*

$\frac{\overset{1}{\cancel{3}} \cdot x}{\underset{1}{\cancel{3}}} = \frac{12}{3}$

$x = 4$

We will check our answers by showing that the cross products are equal.

Check: $1 \cdot 12 = 12 = 3 \cdot 4$

31. $\frac{15}{10} = \frac{3}{x}$ **OR** $\frac{3}{2} = \frac{3}{x}$

Since the numerators are equal, the denominators must be equal, so $x = 2$.

33. $\frac{x}{11} = \frac{32}{4}$ **OR** $\frac{x}{11} = \frac{8}{1}$

$1 \cdot x = 11 \cdot 8$ *Cross products are equivalent*

$x = 88$

Check: $88 \cdot 1 = 88 = 11 \cdot 8$

35. $\frac{42}{x} = \frac{18}{39}$

$x \cdot 18 = 42 \cdot 39$ *Cross products are equivalent*

$\frac{x \cdot \overset{1}{\cancel{18}}}{\underset{1}{\cancel{18}}} = \frac{\overset{7}{\cancel{42}} \cdot \overset{13}{\cancel{39}}}{\underset{1}{\underset{\cancel{6}}{\cancel{18}}}}$

$x = 91$

Check: $42 \cdot 39 = 1638 = 91 \cdot 18$

37. $\frac{x}{25} = \frac{4}{20}$ **OR** $\frac{x}{25} = \frac{1}{5}$

$5 \cdot x = 25 \cdot 1$ *Cross products are equivalent*

$\frac{\overset{1}{\cancel{5}} \cdot x}{\underset{1}{\cancel{5}}} = \frac{\overset{5}{\cancel{25}}}{\underset{1}{\cancel{5}}}$

$x = 5$

Check: $5 \cdot 5 = 25 = 25 \cdot 1$

39. $\frac{8}{x} = \frac{24}{30}$ **OR** $\frac{8}{x} = \frac{4}{5}$

$4 \cdot x = 8 \cdot 5$ *Cross products are equivalent*

$$\frac{\overset{1}{\cancel{4}} \cdot x}{\underset{1}{\cancel{4}}} = \frac{\overset{2}{\cancel{8}} \cdot 5}{\underset{1}{\cancel{4}}}$$

$x = 10$

Check: $8 \cdot 5 = 40 = 10 \cdot 4$

41. $\frac{99}{55} = \frac{44}{x}$ **OR** $\frac{9}{5} = \frac{44}{x}$

$9 \cdot x = 5 \cdot 44$ *Cross products are equivalent*

$$\frac{\overset{1}{\cancel{9}} \cdot x}{\underset{1}{\cancel{9}}} = \frac{220}{9}$$

$$x = \frac{220}{9}$$

≈ 24.44

Check: $9 \cdot \frac{220}{9} = 220 = 5 \cdot 44$

43. $\frac{0.7}{9.8} = \frac{3.6}{x}$

$0.7 \cdot x = 9.8\,(3.6)$ *Cross products are equivalent*

$$\frac{\overset{1}{\cancel{0.7}} \cdot x}{\underset{1}{\cancel{0.7}}} = \frac{35.28}{0.7}$$

$x = 50.4$

Check: $0.7(50.4) = 35.28 = 9.8(3.6)$

45. $\frac{250}{24.8} = \frac{x}{1.75}$

$24.8 \cdot x = 250(1.75)$ *Cross products are equivalent*

$$\frac{\overset{1}{\cancel{24.8}} \cdot x}{\underset{1}{\cancel{24.8}}} = \frac{437.5}{24.8}$$

≈ 17.64

Check: $250(1.75) = 437.5 = 24.8 \cdot \frac{437.5}{24.8}$

47. $\frac{15}{1\frac{2}{3}} = \frac{9}{x}$

$15 \cdot x = 1\frac{2}{3} \cdot 9$ *Cross products are equivalent*

$$15 \cdot x = \frac{5}{\underset{1}{\cancel{3}}} \cdot \frac{\overset{3}{\cancel{9}}}{1}$$

$$\frac{\overset{1}{\cancel{15}} \cdot x}{\underset{1}{\cancel{15}}} = \frac{15}{15}$$

$x = 1$

49. $\frac{2\frac{1}{3}}{1\frac{1}{2}} = \frac{x}{2\frac{1}{4}}$

$$1\frac{1}{2} \cdot x = 2\frac{1}{3} \cdot 2\frac{1}{4}$$

$$\frac{3}{2} \cdot x = \frac{7}{\underset{1}{\cancel{3}}} \cdot \frac{\overset{3}{\cancel{9}}}{4}$$

$$\frac{3}{2} \cdot x = \frac{21}{4}$$

$$\frac{\frac{3}{2} \cdot x}{\frac{3}{2}} = \frac{\frac{21}{4}}{\frac{3}{2}}$$

$$x = \frac{21}{4} \div \frac{3}{2} = \frac{\overset{7}{\cancel{21}}}{\underset{2}{\cancel{4}}} \cdot \frac{\overset{1}{\cancel{2}}}{\underset{1}{\cancel{3}}} = \frac{7}{2} = 3\frac{1}{2}$$

51. Change $\frac{1}{2}$ to a decimal by dividing: $1 \div 2 = 0.5$

$$\frac{\frac{1}{2}}{x} = \frac{2}{0.8}$$

$$\frac{0.5}{x} = \frac{2}{0.8}$$

$x \cdot 2 = 0.5(0.8)$

$x \cdot 2 = 0.4$

$$\frac{x \cdot \overset{1}{\cancel{2}}}{\underset{1}{\cancel{2}}} = \frac{0.4}{2}$$

$x = 0.2$

Now change 0.8 to a fraction and write it in lowest terms:

$$0.8 = \frac{8 \div 2}{10 \div 2} = \frac{4}{5}$$

$$\frac{\frac{1}{2}}{x} = \frac{2}{\frac{4}{5}}$$

$$x \cdot 2 = \frac{1}{\underset{1}{\cancel{2}}} \cdot \frac{\overset{2}{\cancel{4}}}{5}$$

$$x \cdot 2 = \frac{2}{5}$$

$$\frac{x \cdot \overset{1}{\cancel{2}}}{\underset{1}{\cancel{2}}} = \frac{\frac{2}{5}}{2}$$

$$x = \frac{2}{5} \div 2 = \frac{\overset{1}{\cancel{2}}}{5} \cdot \frac{1}{\underset{1}{\cancel{2}}} = \frac{1}{5}$$

53. $\frac{x}{\frac{3}{50}} = \frac{0.15}{1\frac{4}{5}}$

Change to decimals.

$$\frac{x}{0.06} = \frac{0.15}{1.8}$$
$$x \cdot 1.8 = 0.15(0.06)$$
$$\frac{x \cdot 1.8}{1.8} = \frac{0.009}{1.8}$$
$$x = 0.005$$

Change to fractions.

$$\frac{x}{\frac{3}{50}} = \frac{\frac{3}{20}}{1\frac{4}{5}}$$
$$1\frac{4}{5} \cdot x = \frac{3}{50} \cdot \frac{3}{20}$$
$$\frac{1\frac{4}{5} \cdot x}{1\frac{4}{5}} = \frac{\frac{9}{1000}}{1\frac{4}{5}}$$
$$x = \frac{9}{1000} \div 1\frac{4}{5} = \frac{\overset{1}{\cancel{9}}}{\underset{200}{\cancel{1000}}} \cdot \frac{\overset{1}{\cancel{5}}}{\underset{1}{\cancel{9}}} = \frac{1}{200}$$

55. $\frac{10}{4} = \frac{5}{3}$

Find the cross products.

$10 \cdot 3 = 30$

$4 \cdot 5 = 20$

The cross products are *unequal,* so the proportion is *not* true.

Replace 10 with x.

$$\frac{x}{4} = \frac{5}{3}$$
$$3 \cdot x = 4 \cdot 5$$
$$x = \frac{20}{3} = 6\frac{2}{3}, \text{ so}$$

$\frac{6\frac{2}{3}}{4} = \frac{5}{3}$ is a true proportion.

Replace 4 with x.

$$\frac{10}{x} = \frac{5}{3}$$
$$5 \cdot x = 10 \cdot 3$$
$$x = \frac{30}{5} = 6, \text{ so}$$

$\frac{10}{6} = \frac{5}{3}$ is a true proportion.

Replace 5 with x.

$$\frac{10}{4} = \frac{x}{3}$$
$$4 \cdot x = 10 \cdot 3$$
$$x = \frac{30}{4} = \frac{15}{2} = 7.5, \text{ so}$$

$\frac{10}{4} = \frac{7.5}{3}$ is a true proportion.

Replace 3 with x.

$$\frac{10}{4} = \frac{5}{x}$$
$$10 \cdot x = 4 \cdot 5$$
$$x = \frac{20}{10} = 2, \text{ so}$$

$\frac{10}{4} = \frac{5}{2}$ is a true proportion.

56. $\frac{6}{8} = \frac{24}{30}$

Find the cross products.

$6 \cdot 30 = 180$

$8 \cdot 24 = 192$

The cross products are *unequal,* so the proportion is *not* true.

Replace 6 with x.

$$\frac{x}{8} = \frac{24}{30}$$
$$30 \cdot x = 8 \cdot 24$$
$$x = \frac{8 \cdot \overset{4}{\cancel{24}}}{\underset{5}{\cancel{30}}} = \frac{32}{5} = 6.4, \text{ so}$$

$\frac{6.4}{8} = \frac{24}{30}$ is a true proportion.

Replace 8 with x.

$$\frac{6}{x} = \frac{24}{30}$$
$$24 \cdot x = 6 \cdot 30$$
$$x = \frac{6 \cdot \overset{5}{\cancel{30}}}{\underset{4}{\cancel{24}}} = \frac{30}{4} = 7.5, \text{ so}$$

$\frac{6}{7.5} = \frac{24}{30}$ is a true proportion.

Replace 24 with x.

$$\frac{6}{8} = \frac{x}{30}$$
$$8 \cdot x = 6 \cdot 30$$
$$x = \frac{\overset{3}{\cancel{6}} \cdot \overset{15}{\cancel{30}}}{\underset{2}{\cancel{\underset{4}{\cancel{8}}}}} = \frac{45}{2} = 22.5, \text{ so}$$

$\frac{6}{8} = \frac{22.5}{30}$ is a true proportion.

continued

Replace 30 with x.

$$\frac{6}{8}=\frac{24}{x}$$

$$6 \cdot x = 8 \cdot 24$$

$$x = \frac{8 \cdot \overset{4}{\cancel{24}}}{\underset{1}{\cancel{6}}} = 32, \text{ so}$$

$\frac{6}{8}=\frac{24}{32}$ is a true proportion.

Summary Exercises on Ratios, Rates, and Proportions

1. 2400 freshmen to 2000 juniors

$$\frac{2400}{2000}=\frac{2400 \div 400}{2000 \div 400}=\frac{6}{5}$$

3. 1850 seniors and 2150 sophomores to 2000 juniors

$$\frac{1850+2150}{2000}=\frac{4000}{2000}=\frac{4000 \div 2000}{2000 \div 2000}=\frac{2}{1}$$

5. **(i)** 2 million violin players to 22 million piano players

$$\frac{2 \text{ million}}{22 \text{ million}}=\frac{2 \div 2}{22 \div 2}=\frac{1}{11}$$

(ii) 2 million violin players to 20 million guitar players

$$\frac{2 \text{ million}}{20 \text{ million}}=\frac{2 \div 2}{20 \div 2}=\frac{1}{10}$$

(iii) 2 million violin players to 6 million organ players

$$\frac{2 \text{ million}}{6 \text{ million}}=\frac{2 \div 2}{6 \div 2}=\frac{1}{3}$$

(iv) 2 million violin players to 4 million clarinet players

$$\frac{2 \text{ million}}{4 \text{ million}}=\frac{2 \div 2}{4 \div 2}=\frac{1}{2}$$

(v) 2 million violin players to 3 million drum players

$$\frac{2 \text{ million}}{3 \text{ million}}=\frac{2}{3}$$

7. $$\frac{100 \text{ points}}{48 \text{ minutes}}=\frac{100 \div 48 \text{ points}}{48 \div 48 \text{ minutes}} = 2.08\overline{3} \approx 2.1 \text{ points/minute}$$

$$\frac{48 \text{ minutes}}{100 \text{ points}}=\frac{48 \div 100 \text{ minutes}}{100 \div 100 \text{ points}} = 0.48 \approx 0.5 \text{ minute/point}$$

9. $$\frac{\$652.80}{40 \text{ hours}}=\frac{\$652.80 \div 40}{40 \div 40 \text{ hours}}=\$16.32/\text{hour}$$

Her regular hourly pay rate is $16.32/hour.

$$\frac{\$195.84}{8 \text{ hours}}=\frac{\$195.84 \div 8}{8 \div 8 \text{ hours}}=\$24.48/\text{hour}$$

Her overtime hourly pay rate is $24.48/hour.

11.

Size	Cost per Unit
11 ounces	$\frac{\$6.79}{11 \text{ ounces}} \approx \$0.62$
12 ounces	$\frac{\$7.24}{12 \text{ ounces}} = \$0.60\ (*)$
16 ounces	$\frac{\$10.99}{16 \text{ ounces}} \approx \$0.69$

The best buy is 12 ounces for $7.24.

13. $$\frac{28}{21}=\frac{44}{33}$$

$$\frac{28 \div 7}{21 \div 7}=\frac{4}{3} \text{ and } \frac{44 \div 11}{33 \div 11}=\frac{4}{3}$$

Both ratios are equivalent to $\frac{4}{3}$, so the proportion is *true*.

15. $$\frac{2\frac{5}{8}}{3\frac{1}{4}}=\frac{21}{26}$$

Cross products:

$$2\frac{5}{8} \cdot 26 = \frac{21}{\underset{4}{\cancel{8}}} \cdot \frac{\overset{13}{\cancel{26}}}{1} = \frac{273}{4}$$

$$3\frac{1}{4} \cdot 21 = \frac{13}{4} \cdot \frac{21}{1} = \frac{273}{4}$$

The cross products are *equal*, so the proportion is *true*.

17. $$\frac{15}{8}=\frac{6}{x}$$

$15 \cdot x = 8 \cdot 6$ *Cross products are equivalent*

$$\frac{\overset{1}{\cancel{15}} \cdot x}{\underset{1}{\cancel{15}}} = \frac{8 \cdot \overset{2}{\cancel{6}}}{\underset{5}{\cancel{15}}} = \frac{16}{5} = 3.2$$

Check: $15(3.2) = 48 = 8 \cdot 6$

19. $$\frac{10}{11}=\frac{x}{4}$$

$11 \cdot x = 10 \cdot 4$ *Cross products are equivalent*

$$\frac{\overset{1}{\cancel{11}} \cdot x}{\underset{1}{\cancel{11}}} = \frac{40}{11} \approx 3.64$$

Check: $10 \cdot 4 = 40 = 11 \cdot \frac{40}{11}$

21. $\frac{2.6}{x} = \frac{13}{7.8}$

$13 \cdot x = 2.6(7.8)$ *Cross products are equivalent*

$\frac{\overset{1}{\cancel{13}} \cdot x}{\underset{1}{\cancel{13}}} = \frac{20.28}{13} = 1.56$

Check: $2.6(7.8) = 20.28 = 1.56(13)$

23. $\frac{\frac{1}{3}}{8} = \frac{x}{24}$

$8 \cdot x = \frac{1}{3} \cdot 24$ *Cross products are equivalent*

$\frac{\overset{1}{\cancel{8}} \cdot x}{\underset{1}{\cancel{8}}} = \frac{8}{8} = 1$

Check: $\frac{1}{3}(24) = 8 = 8 \cdot 1$

6.4 Problem Solving with Proportions

6.4 Section Exercises

1. $\frac{5 \text{ hours}}{4 \text{ cartoon strips}} = \frac{x \text{ hours}}{18 \text{ cartoon strips}}$

$$\frac{5}{4} = \frac{x}{18}$$
$$4 \cdot x = 5 \cdot 18$$
$$4 \cdot x = 90$$
$$\frac{\overset{1}{\cancel{4}} \cdot x}{\underset{1}{\cancel{4}}} = \frac{90}{4}$$
$$x = 22.5$$

It will take 22.5 hours to sketch 18 cartoon strips.

3. $\frac{60 \text{ newspapers}}{\$27} = \frac{16 \text{ newspapers}}{x}$

$$60 \cdot x = 27 \cdot 16$$
$$\frac{60 \cdot x}{60} = \frac{432}{60}$$
$$x = 7.2$$

The cost of 16 newspapers is \$7.20.

5. $\frac{3 \text{ pounds}}{350 \text{ square feet}} = \frac{x}{4900 \text{ square feet}}$

$$350 \cdot x = 3 \cdot 4900$$
$$\frac{350 \cdot x}{350} = \frac{14{,}700}{350}$$
$$x = 42$$

42 pounds are needed to cover 4900 square feet.

7. $\frac{\$455.75}{5 \text{ days}} = \frac{x}{3 \text{ days}}$

$$5 \cdot x = (455.75)(3)$$
$$\frac{5 \cdot x}{5} = \frac{1367.25}{5}$$
$$x = 273.45$$

In 3 days Tom makes \$273.45.

9. $\frac{6 \text{ ounces}}{7 \text{ servings}} = \frac{x}{12 \text{ servings}}$

$$7 \cdot x = 6 \cdot 12$$
$$\frac{7 \cdot x}{7} = \frac{72}{7}$$
$$x = 10\frac{2}{7} \approx 10$$

You need about 10 ounces for 12 servings.

11. $\frac{3 \text{ quarts}}{270 \text{ square feet}} = \frac{x}{(350 + 100) \text{ square feet}}$

$$\frac{3}{270} = \frac{x}{450}$$
$$270 \cdot x = 3 \cdot 450$$
$$\frac{270 \cdot x}{270} = \frac{1350}{270}$$
$$x = 5$$

You will need 5 quarts.

13. First find the length.

$$\frac{1 \text{ inch}}{4 \text{ feet}} = \frac{3.5 \text{ inches}}{x}$$
$$1 \cdot x = 4(3.5)$$
$$x = 14$$

The kitchen is 14 feet long.

Then find the width.

$$\frac{1 \text{ inch}}{4 \text{ feet}} = \frac{2.5 \text{ inches}}{x \text{ feet}}$$
$$1 \cdot x = 4(2.5)$$
$$x = 10$$

The kitchen is 10 feet wide.

15. The length of the dining area is the same as the length of the kitchen, which is 14 feet by Exercise 13.

Find the width of the dining area.

$$4.5 - 2.5 \text{ inches} = 2 \text{ inches}$$
$$\frac{1 \text{ inch}}{4 \text{ feet}} = \frac{2 \text{ inches}}{x}$$
$$1 \cdot x = 4 \cdot 2$$
$$x = 8$$

The dining area is 8 feet wide.

17. Set up and solve a proportion with pieces of chicken and number of guests.

$$\frac{40 \text{ pieces}}{25 \text{ guests}} = \frac{x}{60 \text{ guests}}$$
$$25 \cdot x = 40 \cdot 60$$
$$\frac{25 \cdot x}{25} = \frac{40 \cdot 60}{25}$$
$$x = \frac{40 \cdot 60}{25} = 96 \text{ pieces of chicken}$$

continued

For the other food items, the proportions are similar, so we simply replace "40" in the last step with the appropriate value.

$\frac{14 \cdot 60}{25} = 33.6$ pounds of lasagna

$\frac{4.5 \cdot 60}{25} = 10.8$ pounds of deli meats

$\frac{\frac{7}{3} \cdot 60}{25} = \frac{28}{5} = 5\frac{3}{5}$ pounds of cheese

$\frac{3 \text{ dozen} \cdot 60}{25} = 7.2$ dozen (about 86) buns

$\frac{6 \cdot 60}{25} = 14.4$ pounds of salad

19. $\frac{7 \text{ refresher}}{10 \text{ entering}} = \frac{x}{2950 \text{ entering}}$

$$10 \cdot x = 7 \cdot 2950$$
$$\frac{10 \cdot x}{10} = \frac{20{,}650}{10}$$
$$x = 2065$$

2065 students will probably need a refresher course. This is a reasonable answer because it's more than half the students, but not all the students.

Incorrect setup

$$\frac{10 \text{ entering}}{7 \text{ refresher}} = \frac{x}{2950 \text{ entering}}$$
$$7 \cdot x = 10 \cdot 2950$$
$$\frac{7 \cdot x}{7} = \frac{29{,}500}{7}$$
$$x \approx 4214$$

The incorrect setup gives an unreasonable estimate of 4214 entering students since there are only 2950 entering students.

21. $\frac{1 \text{ chooses vanilla}}{8 \text{ people}} = \frac{x \text{ choose vanilla}}{238 \text{ people}}$

$$8 \cdot x = 1 \cdot 238$$
$$\frac{8 \cdot x}{8} = \frac{238}{8}$$
$$x = 29.75 \approx 30$$

You would expect about 30 people to choose vanilla ice cream. This is a reasonable answer.

Incorrect setup

$$\frac{8 \text{ people}}{1 \text{ chooses vanilla}} = \frac{x \text{ choose vanilla}}{238 \text{ people}}$$
$$1 \cdot x = 8 \cdot 238$$
$$x = 1904$$

With an incorrect setup, 1904 people choose vanilla ice cream. This is unreasonable because only 238 people attended the ice cream social.

23. $\frac{98}{100} = \frac{x}{109{,}300{,}000}$

$$100 \cdot x = 98 \cdot 109{,}300{,}000$$
$$\frac{100 \cdot x}{100} = \frac{10{,}711{,}400{,}000}{100}$$
$$x = 107{,}114{,}000$$

107,114,000 U.S. households have one or more TVs.

Incorrect setup

$$\frac{100}{98} = \frac{x}{109{,}300{,}000}$$
$$98 \cdot x = 100 \cdot 109{,}300{,}000$$
$$\frac{98 \cdot x}{98} = \frac{10{,}930{,}000{,}000}{98}$$
$$x \approx 111{,}530{,}612$$

The incorrect setup gives about 111,530,612 U.S. households with one or more TVs, but there are only 109,300,000 U.S. households.

25. $\frac{5 \text{ stocks up}}{6 \text{ stocks down}} = \frac{x \text{ stocks up}}{750 \text{ stocks down}}$

$$\frac{5}{6} = \frac{x}{750}$$
$$6 \cdot x = 5 \cdot 750$$
$$\frac{6 \cdot x}{6} = \frac{3750}{6}$$
$$x = 625$$

625 stocks went up.

27. $\frac{8 \text{ length}}{1 \text{ width}} = \frac{32.5 \text{ meters length}}{x \text{ meters width}}$

$$8 \cdot x = 1 \cdot 32.5$$
$$\frac{8 \cdot x}{8} = \frac{32.5}{8}$$
$$x = 4.0625 \approx 4.06$$

The wing must be about 4.06 meters wide.

29. $\frac{150 \text{ pounds}}{222 \text{ calories}} = \frac{210 \text{ pounds}}{x}$

$$150 \cdot x = 222 \cdot 210$$
$$\frac{150 \cdot x}{150} = \frac{46{,}620}{150}$$
$$x = 310.8 \approx 311$$

A 210-pound person would burn about 311 calories.

31. $\frac{1.05 \text{ meters}}{1.68 \text{ meters}} = \frac{6.58 \text{ meters}}{x}$

$$1.05 \cdot x = 1.68(6.58)$$
$$\frac{1.05 \cdot x}{1.05} = \frac{11.0544}{1.05}$$
$$x = 10.528 \approx 10.53$$

The height of the tree is about 10.53 meters.

33. You cannot solve this problem using a proportion because the ratio of age to weight is not constant. As Jim's age increases from 25 to 50 years old, his weight may decrease, stay the same, or increase.

35. First find the number of coffee drinkers.

$$\frac{4}{5} = \frac{x}{50{,}700}$$
$$5 \cdot x = 4 \cdot 50{,}700$$
$$\frac{5 \cdot x}{5} = \frac{202{,}000}{5}$$
$$x = 40{,}560$$

The survey showed that 40,560 students drink coffee. Now find the number of coffee drinkers who use cream.

$$\frac{1}{8} = \frac{x}{40{,}560}$$
$$8 \cdot x = 40{,}560$$
$$\frac{8 \cdot x}{8} = \frac{40{,}560}{8}$$
$$x = 5070$$

According to the survey, 5070 students use cream.

37. First find the number of calories in a $\frac{1}{2}$-cup serving of bran cereal.

$$\frac{\frac{1}{3}\text{ cup}}{80\text{ calories}} = \frac{\frac{1}{2}\text{ cup}}{x}$$
$$\frac{1}{3} \cdot x = 80 \cdot \frac{1}{2}$$
$$\frac{\frac{1}{3} \cdot x}{\frac{1}{3}} = \frac{40}{\frac{1}{3}} = \frac{40}{1} \cdot \frac{3}{1}$$
$$x = 120$$

Then find the number of grams of fiber in a $\frac{1}{2}$-cup serving of bran cereal.

$$\frac{\frac{1}{3}\text{ cup}}{8\text{ grams of fiber}} = \frac{\frac{1}{2}\text{ cup}}{x}$$
$$\frac{1}{3} \cdot x = 8 \cdot \frac{1}{2}$$
$$\frac{\frac{1}{3} \cdot x}{\frac{1}{3}} = \frac{4}{\frac{1}{3}} = \frac{4}{1} \cdot \frac{3}{1}$$
$$x = 12$$

A $\frac{1}{2}$-cup serving of bran cereal provides 120 calories and 12 grams of fiber.

39. *Use proportions.*

Water:
$$\frac{3\frac{1}{2}\text{ cups}}{12\text{ servings}} = \frac{x}{6\text{ servings}}$$
$$12 \cdot x = 3\frac{1}{2} \cdot 6$$
$$12 \cdot x = \frac{7}{2} \cdot 6$$
$$\frac{12 \cdot x}{12} = \frac{21}{12}$$
$$x = \frac{7}{4} = 1\frac{3}{4}$$

Margarine:
$$\frac{6\text{ Tbsp}}{12\text{ servings}} = \frac{x}{6\text{ servings}}$$
$$12 \cdot x = 6 \cdot 6$$
$$\frac{12 \cdot x}{12} = \frac{36}{12}$$
$$x = 3$$

Milk:
$$\frac{1\frac{1}{2}\text{ cups}}{12\text{ servings}} = \frac{x}{6\text{ servings}}$$
$$12 \cdot x = 1\frac{1}{2} \cdot 6$$
$$12 \cdot x = \frac{3}{2} \cdot 6$$
$$\frac{12 \cdot x}{12} = \frac{9}{12}$$
$$x = \frac{3}{4}$$

Potato flakes:
$$\frac{4\text{ cups}}{12\text{ servings}} = \frac{x}{6\text{ servings}}$$
$$12 \cdot x = 4 \cdot 6$$
$$\frac{12 \cdot x}{12} = \frac{24}{12}$$
$$x = 2$$

Multiply the quantities by $\frac{1}{2}$ (or divide by 2), since *6 servings is $\frac{1}{2}$ of 12 servings.*

Water:
$$\frac{3\frac{1}{2}}{2} = \frac{\frac{7}{2}}{2} = \frac{7}{2} \div 2 = \frac{7}{2} \cdot \frac{1}{2}$$
$$= \frac{7}{4} = 1\frac{3}{4}$$

Margarine:
$$\frac{6}{2} = 3$$

Milk:
$$\frac{1\frac{1}{2}}{2} = \frac{\frac{3}{2}}{2} = \frac{3}{2} \div 2$$
$$= \frac{3}{2} \cdot \frac{1}{2} = \frac{3}{4}$$

Potato flakes:
$$\frac{4}{2} = 2$$

For 6 servings, use $1\frac{3}{4}$ cups water, 3 Tbsp margarine, $\frac{3}{4}$ cup milk, and 2 cups potato flakes.

40. *Use proportions.*

Water: $\frac{3\frac{1}{2}\text{ cups}}{12\text{ servings}} = \frac{x}{18\text{ servings}}$

$12 \cdot x = 3\frac{1}{2} \cdot 18$

$12 \cdot x = \frac{7}{2} \cdot 18$

$\frac{12 \cdot x}{12} = \frac{63}{12}$

$x = 5\frac{1}{4}$

Margarine: $\frac{6\text{ Tbsp}}{12\text{ servings}} = \frac{x}{18\text{ servings}}$

$12 \cdot x = 6 \cdot 18$

$\frac{12 \cdot x}{12} = \frac{108}{12}$

$x = 9$

Milk: $\frac{1\frac{1}{2}\text{ cups}}{12\text{ servings}} = \frac{x}{18\text{ servings}}$

$12 \cdot x = 1\frac{1}{2} \cdot 18$

$12 \cdot x = \frac{3}{2} \cdot 18$

$\frac{12 \cdot x}{12} = \frac{27}{12}$

$x = 2\frac{1}{4}$

Potato flakes: $\frac{4\text{ cups}}{12\text{ servings}} = \frac{x}{18\text{ servings}}$

$12 \cdot x = 4 \cdot 18$

$\frac{12 \cdot x}{12} = \frac{72}{12}$

$x = 6$

Multiply the quantities in Exercise 39 by 3, since 18 servings is 3 times 6 servings.

Water: $1\frac{3}{4} \cdot 3 = \frac{7}{4} \cdot \frac{3}{1} = \frac{21}{4} = 5\frac{1}{4}$

Margarine: $3 \cdot 3 = 9$

Milk: $\frac{3}{4} \cdot 3 = \frac{3}{4} \cdot \frac{3}{1} = \frac{9}{4} = 2\frac{1}{4}$

Potato flakes: $2 \cdot 3 = 6$

For 18 servings, use $5\frac{1}{4}$ cups water, 9 Tbsp margarine, $2\frac{1}{4}$ cups milk, and 6 cups potato flakes.

41. Since 3 servings is $\frac{1}{2}$ of 6 servings, multiply each quantity in Exercise 39 by $\frac{1}{2}$.

Water: $1\frac{3}{4} \cdot \frac{1}{2} = \frac{7}{4} \cdot \frac{1}{2} = \frac{7}{8}$

Margarine: $3 \cdot \frac{1}{2} = \frac{3}{1} \cdot \frac{1}{2} = \frac{3}{2} = 1\frac{1}{2}$

Milk: $\frac{3}{4} \cdot \frac{1}{2} = \frac{3}{8}$

Potato flakes: $2 \cdot \frac{1}{2} = \frac{2}{1} \cdot \frac{1}{2} = 1$

For 3 servings, use $\frac{7}{8}$ cup water, $1\frac{1}{2}$ Tbsp margarine, $\frac{3}{8}$ cup milk, and 1 cup potato flakes.

42. Since 9 servings is 3 times 3 servings, multiply each quantity in Exercise 41 by 3.

Water: $\frac{7}{8} \cdot 3 = \frac{7}{8} \cdot \frac{3}{1} = \frac{21}{8} = 2\frac{5}{8}$

Margarine: $\frac{3}{2} \cdot 3 = \frac{3}{2} \cdot \frac{3}{1} = \frac{9}{2} = 4\frac{1}{2}$

Milk: $\frac{3}{8} \cdot 3 = \frac{3}{8} \cdot \frac{3}{1} = \frac{9}{8} = 1\frac{1}{8}$

Potato flakes: $1 \cdot 3 = 3$

For 9 servings, use $2\frac{5}{8}$ cups water, $4\frac{1}{2}$ Tbsp margarine, $1\frac{1}{8}$ cups milk, and 3 cups potato flakes.

6.5 Geometry: Lines and Angles

6.5 Section Exercises

1. This is a *line* named $\overleftrightarrow{CD}$ or $\overleftrightarrow{DC}$. A line is a straight row of points that goes on forever in both directions.

3. The figure has two endpoints so it is a *line segment* named $\overline{GF}$ or $\overline{FG}$.

5. This is a *ray* named $\overrightarrow{PQ}$. A ray is a part of a line that has only one endpoint and goes on forever in one direction.

7. The lines are *perpendicular* because they intersect at right angles.

9. These lines appear to be *parallel* lines. Parallel lines are lines in the same plane that never intersect (cross).

11. The lines intersect so they are *not* parallel. At their intersection they *do not* form a right angle so they are not perpendicular. The lines are *intersecting*.

13. The angle can be named $\angle AOS$ or $\angle SOA$. The middle letter, *O*, identifies the vertex.

15. The angle can be named $\angle CRT$ or $\angle TRC$. The middle letter, *R*, identifies the vertex.

17. The angle can be named $\angle AQC$ or $\angle CQA$. The middle letter, *Q*, identifies the vertex.

19. The angle is a *right angle*, as indicated by the small square at the vertex. Right angles measure exactly 90°.

21. The measure of the angle is between 0° and 90°, so it is an *acute angle*.

23. Two rays in a straight line pointing opposite directions measure 180°. An angle that measures 180° is called a *straight angle*.

25. The pairs of complementary angles are:
$\angle EOD$ and $\angle COD$ because $75° + 15° = 90°$;
$\angle AOB$ and $\angle BOC$ because $25° + 65° = 90°$.

27. The pairs of supplementary angles are:
$\angle HNE$ and $\angle ENF$ because $77° + 103° = 180°$;
$\angle ACB$ and $\angle KOL$ because $103° + 77° = 180°$;
$\angle HNE$ and $\angle ACB$ because $77° + 103° = 180°$;
$\angle ENF$ and $\angle KOL$ because $103° + 77° = 180°$.

29. The complement of 40° is 50°, because $90° - 40° = 50°$.

31. The complement of 86° is 4°, because $90° - 86° = 4°$.

33. The supplement of 130° is 50°, because $180° - 130° = 50°$.

35. The supplement of 90° is 90°, because $180° - 90° = 90°$.

37. $\angle SON \cong \angle TOM$ because they are vertical angles.
$\angle TOS \cong \angle MON$ because they are vertical angles.

39. Because $\angle COE$ and $\angle GOH$ are vertical angles, they are also congruent. This means they have the same measure. $\angle COE$ measures 63°, so $\angle GOH$ measures 63°.

The sum of the measures of $\angle COE$, $\angle AOC$, and $\angle AOH$ equals 180°. Therefore, $\angle AOC$ measures $180° - (63° + 37°) = 180° - 100° = 80°$. Since $\angle AOC$ and $\angle GOF$ are vertical, they are congruent, so $\angle GOF$ measures 80°. Since $\angle AOH$ and $\angle EOF$ are vertical, they are congruent, so $\angle EOF$ measures 37°.

41. "$\angle UST$ is 90°" is *true* because $\overleftrightarrow{UQ}$ is perpendicular to $\overleftrightarrow{ST}$.

42. "$\overleftrightarrow{SQ}$ and $\overleftrightarrow{PQ}$ are perpendicular" is *true* because they form a 90° angle, as indicated by the small red square.

43. "The measure of $\angle USQ$ is less than the measure of $\angle PQR$" is *false*. $\angle USQ$ is a straight angle and so is $\angle PQR$, therefore each measures 180°. A true statement would be: $\angle USQ$ has the same measure as $\angle PQR$.

44. "$\overleftrightarrow{ST}$ and $\overleftrightarrow{PR}$ are intersecting" is *false*. $\overleftrightarrow{ST}$ and $\overleftrightarrow{PR}$ are parallel and will never intersect.

45. "$\overleftrightarrow{QU}$ and $\overleftrightarrow{TS}$ are parallel" is *false*. $\overleftrightarrow{QU}$ and $\overleftrightarrow{TS}$ are perpendicular because they intersect at right angles.

46. "$\angle UST$ and $\angle UQR$ measure the same number of degrees" is *true* because both angles are formed by perpendicular lines, so they both measure 90°.

47. There are four pairs of corresponding angles:

$$\angle 1 \text{ and } \angle 8,\quad \angle 2 \text{ and } \angle 5,$$
$$\angle 3 \text{ and } \angle 6,\quad \angle 4 \text{ and } \angle 7$$

There are two pairs of alternate interior angles:

$$\angle 4 \text{ and } \angle 5,\quad \angle 3 \text{ and } \angle 8$$

49. $\angle 8 \cong \angle 6 \cong \angle 2 \cong \angle 4$, so each measures 130°.

$\angle 5$ measures $180° - 130° = 50°$.

$\angle 5 \cong \angle 7 \cong \angle 1 \cong \angle 3$, so each measures 50°.

51. $\angle 6 \cong \angle 1 \cong \angle 8 \cong \angle 3$, so each measures 47°.

$\angle 5$ measures $180° - 47° = 133°$.

$\angle 5 \cong \angle 2 \cong \angle 7 \cong \angle 4$, so each measures 133°.

53. $\angle 6 \cong \angle 8 \cong \angle 4 \cong \angle 2$, so each measures 114°.

$\angle 7$ measures $180° - 114° = 66°$.

$\angle 7 \cong \angle 5 \cong \angle 3 \cong \angle 1$, so each measures 66°.

55. $\angle 2$ and $\angle ABC$ are alternate interior angles, so they have the same measure, 42°.
$\angle 1$ and $\angle ABC$ are supplementary angles, so $\angle 1 = 180° - 42° = 138°$.
$\angle 1$ and $\angle 3$ are supplements of alternate interior angles, so they have the same measure, 138°.

6.6 Geometry Applications: Congruent and Similar Triangles

6.6 Section Exercises

1. If you picked up $\triangle ABC$ and slid it over on top of $\triangle DEF$, the two triangles would match.

The corresponding angles are

$$\angle 1 \text{ and } \angle 4,\quad \angle 2 \text{ and } \angle 5,\quad \angle 3 \text{ and } \angle 6.$$

The corresponding sides are

$$\overline{AB} \text{ and } \overline{DE},\quad \overline{BC} \text{ and } \overline{EF},\quad \overline{AC} \text{ and } \overline{DF}.$$

3. If you rotate $\triangle TUS$, then slide it on top of $\triangle WXY$, the two triangles would match.

The corresponding angles are

$$\angle 1 \text{ and } \angle 6,\quad \angle 2 \text{ and } \angle 4,\quad \angle 3 \text{ and } \angle 5.$$

The corresponding sides are

$$\overline{ST} \text{ and } \overline{YW},\quad \overline{TU} \text{ and } \overline{WX},\quad \overline{SU} \text{ and } \overline{YX}.$$

5. If you flipped $\triangle MNL$ over, then slid it on top of $\triangle SRT$, the two triangles would match.

The corresponding angles are

$$\angle 1 \text{ and } \angle 6,\quad \angle 2 \text{ and } \angle 5,\quad \angle 3 \text{ and } \angle 4.$$

The corresponding sides are

$$\overline{LM} \text{ and } \overline{TS},\quad \overline{LN} \text{ and } \overline{TR},\quad \overline{MN} \text{ and } \overline{SR}.$$

7. On both triangles, two corresponding sides and the angle between them measure the same, so the Side-Angle-Side (SAS) method can be used to prove that the triangles are congruent.

9. Each pair of corresponding sides has the same length, so the Side-Side-Side (SSS) method can be used to prove that the triangles are congruent.

11. On both triangles, two corresponding angles and the side that connects them measure the same, so the Angle-Side-Angle (ASA) method can be used to prove that the triangles are congruent.

13. use SAS: $BC = CE$, $\angle ABC \cong \angle DCE$, $BA = CD$

14. use SSS: $WP = YP$, $ZP = XP$, $WZ = YX$

15. use SAS: $PS = SR$, $m\angle QSP = m\angle QSR = 90°$, $QS = QS$ (common side)

16. use SAS: $LM = OM$, $PM = NM$, $\angle LMP \cong \angle OMN$ (vertical angles)

17. Set up a ratio of corresponding sides.

$$\frac{6 \text{ cm}}{12 \text{ cm}} = \frac{6}{12} = \frac{1}{2}$$

Write a proportion to find a.

$$\frac{a}{12} = \frac{1}{2}$$
$$2 \cdot a = 12 \cdot 1$$
$$\frac{2 \cdot a}{2} = \frac{12}{2}$$
$$a = 6 \text{ cm}$$

Write a proportion to find b.

$$\frac{7.5}{b} = \frac{1}{2}$$
$$1 \cdot b = 2 \cdot 7.5$$
$$b = 15 \text{ cm}$$

19. Set up a ratio of corresponding sides.

$$\frac{6 \text{ mm}}{12 \text{ mm}} = \frac{6}{12} = \frac{1}{2}$$

Write a proportion to find a.

$$\frac{a}{10} = \frac{1}{2}$$
$$a \cdot 2 = 10 \cdot 1$$
$$\frac{a \cdot 2}{2} = \frac{10}{2}$$
$$a = 5 \text{ mm}$$

Write a proportion to find b.

$$\frac{b}{6} = \frac{1}{2}$$
$$b \cdot 2 = 6 \cdot 1$$
$$\frac{b \cdot 2}{2} = \frac{6}{2}$$
$$b = 3 \text{ mm}$$

21. Set up a ratio of corresponding sides.

$$\frac{18 \text{ in.}}{12 \text{ in.}} = \frac{18}{12} = \frac{3}{2}$$

Write a proportion to find a.

$$\frac{a}{16} = \frac{3}{2}$$
$$a \cdot 2 = 16 \cdot 3$$
$$\frac{a \cdot 2}{2} = \frac{48}{2}$$
$$a = 24 \text{ inches}$$

Write a proportion to find b.

$$\frac{30}{b} = \frac{3}{2}$$
$$b \cdot 3 = 30 \cdot 2$$
$$\frac{b \cdot 3}{3} = \frac{60}{3}$$
$$b = 20 \text{ inches}$$

23. Write a proportion to find x.

$$\frac{x}{18.6} = \frac{28}{21} \quad \textbf{OR} \quad \frac{x}{18.6} = \frac{4}{3}$$
$$3 \cdot x = 4 \cdot 18.6$$
$$\frac{3 \cdot x}{3} = \frac{74.4}{3}$$
$$x = 24.8 \text{ m}$$

$$P = 24.8 \text{ m} + 28 \text{ m} + 20 \text{ m} = 72.8 \text{ m}$$

Write a proportion to find y.

$$\frac{y}{20} = \frac{21}{28} \quad \textbf{OR} \quad \frac{y}{20} = \frac{3}{4}$$
$$4 \cdot y = 20 \cdot 3$$
$$\frac{4 \cdot y}{4} = \frac{60}{4}$$
$$y = 15 \text{ m}$$

$$P = 15 \text{ m} + 21 \text{ m} + 18.6 \text{ m} = 54.6 \text{ m}$$

25. Since triangles CDE and FGH are similar, each missing side of triangle FGH is 8 cm.

Perimeter of triangle FGH

$$= 8 \text{ cm} + 8 \text{ cm} + 8 \text{ cm}$$
$$= 24 \text{ cm}$$

Set up a ratio of corresponding sides to find the height h of triangle FGH.

$$\frac{10.4}{12} = \frac{h}{8}$$
$$12 \cdot h = 8 \cdot 10.4$$
$$\frac{12 \cdot h}{12} = \frac{83.2}{12}$$
$$h \approx 6.9 \text{ cm}$$

Area of triangle FGH
$= 0.5 \cdot b \cdot h$
$\approx 0.5 \cdot 8 \text{ cm} \cdot 6.9 \text{ cm}$
$= 27.6 \text{ cm}^2$

27. Write a proportion to find h.

$$\frac{2}{16} = \frac{3}{h}$$
$$2 \cdot h = 3 \cdot 16$$
$$\frac{2 \cdot h}{2} = \frac{48}{2}$$
$$h = 24 \text{ ft}$$

The height of the house is 24 ft.

29. One dictionary definition is "resembling, but not identical." Examples of similar objects are sets of different size pots or measuring cups; small and large size cans of beans; child's tennis shoe and adult's tennis shoe.

31. Using the hint, we can write a proportion to find x.

$$\frac{x}{120} = \frac{100}{100 + 140} = \frac{100}{240} \quad \textbf{OR}$$
$$\frac{x}{120} = \frac{5}{12}$$
$$x \cdot 12 = 120 \cdot 5$$
$$\frac{x \cdot 12}{12} = \frac{600}{12}$$
$$x = 50 \text{ m}$$

33. Write a proportion to find n.

$$\frac{50}{n} = \frac{100}{100 + 120} = \frac{100}{220} \quad \textbf{OR}$$
$$\frac{50}{n} = \frac{5}{11}$$
$$5 \cdot n = 50 \cdot 11$$
$$\frac{5 \cdot n}{5} = \frac{550}{5}$$
$$n = 110 \text{ m}$$

The length of the lake is 110 m.

Chapter 6 Review Exercises

1. orca whale's length to whale shark's length

$$\frac{30 \text{ ft}}{40 \text{ ft}} = \frac{30}{40} = \frac{30 \div 10}{40 \div 10} = \frac{3}{4}$$

2. blue whale's length to great white shark's length

$$\frac{80 \text{ ft}}{20 \text{ ft}} = \frac{80}{20} = \frac{80 \div 20}{20 \div 20} = \frac{4}{1}$$

3. To get a ratio of $\frac{1}{2}$, we can start with the smallest value in the table and see if there is a value that is two times the smallest. $2 \times 20 = 40$, so the ratio of the *great white shark's* length to the *whale shark's* length is $\frac{20}{40} = \frac{1}{2}$. The length of the orca whale is 30 ft, but $2 \times 30 = 60$ is not in the table. $2 \times 40 = 80$, so the ratio of the *whale shark's* length to the *blue whale's* length is $\frac{40}{80} = \frac{1}{2}$.

4. \$2.50 to \$1.25

$$\frac{\$2.50}{\$1.25} = \frac{2.50}{1.25} = \frac{2.50 \div 1.25}{1.25 \div 1.25} = \frac{2}{1}$$

5. \$0.30 to \$0.45

$$\frac{\$0.30}{\$0.45} = \frac{0.30}{0.45} = \frac{0.30 \div .15}{0.45 \div .15} = \frac{2}{3}$$

6. $1\frac{2}{3}$ cups to $\frac{2}{3}$ cup

$$\frac{1\frac{2}{3} \text{ cups}}{\frac{2}{3} \text{ cup}} = \frac{1\frac{2}{3}}{\frac{2}{3}} = \frac{5}{3} \div \frac{2}{3}$$
$$= \frac{5}{\cancel{3}_1} \cdot \frac{\cancel{3}^1}{2} = \frac{5}{2}$$

7. $2\frac{3}{4}$ miles to $16\frac{1}{2}$ miles

$$\frac{2\frac{3}{4} \text{ miles}}{16\frac{1}{2} \text{ miles}} = \frac{2\frac{3}{4}}{16\frac{1}{2}} = \frac{\frac{11}{4}}{\frac{33}{2}} = \frac{11}{4} \div \frac{33}{2}$$
$$= \frac{\cancel{11}^1}{\cancel{4}_2} \cdot \frac{\cancel{2}^1}{\cancel{33}_3} = \frac{1}{6}$$

8. 5 hours to 100 minutes

5 hours $= 5 \cdot 60$ minutes $= 300$ minutes

$$\frac{300 \text{ minutes}}{100 \text{ minutes}} = \frac{300}{100} = \frac{300 \div 100}{100 \div 100} = \frac{3}{1}$$

9. 9 inches to 2 feet

2 feet $= 24$ inches

$$\frac{9 \text{ inches}}{24 \text{ inches}} = \frac{9}{24} = \frac{9 \div 3}{24 \div 3} = \frac{3}{8}$$

10. 1 ton to 1500 pounds

1 ton $= 2000$ pounds

$$\frac{2000 \text{ pounds}}{1500 \text{ pounds}} = \frac{2000}{1500} = \frac{2000 \div 500}{1500 \div 500} = \frac{4}{3}$$

11. 8 hours to 3 days

3 days $= 3 \cdot 24$ hours $= 72$ hours

$$\frac{8 \text{ hours}}{72 \text{ hours}} = \frac{8}{72} = \frac{8 \div 8}{72 \div 8} = \frac{1}{9}$$

12. \$500 to \$350

$$\frac{\$500}{\$350} = \frac{500}{350} = \frac{500 \div 50}{350 \div 50} = \frac{10}{7}$$

The ratio of her sales to his sales is $\frac{10}{7}$.

13. $$\frac{35 \text{ miles per gallon}}{25 \text{ miles per gallon}} = \frac{35}{25} = \frac{35 \div 5}{25 \div 5} = \frac{7}{5}$$

The ratio of the new car's mileage to the old car's mileage is $\frac{7}{5}$.

14. $\dfrac{6000 \text{ students}}{7200 \text{ students}} = \dfrac{6000}{7200} = \dfrac{6000 \div 1200}{7200 \div 1200} = \dfrac{5}{6}$

The ratio of the math students to the English students is $\frac{5}{6}$.

15. \$88 for 8 dozen

$$\frac{\$88 \div 8}{8 \text{ dozen} \div 8} = \frac{\$11}{1 \text{ dozen}}$$

16. 96 children in 40 families

$$\frac{96 \text{ children} \div 8}{40 \text{ families} \div 8} = \frac{12 \text{ children}}{5 \text{ families}}$$

17. 4 pages in 20 minutes

(i) $\dfrac{4 \text{ pages} \div 20}{20 \text{ minutes} \div 20} = \dfrac{0.2 \text{ page}}{1 \text{ minute}}$
$= 0.2 \text{ page/minute}$
or $\dfrac{1}{5}$ page/minute

(ii) $\dfrac{20 \text{ minutes} \div 4}{4 \text{ pages} \div 4} = \dfrac{5 \text{ minutes}}{1 \text{ page}}$
$= 5 \text{ minutes/page}$

18. \$24 in 3 hours

(i) $\dfrac{\$24 \div 3}{3 \text{ hours} \div 3} = \dfrac{\$8}{1 \text{ hour}} = \$8/\text{hour}$

(ii) $\dfrac{3 \text{ hours} \div 24}{\$24 \div 24} = \dfrac{0.125 \text{ hour}}{\$1}$
$= 0.125 \text{ hour/dollar}$
or $\dfrac{1}{8}$ hour/dollar

19.

Size	Cost per Unit
13 ounces	$\dfrac{\$2.29}{13 \text{ ounces}} \approx \$0.176\ (*)$
8 ounces	$\dfrac{\$1.45}{8 \text{ ounces}} \approx \$0.181$
3 ounces	$\dfrac{\$0.95}{3 \text{ ounces}} \approx \$0.317$

The best buy is 13 ounces for \$2.29.

20. 50 pounds for \$19.95 − \$1.00 (coupon) = \$18.95

$$\frac{\$18.95}{50 \text{ pounds}} = \$0.379$$

25 pounds for \$10.40 − \$1.00 (coupon) = \$9.40

$$\frac{\$9.40}{25 \text{ pounds}} = \$0.376\ (*)$$

8 pounds for \$3.40

$$\frac{\$3.40}{8 \text{ pounds}} = \$0.425$$

The best buy is 25 pounds for \$10.40 with the \$1 coupon.

21. $\dfrac{6}{10} = \dfrac{9}{15}$

$\dfrac{6 \div 2}{10 \div 2} = \dfrac{3}{5}$ and $\dfrac{9 \div 3}{15 \div 3} = \dfrac{3}{5}$

Both ratios are equivalent to $\frac{3}{5}$, so the proportion is *true.*

22. $\dfrac{6}{48} = \dfrac{9}{36}$

Cross products: $6 \cdot 36 = 216$; $48 \cdot 9 = 432$

The cross products are *unequal,* so the proportion is *false.*

23. $\dfrac{47}{10} = \dfrac{98}{20}$

Cross products: $47 \cdot 20 = 940$; $10 \cdot 98 = 980$

The cross products are *unequal,* so the proportion is *false.*

24. $\dfrac{1.5}{2.4} = \dfrac{2}{3.2}$

Cross products: $1.5(3.2) = 4.8$; $2.4 \cdot 2 = 4.8$

The cross products are *equal,* so the proportion is *true.*

25. $\dfrac{3\frac{1}{2}}{2\frac{1}{3}} = \dfrac{6}{4}$

Cross products:

$$3\frac{1}{2} \cdot 4 = \frac{7}{\cancel{2}_1} \cdot \frac{\cancel{4}^2}{1} = 14$$

$$2\frac{1}{3} \cdot 6 = \frac{7}{\cancel{3}_1} \cdot \frac{\cancel{6}^2}{1} = 14$$

The cross products are *equal,* so the proportion is *true.*

26. $\dfrac{4}{42} = \dfrac{150}{x}$ **OR** $\dfrac{2}{21} = \dfrac{150}{x}$

$2 \cdot x = 21 \cdot 150$ *Cross products are equivalent*

$\dfrac{2 \cdot x}{2} = \dfrac{3150}{2}$

$x = 1575$

Check: $2 \cdot 1575 = 3150 = 21 \cdot 150$

27. $\dfrac{16}{x} = \dfrac{12}{15}$ **OR** $\dfrac{16}{x} = \dfrac{4}{5}$

$4 \cdot x = 5 \cdot 16$ *Cross products are equivalent*

$\dfrac{4 \cdot x}{4} = \dfrac{80}{4}$

$x = 20$

Check: $16 \cdot 5 = 80 = 20 \cdot 4$

28. $\frac{100}{14} = \frac{x}{56}$ **OR** $\frac{50}{7} = \frac{x}{56}$

$7 \cdot x = 50 \cdot 56$ *Cross products are equivalent*

$\frac{7 \cdot x}{7} = \frac{2800}{7}$

$x = 400$

Check: $50 \cdot 56 = 2800 = 7 \cdot 400$

29. $\frac{5}{8} = \frac{x}{20}$

$8 \cdot x = 5 \cdot 20$ *Cross products are equivalent*

$\frac{8 \cdot x}{8} = \frac{100}{8}$

$x = 12.5$

Check: $5 \cdot 20 = 100 = 8(12.5)$

30. $\frac{x}{24} = \frac{11}{18}$

$18 \cdot x = 24 \cdot 11$ *Cross products are equivalent*

$\frac{18 \cdot x}{18} = \frac{264}{18}$

$x = \frac{44}{3} \approx 14.67$

Check: $\frac{44}{3} \cdot 18 = 264 = 24 \cdot 11$

31. $\frac{7}{x} = \frac{18}{21}$ **OR** $\frac{7}{x} = \frac{6}{7}$

$6 \cdot x = 7 \cdot 7$ *Cross products are equivalent*

$\frac{6 \cdot x}{6} = \frac{49}{6}$

$x = \frac{49}{6} \approx 8.17$

Check: $7 \cdot 7 = 49 = \frac{49}{6} \cdot 6$

32. $\frac{x}{3.6} = \frac{9.8}{0.7}$

$0.7 \cdot x = 9.8(3.6)$ *Cross products are equivalent*

$\frac{0.7 \cdot x}{0.7} = \frac{35.28}{0.7}$

$x = 50.4$

Check: $50.4(0.7) = 35.28 = 3.6(9.8)$

33. $\frac{13.5}{1.7} = \frac{4.5}{x}$

$13.5 \cdot x = 1.7(4.5)$ *Cross products are equivalent*

$\frac{13.5 \cdot x}{13.5} = \frac{7.65}{13.5}$

$x \approx 0.57$

Check: $13.5\left(\frac{7.65}{13.5}\right) = 7.65 = 1.7(4.5)$

34. $\frac{0.82}{1.89} = \frac{x}{5.7}$

$1.89 \cdot x = 0.82(5.7)$ *Cross products are equivalent*

$\frac{1.89 \cdot x}{1.89} = \frac{4.674}{1.89}$

$x \approx 2.47$

Check: $0.82(5.7) = 4.674 = 1.89\left(\frac{4.674}{1.89}\right)$

35. $\frac{3 \text{ cats}}{5 \text{ dogs}} = \frac{x}{45 \text{ dogs}}$

$5 \cdot x = 3 \cdot 45$

$\frac{5 \cdot x}{5} = \frac{135}{5}$

$x = 27$

There are 27 cats.

36. $\frac{8 \text{ hits}}{28 \text{ at bats}} = \frac{x}{161 \text{ at bats}}$

$28 \cdot x = 8 \cdot 161$

$\frac{28 \cdot x}{28} = \frac{1288}{28}$

$x = 46$

She will get 46 hits.

37. $\frac{3.5 \text{ pounds}}{\$9.77} = \frac{5.6 \text{ pounds}}{x}$

$3.5 \cdot x = 9.77(5.6)$

$\frac{3.5 \cdot x}{3.5} = \frac{54.712}{3.5}$

$x \approx 15.63$

The cost for 5.6 pounds of ground beef is $15.63 (rounded).

38. $\frac{4 \text{ voting students}}{10 \text{ students}} = \frac{x}{8247 \text{ students}}$

$10 \cdot x = 4 \cdot 8247$

$\frac{10 \cdot x}{10} = \frac{32{,}988}{10}$

$x \approx 3299$

They should expect about 3299 students to vote.

39. $\frac{1 \text{ inch}}{16 \text{ feet}} = \frac{4.25 \text{ inches}}{x}$

$1 \cdot x = 16(4.25)$

$x = 68$

The length of the real boxcar is 68 feet.

40. 2 dozen necklaces $= 2 \cdot 12 = 24$ necklaces

$$\frac{24 \text{ necklaces}}{16\frac{1}{2} \text{ hours}} = \frac{40 \text{ necklaces}}{x}$$

$$24 \cdot x = 16\frac{1}{2} \cdot 40 = \frac{33}{2} \cdot \frac{40}{1}$$

$$\frac{24 \cdot x}{24} = \frac{660}{24}$$

$$x = 27.5 = 27\frac{1}{2}$$

It will take Marvette $27\frac{1}{2}$ hours or 27.5 hours to make 40 necklaces.

41. $\frac{284 \text{ calories}}{25 \text{ minutes}} = \frac{x}{45 \text{ minutes}}$

$$25 \cdot x = 284 \cdot 45$$

$$\frac{25 \cdot x}{25} = \frac{12{,}780}{25}$$

$$x = 511.2 \approx 511$$

A 180-pound person would burn about 511 calories in 45 minutes.

42. $\frac{3.5 \text{ milligrams}}{50 \text{ pounds}} = \frac{x}{210 \text{ pounds}}$

$$50 \cdot x = 3.5 \cdot 210$$

$$\frac{50 \cdot x}{50} = \frac{735}{50}$$

$$x = 14.7$$

A patient who weighs 210 pounds should be given 14.7 milligrams of the medicine.

43. The figure has two endpoints so it is a *line segment* named $\overline{AB}$ or $\overline{BA}$.

44. This is a *line* named $\overleftrightarrow{CD}$ or $\overleftrightarrow{DC}$. A line is a straight row of points that goes on forever in both directions.

45. This is a *ray* named $\overrightarrow{OP}$. A ray is a part of a line that has only one endpoint and goes on forever in one direction.

46. These lines appear to be *parallel* lines. Parallel lines are lines in the same plane that never intersect (cross).

47. The lines are *perpendicular* because they intersect at right angles.

48. The lines intersect so they are *not* parallel. At their intersection they *do not* form a right angle so they are not perpendicular. The lines are *intersecting*.

49. The measure of the angle is between 0° and 90°, so it is an *acute angle*.

50. The measure of the angle is between 90° and 180°, so it is an *obtuse angle*.

51. Two rays in a straight line pointing opposite directions measure 180°. An angle that measures 180° is called a *straight angle*.

52. The angle is a *right angle*, as indicated by the small square at the vertex. Right angles measure exactly 90°.

53. **(a)** The complement of 80° is 10°, because $90° - 80° = 10°$.

(b) The complement of 45° is 45°, because $90° - 45° = 45°$.

(c) The complement of 7° is 83°, because $90° - 7° = 83°$.

54. **(a)** The supplement of 155° is 25°, because $180° - 155° = 25°$.

(b) The supplement of 90° is 90°, because $180° - 90° = 90°$.

(c) The supplement of 33° is 147°, because $180° - 33° = 147°$.

55. $\angle 2 \cong \angle 5$, so $\angle 5$ measures 60°.
$\angle 6 \cong \angle 3$, so $\angle 6$ and $\angle 3$ measure 90°.
$\angle 1$ measures $90° - 60° = 30°$.
$\angle 1 \cong \angle 4$, so $\angle 4$ measures 30°.

56. $\angle 8 \cong \angle 3 \cong \angle 6 \cong \angle 1$, so each measures 160°.
$\angle 4$ measures $180° - 160° = 20°$.
$\angle 4 \cong \angle 7 \cong \angle 2 \cong \angle 5$, so each measures 20°.

57. Each pair of corresponding sides has the same length, so the Side-Side-Side (SSS) method can be used to prove that the triangles are congruent.

58. On both triangles, two corresponding sides and the angle between them measure the same, so the Side-Angle-Side (SAS) method can be used to prove that the triangles are congruent.

59. On both triangles, two corresponding angles and the side that connects them measure the same, so the Angle-Side-Angle (ASA) method can be used to prove that the triangles are congruent.

60. Set up a ratio of corresponding sides.

$$\frac{40 \text{ ft}}{20 \text{ ft}} = \frac{40}{20} = \frac{2}{1}$$

Write a proportion to find y.

$$\frac{y}{15} = \frac{2}{1}$$

$$y \cdot 1 = 2 \cdot 15$$

$$y = 30 \text{ ft}$$

Write a proportion to find x.

$$\frac{x}{17} = \frac{2}{1}$$
$$x \cdot 1 = 17 \cdot 2$$
$$x = 34 \text{ ft}$$

$$P = 34 \text{ ft} + 30 \text{ ft} + 40 \text{ ft}$$
$$= 104 \text{ ft}$$

61. Set up a ratio of corresponding sides.

$$\frac{4 \text{ m}}{6 \text{ m}} = \frac{4}{6} = \frac{2}{3}$$

Write a proportion to find x.

$$\frac{6}{x} = \frac{2}{3}$$
$$2 \cdot x = 3 \cdot 6$$
$$\frac{2 \cdot x}{2} = \frac{18}{2}$$
$$x = 9 \text{ m}$$

Write a proportion to find y.

$$\frac{5}{y} = \frac{2}{3}$$
$$2 \cdot y = 5 \cdot 3$$
$$\frac{2 \cdot y}{2} = \frac{15}{2}$$
$$y = 7.5 \text{ m}$$

$$P = 9 \text{ m} + 7.5 \text{ m} + 6 \text{ m}$$
$$= 22.5 \text{ m}$$

62. Set up a ratio of corresponding sides.

$$\frac{16 \text{ mm}}{12 \text{ mm}} = \frac{16}{12} = \frac{4}{3}$$

Write a proportion to find x.

$$\frac{x}{9} = \frac{4}{3}$$
$$3 \cdot x = 9 \cdot 4$$
$$\frac{3 \cdot x}{3} = \frac{36}{3}$$
$$x = 12 \text{ mm}$$

Write a proportion to find y.

$$\frac{10}{y} = \frac{4}{3}$$
$$y \cdot 4 = 10 \cdot 3$$
$$\frac{y \cdot 4}{4} = \frac{30}{4}$$
$$y = 7.5 \text{ mm}$$

$$P = 12 \text{ mm} + 10 \text{ mm} + 16 \text{ mm}$$
$$= 38 \text{ mm}$$

63. **[6.3]** $\frac{x}{45} = \frac{70}{30}$ **OR** $\frac{x}{45} = \frac{7}{3}$

$3 \cdot x = 7 \cdot 45$ *Cross products are equivalent*

$$\frac{3 \cdot x}{3} = \frac{315}{3}$$
$$x = 105$$

Check: $105 \cdot 3 = 315 = 45 \cdot 7$

64. **[6.3]** $\frac{x}{52} = \frac{0}{20}$

Since $\frac{0}{20} = 0$, x *must* equal 0. The denominator 52 could be any number (except 0).

65. **[6.3]** $\frac{64}{10} = \frac{x}{20}$ **OR** $\frac{32}{5} = \frac{x}{20}$

$5 \cdot x = 32 \cdot 20$ *Cross products are equivalent*

$$\frac{5 \cdot x}{5} = \frac{640}{5}$$
$$x = 128$$

Check: $32 \cdot 20 = 640 = 5 \cdot 128$

66. **[6.3]** $\frac{15}{x} = \frac{65}{100}$ **OR** $\frac{15}{x} = \frac{13}{20}$

$13 \cdot x = 15 \cdot 20$ *Cross products are equivalent*

$$\frac{13 \cdot x}{13} = \frac{300}{13}$$
$$x = \frac{300}{13} \approx 23.08$$

Check: $15 \cdot 20 = 300 = \frac{300}{13} \cdot 13$

67. **[6.3]** $\frac{7.8}{3.9} = \frac{13}{x}$ **OR** $\frac{2}{1} = \frac{13}{x}$

$$2 \cdot x = 1 \cdot 13$$
$$\frac{2 \cdot x}{2} = \frac{13}{2}$$
$$x = 6.5$$

Check: $2(6.5) = 13 = 1 \cdot 13$

68. **[6.3]** $\frac{34.1}{x} = \frac{0.77}{2.65}$

$$0.77 \cdot x = 34.1(2.65)$$
$$\frac{0.77 \cdot x}{0.77} = \frac{90.365}{0.77}$$
$$x \approx 117.36$$

Check: $34.1(2.65) = 90.365 = \left(\frac{90.365}{0.77}\right)(0.77)$

69. **[6.1]** 4 dollars to 10 quarters

4 dollars $= 4 \cdot 4 = 16$ quarters

$$\frac{16 \text{ quarters} \div 2}{10 \text{ quarters} \div 2} = \frac{8}{5}$$

70. **[6.1]** $4\frac{1}{8}$ inches to 10 inches

$$\frac{4\frac{1}{8}\text{ inches}}{10\text{ inches}} = \frac{\frac{33}{8}}{10} = \frac{33}{8} \div 10 = \frac{33}{8} \cdot \frac{1}{10} = \frac{33}{80}$$

71. **[6.1]** 10 yards to 8 feet

10 yards $= 10 \cdot 3 = 30$ feet

$$\frac{30\text{ feet}}{8\text{ feet}} = \frac{30 \div 2}{8 \div 2} = \frac{15}{4}$$

72. **[6.1]** \$3.60 to \$0.90

$$\frac{\$3.60}{\$0.90} = \frac{3.60 \div 0.90}{0.90 \div 0.90} = \frac{4}{1}$$

73. **[6.1]** 12 eggs to 15 eggs

$$\frac{12\text{ eggs}}{15\text{ eggs}} = \frac{12 \div 3}{15 \div 3} = \frac{4}{5}$$

74. **[6.1]** 37 meters to 7 meters

$$\frac{37\text{ meters}}{7\text{ meters}} = \frac{37}{7}$$

75. **[6.1]** 3 pints to 4 quarts

4 quarts $= 4 \cdot 2 = 8$ pints

$$\frac{3\text{ pints}}{8\text{ pints}} = \frac{3}{8}$$

76. **[6.1]** 15 minutes to 3 hours

3 hours $= 3 \cdot 60 = 180$ minutes

$$\frac{15\text{ minutes}}{180\text{ minutes}} = \frac{15 \div 15}{180 \div 15} = \frac{1}{12}$$

77. **[6.1]** $4\frac{1}{2}$ miles to $1\frac{3}{10}$ miles

$$\frac{4\frac{1}{2}\text{ miles}}{1\frac{3}{10}\text{ miles}} = \frac{4\frac{1}{2}}{1\frac{3}{10}} = 4\frac{1}{2} \div 1\frac{3}{10} = \frac{9}{2} \div \frac{13}{10} = \frac{9}{\cancel{2}_1} \cdot \frac{\cancel{10}^5}{13} = \frac{45}{13}$$

78. **[6.4]**

$$\frac{7\text{ buying fans}}{8\text{ fans}} = \frac{x}{28{,}500\text{ fans}}$$
$$8 \cdot x = 7 \cdot 28{,}500$$
$$\frac{8 \cdot x}{8} = \frac{199{,}500}{8}$$
$$x \approx 24{,}937.5$$

or 24,900
(*rounded to the nearest hundred*)

At today's concert, about 24,900 fans can be expected to buy a beverage.

79. **[6.1]**

$$\frac{\$400\text{ spent on car insurance}}{\$150\text{ spent on repairs}} = \frac{400 \div 50}{150 \div 50} = \frac{8}{3}$$

The ratio of the amount spent on insurance to the amount spent on repairs is $\frac{8}{3}$.

80. **[6.2]** 25 feet for \$0.78

$$\frac{\$0.78}{25\text{ feet}} \approx \$0.031\text{ per foot}$$

75 feet for \$1.99 – \$0.50 (coupon) = \$1.49

$$\frac{\$1.49}{75\text{ feet}} \approx \$0.020\text{ per foot } (*)$$

100 feet for \$2.59 – \$0.50 (coupon) = \$2.09

$$\frac{\$2.09}{100\text{ feet}} \approx \$0.021\text{ per foot}$$

The best buy is 75 feet for \$1.99 with a 50¢ coupon.

81. **[6.4]** First find the length.

$$\frac{0.5\text{ inch}}{6\text{ feet}} = \frac{1.75\text{ inches}}{x}$$
$$0.5 \cdot x = 6(1.75)$$
$$\frac{0.5 \cdot x}{0.5} = \frac{10.5}{0.5}$$
$$x = 21$$

When it is built, the actual length of the patio will be 21 feet.

Then find the width.

$$\frac{0.5\text{ inch}}{6\text{ feet}} = \frac{1.25\text{ inches}}{x}$$
$$0.5 \cdot x = 6(1.25)$$
$$\frac{0.5 \cdot x}{0.5} = \frac{7.5}{0.5}$$
$$x = 15$$

When it is built, the actual width of the patio will be 15 feet.

82. **[6.4]** **(a)**

$$\frac{1000\text{ milligrams}}{5\text{ pounds}} = \frac{x}{7\text{ pounds}}$$
$$5 \cdot x = 1000 \cdot 7$$
$$\frac{5 \cdot x}{5} = \frac{7000}{5}$$
$$x = 1400$$

A 7-pound cat should be given 1400 milligrams.

(b) 8 ounces $= \frac{8}{16}$ pound $= 0.5$ pound

$$\frac{1000\text{ milligrams}}{5\text{ pounds}} = \frac{x}{0.5\text{ pounds}}$$
$$5 \cdot x = 1000(0.5)$$
$$\frac{5 \cdot x}{5} = \frac{500}{5}$$
$$x = 100$$

An 8-ounce kitten should be given 100 milligrams.

83. **[6.4]** $\frac{251 \text{ points}}{169 \text{ minutes}} = \frac{x}{14 \text{ minutes}}$

$$169 \cdot x = 251 \cdot 14$$
$$\frac{169 \cdot x}{169} = \frac{3514}{169}$$
$$x \approx 20.792899$$

Charles should score 21 points (rounded).

84. **[6.4]** $\frac{1\frac{1}{2} \text{ teaspoons}}{24 \text{ pounds}} = \frac{x}{8 \text{ pounds}}$

$$24 \cdot x = 1\frac{1}{2} \cdot 8 = \frac{3}{2} \cdot \frac{8}{1} = \frac{24}{2} = 12$$
$$\frac{24 \cdot x}{24} = \frac{12}{24}$$
$$x = \frac{1}{2} \text{ or } 0.5$$

The infant should be given $\frac{1}{2}$ or 0.5 teaspoon.

85. **[6.5]** $\overleftrightarrow{WX}$ and $\overleftrightarrow{YZ}$ are parallel lines.

86. **[6.5]** $\overline{QR}$ is a line segment.

87. **[6.5]** $\angle CQD$ is an acute angle.

88. **[6.5]** $\overleftrightarrow{PQ}$ and $\overleftrightarrow{NO}$ are intersecting lines.

89. **[6.5]** $\angle APB$ is a right angle measuring 90°.

90. **[6.5]** $\overrightarrow{AB}$ is a ray.

91. **[6.5]** $\overleftrightarrow{T}$ is a straight angle measuring 180°.

92. **[6.5]** $\angle FEG$ is an obtuse angle.

93. **[6.5]** $\overleftrightarrow{LM}$ and $\overleftrightarrow{JK}$ are perpendicular lines.

94. **[6.5]** **(a)** If your car "did a 360," the car turned around in a complete circle.

(b) If the governor's view on taxes "took a 180° turn," he or she took the opposite view. For example, he or she may have opposed taxes but now supports them.

95. **[6.5]** **(a)** No; obtuse angles are $> 90°$, so their sum would be $> 180°$.

(b) Yes; acute angles are $< 90°$, so their sum could equal 90°.

96. **[6.5]** $\angle 1$ measures $90° - 45° = 45°$.
$\angle 7 \cong \angle 4$, so each measures 55°.
$\angle 3$ measures $90° - 55° = 35°$.
$\angle 3 \cong \angle 6$, so $\angle 6$ measures 35°.
$\angle 5$ measures 90°.

97. **[6.5]** $\angle 5 \cong \angle 2 \cong \angle 7 \cong \angle 4$, so each measures 75°.
$\angle 6$ measures $180° - 75° = 105°$.
$\angle 6 \cong \angle 1 \cong \angle 8 \cong \angle 3$, so each measures 105°.

Chapter 6 Test

1. \$15 for 75 minutes

$$\frac{\$15 \div 15}{75 \text{ minutes} \div 15} = \frac{\$1}{5 \text{ minutes}}$$

2. 3 hours to 40 minutes

3 hours $= 3 \cdot 60$ minutes $= 180$ minutes

$$\frac{180 \text{ minutes}}{40 \text{ minutes}} = \frac{180 \div 20}{40 \div 20} = \frac{9}{2}$$

3. $\frac{1200 \text{ seats}}{320 \text{ seats}} = \frac{1200 \div 80}{320 \div 80} = \frac{15}{4}$

4. 28 ounces of Brand X for
\$3.89 – \$0.75 (coupon) = \$3.14

$$\frac{\$3.14}{28 \text{ ounces}} \approx \$0.112 \text{ per ounce}$$

18 ounces of Brand Y for
\$1.89 – \$0.25 (coupon) = \$1.64

$$\frac{\$1.64}{18 \text{ ounces}} \approx \$0.091 \text{ per ounce } (*)$$

13 ounces of Brand Z for \$1.29.

$$\frac{\$1.29}{13 \text{ ounces}} \approx \$0.099 \text{ per ounce}$$

The best buy is 18 ounces of Brand Y for \$1.89 with a 25¢ coupon.

5. $\frac{5}{9} = \frac{x}{45}$

$9 \cdot x = 5 \cdot 45$ *Cross products are equivalent*

$$\frac{9 \cdot x}{9} = \frac{225}{9}$$
$$x = 25$$

Check: $5 \cdot 45 = 225 = 9 \cdot 25$

6. $\frac{3}{1} = \frac{8}{x}$

$3 \cdot x = 8 \cdot 1$ *Cross products are equivalent*

$$\frac{3 \cdot x}{3} = \frac{8}{3}$$
$$x \approx 2.67$$

Check: $3\left(\frac{8}{3}\right) = 8 = 1 \cdot 8$

7. $\frac{x}{20} = \frac{6.5}{0.4}$

$0.4 \cdot x = 20(6.5)$ *Cross products are equivalent*

$$\frac{0.4 \cdot x}{0.4} = \frac{130}{0.4}$$
$$x = 325$$

Check: $325(0.4) = 130 = 20(6.5)$

8. $\frac{2\frac{1}{3}}{x} = \frac{\frac{8}{9}}{4}$

$\frac{8}{9} \cdot x = 2\frac{1}{3} \cdot 4 = \frac{7}{3} \cdot \frac{4}{1} = \frac{28}{3}$

$\frac{\frac{8}{9} \cdot x}{\frac{8}{9}} = \frac{\frac{28}{3}}{\frac{8}{9}}$

$x = \frac{28}{3} \div \frac{8}{9} = \frac{\overset{7}{\cancel{28}}}{\underset{1}{\cancel{3}}} \cdot \frac{\overset{3}{\cancel{9}}}{\underset{2}{\cancel{8}}} = \frac{21}{2} = 10\frac{1}{2}$

Check: $2\frac{1}{3} \cdot 4 = \frac{28}{3} = \frac{21}{2} \cdot \frac{8}{9}$

9. $\frac{240 \text{ words}}{5 \text{ minutes}} = \frac{x}{12 \text{ minutes}}$

$5 \cdot x = 240 \cdot 12$

$\frac{5 \cdot x}{5} = \frac{2880}{5}$

$x = 576$

Pedro could type 576 words in 12 minutes.

10. $\frac{2 \text{ left-handed people}}{15 \text{ people}} = \frac{x}{650 \text{ students}}$

$15 \cdot x = 2 \cdot 650$

$\frac{15 \cdot x}{15} = \frac{1300}{15}$

$x = \frac{260}{3} \approx 87$

You could expect about 87 students to be left-handed.

11. $\frac{8.2 \text{ grams}}{50 \text{ pounds}} = \frac{x}{145 \text{ pounds}}$

$50 \cdot x = 8.2(145)$

$\frac{50 \cdot x}{50} = \frac{1189}{50}$

$x = 23.78 \approx 23.8$

A 145-pound person should be given 23.8 grams (rounded).

12. $\frac{1 \text{ inch}}{8 \text{ feet}} = \frac{7.5 \text{ inches}}{x}$

$1 \cdot x = 8(7.5)$

$x = 60$

The actual height of the building is 60 feet.

13. $\angle LOM$ is an acute angle, so the answer is (e).

14. $\angle YOX$ is a right angle, so the answer is (a). Its measure is 90°.

15. $\overrightarrow{GH}$ is a ray, so the answer is (d).

16. $\overleftrightarrow{W}$ is a straight angle, so the answer is (g). Its measure is 180°.

17. Parallel lines are lines in the same plane that never intersect.
Perpendicular lines intersect to form a right angle.

Parallel lines

Perpendicular lines

18. The complement of an angle measuring 81° is 90° – 81° = 9°.

19. The supplement of an angle measuring 20° is 180° – 20° = 160°.

20. $\angle 4 \cong \angle 1$, so each measures 50°.
$\angle 6 \cong \angle 3$, so each measures 95°.
$\angle 2$ measures 180° – 50° – 95° = 35°.
$\angle 2 \cong \angle 5$, so each measures 35°.

21. $\angle 3 \cong \angle 1 \cong \angle 5 \cong \angle 7$, so each measures 65°.
$\angle 4$ measures 180° – 65° = 115°.
$\angle 4 \cong \angle 2 \cong \angle 6 \cong \angle 8$, so each measures 115°.

22. On both triangles, two corresponding angles and the side that connects them measure the same, so the Angle-Side-Angle (ASA) method can be used to prove that the triangles are congruent.

23. On both triangles, two corresponding sides and the angle between them measure the same, so the Side-Angle-Side (SAS) method can be used to prove that the triangles are congruent.

24. Set up a ratio of corresponding sides.

$\frac{10 \text{ cm}}{15 \text{ cm}} = \frac{10}{15} = \frac{2}{3}$

Write a proportion to find y.

$\frac{y}{18} = \frac{2}{3}$

$y \cdot 3 = 18 \cdot 2$

$\frac{y \cdot 3}{3} = \frac{36}{3}$

$y = 12 \text{ cm}$

Write a proportion to find z.

$\frac{z}{9} = \frac{2}{3}$

$z \cdot 3 = 9 \cdot 2$

$\frac{z \cdot 3}{3} = \frac{18}{3}$

$z = 6 \text{ cm}$

25. Set up a ratio of corresponding sides.

$\frac{18 \text{ mm}}{15 \text{ mm}} = \frac{18}{15} = \frac{6}{5}$

Write a proportion to find x.

$\frac{x}{10} = \frac{6}{5}$

$x \cdot 5 = 10 \cdot 6$

$\frac{x \cdot 5}{5} = \frac{60}{5}$

$x = 12 \text{ mm}$

Write a proportion to find y.

$$\frac{16.8}{y} = \frac{6}{5}$$
$$y \cdot 6 = 16.8 \cdot 5$$
$$\frac{y \cdot 6}{6} = \frac{84}{6}$$
$$y = 14 \text{ mm}$$

The perimeter of the larger triangle:

$$P = 18 \text{ mm} + 12 \text{ mm} + 16.8 \text{ mm}$$
$$P = 46.8 \text{ mm}$$

The perimeter of the smaller triangle:

$$P = 15 \text{ mm} + 10 \text{ mm} + 14 \text{ mm}$$
$$P = 39 \text{ mm}$$

Cumulative Review Exercises (Chapters 1–6)

1. **(a)** In words, 77,001,000,805 is seventy-seven billion, one million, eight hundred five.

(b) In words, 0.02 is two hundredths.

2. **(a)** Three and forty thousandths can be written as 3.040.

(b) Five hundred million, thirty-seven thousand can be written as 500,037,000.

3. $\frac{0}{-16} = 0$

4. $|0| + |-6 - 8|$ Add the opposite.
$= |0| + |-6 + (-8)|$ Add.
$= |0| + |-14|$ Absolute values
$= 0 + 14$ Add.
$= 14$

5. $4\frac{3}{4} - 1\frac{5}{6}$ Change to improper fractions.
$= \frac{19}{4} - \frac{11}{6}$ LCD = 12
$= \frac{19 \cdot 3}{4 \cdot 3} - \frac{11 \cdot 2}{6 \cdot 2}$
$= \frac{57}{12} - \frac{22}{12}$
$= \frac{57 - 22}{12}$
$= \frac{35}{12}$ or $2\frac{11}{12}$

6. $\frac{h}{5} - \frac{3}{10} = \frac{2h}{10} - \frac{3}{10} = \frac{2h - 3}{10}$

7. $100 - 0.0095$

$$\begin{array}{r} \overset{9\;9\;9\;9\;9}{\cancel{1}\,\cancel{0}\,\cancel{0}.\cancel{0}\,\cancel{0}\,\cancel{0}\,\cancel{0}} \\ -\;\;0.0095 \\ \hline 99.9905 \end{array}$$

8. $\frac{-4+7}{9-3^2} = \frac{3}{9-9} = \frac{3}{0}$, which is undefined.

9. $-6 + 3(0 - 4)$ Parentheses
$= -6 + 3(-4)$ Multiply.
$= -6 + (-12)$ Add.
$= -18$

10. $\frac{-8}{\frac{4}{7}} = -\frac{8}{1} \div \frac{4}{7} = -\frac{8}{1} \cdot \frac{7}{4} = -\frac{2 \cdot \cancel{4} \cdot 7}{1 \cdot \cancel{4}}$
$= -\frac{14}{1} = -14$

11. $\frac{5n}{6m^3} \div \frac{10}{3m^2} = \frac{5n}{6m^3} \cdot \frac{3m^2}{10}$
$= \frac{\cancel{5} \cdot n \cdot \cancel{3} \cdot \cancel{m} \cdot \cancel{m}}{2 \cdot \cancel{3} \cdot \cancel{m} \cdot \cancel{m} \cdot m \cdot 2 \cdot \cancel{5}}$
$= \frac{n}{4m}$

12. $(0.06)(-0.007)$ Different signs, negative product

0.007	←	*3 decimal places*
× 0.06	←	*2 decimal places*
0.00042	←	*5 decimal places*

Answer: -0.00042

13. $\frac{x}{14y} \cdot \frac{7}{xy} = \frac{\cancel{x} \cdot \cancel{7}}{2 \cdot \cancel{7} \cdot y \cdot \cancel{x} \cdot y} = \frac{1}{2y^2}$

14. $\frac{9}{n} + \frac{2}{3} = \frac{9 \cdot 3}{n \cdot 3} + \frac{2 \cdot n}{3 \cdot n} = \frac{27}{3n} + \frac{2n}{3n} = \frac{27 + 2n}{3n}$

15. $-40 + 8(-5) + 2^4$ Exponent
$= -40 + 8(-5) + 16$ Multiply.
$= -40 + (-40) + 16$ Add.
$= -80 + 16$ Add.
$= -64$

16. $5.8 - (-0.6)^2 \div 0.9$ Exponent
$= 5.8 - 0.36 \div 0.9$ Divide.
$= 5.8 - 0.4$ Subtract.
$= 5.4$

17. $\left(-\frac{1}{2}\right)^3 \left(\frac{2}{3}\right)^2 = -\frac{1}{2}\left(-\frac{1}{2}\right)\left(-\frac{1}{2}\right)\frac{2}{3} \cdot \frac{2}{3}$
$= -\frac{1 \cdot 1 \cdot 1 \cdot \cancel{2} \cdot \cancel{2}}{2 \cdot \cancel{2} \cdot \cancel{2} \cdot 3 \cdot 3} = -\frac{1}{18}$

18.
$$\begin{aligned} 7y+5 &= -3-y \\ 7y+5 &= -3+(-1y) \\ 1y \quad & \quad\quad 1y \\ 8y+5 &= -3+0 \\ 8y+5 &= -3 \\ -5 \quad & \quad -5 \\ 8y+0 &= -8 \\ \frac{8y}{8} &= \frac{-8}{8} \\ y &= -1 \end{aligned}$$

$-y = -1y$ Add $1y$ to both sides.

Add -5 to both sides.

Divide both sides by 8.

The solution is -1.

19.
$$\begin{aligned} -2+\frac{3}{5}x &= 7 \\ +2 \quad & \quad +2 \\ 0+\frac{3}{5}x &= 9 \\ \frac{5}{3}\left(\frac{3}{5}x\right) &= \frac{5}{3}(9) \\ x &= 15 \end{aligned}$$

Add 2 to both sides.

Multiply both sides by $\frac{5}{3}$.

The solution is 15.

20. $\frac{4}{x} = \frac{14}{35}$ **OR** $\frac{4}{x} = \frac{2}{5}$

$$\begin{aligned} x \cdot 2 &= 4 \cdot 5 \\ \frac{2x}{2} &= \frac{20}{2} \\ x &= 10 \end{aligned}$$

The solution is 10.

21. *Step 1*
Unknown: starting temperature

Known: 15° rise, 23° drop, 5° rise, final temperature 71°.

Step 2
Let t be the starting temperature.

Step 3
$t + 15 + (-23) + 5 = 71$

Step 4
$$\begin{aligned} t+(-3) &= 71 \\ +3 \quad & \quad +3 \\ t+0 &= 74 \\ t &= 74 \end{aligned}$$

Step 5
The starting temperature was 74 degrees.

Step 6
Check: $74 + 15 - 23 + 5 = 71$

22. *Step 1*
Unknown: length and width

Known: perimeter is 100 ft, width is 14 ft less than the length

Step 2
Let l represent the length. Then $l - 14$ represents the width.

Step 3
$P = 2l + 2w$
$2(l) + 2(l - 14) = 100$

Step 4
$$\begin{aligned} 2(l)+2(l-14) &= 100 \\ 2l+2l-28 &= 100 \\ 4l-28 &= 100 \\ +28 \quad & \quad +28 \\ 4l+0 &= 128 \\ \frac{4l}{4} &= \frac{128}{4} \\ l &= 32 \end{aligned}$$

Step 5
The length is 32 ft and the width is 14 ft less than the length, or 18 ft.

Step 6
Check: $P = 2l + 2w$
$P = 2 \cdot 32 \text{ ft} + 2 \cdot 18 \text{ ft}$
$P = 64 \text{ ft} + 36 \text{ ft}$
$P = 100 \text{ ft}$

23. $P = 2l + 2w$
$P = 2 \cdot 2.8 \text{ km} + 2 \cdot 0.7 \text{ km}$
$P = 5.6 \text{ km} + 1.4 \text{ km}$
$P = 7 \text{ km}$

$A = l \cdot w$
$A = 2.8 \text{ km} \cdot 0.7 \text{ km}$
$A = 1.96 \text{ km}^2$
$A \approx 2.0 \text{ km}^2$ (rounded to the nearest tenth)

24. $d = 2r$
$d = 2 \cdot 8.5 \text{ m}$
$d = 17 \text{ m}$

The diameter is 17 m.

$C = \pi \cdot d$
$C \approx 3.14 \cdot 17 \text{ m}$
$C \approx 53.38 \text{ m}$

The circumference is about 53.4 m.

$A = \pi r^2$
$A = \pi \cdot r \cdot r$
$A \approx 3.14 \cdot 8.5 \text{ m} \cdot 8.5 \text{ m}$
$A \approx 226.865 \text{ m}^2$

The area is about 226.9 m^2.

25. $V = \pi r^2 h$
$V = \pi \cdot r \cdot r \cdot h$
$V \approx 3.14 \cdot 3 \text{ ft} \cdot 3 \text{ ft} \cdot 12 \text{ ft}$
$V \approx 339.12 \text{ ft}^3$

The volume is about 339.1 ft^3.

$SA = 2\pi rh + 2\pi r^2$
$SA \approx 2 \cdot 3.14 \cdot 3 \text{ ft} \cdot 12 \text{ ft} + 2 \cdot 3.14 \cdot 3 \text{ ft} \cdot 3 \text{ ft}$
$SA \approx 282.6 \text{ ft}^2$

The surface area is about 282.6 ft^2.

26. Find the amount already collected.

$$\frac{5}{\not{6}_1} \cdot \frac{\overset{250}{\not{1500}}}{1} = 1250$$

$1500 - 1250 = 250$

They need to collect 250 pounds.

27. $$\frac{\frac{1}{2} \text{ teaspoon}}{2 \text{ quarts water}} = \frac{x}{5 \text{ quarts water}}$$

$$2 \cdot x = \frac{1}{2} \cdot 5$$

$$\frac{2 \cdot x}{2} = \frac{\frac{5}{2}}{2}$$

$$x = \frac{5}{2} \div 2 = \frac{5}{2} \cdot \frac{1}{2} = \frac{5}{4} = 1\frac{1}{4}$$

Use $1\frac{1}{4}$ teaspoons of plant food.

28. First add the number of times Norma ran in the morning and in the afternoon.

$$4 + 2\frac{1}{2} = 6\frac{1}{2}$$

Then multiply the sum by the distance around Dunning Pond.

$$1\frac{1}{10} \cdot 6\frac{1}{2} = \frac{11}{10} \cdot \frac{13}{2} = \frac{143}{20} = 7\frac{3}{20}$$

Norma ran $7\frac{3}{20}$ miles.

29. Divide the number of miles driven by the number of gallons of gas purchased.

```
         1 8. 0 0
49.8^)8 9 6. 5^ 0 0
      4 9 8
      3 9 8 5
      3 9 8 4
            1 0
              0
            1 0 0
                0
            1 0 0
```

Rodney's SUV got 18.0 miles per gallon (rounded) on the vacation.

30. The cost per minute rates (in dollars) from least to greatest are:
0.08, 0.10, 0.104, 0.111, 0.14, 0.172, 0.333

31. $$\frac{\$10}{\$0.14/\text{minute}} = \frac{10}{0.14} \approx 71.4 \text{ minutes}$$

You could make a call to Japan for 71 minutes (rounded).

32. $$\frac{\$10}{\$0.333/\text{minute}} \approx 30.03 \text{ minutes}$$

You get 30 minutes (rounded) per \$10 card, so buy 4 cards for 120 minutes (2 hours).

33. **For Canada:**

$$\frac{\$10}{\$0.08/\text{minute}} = 125 \text{ minutes}$$

For Mexico City:

$$\frac{\$10}{\$0.10/\text{minute}} = 100 \text{ minutes}$$

Ratio of minutes to Canada to minutes to Mexico City

$$\frac{125 \div 25}{100 \div 25} = \frac{5}{4}$$

The ratio of *minutes for a \$10 call to Canada* to *minutes for a \$10 call to Mexico* is $\frac{5}{4}$.

CHAPTER 7 PERCENT

7.1 The Basics of Percent

7.1 Section Exercises

1. $25\% = 25.\% = 0.25$

Drop the percent sign and move the decimal point two places to the left.

3. $30\% = 30.\% = 0.30$ or 0.3

Drop the percent sign and move the decimal point two places to the left.

5. $6\% = 06.\% = 0.06$

0 is attached so the decimal point can be moved two places to the left.

7. $140\% = 140.\% = 1.40$ or 1.4

Drop the percent sign and move the decimal point two places to the left.

9. $7.8\% = 07.8\% = 0.078$

0 is attached so the decimal point can be moved two places to the left.

11. $100\% = 100.\% = 1.00$ or 1

13. $0.5\% = 00.5\% = 0.005$

0 is attached so the decimal point can be moved two places to the left.

15. $0.35\% = 00.35\% = 0.0035$

0 is attached so the decimal point can be moved two places to the left.

17. $0.5 = 0.50 = 50\%$

0 is attached so the decimal point can be moved two places to the right. Attach a percent sign. (The decimal point is not written with whole numbers.)

19. $0.62 = 62\%$

Move the decimal point two places to the right and attach a percent sign.

21. $0.03 = 3\%$

Move the decimal point two places to the right and attach a percent sign.

23. $0.125 = 12.5\%$

Move the decimal point two places to the right and attach a percent sign.

25. $0.629 = 62.9\%$

Move the decimal point two places to the right and attach a percent sign.

27. $2 = 2.00 = 200\%$

Two zeros are attached so the decimal point can be moved two places to the right. Attach a percent sign.

29. $2.6 = 2.60 = 260\%$

One zero is attached so the decimal point can be moved two places to the right. Attach a percent sign.

31. $0.0312 = 3.12\%$

Move the decimal point two places to the right and attach a percent sign.

33. $20\% = \frac{20}{100} = \frac{20 \div 20}{100 \div 20} = \frac{1}{5}$

35. $50\% = \frac{50}{100} = \frac{50 \div 50}{100 \div 50} = \frac{1}{2}$

37. $55\% = \frac{55}{100} = \frac{55 \div 5}{100 \div 5} = \frac{11}{20}$

39. $37.5\% = \frac{37.5}{100} = \frac{37.5(10)}{100(10)} = \frac{375 \div 125}{1000 \div 125} = \frac{3}{8}$

41. $6.25\% = \frac{6.25}{100} = \frac{6.25(100)}{100(100)} = \frac{625 \div 625}{10{,}000 \div 625} = \frac{1}{16}$

43. $16\frac{2}{3}\% = \frac{16\frac{2}{3}}{100} = 16\frac{2}{3} \div \frac{100}{1} = \frac{\overset{1}{\cancel{50}}}{3} \cdot \frac{1}{\underset{2}{\cancel{100}}} = \frac{1}{6}$

45. $130\% = \frac{130}{100} = \frac{130 \div 10}{100 \div 10} = \frac{13}{10}$ or $1\frac{3}{10}$

47. $250\% = \frac{250}{100} = \frac{250 \div 50}{100 \div 50} = \frac{5}{2} = 2\frac{1}{2}$

49. $\frac{1}{4} \cdot 100\% = \frac{1}{4} \cdot \frac{100}{1}\% = \frac{1 \cdot \overset{1}{\cancel{4}} \cdot 25}{\underset{1}{\cancel{4}} \cdot 1}\% = \frac{25}{1}\% = 25\%$

51. $\frac{3}{10} \cdot 100\% = \frac{3}{10} \cdot \frac{100}{1}\% = \frac{3 \cdot \overset{1}{\cancel{10}} \cdot 10}{\underset{1}{\cancel{10}} \cdot 1}\% = \frac{30}{1}\%$

$= 30\%$

53. $\frac{3}{5} \cdot 100\% = \frac{3}{5} \cdot \frac{100}{1}\% = \frac{3 \cdot \overset{1}{\cancel{5}} \cdot 20}{\underset{1}{\cancel{5}} \cdot 1}\% = \frac{60}{1}\%$

$= 60\%$

55. The denominator is already 100.

$$\frac{37}{100} = 37\%$$

57. $\frac{3}{8} \cdot 100\% = \frac{3}{8} \cdot \frac{100}{1}\% = \frac{3 \cdot \overset{1}{\cancel{4}} \cdot 25}{2 \cdot \underset{1}{\cancel{4}} \cdot 1}\% = \frac{75}{2}\%$

$= 37\frac{1}{2}\%$ or 37.5%

59. $\frac{1}{20} \cdot 100\% = \frac{1}{20} \cdot \frac{100}{1}\% = \frac{1 \cdot \overset{1}{\cancel{20}} \cdot 5}{\underset{1}{\cancel{20}} \cdot 1}\% = \frac{5}{1}\%$

$= 5\%$

61. $\frac{5}{9} \cdot 100\% = \frac{5}{9} \cdot \frac{100}{1}\% = \frac{500}{9}\%$

$= 55\frac{5}{9}\%$, or 55.6% (rounded)

63. $\frac{1}{7} \cdot 100\% = \frac{1}{7} \cdot \frac{100}{1}\% = \frac{100}{7}\%$

$= 14\frac{2}{7}\%$, or 14.3% (rounded)

65. Drop the percent sign and move the decimal point two places to the left.

8% of U.S. homes = 0.08

67. Drop the percent sign and move the decimal point two places to the left.

42% of tornadoes = 0.42

69. Move the decimal point two places to the right and attach a percent sign.

0.035 property tax rate = 3.5%

71. Attach two 0's so that the decimal point can be moved two places to the right. Attach a percent sign.

2 times that of the last session = 200%

73. Ninety-five parts of the one hundred parts are shaded.

$\frac{95}{100} = 0.95 = 95\%$

Five parts of the one hundred parts are unshaded.

$\frac{5}{100} = 0.05 = 5\%$

75. Three of the ten parts are shaded.

$\frac{3}{10} = \frac{30}{100} = 30\%$

Seven of the ten parts are unshaded.

$\frac{7}{10} = \frac{70}{100} = 70\%$

77. Three parts of the four parts are shaded.

$\frac{3}{4} = \frac{3(25)}{4(25)} = \frac{75}{100} = 75\%$

One part of the four parts is unshaded.

$\frac{1}{4} = \frac{1(25)}{4(25)} = \frac{25}{100} = 25\%$

79. $\frac{1}{100} = 0.01 \leftarrow$ Decimal

$= 1\% \leftarrow$ Percent

81. $0.2 = 0.20 = \frac{20}{100} = \frac{1}{5} \leftarrow$ Fraction

$= 20\% \leftarrow$ Percent

83. $30\% = \frac{30}{100} = \frac{3}{10} \leftarrow$ Fraction

$= 0.30 = 0.3 \leftarrow$ Decimal

85. $\frac{1}{2} = \frac{50}{100} = 0.50 \leftarrow$ Decimal

$= 50\% \leftarrow$ Percent

87. $90\% = \frac{90}{100} = \frac{9}{10} \leftarrow$ Fraction

$= 0.90 = 0.9 \leftarrow$ Decimal

89. $1.5 = \frac{150}{100} = \frac{3}{2} = 1\frac{1}{2} \leftarrow$ Fraction

$= 150\% \leftarrow$ Percent

91. **(a)** The fourth-most-common perk was employee entertainment/company product discounts.

(b) $40\% = \frac{40}{100} = \frac{4}{10} = \frac{2}{5}$

93. **(a)** The perk offered by *about* $\frac{1}{2}$ of the companies, at 49% (closest to 50%), was personal development training.

(b) The perk offered by *about* $\frac{1}{4}$ of the companies, at 27% (closest to 25%), was telecommuting.

95. There are 8 incisors and 32 total teeth.

$\frac{8}{32} = \frac{8 \div 8}{32 \div 8} = \frac{1}{4}$

$\frac{1}{4} = \frac{1(25)}{4(25)} = \frac{25}{100} = 0.25$

$= 25\%$

97. There are 12 molars and 32 total teeth.

$\frac{12}{32} = \frac{12 \div 4}{32 \div 4} = \frac{3}{8}$

$\frac{3}{8} = \frac{3(125)}{8(125)} = \frac{375}{1000} = 0.375$

$= 37.5\%$

99. There are 4 canines and 28 total teeth.

$\frac{4}{28} = \frac{4 \div 4}{28 \div 4} = \frac{1}{7}$

$\frac{1}{7} = \frac{1}{7} \cdot 100\% = \frac{100}{7}\% = 14\frac{2}{7}\% \approx 14.3\%$

101. **(a)** The student forgot to move the decimal point in 0.35 two places to the right. So $\frac{7}{20} = 35\%$.

(b) The student did the division in the wrong order. Enter $16 \div 25$ to get 0.64 and then move the decimal point two places to the right. So $\frac{16}{25} = 0.64 = 64\%$.

103. **(a)** 100% of \$78 is <u>\$78</u>.

(b) 50% of \$78 is half of \$78 or <u>\$39</u>.

105. **(a)** 100% of 15 inches is <u>15 inches</u>.

(b) 50% of 15 inches is half of 15 inches or <u>$7\frac{1}{2}$ inches</u>.

107. 100% of 20 children is 20 children. 20 children are served both meals.

109. **(a)** 100% of 2.8 miles is <u>2.8 miles</u>.

(b) 50% of 2.8 miles is half of 2.8 miles or <u>1.4 miles</u>.

111. **(a)** 50% of \$285 is half of \$285 or <u>\$142.50</u>.

(b) John will have to pay the remaining 50% of the tuition (100% − 50% = 50%).

113. **(a)** 50% of 8200 is 4100, so the number of students who work more than 20 hours per week is *about* 4100.

(b) The percent of students who work 20 hours or less per week is 100% − 50% = 50%.

115. **(a)** 100% of 35 problems is 35 problems.

(b) He correctly worked all 35 problems, so he missed 0 problems.

117. 50% means 50 parts out of 100 parts. That's half of the number. A shortcut for finding 50% of a number is to divide the number by 2. Examples will vary.

7.2 The Percent Proportion

7.2 Section Exercises

1. **(a)** Percent: 10%
(b) Whole: 3000 runners
(c) Part: unknown (n)

(d) *Step 1* $\frac{10}{100} = \frac{n}{3000}$

Step 2 $100 \cdot n = 10 \cdot 3000$

Step 3 $\frac{100n}{100} = \frac{30{,}000}{100}$

$n = 300$

10% of 3000 runners is 300 runners.

3. **(a)** Percent: 4%
(b) Whole: 120 ft
(c) Part: unknown (n)

(d) *Step 1* $\frac{4}{100} = \frac{n}{120}$

Step 2 $100 \cdot n = 4 \cdot 120$

Step 3 $\frac{100n}{100} = \frac{480}{100}$

$n = 4.8$

4% of 120 feet is 4.8 feet.

5. **(a)** Percent: unknown (p)
(b) Whole: 32 pizzas
(c) Part: 16 pizzas

(d) *Step 1* $\frac{p}{100} = \frac{16}{32}$

Step 2 $32 \cdot p = 100 \cdot 16$

Step 3 $\frac{32p}{32} = \frac{1600}{32}$

$p = 50$

16 pizzas is 50% of 32 pizzas.

7. **(a)** Percent: unknown (p)
(b) Whole: 200 calories
(c) Part: 16 calories

(d) *Step 1* $\frac{p}{100} = \frac{16}{200}$

Step 2 $200 \cdot p = 100 \cdot 16$

Step 3 $\frac{200p}{200} = \frac{1600}{200}$

$p = 8$

8% of 200 calories is 16 calories.

9. **(a)** Percent: 90%
(b) Whole: unknown (n)
(c) Part: 495 students

(d) *Step 1* $\frac{90}{100} = \frac{495}{n}$

Step 2 $90 \cdot n = 100 \cdot 495$

Step 3 $\frac{90n}{90} = \frac{49{,}500}{90}$

$n = 550$

495 students is 90% of 550 students.

11. **(a)** Percent: $12\frac{1}{2}\% = 12.5\%$
(b) Whole: unknown (n)
(c) Part: \$3.50

(d) *Step 1* $\frac{12.5}{100} = \frac{3.50}{n}$

Step 2 $12.5 \cdot n = 100 \cdot 3.50$

Step 3 $\frac{12.5n}{12.5} = \frac{350}{12.5}$

$n = 28$

$12\frac{1}{2}\%$ of \$28 is \$3.50.

13. Part: unknown (n)
Whole: 7
Percent: 250%

Step 1 $\frac{250}{100} = \frac{n}{7}$

Step 2 $100 \cdot n = 250 \cdot 7$

Step 3 $\frac{100n}{100} = \frac{1750}{100}$

$n = 17.5$

250% of 7 hours is 17.5 hours.

15. Part: 32
Whole: 172
Percent: unknown (p)

Step 1 $\frac{p}{100} = \frac{32}{172}$
Step 2 $172 \cdot p = 100 \cdot 32$
Step 3 $\frac{172p}{172} = \frac{3200}{172}$
$p \approx 18.6$

$32 is 18.6% (rounded) of $172.

17. Part: 748
Whole: unknown (n)
Percent: 110%

Step 1 $\frac{110}{100} = \frac{748}{n}$
Step 2 $110 \cdot n = 100 \cdot 748$
Step 3 $\frac{110n}{110} = \frac{74{,}800}{110}$
$n = 680$

748 books is 110% of 680 books.

19. Part: unknown (n)
Whole: $274
Percent: 14.7%

Step 1 $\frac{14.7}{100} = \frac{n}{274}$
Step 2 $100 \cdot n = 14.7 \cdot 274$
Step 3 $\frac{100n}{100} = \frac{4027.8}{100}$
$n \approx 40.28$

14.7% of $274 is $40.28 (rounded).

21. Part: 105
Whole: 54
Percent: unknown (p)

Step 1 $\frac{p}{100} = \frac{105}{54}$
Step 2 $54 \cdot p = 100 \cdot 105$
Step 3 $\frac{54p}{54} = \frac{10{,}500}{54}$
$p \approx 194.4$

105 is 194.4% (rounded) of 54.

23. Part: $0.33
Whole: unknown (n)
Percent: 4%

Step 1 $\frac{4}{100} = \frac{0.33}{n}$
Step 2 $4 \cdot n = 100 \cdot 0.33$
Step 3 $\frac{4n}{4} = \frac{33}{4}$
$n = 8.25$

$0.33 is 4% of $8.25.

25. 150% of $30 cannot be less than $30 because 150% is greater than 100%. The answer must be greater than $30.

25% of $16 cannot be greater than $16 because 25% is less than 100%. The answer must be less than $16.

27. The correct proportion is:

$\frac{p}{100} = \frac{14}{8}$
$8 \cdot p = 100 \cdot 14$
$\frac{8p}{8} = \frac{1400}{8}$
$p = 175$

The answer should be labeled with the % symbol. Correct answer is 175%.

7.3 The Percent Equation

7.3 Section Exercises

1. 50% of 3000 is the same as half of 3000. Choose 1500 patients.

3. 25% of $60 is the same as dividing $60 by 4.

25% of $60 $= \frac{\$60}{4} = \$15$. Choose $15.

5. 10% of 45 pounds is the same as dividing by 10. Move the decimal one place to the left.

10% of 45. = 4.5 pounds. Choose 4.5 pounds.

7. 200% of $3.50 is the same as multiplying $3.50 by 2.

200% of $3.50 $= 2 \cdot \$3.50 = \$7.00$.
Choose $7.00.

9. 1% of 5200 students is the same as dividing by 100. Move the decimal two places to the left.

1% of 5200. = 52 students. Choose 52 students.

11. 10% of 8700 cell phones is the same as dividing by 10. Move the decimal one place to the left.

10% of 8700. = 870 cell phones.
Choose 870 phones.

13. 25% of 19 hours is the same as dividing 19 by 4.

25% of 19 hours $= \frac{19}{4} = 4\frac{3}{4} = 4.75$ hours.

Choose 4.75 hours.

15. **(a)** 10% means $\frac{10}{100}$ or $\frac{1}{10}$. The denominator tells you to divide the whole by 10. The shortcut for dividing by 10 is to move the decimal point one place to the left.

(b) Once you find 10% of a number, multiply the result by 2 for 20% and by 3 for 30%.

17. 35% of 660 programs is how many programs?

$$0.35 \cdot 660 = n$$
$$231 = n$$

35% of 660 programs is 231 programs.

19. 70 truckloads is what percent of 140 truckloads?

$$70 = p \cdot 140$$
$$\frac{70}{140} = \frac{140p}{140}$$
$$0.5 = p \quad (50\%)$$

70 truckloads is 50% of 140 truckloads.

21. 476 circuits is 70% of what number of circuits?

$$476 = 0.70 \cdot n$$
$$\frac{476}{0.70} = \frac{0.70n}{0.70}$$
$$680 = n$$

476 circuits is 70% of 680 circuits.

23. $12\frac{1}{2}$% of what number of people is 135 people?

$$0.125 \cdot n = 135$$
$$\frac{0.125n}{0.125} = \frac{135}{0.125}$$
$$n = 1080$$

$12\frac{1}{2}$% of 1080 people is 135 people.

25. What is 65% of 1300 species?

$$n = 0.65 \cdot 1300$$
$$n = 845$$

845 species is 65% of 1300 species.

27. 4% of \$520 is how much?

$$0.04 \cdot 520 = n$$
$$20.8 = n$$

4% of \$520 is \$20.80.

29. 38 styles is what percent of 50 styles?

$$38 = p \cdot 50$$
$$\frac{38}{50} = \frac{50p}{50}$$
$$0.76 = p \quad (76\%)$$

38 styles is 76% of 50 styles.

31. What percent of \$264 is \$330?

$$p \cdot 264 = 330$$
$$\frac{264p}{264} = \frac{330}{264}$$
$$p = 1.25 \quad (125\%)$$

125% of \$264 is \$330.

33. 141 employees is 3% of what number of employees?

$$141 = 0.03 \cdot n$$
$$\frac{141}{0.03} = \frac{0.03n}{0.03}$$
$$4700 = n$$

141 employees is 3% of 4700 employees.

35. 32% of 260 quarts is how many quarts?

$$0.32 \cdot 260 = n$$
$$83.2 = n$$

32% of 260 quarts is 83.2 quarts.

37. \$1.48 is what percent of \$74?

$$1.48 = p \cdot 74$$
$$\frac{1.48}{74} = \frac{74p}{74}$$
$$0.02 = p \quad (2\%)$$

\$1.48 is 2% of \$74.

39. How many tablets is 140% of 500 tablets?

$$n = 1.40 \cdot 500$$
$$n = 700$$

700 tablets is 140% of 500 tablets.

41. 40% of what number of salads is 130 salads?

$$\begin{aligned} 0.40 \cdot n &= 130 \\ \frac{0.40n}{0.40} &= \frac{130}{0.40} \\ n &= 325 \end{aligned}$$

40% of 325 salads is 130 salads.

43. What percent of 160 liters is 2.4 liters?

$$\begin{aligned} p \cdot 160 &= 2.4 \\ \frac{160p}{160} &= \frac{2.4}{160} \\ p &= 0.015\ (1.5\%) \end{aligned}$$

1.5% of 160 liters is 2.4 liters.

45. 225% of what number of gallons is 11.25 gallons?

$$\begin{aligned} 2.25 \cdot n &= 11.25 \\ \frac{2.25n}{2.25} &= \frac{11.25}{2.25} \\ n &= 5 \end{aligned}$$

225% of 5 gallons is 11.25 gallons.

47. What is 12.4% of 8300 meters?

$$\begin{aligned} n &= 0.124 \cdot 8300 \\ n &= 1029.2 \end{aligned}$$

1029.2 meters is 12.4% of 8300 meters.

49. **(a)** Multiply 0.2 by 100 to change it from a decimal to a percent. So, $0.20 = 20\%$.

(b) The correct equation is $50 = p \cdot 20$.

$$\begin{aligned} \frac{50}{20} &= \frac{20p}{20} \\ 2.5 &= p \end{aligned}$$

The solution is 250%.

51. **(a)** $33\frac{1}{3}\% = \frac{100}{3}\% = \frac{\frac{100}{3}}{100} = \frac{100}{3} \div 100$

$$= \frac{100}{3} \div \frac{100}{1}$$

$$= \frac{100}{3} \cdot \frac{1}{100} = \frac{1}{3}$$

Thus, $33\frac{1}{3}\% = \frac{1}{3}$.

$33\frac{1}{3}\%$ of \$162 is how much?

$$\begin{aligned} \frac{1}{3} \cdot 162 &= n \\ 54 &= n \end{aligned}$$

The solution is \$54.

(b) $33\frac{1}{3}\% = \frac{1}{3} \approx 0.333333333$

$$\begin{aligned} (0.333333333)(162) &= n \\ 54 &\approx n \end{aligned}$$

Depending upon how your calculator rounds numbers, the solution is either \$54 or \$53.99999995.

(c) There is no difference or the difference is insignificant. The small variation in the solutions is due to truncating or rounding.

52. **(a)** $66\frac{2}{3}\% = \frac{200}{3}\% = \frac{\frac{200}{3}}{100} = \frac{200}{3} \div 100$

$$= \frac{200}{3} \div \frac{100}{1} = \frac{200}{3} \cdot \frac{1}{100} = \frac{2}{3}$$

Thus, $66\frac{2}{3}\% = \frac{2}{3}$.

$$\begin{aligned} \text{part} &= \text{percent} \cdot \text{whole} \\ 22 &= \frac{2}{3} \cdot n \\ \frac{22}{\frac{2}{3}} &= \frac{\frac{2}{3}n}{\frac{2}{3}} \\ 33 &= n \end{aligned}$$

22 cans is $66\frac{2}{3}\%$ of 33 cans.

(b) From part (a), $66\frac{2}{3}\% = \frac{2}{3} = 0.6666\overline{6}$.

Using the truncated decimal form of $\frac{2}{3}$:

$$\begin{aligned} \text{part} &= \text{percent} \cdot \text{whole} \\ 22 &= 0.66666666 \cdot n \\ \frac{22}{0.66666666} &= \frac{0.66666666n}{0.66666666} \\ 33.00000003 &= n \end{aligned}$$

or

Using the rounded decimal form of $\frac{2}{3}$:

$$\begin{aligned} \text{part} &= \text{percent} \cdot \text{whole} \\ 22 &= 0.66666667 \cdot n \\ \frac{22}{0.66666667} &= \frac{0.66666667n}{0.66666667} \\ 32.99999998 &= n \end{aligned}$$

Depending on whether you truncate the decimal form or round it, the solution will be either 33.00000003 or 32.99999998.

(c) There is no difference or the difference is insignificant. The small variation in the solutions is due to truncating or rounding.

Summary Exercises on Percent

1. **(a)** $\frac{3}{100} = 0.03 \leftarrow$ Decimal

$= 3\% \leftarrow$ Percent

(b) $30\% = \frac{30}{100} = \frac{3}{10} \leftarrow$ Fraction

$= 0.30 = 0.3 \leftarrow$ Decimal

(c) $0.375 = \frac{375}{1000} = \frac{375 \div 125}{1000 \div 125} = \frac{3}{8} \leftarrow$ Fraction

$= 37.5\% \leftarrow$ Percent

(d) $160\% = \frac{160}{100} = \frac{8}{5} = 1\frac{3}{5} \leftarrow$ Fraction

$= 1.6 \leftarrow$ Decimal

(e) Since 16 doesn't divide evenly into 100 or 1000, we divide 10,000 by 16 to get 625.

$\frac{1}{16} = \frac{625}{10{,}000} = 0.0625 \leftarrow$ Decimal

$= 6.25\% \leftarrow$ Percent

(f) $5\% = \frac{5}{100} = \frac{1}{20} \leftarrow$ Fraction

$= 0.05 \leftarrow$ Decimal

(g) $2.0 = \frac{200}{100} = \frac{2}{1} = 2 \leftarrow$ Fraction

$= 200\% \leftarrow$ Percent

(h) $\frac{4}{5} = \frac{4(20)}{5(20)} = \frac{80}{100} = 0.8 \leftarrow$ Decimal

$= 80\% \leftarrow$ Percent

(i) $0.072 = \frac{72}{1000} = \frac{72 \div 8}{1000 \div 8} = \frac{9}{125} \leftarrow$ Fraction

$= 7.2\% \leftarrow$ Percent

3. Part: 9
Whole: 72
Percent: unknown (p)

Step 1 $\frac{p}{100} = \frac{9}{72}$

Step 2 $72 \cdot p = 100 \cdot 9$

Step 3 $\frac{72p}{72} = \frac{900}{72}$

$p = 12.5$

9 Web sites is 12.5% of 72 Web sites.

5. Part: unknown (n)
Whole: \$8.79
Percent: 6%

Step 1 $\frac{6}{100} = \frac{n}{8.79}$

Step 2 $100 \cdot n = 6 \cdot 8.79$

Step 3 $\frac{100n}{100} = \frac{52.74}{100}$

$n \approx 0.53$

6% of \$8.79 is \$0.53 (rounded).

7. Part: unknown (n)
Whole: 168
Percent: $3\frac{1}{2}\%$

Step 1 $\frac{3.5}{100} = \frac{n}{168}$

Step 2 $100 \cdot n = 3.5 \cdot 168$

Step 3 $\frac{100n}{100} = \frac{588}{100}$

$n = 5.88$

$3\frac{1}{2}\%$ of 168 pounds is 5.9 pounds (rounded).

9. Part: 40,000
Whole: 80,000
Percent: unknown (p)

Step 1 $\frac{p}{100} = \frac{40{,}000}{80{,}000}$

Step 2 $80{,}000 \cdot p = 100 \cdot 40{,}000$

Step 3 $\frac{80{,}000p}{80{,}000} = \frac{4{,}000{,}000}{80{,}000}$

$p = 50$

50% of 80,000 deer is 40,000 deer.

11. Part: unknown (n)
Whole: 35
Percent: 280%

Step 1 $\frac{280}{100} = \frac{n}{35}$

Step 2 $100 \cdot n = 280 \cdot 35$

Step 3 $\frac{100n}{100} = \frac{9800}{100}$

$n = 98$

98 golf balls is 280% of 35 golf balls.

13. 9% of what number of apartments is 207 apartments?

$$0.09 \cdot n = 207$$

$$\frac{0.09n}{0.09} = \frac{207}{0.09}$$

$$n = 2300$$

9% of 2300 apartments is 207 apartments.

15. $\underbrace{\$1160}$ is $\underbrace{\text{what percent}}$ of $\underbrace{\$800?}$

$$1160 = p \cdot 800$$
$$\frac{1160}{800} = \frac{800p}{800}$$
$$1.45 = p \quad (145\%)$$

$1160 is 145% of $800.

17. $\underbrace{\text{What}}$ is $\underbrace{300\%}$ of $\underbrace{0.007 \text{ inch?}}$

$$n = 3 \cdot 0.007$$
$$n = 0.021$$

0.021 inch is 300% of 0.007 inch.

19. $\underbrace{\text{What percent}}$ of $\underbrace{60 \text{ yards}}$ is $\underbrace{4.8 \text{ yards?}}$

$$p \cdot 60 = 4.8$$
$$\frac{60p}{60} = \frac{4.8}{60}$$
$$p = 0.08 \quad (8\%)$$

8% of 60 yards is 4.8 yards.

21. The smallest percentage is $4\frac{1}{2}\%$ for flowers.
$4\frac{1}{2}\%$ of $20,500 = 0.045($20,500) = $922.50

23. 14.6% of $20,500 = 0.146($20,500) = $2993

7.4 Problem Solving with Percent

7.4 Section Exercises

1. *Step 1*
Unknown: amount withheld
Known: withhold 18% of $210

Step 2
Let n be the amount withheld.

Step 3
percent • whole = part
$18\% \cdot \$210 = n$

Step 4
$(0.18)(210) = n$
$\$37.80 = n$

Step 5
The amount withheld is $37.80.

Step 6
Check: 10% of $210 is $21, so 20% would be $2 \cdot \$21 = \$42$. $37.80 is slightly less than $42, so it is reasonable.

3. **(a)** *Step 1*
Unknown: percent
Known: $20 withdrawal; $2 fee

Step 2
Let p be the percent.

Step 3
percent • whole = part
$p \cdot 20 = 2$

Step 4
$$\frac{20p}{20} = \frac{2}{20}$$
$p = 0.10 = 10\%$

Step 5
The $2 fee is 10% of the $20 withdrawal.

Step 6
Check: 10% of $20. is $2. ✓

(b) $p = \frac{2}{40} = 0.05 = 5\%$

(c) $p = \frac{2}{100} = 0.02 = 2\%$

(d) $p = \frac{2}{200} = 0.01 = 1\%$

5. **(a)** *Step 1*
Unknown: number of pounds of water
Known: water weight is 61.6% of 165 pounds

Step 2
Let n be the number of pounds of water.

Step 3
percent • whole = part
$61.6\% \cdot 165 = n$

Step 4
$(0.616)(165) = n$
$101.64 = n$

Step 5
101.6 pounds (rounded) of the 165 pounds is water.

Step 6
Check: 50% of 165 pounds is 82.5 pounds, so 101.6 pounds is reasonable.

(b) (minerals) $n = 6.1\%$ of 165
$= (0.061)(165)$
$= 10.065$
$= 10.1$ pounds (rounded)

7. Whole: 335 people
Percent: unknown (p)
Part: 44 female

Percent • Whole = Part
$p \cdot 335 = 44$

$$\frac{335p}{335} = \frac{44}{335}$$
$p \approx 0.131$

About 13.1% of the crew is female and 86.9% (100% – 13.1%) is male.

9. Whole: Total U.S. population is unknown (n)
Part: 35.7 million are 65 years of age or older
Percent: 12.7%

$$\text{Percent} \cdot \text{Whole} = \text{Part}$$
$$0.127 \cdot n = 35.7$$
$$\frac{0.127n}{0.127} = \frac{35.7}{0.127}$$
$$n \approx 281$$

There were 281 million people (rounded) in the U.S. at the time of the 2000 census.

11. Whole: $50,000
Percent: unknown (p)
Part: $69,000

$$\text{Percent} \cdot \text{Whole} = \text{Part}$$
$$p \cdot 50{,}000 = 69{,}000$$
$$\frac{50{,}000p}{50{,}000} = \frac{69{,}000}{50{,}000}$$
$$p = 1.38 \qquad (138\%)$$

The society raised 138% of their goal.

13. Whole: Number of problems on the test is unknown (n)
Part: 38 problems done correctly
Percent: 95%

$$\text{Percent} \cdot \text{Whole} = \text{Part}$$
$$0.95 \cdot n = 38$$
$$\frac{0.95n}{0.95} = \frac{38}{0.95}$$
$$n = 40$$

There were 40 problems on the test.

15. Whole: number of shots tried is unknown (n)
Percent: 50.22%
Part: 683 shots made

$$\text{Percent} \cdot \text{Whole} = \text{Part}$$
$$0.5022 \cdot n = 683$$
$$\frac{0.5022n}{0.5022} = \frac{683}{0.5022}$$
$$n \approx 1360$$

He tried 1360 shots (rounded).

17. First calculate her increase in mileage.

Percent	$\cdot$	Original mileage	=	Amount of increase
0.15	$\cdot$	20.6	=	n
		3.09	=	n

The new tires should increase her mileage by 3.09 miles per gallon.

Her new mileage should be
$20.6 + 3.09 = 23.69 \approx 23.7$ miles per gallon (rounded)

19. Lowest sales month (March) sold 7%.

7%	of	350 million cans	is	how many?
↓	↓	↓	↓	↓
0.07	$\cdot$	350	=	n
		24.5	=	n

24.5 million (24,500,000) cans were sold in March.

21. Highest sales month: January, 15%
Second-highest sales month: February, 11%
Whole: 350 million cans sold each year

Highest sales month sold:

$0.15 \cdot 350 = 52.5$ million cans

Second-highest month sold:

$0.11 \cdot 350 = 38.5$ million cans

23. Amount of decrease = $825 – $290 = $535

Let p be the percent of decrease.

percent	of	original value	=	amount of decrease	
↓	↓	↓	↓	↓	
p	$\cdot$	$825	=	$535	
		$\frac{825p}{825}$	=	$\frac{535}{825}$	
		p	$\approx$	0.648	(64.8%)

The percent of decrease was 64.8% (rounded).

25. Original tuition last semester: $1328
New tuition this semester: $1449
Amount of increase = $1449 – $1328 = $121

Let p be the percent of increase.

percent	of	original tuition	=	amount of increase	
↓	↓	↓	↓	↓	
p	$\cdot$	1328	=	121	
		$\frac{1328p}{1328}$	=	$\frac{121}{1328}$	
		p	$\approx$	0.091	(9.1%)

The percent increase is 9.1% (rounded).

27. Original hours = 30 hours
New hours = 18 hours
Amount of decrease = 30 − 18 = 12

Let p be the percent of decrease.

percent of original amount = amount of decrease

$$p \cdot 30 = 12$$
$$\frac{30p}{30} = \frac{12}{30}$$
$$p = 0.4 \quad (40\%)$$

The percent decrease is 40%.

29. Original number = 78
New number = 519
Amount of increase = 519 − 78 = 441

Let p be the percent of increase.

percent of original number = amount of increase

$$p \cdot 78 = 441$$
$$\frac{78p}{78} = \frac{441}{78}$$
$$p \approx 5.65 \quad (565\%)$$

The percent of increase is 565% (rounded).

31. No. 100% is the entire price, so a decrease of 100% would take the price down to 0. Therefore, 100% is the maximum possible decrease in the price of something.

33. George ate more than 65 grams, so the percent must be $> 100\%$. Use $p \cdot 65 = 78$ to get 120%.

34. The team won more than half the games, so the percent must be $> 50\%$. Correct solution is $0.72 = 72\%$.

35. The brain could not weigh 375 pounds, which is more than the person weighs.
$2\frac{1}{2}\% = 2.5\% = 0.025$, so $(0.025)(150) = n$ and $n = 3.75$ pounds.

36. If 80% were absent, then only 20% made it to class. 800 − 640 = 160 students, or use $n = (0.20)(800) = 160$ students.

7.5 Consumer Applications: Sales Tax, Tips, Discounts, and Simple Interest

7.5 Section Exercises

1. Cost of Item = \$100
Tax Rate = 6% = 0.06
Amount of Tax = 0.06(100) = \$6
Total Cost = \$100 + \$6 = \$106

3. Cost of Item = \$68
Tax Rate = unknown (p)
Amount of Tax = \$2.04
Total Cost = \$68 + \$2.04 = \$70.04

Tax Rate · Cost of Item = Amount of Tax

$$p \cdot 68 = 2.04$$
$$\frac{68p}{68} = \frac{2.04}{68}$$
$$p = 0.03 \quad (3\%)$$

The tax rate is 3%.

5. Cost of Item = \$365.98
Tax Rate = 8% = 0.08
Amount of Tax = 0.08(365.98) = \$29.28 (rounded)
Total Cost = \$365.98 + \$29.28 = \$395.26

7. Cost of Item = \$2.10
Tax Rate = $5\frac{1}{2}\%$ = 0.055
Amount of Tax = 0.055(2.10) = \$0.12 (rounded)
Total Cost = \$2.10 + \$0.12 = \$2.22

9. Cost of Item = \$12,600
Tax Rate = unknown (p)
Amount of Tax = \$567
Total Cost = \$12,600 + \$567 = \$13,167

Tax Rate · Cost of Item = Amount of Tax

$$p \cdot 12{,}600 = 567$$
$$\frac{12{,}600p}{12{,}600} = \frac{567}{12{,}600}$$
$$p = 0.045 \quad (4.5\%)$$

The tax rate is 4.5% or $4\frac{1}{2}\%$.

11. The bill of \$32.17 rounds to \$30.

Estimate of 15% tip:

10% of \$30 is \$3. 5% of \$30 is half of \$3, or \$1.50. An estimate is \$3 + \$1.50 = \$4.50.

Exact 15% tip:

$$0.15(\$32.17) = \$4.8255 \approx \$4.83$$

Estimate of 20% tip:

10% of \$30 is \$3. 20% is 2 times \$3, or \$6. An estimate is \$6.

Exact 20% tip:

$$0.20(\$32.17) = \$6.434 \approx \$6.43$$

13. The bill of $78.33 rounds to $80.

Estimate of 15% tip:

10% of $80 is $8. 5% of $80 is half of $8, or $4. An estimate is $8 + $4 = $12.

Exact 15% tip:

$$0.15(\$78.33) = \$11.7495 \approx \$11.75$$

Estimate of 20% tip:

10% of $80 is $8. 20% is 2 times $8, or $16. An estimate is $16.

Exact 20% tip:

$$0.20(\$78.33) = \$15.666 \approx \$15.67$$

15. The bill of $9.55 rounds to $10.

Estimate of 15% tip:

10% of $10 is $1. 5% of $10 is half of $1, or $0.50. An estimate is $1 + $0.50 = $1.50.

Exact 15% tip:

$$0.15(\$9.55) = \$1.4325 \approx \$1.43$$

Estimate of 20% tip:

10% of $10 is $1. 20% is 2 times $1, or $2. An estimate is $2.

Exact 20% tip:

$$0.20(\$9.55) = \$1.91$$

17. Original Price = $100
Rate of Discount = 15% = 0.15
Amount of Discount = 0.15(100) = $15
Sale Price = $100 − $15 = $85

19. Original Price = $180
Rate of Discount = unknown (p)
Amount of Discount = $54
Sale Price = $180 − $54 = $126

amount of discount = rate of discount • original price

$$54 = p \cdot 180$$
$$\frac{54}{180} = \frac{180p}{180}$$
$$0.3 = p \qquad (30\%)$$

The rate of discount is 30%.

21. Original Price = $17.50
Rate of Discount = 25% = 0.25
Amount of
Discount = 0.25(17.50) = $4.375 ≈ $4.38
Sale Price = $17.50 − $4.38 = $13.12

23. Original Price = $37.88
Rate of Discount = 10% = 0.10
Amount of
Discount = 0.10($37.88) = $3.788 ≈ $3.79
Sale Price = $37.88 − $3.79 = $34.09

25. $300 at 14% for 1 year

$$I = p \cdot r \cdot t$$
$$= (300)(0.14)(1)$$
$$= 42$$

The interest is $42.

amount due = principal + interest
= $300 + $42
= $342

The total amount due is $342.

27. $740 at 6% for 9 months

$$I = p \cdot r \cdot t$$
$$= (740)(0.06)(\tfrac{9}{12})$$
$$= 33.30$$

The interest is $33.30.

amount due = principal + interest
= $740 + $33.30
= $773.30

The total amount due is $773.30.

29. $1500 at $9\frac{1}{2}$% for $1\frac{1}{2}$ years

$$I = p \cdot r \cdot t$$
$$= (1500)(0.095)(1.5)$$
$$= 213.75$$

The interest is $213.75.

amount due = principal + interest
= $1500 + $213.75
= $1713.75

The total amount due is $1713.75.

31. $17,800 at $7\frac{3}{4}$% for 8 months

$$I = p \cdot r \cdot t$$
$$= (17{,}800)(0.0775)(\tfrac{8}{12})$$
$$\approx 919.67$$

The interest is $919.67.

amount due = principal + interest
= $17,800 + $919.67
= $18,719.67

The total amount due is $18,719.67.

33. Original Price = $1950
Rate of Discount = 40% = 0.40
Amount of Discount = 0.40(1950) = $780
Sale Price = $1950 − $780 = $1170

The sale price of the ring is $1170.

35. \$7500 at $8\frac{1}{2}\%$ for 9 months

$$
\begin{aligned}
I &= p \cdot r \cdot t \\
&= (7500)(0.085)(\tfrac{9}{12}) \\
&= 478.13 \text{ (rounded)}
\end{aligned}
$$

The interest is \$478.13.

$$
\begin{aligned}
\text{amount due} &= \text{principal} + \text{interest} \\
&= \$7500 + \$478.13 \\
&= \$7978.13
\end{aligned}
$$

Rick will owe his mother \$7978.13.

37. Cost of Item = \$99.99
Tax Rate = $6\frac{1}{2}\% = 0.065$
Amount of Tax = $0.065(99.99) \approx \$6.50$
Total Cost = \$99.99 + \$6.50 = \$106.49

The total cost of the phone is \$106.49.

39. Cost of Item = \$1980
Tax Rate = unknown(p)
Amount of Tax = \$99

$$
\begin{array}{ccccc}
\text{Tax Rate} & \cdot & \text{Cost of Item} & = & \text{Amount of Tax} \\
\downarrow & \downarrow & \downarrow & \downarrow & \downarrow \\
p & \cdot & 1980 & = & 99 \\
 & & \dfrac{1980p}{1980} & = & \dfrac{99}{1980} \\
 & & p & = & 0.05 \quad (5\%)
\end{array}
$$

The sales tax rate is 5%.

41. Original Price = \$135
Rate of Discount = 45%
Amount of Discount = 0.45(135) = \$60.75
Sale Price = \$135 − \$60.75 = \$74.25

The sale price of the parka is \$74.25.

43.
$$
\begin{array}{ccccc}
\text{Tip rate} & \cdot & \text{Original price} & = & \text{Tip amount} \\
\downarrow & \downarrow & \downarrow & \downarrow & \downarrow \\
0.15 & \cdot & \$43.70 & = & n \\
 & & \$6.555 & = & n
\end{array}
$$

The tip amount is \$6.56 (rounded).
Total = \$43.70 + \$6.56 = \$50.26
Amount paid by each = \$50.26 ÷ 2 = \$25.13

45. Original Price = \$590
Rate of Discount = 18%
Amount of Discount = 0.18(590) = \$106.20
Sale Price = \$590 − \$106.20 = \$483.80

The discount is \$106.20 and the sale price is \$483.80.

47. \$1900 at $12\frac{1}{4}\%$ for 6 months

$$
\begin{aligned}
I &= p \cdot r \cdot t \\
&= (1900)(0.1225)(\tfrac{6}{12}) \\
&= 116.375
\end{aligned}
$$

The interest is \$116.38 (rounded).

$$
\begin{aligned}
\text{amount due} &= \text{principal} + \text{interest} \\
&= \$1900 + \$116.38 \\
&= \$2016.38
\end{aligned}
$$

She must pay a total amount of \$2016.38.

49.
$$
\begin{array}{ccccc}
\text{Tip rate} & \cdot & \text{Original price} & = & \text{Tip amount} \\
\downarrow & \downarrow & \downarrow & \downarrow & \downarrow \\
0.15 & \cdot & \$17.98 & = & n \\
 & & \$2.697 & = & n
\end{array}
$$

The 15% tip would be about \$2.70.
Total = \$17.98 + \$2.70 = \$20.68

They would give the delivery person \$21 (rounded to the nearest dollar).

51. **Computer modem:** (\$129, 65% off)

Discount amount = 0.65(\$129) = \$83.85
Sale price = \$129 − \$83.85 = \$45.15
Tax amount = 0.06(\$45.15) ≈ \$2.71
Total = \$45.15 + \$2.71 = **\$47.86**

Earrings: (\$60, 30% off)

Discount amount = 0.30(\$60) = \$18
Sale price = \$60 − \$18 = \$42
Tax amount = 0.06(\$42) = \$2.52
Total = \$42 + \$2.52 = **\$44.52**

Bill = \$47.86 + \$44.52 = **\$92.38**

53. **Television:** (\$287.95, 65% off)

Discount amount = 0.65(\$287.95) ≈ \$187.17
Sale price = \$287.95 − \$187.17 = \$100.78
Tax amount = 0.06(\$100.78) ≈ \$6.05
Total = \$100.78 + \$6.05 = **\$106.83**

Jeans: (2 @ \$48, 45% off)

Discount amount = 0.45(\$48) = \$21.60
Sale price = \$48 − \$21.60 = \$26.40
Tax amount: no tax on clothing
Total for 2 pairs = 2(\$26.40) = **\$52.80**

Ring: (\$95, 30% off)

Discount amount = 0.30(\$95) = \$28.50
Sale price = \$95 − \$28.50 = \$66.50
Tax amount = 0.06(\$66.50) = \$3.99
Total = \$66.50 + \$3.99 = **\$70.49**

Bill = \$106.83 + \$52.80 + \$70.49 = **\$230.12**

55. **(a)** Cost of item = \$18.50
Discount rate = 6% = 0.06
Amount of discount = 0.06(18.50) = \$1.11
Cost of discounted book = \$18.50 − \$1.11
= \$17.39

Tax rate = 6% = 0.06
Amount of tax = 0.06(17.39) ≈ \$1.04
Final price = \$17.39 + \$1.04 = \$18.43

(b) When calculating the discount, the *whole* is \$18.50. But when calculating the sales tax, the whole is only \$17.39 (the discounted price).

56. **(a)** Cost of item = \$398
Discount rate = 7% = 0.07
Amount of discount = 0.07(\$398) = \$27.86
Cost of discounted item
= \$398 − \$27.86 = \$370.14

Tax rate = 7% = 0.07
Amount of tax = 0.07(\$370.14) ≈ \$25.91
Final price = \$370.14 + \$25.91 = \$396.05

(b) Tax amount = rate of tax • price of item

$$27.86 = r \cdot 370.14$$
$$\frac{27.86}{370.14} = \frac{370.14r}{370.14}$$
$$0.0753 \approx r \qquad (7.53\%)$$

A sales tax of 7.53% would give a tax amount of \$27.87 and a final cost of \$398.01.

Chapter 7 Review Exercises

1. $25\% = 25 \div 100 = 0.25$

2. $180\% = 180 \div 100 = 1.80$

3. $12.5\% = 12.5 \div 100 = 0.125$

4. $7\% = 7 \div 100 = 0.07$

5. $2.65 = 2.65 \cdot 100\% = 265\%$

6. $0.02 = 0.02 \cdot 100\% = 2\%$

7. $0.3 = 0.3 \cdot 100\% = 30\%$

8. $0.002 = 0.002 \cdot 100\% = 0.2\%$

9. $12\% = \frac{12}{100} = \frac{12 \div 4}{100 \div 4} = \frac{3}{25}$

10. $37.5\% = \frac{37.5}{100} = \frac{37.5 \cdot 10}{100 \cdot 10} = \frac{375}{1000}$
$= \frac{375 \div 125}{1000 \div 125} = \frac{3}{8}$

11. $250\% = \frac{250}{100} = \frac{250 \div 50}{100 \div 50} = \frac{5}{2} = 2\frac{1}{2}$

12. $5\% = \frac{5}{100} = \frac{5 \div 5}{100 \div 5} = \frac{1}{20}$

13. $\frac{3}{4} = \frac{3}{4} \cdot 100\% = \frac{3}{4} \cdot \frac{100}{1}\% = \frac{3 \cdot \overset{1}{\cancel{4}} \cdot 25}{\underset{1}{\cancel{4}} \cdot 1}\%$
$= \frac{75}{1}\% = 75\%$

14. $\frac{5}{8} = \frac{5}{8} \cdot 100\% = \frac{5}{8} \cdot \frac{100}{1}\% = \frac{5 \cdot \overset{1}{\cancel{4}} \cdot 25}{2 \cdot \underset{1}{\cancel{4}} \cdot 1}\%$
$= \frac{125}{2}\% = 62\frac{1}{2}\%$ or 62.5%

15. $3\frac{1}{4} = \frac{13}{4} = \frac{13}{4} \cdot \frac{100}{1}\% = \frac{13 \cdot \overset{1}{\cancel{4}} \cdot 25}{\underset{1}{\cancel{4}} \cdot 1}\%$
$= \frac{325}{1}\% = 325\%$

16. $\frac{3}{50} = \frac{3}{50} \cdot \frac{100}{1}\% = \frac{3 \cdot 2 \cdot \overset{1}{\cancel{50}}}{\underset{1}{\cancel{50}} \cdot 1}\% = \frac{6}{1}\% = 6\%$

17. $\frac{1}{8} = 1 \div 8 = 0.125$

18. $0.125 = 0.125 \cdot 100\% = 12.5\%$

19. $0.15 = \frac{15}{100} = \frac{15 \div 5}{100 \div 5} = \frac{3}{20}$

20. $0.15 = 0.15 \cdot 100\% = 15\%$

21. $180\% = \frac{180}{100} = \frac{18}{10} = \frac{9}{5}$ or $1\frac{4}{5}$

22. $180\% = 180 \div 100 = 1.80$ or 1.8

23. 100% of \$46 is all of the money, or \$46.

24. 50% of \$46 is half of the money, or \$23.

25. 100% of 9 hours is all of the hours, or 9 hours.

26. 50% of 9 hours is half of the hours, which is $4\frac{1}{2}$ or 4.5 hours.

27. Part: 338.8
Whole: unknown (n)
Percent: 140%

Step 1 $\frac{140}{100} = \frac{338.8}{n}$

Step 2 $140 \cdot n = 100 \cdot 338.8$

Step 3 $\frac{140n}{140} = \frac{33{,}880}{140}$
$n = 242$

338.8 meters is 140% of 242 meters.

28. Part: 425
Whole: unknown (n)
Percent: 2.5%

Step 1 $\frac{2.5}{100} = \frac{425}{n}$

Step 2 $2.5 \cdot n = 100 \cdot 425$

Step 3 $\frac{2.5n}{2.5} = \frac{42{,}500}{2.5}$

$n = 17{,}000$

2.5% of 17,000 cases is 425 cases.

29. Part: unknown (n)
Whole: 450
Percent: 6%

Step 1 $\frac{6}{100} = \frac{n}{450}$

Step 2 $100 \cdot n = 6 \cdot 450$

Step 3 $\frac{100n}{100} = \frac{2700}{100}$

$n = 27$

6% of 450 cellular phones is 27 cellular phones.

30. Part: unknown (n)
Whole: 1450
Percent: 60%

Step 1 $\frac{60}{100} = \frac{n}{1450}$

Step 2 $100 \cdot n = 60 \cdot 1450$

Step 3 $\frac{100n}{100} = \frac{87{,}000}{100}$

$n = 870$

60% of 1450 reference books is 870 reference books.

31. Part: 36
Whole: 380
Percent: unknown (p)

Step 1 $\frac{p}{100} = \frac{36}{380}$

Step 2 $380 \cdot p = 100 \cdot 36$

Step 3 $\frac{380p}{380} = \frac{3600}{380}$

$p \approx 9.5$

36 pairs is 9.5% (rounded) of 380 pairs.

32. Part: 1440
Whole: 640
Percent: unknown (p)

Step 1 $\frac{p}{100} = \frac{1440}{640}$

Step 2 $640 \cdot p = 100 \cdot 1440$

Step 3 $\frac{640p}{640} = \frac{144{,}000}{640}$

$p = 225$

1440 cans is 225% of 640 cans.

33. 11% of $23.60 is $\underbrace{\text{how much?}}$

$$0.11 \cdot 23.60 = n$$
$$2.596 = n$$
$$\$2.60 \text{ (rounded)} = n$$

34. $\underbrace{\text{What}}$ is 125% of 64 days?

$$n = 1.25 \cdot 64$$
$$n = 80 \text{ days}$$

35. 1.28 ounces is $\underbrace{\text{what percent}}$ of 32 ounces?

$$1.28 = p \cdot 32$$
$$1.28 = 32p$$
$$\frac{1.28}{32} = \frac{32p}{32}$$
$$0.04 = p$$
$$4\% = p$$

36. $46 is 8% of $\underbrace{\text{what number of dollars?}}$

$$46 = 0.08 \cdot n$$
$$46 = 0.08n$$
$$\frac{46}{0.08} = \frac{0.08n}{0.08}$$
$$\$575 = n$$

37. 8 people is 40% of $\underbrace{\text{what number of people?}}$

$$8 = 0.40 \cdot n$$
$$8 = 0.40n$$
$$\frac{8}{0.40} = \frac{0.40n}{0.40}$$
$$20 \text{ people} = n$$

38. $\underbrace{\text{What percent}}$ of 174 ft is 304.5 ft?

$$p \cdot 174 = 304.5$$
$$174p = 304.5$$
$$\frac{174p}{174} = \frac{304.5}{174}$$
$$p = 1.75 = 175\%$$

39. **(a)** Part: 504 late patients
Whole: total number of patients, unknown (n)
Percent: 16.8%

Percent • Whole = Part

$$0.168 \cdot n = 504$$
$$\frac{0.168n}{0.168} = \frac{504}{0.168}$$
$$n = 3000$$

There were 3000 patients in January.

(b) Amount of decrease $= 504 - 345 = 159$

Let p be the percent of decrease.

$$\underbrace{\text{percent}}_{p} \quad \underbrace{\text{of}}_{\cdot} \quad \underbrace{\text{original value}}_{504} \quad = \quad \underbrace{\text{amount of decrease}}_{159}$$

$$\frac{504p}{504} = \frac{159}{504}$$

$$p \approx 0.315 \quad (31.5\%)$$

The percent of decrease was 31.5% (rounded).

40. **(a)** Part: actual amount spent, unknown (n)
Whole: \$280 budgeted for food
Percent: 130%

$$\text{Percent} \cdot \text{Whole} = \text{Part}$$
$$1.30 \cdot 280 = n$$
$$364 = n$$

\$364 was actually spent on food.

(b) Part: amount spent, \$112
Whole: amount budgeted, \$50
Percent: unknown (p)

$$\text{Percent} \cdot \text{Whole} = \text{Part}$$
$$p \cdot 50 = 112$$
$$\frac{50p}{50} = \frac{112}{50}$$
$$p = 2.24 \quad (224\%)$$

She spent 224% of the budgeted amount.

41. **(a)** Part: 640 trees still living after 1 year
Whole: 800 trees planted
Percent: unknown (p)

$$\text{Percent} \cdot \text{Whole} = \text{Part}$$
$$p \cdot 800 = 640$$
$$\frac{800p}{800} = \frac{640}{800}$$
$$p = 0.8 \quad (80\%)$$

80% of the trees were still living.

(b) Amount of increase $= 850 - 800 = 50$

Let p be the percent of increase.

$$\underbrace{\text{percent}}_{p} \quad \underbrace{\text{of}}_{\cdot} \quad \underbrace{\text{original value}}_{800} \quad = \quad \underbrace{\text{amount of increase}}_{50}$$

$$\frac{800p}{800} = \frac{50}{800}$$

$$p = 0.0625 \quad (6.25\%)$$

The percent of increase was 6.3% (rounded).

42. Part: 1000 species in danger of extinction
Whole: 9600 species
Percent: unknown (p)

$$\text{Percent} \cdot \text{Whole} = \text{Part}$$
$$p \cdot 9600 = 1000$$
$$\frac{9600p}{9600} = \frac{1000}{9600}$$
$$p \approx 0.104 \quad (10.4\%)$$

10.4% (rounded) of the bird species are in danger of extinction.

43. Part: Sales tax = unknown (n)
Whole: Cost of item = \$2.79
Percent: Tax rate = 4%

$$\text{Percent} \cdot \text{Whole} = \text{Part}$$
$$0.04 \cdot 2.79 = n$$
$$0.1116 = n$$

Amount of tax = \$0.11 (rounded)
Total cost = \$2.79 + \$0.11 = \$2.90

44. Part: Sales tax = \$58.50
Whole: Cost of item = \$780
Percent: Tax rate = unknown (p)

$$\text{Percent} \cdot \text{Whole} = \text{Part}$$
$$p \cdot 780 = 58.50$$
$$\frac{780p}{780} = \frac{58.50}{780}$$
$$p = 0.075 \quad (7.5\%)$$

The tax rate is 7.5% or $7\frac{1}{2}\%$.
Total cost = \$780 + \$58.50 = \$838.50

45. The bill of \$42.73 rounds to \$40.

Estimate of 15% tip:

10% of \$40 is \$4. 5% of \$40 is half of \$4, or \$2. An estimate is \$4 + \$2 = \$6.

Exact 15% tip:

$$0.15(\$42.73) = \$6.4095 \approx \$6.41$$

Estimate of 20% tip:

10% of \$40 is \$4. 20% is 2 times \$4, or \$8. An estimate is \$8.

Exact 20% tip:

$$0.20(\$42.73) = \$8.546 \approx \$8.55$$

46. The bill of \$8.05 rounds to \$8.

Estimate of 15% tip:

10% of \$8 is \$0.80. 5% of \$8 is half of \$0.80, or \$0.40. An estimate is $\$0.80 + \$0.40 = \$1.20$.

Exact 15% tip:

$$0.15(\$8.05) = \$1.2075 \approx \$1.21$$

Estimate of 20% tip:

10% of \$8 is \$0.80. 20% is 2 times \$0.80, or \$1.60. An estimate is \$1.60.

Exact 20% tip:

$$0.20(\$8.05) = \$1.61$$

47. Original Price = \$37.50
Rate of Discount = 10% = 0.10
Amount of Discount = 0.10(37.50) = \$3.75
Sale Price = \$37.50 − \$3.75 = \$33.75

48. Original Price = \$252
Rate of Discount = unknown (p)
Amount of Discount = \$63
Sale Price = \$252 − \$63 = \$189

amount of discount = rate of discount • original price

$$63 = p \cdot 252$$
$$\frac{63}{252} = \frac{252p}{252}$$
$$0.25 = p \qquad (25\%)$$

The rate of discount is 25%.

49. \$350 at $6\frac{1}{2}\%$ for 3 years

$$\begin{aligned} I &= p \cdot r \cdot t \\ &= (350)(0.065)(3) \\ &= 68.25 \end{aligned}$$

The interest is \$68.25.

$$\begin{aligned} \text{amount due} &= \text{principal} + \text{interest} \\ &= \$350 + \$68.25 \\ &= \$418.25 \end{aligned}$$

The total amount due is \$418.25.

50. \$1530 at 16% for 9 months

$$\begin{aligned} I &= p \cdot r \cdot t \\ &= (1530)(0.16)(\tfrac{9}{12}) \\ &= 183.60 \end{aligned}$$

The interest is \$183.60.

$$\begin{aligned} \text{amount due} &= \text{principal} + \text{interest} \\ &= \$1530 + \$183.60 \\ &= \$1713.60 \end{aligned}$$

The total amount due is \$1713.60.

51. **[7.1]** $\frac{1}{3} = \frac{1}{3} \cdot 100\% = \frac{100}{3}\% = 33\frac{1}{3}\%$ (exact)

$33\frac{1}{3}\% \approx 0.333 = 33.3\%$ (rounded)

52. **[7.1]** The largest percent is 68%, which represents cards, so cards is the most popular game.

$$68\% = 0.68 = \frac{68}{100} = \frac{68 \div 4}{100 \div 4} = \frac{17}{25}$$

53. **[7.3]** $\frac{1}{3} \cdot 830 \approx 276.7 \approx 277$

277 adults (rounded) play electronic/computer games.

54. **[7.3]** Most popular: cards, 68%

$0.68(277) = 188.36 \approx 188$ adults (rounded)

Least popular: sci-fi/simulation, 37%

$0.37(277) = 102.49 \approx 102$ adults (rounded)

55. **[7.4]** Part: 2599 dogs placed
Whole: 5371 dogs received
Percent: unknown (p)

Percent • Whole = Part

$$p \cdot 5371 = 2599$$
$$\frac{5371p}{5371} = \frac{2599}{5371}$$
$$p \approx 0.484 \quad (48.4\%)$$

48.4% (rounded) of the dogs were placed in new homes.

56. **[7.4]** Part: 5371 received so far this year
Whole: number expected, unknown (n)
Percent: 75%

Percent • Whole = Part

$$0.75 \cdot n = 5371$$
$$\frac{0.75n}{0.75} = \frac{5371}{0.75}$$
$$n \approx 7161$$

They expect to receive 7161 dogs (rounded).

57. **[7.4]** Part: 2346 placed in new homes
Whole: 6447 received
Percent: unknown (p)

Percent • Whole = Part

$$p \cdot 6447 = 2346$$
$$\frac{6447p}{6447} = \frac{2346}{6447}$$
$$p \approx 0.364 \quad (36.4\%)$$

About 36.4% of the cats were placed in new homes.

58. **[7.4]** During the first 9 months:

Last year = 2300 cats received
This year = 6447 cats received

Amount of increase = 6447 − 2300
= 4147 cats

$$\begin{array}{ccccc} \text{Percent increase} & \cdot & \text{Original amount} & = & \text{Amount of increase} \\ p & \cdot & 2300 & = & 4147 \\ & & \dfrac{2300p}{2300} & = & \dfrac{4147}{2300} \\ & & p & \approx & 1.803 \quad (180.3\%) \end{array}$$

The percent increase is 180.3% (rounded).

59. **[7.4]** Total number of animals received
= 5371 + 6447 + 2223 = 14,041
Total number of animals placed
= 2599 + 2346 + 406 = 5351

Percent • Whole = Part

$$p \cdot 14{,}041 = 5351$$
$$\frac{14{,}041p}{14{,}041} = \frac{5351}{14{,}041}$$
$$p \approx 0.381 \quad (38.1\%)$$

About 38.1% of the animals received were placed in new homes.

60. **[7.4]** 40% of all animals received:

$$0.40(14{,}041) \approx 5616$$

Since they placed 5351 animals,
5616 − 5351 = 265 (rounded) animals need to be placed in a home in order to reach their goal.

Chapter 7 Test

1. $75\% = 75 \div 100 = 0.75$

2. $0.6 = 0.60 = 60\%$

3. $1.8 = 1.80 = 180\%$

4. $0.075 = 7.5\%$ or $7\frac{1}{2}\%$

5. $300\% = 300 \div 100 = 3.00$ or 3

6. $2\% = 2 \div 100 = 0.02$

7. $$62.5\% = \frac{62.5}{100} = \frac{62.5 \cdot 10}{100 \cdot 10} = \frac{625}{1000} = \frac{625 \div 125}{1000 \div 125} = \frac{5}{8}$$

8. $$240\% = \frac{240}{100} = \frac{24}{10} = \frac{12}{5} = 2\frac{2}{5}$$

9. $$\frac{1}{20} = \frac{1}{20} \cdot \frac{100}{1}\% = \frac{1 \cdot \cancel{20}^{1} \cdot 5}{\cancel{20}_{1} \cdot 1}\% = \frac{5}{1}\% = 5\%$$

10. $$\frac{7}{8} = \frac{7}{8} \cdot \frac{100}{1}\% = \frac{7 \cdot \cancel{4}^{1} \cdot 25}{2 \cdot \cancel{4}_{1}}\% = \frac{175}{2}\% = 87\frac{1}{2}\% \text{ or } 87.5\%$$

11. $$1\frac{3}{4} = \frac{7}{4} = \frac{7}{4} \cdot \frac{100}{1}\% = \frac{7 \cdot \cancel{4}^{1} \cdot 25}{\cancel{4}_{1} \cdot 1}\% = \frac{175}{1}\% = 175\%$$

12. 16 laptops is 5% of what number of laptops?

$$16 = 0.05 \cdot n$$
$$\frac{16}{0.05} = \frac{0.05n}{0.05}$$
$$320 = n$$

16 laptops is 5% of 320 laptops.

13. \$192 is what percent of \$48?

$$192 = p \cdot 48$$
$$\frac{192}{48} = \frac{48p}{48}$$
$$4 = p \quad (400\%)$$

\$192 is 400% of \$48.

14. Part: \$14,625
Whole: Amount needed = unknown (n)
Percent: 75%

Percent • Whole = Part

$$0.75 \cdot n = 14{,}625$$
$$\frac{0.75n}{0.75} = \frac{14{,}625}{0.75}$$
$$n = 19{,}500$$

\$19,500 is needed for a down payment.

15. Part: Sales tax = unknown (n)
Whole: Cost of item = \$7950
Percent: Tax rate = $6\frac{1}{2}\% = 6.5\% = 0.065$

Percent • Whole = Part

$$0.065 \cdot 7950 = n$$
$$516.75 = n$$

The sales tax is \$516.75.

Total cost of the car = \$7950 + \$516.75
= \$8466.75

16. Last semester = 1440 students
Current semester = 1925 students
Amount of increase
= 1925 − 1440 = 485 students

Percent • Original = Amount of increase

$p \cdot 1440 = 485$

$$\frac{1440p}{1440} = \frac{485}{1440}$$

$p \approx 0.34\ (34\%)$

The percent of increase is 34% (rounded).

17. To find 50% of a number, divide the number by 2. To find 25% of a number, divide the number by 4. Examples will vary.

18. Round \$31.94 to \$30. Then 10% of \$30. is \$3 and 5% of \$30 is half of \$3 or \$1.50, so a 15% tip estimate is \$3 + \$1.50 = \$4.50. A 20% tip estimate is 2(\$3) = \$6.

19. Percent • Whole = Part
0.15 • \$31.94 = Tip amount
\$4.791 = Tip amount

The tip amount would be \$4.79 (rounded).

Total expense = \$31.94 + \$4.79
= \$36.73

If the total expense is shared by 3 friends, each person will pay \$36.73 ÷ 3 ≈ \$12.24.

20. Discount rate • Original price = Discount amount
0.08 • \$48 = Discount amount
\$3.84 = Discount amount

Sale price = Original price − Discount amount
= \$48 − \$3.84
= \$44.16

21. Discount rate • Original price = Discount amount
0.18 • \$229.95 = Discount amount
\$41.391 = Discount amount

Rounded to the nearest cent, the discount amount is \$41.39.

Sale price = \$229.95 − \$41.39
= \$188.56

22. Discount rate • Original price = Discount amount
0.30 • \$1089 = Discount amount
\$326.70 = Discount amount

Discounted price = \$1089 − \$326.70 = \$762.30

Tax rate • Price = Sales tax
0.07 • \$762.30 = Sales tax
\$53.361 = Sales tax
\$53.36 (rounded) = Sales tax

Total bill = \$762.30 + \$53.36
= \$815.66

Since the payments will be spread out over 6 months, Jamal will need to pay \$815.66 ÷ 6 ≈ \$135.94 per month.

23. \$5000 at $8\frac{1}{4}\%$ for 4 years

$I = p \cdot r \cdot t$ $\qquad 8\frac{1}{4}\% = 0.0825$
$= (5000)(0.0825)(4)$
$= 1650$

The interest is \$1650.

Total amount due = principal + interest
= \$5000 + \$1650
= \$6650

24. \$860 at 12% for 6 months

$I = p \cdot r \cdot t$
$= (860)(0.12)(\frac{6}{12})$
$= 51.60$

The interest due on the loan is \$51.60.

Total amount due = principal + interest
= \$860 + \$51.60
= \$911.60

Cumulative Review Exercises (Chapters 1–7)

1. **(a)** In words, 90.105 is

ninety and one hundred five thousandths.

(b) In words, 125,000,670 is

one hundred twenty-five million, six hundred seventy.

2. **(a)** Written in digits, thirty billion, five million is

30,005,000,000.

(b) Written in digits, seventy-eight ten-thousandths is

0.0078.

3. **(a) 49,617 to the nearest thousand**

Underline the thousands place: 4<u>9</u>,617
The next digit is 5 or more. Add 1 to 9. Write 0 and regroup the 1 to the ten-thousands place. Change 6, 1, and 7 to 0. **50,000**

(b) 0.7039 to the nearest hundredth

Underline the hundredths place: 0.7$\underline{0}$39
The next digit is 4 or less. Leave 0. Drop all digits to the right of the underlined place. **0.70**

(c) 8945 to the nearest hundred

Underline the hundreds place: 8$\underline{9}$45
The next digit is 4 or less. Leave 9. Change 4 and 5 to 0. **8900**

4. **(a)** $(-7)(0.8) = (0.8)(-7)$

commutative property of multiplication

(b) $\left(\frac{2}{3}+\frac{3}{4}\right)+\frac{1}{2}=\frac{2}{3}+\left(\frac{3}{4}+\frac{1}{2}\right)$

associative property of addition

5. $-18\boxed{<}-8$ because -18 lies to the left of -8 on the number line.

$0\boxed{>}-5$ because 0 lies to the right of -5 on the number line.

6. $0.705 = 0.7050$
$0.755 = 0.7550$
$\frac{3}{4} = 0.75 = 0.7500$
$0.7005 = 0.7005$

Write each number in decimal form. Fill in zeros so that each number has an equal number of decimal places.

From smallest to largest: 0.7005; 0.705; $\frac{3}{4}$; 0.755

7. Mean

$$= \frac{\text{sum of all values}}{\text{number of values}}$$

$$= \frac{(\$710+\$780+\$650+\$785 + \$1125+\$695+\$740+\$685)}{8}$$

$$= \frac{\$6170}{8} = \$771.25$$

Median: arrange the data from smallest to largest

\$650, \$685, \$695, $\underline{\$710}$, $\underline{\$740}$, \$780, \$785, \$1125

Since there is an even number of values, the median is the average of the middle values.

$$\text{Median} = \frac{\$710+\$740}{2} = \$725$$

8. Brand A: $\frac{\$2.89}{17\text{ ounces}} = \$0.170$ per ounce (*)

Brand B: $\frac{\$3.59}{21\text{ ounces}} \approx \$0.171$ per ounce

Brand C: $\frac{\$2.79}{15\text{ ounces}} = \$0.186$ per ounce

The best buy is Brand A since it is the least expensive per ounce.

9. $50 - 1.099$

$$\begin{array}{r} \overset{4}{\cancel{5}}\overset{\overset{9}{\cancel{10}}}{\cancel{0}}.\overset{\overset{9}{\cancel{10}}}{\cancel{0}}\overset{\overset{9}{\cancel{10}}}{\cancel{0}}\overset{10}{\cancel{0}} \\ -\ 1.0\ 9\ 9 \\ \hline 48.9\ 0\ 1 \end{array}$$

10. $(-3)^2 + 2^3$
$= (-3)(-3) + 2\cdot 2\cdot 2$
$= 9 + 8$
$= 17$

11. $\frac{3b}{10a}\cdot\frac{15ab}{4} = \frac{3\cdot b\cdot 3\cdot \cancel{5}\cdot \cancel{a}\cdot b}{2\cdot\cancel{5}\cdot\cancel{a}\cdot 4} = \frac{9b^2}{8}$

12. $3\frac{3}{10} - 2\frac{4}{5} = \frac{33}{10} - \frac{14}{5}$
$= \frac{33}{10} - \frac{28}{10}$
$= \frac{33-28}{10} = \frac{5}{10} = \frac{1}{2}$

13. $0 + 2(-6+1)$
$= 0 + 2(-5)$
$= 0 + (-10)$
$= -10$

14. $\frac{4}{5}+\frac{x}{4} = \frac{4\cdot 4}{5\cdot 4}+\frac{x\cdot 5}{4\cdot 5} = \frac{16}{20}+\frac{5x}{20} = \frac{16+5x}{20}$

15. $-20 - 20$
$= -20 + (-20)$
$= -40$

16. $(-0.5)(0.002)$

0.002	←	*3 decimal places*
× 0.5	←	*1 decimal place*
0.0010	←	*4 decimal places*

The signs are opposite, so the product is negative.

$$(-0.5)(0.002) = -0.001$$

17. $\frac{-\frac{10}{11}}{-\frac{5}{6}} = -\frac{10}{11} \div \left(-\frac{5}{6}\right)$
$= -\frac{10}{11}\left(-\frac{6}{5}\right)$
$= \frac{2\cdot\cancel{5}\cdot 6}{11\cdot\cancel{5}}$
$= \frac{12}{11}$ or $1\frac{1}{11}$

18. $\frac{3}{8}$ of $328 = \frac{3}{8}\cdot\frac{328}{1} = \frac{3\cdot\cancel{8}\cdot 41}{\cancel{8}\cdot 1}$
$= \frac{123}{1} = 123$

19. $\frac{-16+2^4}{-3-2} = \frac{-16+16}{-3+(-2)} = \frac{0}{-5} = 0$

20. $\frac{7}{8} - \frac{2}{m} = \frac{7 \cdot m}{8 \cdot m} - \frac{2 \cdot 8}{m \cdot 8} = \frac{7m}{8m} - \frac{16}{8m}$

$= \frac{7m-16}{8m}$

21. $\frac{4.8}{-0.16}$ Different signs, quotient is negative.

$0.16\overline{)4.80} \rightarrow 16\overline{)480}$ quotient 30; 48, remainder 0

Answer: -30

22. $\frac{8}{9} \div 2n = \frac{8}{9} \div \frac{2n}{1} = \frac{8}{9} \cdot \frac{1}{2n}$

$= \frac{\overset{1}{\cancel{2}} \cdot 4 \cdot 1}{9 \cdot \underset{1}{\cancel{2}} \cdot n} = \frac{4}{9n}$

23. $1\frac{5}{6} + 1\frac{2}{3} = \frac{11}{6} + \frac{5}{3} = \frac{11}{6} + \frac{10}{6}$

$= \frac{11+10}{6} = \frac{21}{6}$

$= \frac{\overset{1}{\cancel{3}} \cdot 7}{\underset{1}{\cancel{3}} \cdot 2} = \frac{7}{2} = 3\frac{1}{2}$

24. $5 - 1\frac{7}{9} = \frac{5}{1} - \frac{16}{9} = \frac{45}{9} - \frac{16}{9}$

$= \frac{45-16}{9}$

$= \frac{29}{9} = 3\frac{2}{9}$

25. $|10-30| + (-4)^3$
$= |10+(-30)| + (-4)(-4)(-4)$
$= |-20| + (-64)$
$= 20 + (-64)$
$= -44$

26. $0.6 \div 12(3.6-4)$ Parentheses
$= 0.6 \div 12(-0.4)$ Divide.
$= 0.05(-0.4)$ Multiply.
$= -0.02$

27. $\frac{3}{10} - \left(\frac{1}{4} - \frac{3}{4}\right)^2 + \left(\frac{1}{2}\right)^2$

$= \frac{3}{10} - \left(\frac{1-3}{4}\right)^2 + \left(\frac{1}{2}\right)^2$

$= \frac{3}{10} - \left(\frac{-2}{4}\right)^2 + \left(\frac{1}{2}\right)^2$

$= \frac{3}{10} - \left(\frac{-1}{2}\right)^2 + \left(\frac{1}{2}\right)^2$

$= \frac{3}{10} - \frac{1}{4} + \frac{1}{4}$

$= \frac{3}{10} + \left(-\frac{1}{4} + \frac{1}{4}\right)$

$= \frac{3}{10} + 0$

$= \frac{3}{10}$

28. $-6w - 5$ Replace w with -4.
$= -6(-4) - 5$
$= 24 - 5$
$= 19$

29. $5x + 3y$ Replace x with -2 and y with 3.
$= 5(-2) + 3(3)$
$= -10 + 9$
$= -1$

30. x^3y
$= x \cdot x \cdot x \cdot y$ Replace x with -2 and y with 3.
$= (-2)(-2)(-2)(3)$
$= (4)(-2)(3)$
$= (-8)(3)$
$= -24$

31. $-5w^2x$
$= -5 \cdot w \cdot w \cdot x$ Replace x with -2 and w with -4.
$= (-5)(-4)(-4)(-2)$
$= (20)(-4)(-2)$
$= (-80)(-2)$
$= 160$

32. $-2x^2 + 5x - 7x^2$
$= -2x^2 + 5x + (-7x^2)$ Combine like terms.
$= -9x^2 + 5x$

33. $ab - ab$
$= 1ab - 1ab$
$= 1ab + (-1ab)$
$= 0ab$
$= 0$

34. $-10(4w^3)$
$= (-10 \cdot 4)w^3$
$= -40w^3$

35. $3(h-4) + 2$ Distribute.
$= 3h - 12 + 2$
$= 3h - 10$

36. $2n - 3n = 0 - 5$
$2n + (-3n) = 0 + (-5)$
$-1n = -5$
$\frac{-1n}{-1} = \frac{-5}{-1}$
$n = 5$

The solution is 5.

37.
$$\begin{aligned} 12-h&=-3h\\ 12+(-1h)&=-3h\\ 12+(-1h)+1h&=-3h+1h\\ 12&=-2h\\ \frac{12}{-2}&=\frac{-2h}{-2}\\ -6&=h \end{aligned}$$

The solution is -6.

38.
$$\begin{aligned} 5a-0.6&=10.4\\ 5a-0.6+0.6&=10.4+0.6\\ 5a&=11\\ \frac{5a}{5}&=\frac{11}{5}\\ a&=2.2 \end{aligned}$$

The solution is 2.2.

39.
$$\begin{aligned} -\frac{7}{8}&=\frac{3}{16}y\\ \frac{16}{3}\left(-\frac{7}{8}\right)&=\frac{16}{3}\left(\frac{3}{16}y\right)\\ -\frac{2\cdot \overset{1}{\cancel{8}}\cdot 7}{3\cdot \underset{1}{\cancel{8}}}&=y\\ -\frac{14}{3}\text{ or }-4\frac{2}{3}&=y \end{aligned}$$

The solution is $-\frac{14}{3}$.

40.
$$\begin{aligned} 3+\frac{1}{10}b&=5 \qquad \text{Add } -3 \text{ to both sides.}\\ \frac{1}{10}b&=2\\ \frac{10}{1}\left(\frac{1}{10}b\right)&=\frac{10}{1}(2)\\ b&=20 \end{aligned}$$

The solution is 20.

41.
$$\begin{aligned} \frac{0.2}{3.25}&=\frac{10}{x}\\ 0.2\cdot x&=3.25\cdot 10 \quad \text{Cross products}\\ \frac{0.2x}{0.2}&=\frac{32.5}{0.2}\\ x&=162.5 \end{aligned}$$

The solution is 162.5.

42.
$$\begin{aligned} 32-3h&=5h+8\\ 32-3h+3h&=5h+8+3h\\ 32&=8h+8\\ 32-8&=8h+8-8\\ 24&=8h\\ \frac{24}{8}&=\frac{8h}{8}\\ 3&=h \end{aligned}$$

The solution is 3.

43.
$$\begin{aligned} 3+2(x+4)&=-4x+7+2x \quad \text{Distribute.}\\ 3+2x+8&=-4x+7+2x \quad \text{Add like terms.}\\ 2x+11&=-2x+7\\ 2x+11+2x&=-2x+7+2x\\ 4x+11&=7\\ 4x+11-11&=7-11\\ 4x&=-4\\ \frac{4x}{4}&=\frac{-4}{4}\\ x&=-1 \end{aligned}$$

The solution is -1.

44. Let n represent the unknown number.

$$\begin{aligned} 4n-5&=-17\\ 4n-5+5&=-17+5\\ 4n&=-12\\ \frac{4n}{4}&=\frac{-12}{4}\\ n&=-3 \end{aligned}$$

The solution is -3.

45. Let n represent the number.

$$\begin{aligned} n+31&=3n+1\\ n+31-n&=3n+1-n\\ 31&=2n+1\\ 31-1&=2n+1-1\\ \frac{30}{2}&=\frac{2n}{2}\\ 15&=n \end{aligned}$$

The solution is 15.

46. *Step 1*
Unknown: Number of diapers in one package
Known: 17 used, 19 used, 12 left, 3 packages originally purchased

Step 2
Let d represent the number of diapers in each package.

Step 3
$3d-17-19=12$

Step 4
$$\begin{aligned} 3d+(-17)+(-19)&=12\\ 3d+(-36)&=12\\ 3d+(-36)+36&=12+36\\ 3d&=48\\ \frac{3d}{3}&=\frac{48}{3}\\ d&=16 \end{aligned}$$

Step 5
There were 16 diapers in each package.

Step 6
Check: 3 packages with 16 diapers in each = 48 diapers.

$48 - 17 = 31$ First day
$31 - 19 = 12$ Second day (matches 12 left)

47. *Step 1*
Unknown: How much Susanna made and how much Neoka made
Known: $1620 total, Neoka earned twice as much

Step 2
Let p be Susanna's pay. Then $2p$ would be Neoka's pay.

Step 3
Susanna's pay + Neoka's pay = Total pay

Step 4
$$p + 2p = 1620$$
$$3p = 1620$$
$$\frac{3p}{3} = \frac{1620}{3}$$
$$p = 540$$

Step 5
Susanna made $540 and Neoka made $1080.

Step 6
Check: $1080 is twice $540 and $1080 + $540 = $1620.

48. $P = 10\text{ yd} + 10\text{ yd} + 10\text{ yd} + 10\text{ yd} + 10\text{ yd}$
$P = 50\text{ yards}$

The perimeter is 50 yards.

A = area of square + area of triangle
$$A = s \cdot s + \frac{1}{2}bh$$
$$A = 10\text{ yd} \cdot 10\text{ yd} + \frac{1}{2} \cdot 10\text{ yd} \cdot 8.4\text{ yd}$$
$$A = 100\text{ yd}^2 + 42\text{ yd}^2$$
$$A = 142\text{ yd}^2$$

The area is 142 yd^2.

49. $C = 2\pi r$
$C \approx 2 \cdot 3.14 \cdot 5\text{ ft}$
$C = 31.4\text{ ft}$

The circumference is about 31.4 ft.

$$A = \pi r^2$$
$$A = \pi \cdot r \cdot r$$
$$A \approx 3.14 \cdot 5\text{ ft} \cdot 5\text{ ft}$$
$$A = 78.5\text{ ft}^2$$

The area is about 78.5 ft^2.

50.
$$\frac{x}{22} = \frac{5.5}{15}$$
$$15 \cdot x = 22 \cdot 5.5$$
$$\frac{15x}{15} = \frac{121}{15}$$
$$x \approx 8.1\text{ ft}$$

51. $V = \ell \cdot w \cdot h$
$V = 3.5\text{ m} \cdot 3\text{ m} \cdot 0.7\text{ m}$
$V = 7.35\text{ m}^3$
$\approx 7.4\text{ m}^3$

$SA = 2\ell w + 2\ell h + 2wh$
$SA = 2(3.5\text{ m})(3\text{ m}) + 2(3.5\text{ m})(0.7\text{ m}) + 2(3\text{ m})(0.7\text{ m})$
$SA = 30.1\text{ m}^2$

52. Use the formula for the volume of a cylinder.
$$V = \pi \cdot r^2 \cdot h\left(r = \frac{d}{2} = \frac{13\text{ cm}}{2} = 6.5\text{ cm}\right)$$
$$\approx 3.14 \cdot 6.5\text{ cm} \cdot 6.5\text{ cm} \cdot 17\text{ cm}$$
$$= 2255.305$$
$$\approx 2255.3\text{ cm}^3$$

53. hypotenuse $= \sqrt{(\text{leg})^2 + (\text{leg})^2}$
$$y = \sqrt{(15)^2 + (19)^2}$$
$$y = \sqrt{225 + 361}$$
$$y = \sqrt{586}$$
$$y \approx 24.2\text{ mm}$$

$$A = \frac{1}{2}bh$$
$$A = \frac{1}{2}(19\text{ mm})(15\text{ mm})$$
$$A = 142.5\text{ mm}^2$$

54. $\frac{3}{8}$ of 5600 students

$$\frac{3}{8} \cdot 5600 = \frac{3}{\cancel{8}_1} \cdot \frac{\cancel{5600}^{700}}{1} = 2100$$

The survey showed that 2100 students work 20 hours or more per week.

55. The length of the parking space is 18 ft and the width is 9 ft + 5 ft = 14 ft.

$$P = 2\ell + 2w$$
$$P = 2 \cdot 18\text{ ft} + 2 \cdot 14\text{ ft}$$
$$P = 36\text{ ft} + 28\text{ ft}$$
$$P = 64\text{ ft}$$

$$A = \ell \cdot w$$
$$A = 18\text{ ft} \cdot 14\text{ ft}$$
$$A = 252\text{ ft}^2$$

56. Add the amounts needed for the recipes.

$$\begin{array}{rcl} 2\frac{1}{4} & = & 2\frac{1}{4} \\ 1\frac{1}{2} & = & 1\frac{2}{4} \\ +\frac{3}{4} & = & \frac{3}{4} \\ \hline & & 3\frac{6}{4} = 4\frac{2}{4} = 4\frac{1}{2} \text{ cups} \end{array}$$

Find the total amount of brown sugar.

$$\begin{array}{r} 2\frac{1}{3} \\ +2\frac{1}{3} \\ \hline 4\frac{2}{3} \text{ cups} \end{array}$$

Subtract the amount to be used from the total amount.

$$\begin{array}{rcl} 4\frac{2}{3} & = & 4\frac{4}{6} \\ -4\frac{1}{2} & = & 4\frac{3}{6} \\ \hline & & \frac{1}{6} \end{array}$$

The Jackson family will have $\frac{1}{6}$ cup more than the amount needed.

57.

$$\begin{array}{rcl} \text{Discount amount} & = & \text{Discount rate} \cdot \text{Original price} \\ & = & 0.30 \cdot \$189 \\ & = & \$56.70 \end{array}$$

$$\begin{array}{rcl} \text{Sale price} & = & \text{Original price} - \text{Discount amount} \\ & = & \$189 - \$56.70 \\ & = & \$132.30 \end{array}$$

58. Let x represent the actual distance from Springfield to Bloomington. Set up a proportion to solve.

$$\frac{1 \text{ cm}}{12 \text{ km}} = \frac{7.8 \text{ cm}}{x \text{ km}} \quad \text{Units must agree.}$$

$$1 \cdot x = 12 \cdot 7.8 \quad \text{Cross products}$$

$$x = 93.6$$

The actual distance is 93.6 km.

59. Use the percent proportion.
part is 31; whole is 35; percent is unknown

$$\frac{31}{35} = \frac{p}{100}$$

$$35 \cdot p = 31 \cdot 100$$

$$\frac{35 \cdot p}{35} = \frac{3100}{35}$$

$$p \approx 88.57 \quad (88.57\%)$$

The percent of the problems that were correct is 88.6% (rounded).

60. The cost of the 18 prints is

$$18(\$0.12) = \$2.16.$$

Adding the shipping charge gives

$$\$2.16 + \$1.97 = \$4.13.$$

So the cost per print is

$$\frac{\$4.13}{18} \approx \$0.2294 \approx \$0.23.$$

Each print cost \$0.23 (rounded).

61. **(a)** (5 days/week)(6 weeks) = 30 days, so Britain averages six weeks of vacation.

(b) (5 days/week)(8 weeks) = 40 days, so Brazil averages eight weeks of vacation.

62. **(a)** $\frac{\text{Brazil}}{\text{U.S.}} = \frac{40 \text{ days}}{20 \text{ days}} = \frac{40 \div 20}{20 \div 20} = \frac{2}{1}$

(b) $\frac{\text{Canada}}{\text{France}} = \frac{24 \text{ days}}{36 \text{ days}} = \frac{24 \div 12}{36 \div 12} = \frac{2}{3}$

63. **(a)** $\frac{\text{U.S.}}{\text{Italy}} = \frac{20 \text{ days}}{42 \text{ days}} = \frac{10}{21} \approx 0.476 = 47.6\%$

(b) $\frac{\text{France}}{\text{U.S.}} = \frac{36 \text{ days}}{20 \text{ days}} = \frac{9}{5} = 1.8 = 180\%$

64. **(a)** Canada days – U.S. days = 24 – 20 = 4

4 is what percent of 20?

$$\frac{4}{20} = \frac{4 \div 4}{20 \div 4} = \frac{1}{5} = 0.2 = 20\%$$

(b) Britain days – U.S. days = 30 – 20 = 10

10 is what percent of 20?

$$\frac{10}{20} = \frac{10 \div 10}{20 \div 10} = \frac{1}{2} = 0.5 = 50\%$$

(c) Brazil days – U.S. days = 40 – 20 = 20

20 is what percent of 20?

$$\frac{20}{20} = 1 = 100\%$$

CHAPTER 8 MEASUREMENT

8.1 Problem Solving with English Measurement

8.1 Section Exercises

1. **(a)** 1 yd = 3 ft

(b) 12 in. = 1 ft

3. **(a)** 8 fl oz = 1 c

(b) 1 qt = 2 pt

5. **(a)** 1 mi = 5280 ft

(b) 3 ft = 1 yd

7. **(a)** 2000 lb = 1 T

(b) 1 lb = 16 oz

9. **(a)** 1 min = 60 sec

(b) 60 min = 1 hr

11. **(a)** 120 sec to minutes

$$120\text{ sec} = \frac{\overset{2}{\cancel{120}}\ \cancel{\text{sec}}}{1} \cdot \frac{1\text{ min}}{\underset{1}{\cancel{60}}\ \cancel{\text{sec}}} = 2\text{ min}$$

(b) 4 hours to minutes

$$4\text{ hr} = \frac{4\ \cancel{\text{hr}}}{1} \cdot \frac{60\text{ min}}{1\ \cancel{\text{hr}}} = 240\text{ min}$$

13. **(a)** 2 quarts to gallons

$$2\text{ qt} = \frac{\overset{1}{\cancel{2}}\ \cancel{\text{qt}}}{1} \cdot \frac{1\text{ gal}}{\underset{2}{\cancel{4}}\ \cancel{\text{qt}}} = \frac{1}{2}\text{ gal}$$

(b) $6\frac{1}{2}$ feet to inches

$$6\frac{1}{2}\text{ ft} = \frac{\frac{13}{2}\ \cancel{\text{ft}}}{1} \cdot \frac{12\text{ in.}}{1\ \cancel{\text{ft}}} = \frac{13}{\underset{1}{\cancel{2}}} \cdot \frac{\overset{6}{\cancel{12}}}{1}\text{ in.} = 78\text{ in.}$$

15. 7 to 8 tons to pounds

$$\frac{7\ \cancel{\text{T}}}{1} \cdot \frac{2000\text{ lb}}{1\ \cancel{\text{T}}} = 14{,}000\text{ lb}$$

$$\frac{8\ \cancel{\text{T}}}{1} \cdot \frac{2000\text{ lb}}{1\ \cancel{\text{T}}} = 16{,}000\text{ lb}$$

An adult African elephant may weigh 14,000 to 16,000 lb.

17. 9 yd to feet

$$\frac{9\ \cancel{\text{yd}}}{1} \cdot \frac{3\text{ ft}}{1\ \cancel{\text{yd}}} = 9 \cdot 3\text{ ft} = 27\text{ ft}$$

19. 7 lb to ounces

$$\frac{7\ \cancel{\text{lb}}}{1} \cdot \frac{16\text{ oz}}{1\ \cancel{\text{lb}}} = 7 \cdot 16\text{ oz} = 112\text{ oz}$$

21. 5 qt to pints

$$\frac{5\ \cancel{\text{qt}}}{1} \cdot \frac{2\text{ pt}}{1\ \cancel{\text{qt}}} = 5 \cdot 2\text{ pt} = 10\text{ pt}$$

23. 90 min to hours

$$\frac{\overset{3}{\cancel{90}}\ \cancel{\text{min}}}{1} \cdot \frac{1\text{ hr}}{\underset{2}{\cancel{60}}\ \cancel{\text{min}}} = \frac{3}{2}\text{ hr} = 1\frac{1}{2}\text{ or }1.5\text{ hr}$$

25. 3 in. to feet

$$\frac{\overset{1}{\cancel{3}}\ \cancel{\text{in.}}}{1} \cdot \frac{1\text{ ft}}{\underset{4}{\cancel{12}}\ \cancel{\text{in.}}} = \frac{1}{4}\text{ or }0.25\text{ ft}$$

27. 24 oz to pounds

$$\frac{\overset{3}{\cancel{24}}\ \cancel{\text{oz}}}{1} \cdot \frac{1\text{ lb}}{\underset{2}{\cancel{16}}\ \cancel{\text{oz}}} = \frac{3}{2}\text{ lb} = 1\frac{1}{2}\text{ or }1.5\text{ lb}$$

29. 5 c to pints

$$\frac{5\ \cancel{\text{c}}}{1} \cdot \frac{1\text{ pt}}{2\ \cancel{\text{c}}} = \frac{5}{2}\text{ pt} = 2\frac{1}{2}\text{ or }2.5\text{ pt}$$

31. $\frac{1}{2}\text{ ft} = \frac{\frac{1}{2}\ \cancel{\text{ft}}}{1} \cdot \frac{12\text{ in.}}{1\ \cancel{\text{ft}}} = \frac{1}{2} \cdot 12\text{ in.} = 6\text{ in.}$

The ice will safely support a snowmobile or ATV or a person walking.

33. $2\frac{1}{2}$ T to lb

$$\frac{2\frac{1}{2}\ \cancel{\text{T}}}{1} \cdot \frac{2000\text{ lb}}{1\ \cancel{\text{T}}} = \frac{5}{\underset{1}{\cancel{2}}} \cdot \frac{\overset{1000}{\cancel{2000}}}{1}\text{ lb} = 5000\text{ lb}$$

35. $4\frac{1}{4}$ gal to quarts

$$\frac{4\frac{1}{4}\ \cancel{\text{gal}}}{1} \cdot \frac{4\text{ qt}}{1\ \cancel{\text{gal}}} = \frac{17}{\underset{1}{\cancel{4}}} \cdot \frac{\overset{1}{\cancel{4}}}{1}\text{ qt} = 17\text{ qt}$$

37. $\frac{1}{3}\text{ ft} = \frac{\frac{1}{3}\ \cancel{\text{ft}}}{1} \cdot \frac{12\text{ in.}}{1\ \cancel{\text{ft}}} = \frac{1}{\underset{1}{\cancel{3}}} \cdot \frac{\overset{4}{\cancel{12}}}{1}\text{ in.} = 4\text{ in.}$

Two-thirds of a foot would be twice as high; that is, 2(4 in.) = 8 in. The cactus could be 4 to 8 in. tall.

39. 6 yd to inches

$$\frac{6\ \cancel{\text{yd}}}{1} \cdot \frac{3\ \cancel{\text{ft}}}{1\ \cancel{\text{yd}}} \cdot \frac{12\text{ in.}}{1\ \cancel{\text{ft}}} = 6 \cdot 3 \cdot 12\text{ in.} = 216\text{ in.}$$

41. 112 c to quarts

$$\frac{112\text{ c}}{1}\cdot\frac{1\text{ pt}}{2\text{ c}}\cdot\frac{1\text{ qt}}{2\text{ pt}} = 28\text{ qt}$$

43. 6 days to seconds

$$\frac{6\text{ days}}{1}\cdot\frac{24\text{ hr}}{1\text{ day}}\cdot\frac{60\text{ min}}{1\text{ hr}}\cdot\frac{60\text{ sec}}{1\text{ min}}$$
$$= 6\cdot 24\cdot 60\cdot 60\text{ sec}$$
$$= 518{,}400\text{ sec}$$

45. $1\frac{1}{2}$ T to ounces

$$\frac{1\frac{1}{2}\text{ T}}{1}\cdot\frac{2000\text{ lb}}{1\text{ T}}\cdot\frac{16\text{ oz}}{1\text{ lb}} = \frac{3}{2}\cdot\frac{2000}{1}\cdot\frac{16}{1}\text{ oz}$$
$$= 48{,}000\text{ oz}$$

47. **(a)** There is only one relationship that uses 1 to 16: 1 pound = 16 ounces

(b) 10 to 20 is the same as 1 to 2. There are 2 relationships like this: 1 pint = 2 cups and 1 quart = 2 pints.
10 quarts = 20 pints or 10 pints = 20 cups

(c) 120 to 2 is the same as 60 to 1. There are 2 relationships like this: 60 minutes = 1 hour and 60 seconds = 1 minute.
120 minutes = 2 hours or
120 seconds = 2 minutes

(d) 2 to 24 is the same as 1 to 12. Use 1 foot = 12 inches.
2 feet = 24 inches

(e) 6000 to 3 is the same as 2000 to 1. Use 2000 pounds = 1 ton.
6000 pounds = 3 tons

(f) 35 to 5 is the same as 7 to 1. Use 7 days = 1 week.
35 days = 5 weeks

49. $2\frac{3}{4}$ miles to inches

$$\frac{2\frac{3}{4}\text{ mi}}{1}\cdot\frac{5280\text{ ft}}{1\text{ mi}}\cdot\frac{12\text{ in.}}{1\text{ ft}} = \frac{11}{4}\cdot\frac{5280}{1}\cdot\frac{12}{1}\text{ in.}$$
$$= 174{,}240\text{ in.}$$

51. $6\frac{1}{4}$ gal to fluid ounces

$$\frac{6\frac{1}{4}\text{ gal}}{1}\cdot\frac{4\text{ qt}}{1\text{ gal}}\cdot\frac{32\text{ oz}}{1\text{ qt}} = \frac{25}{4}\cdot 4\cdot 32\text{ oz}$$
$$= 800\text{ fl oz}$$

53. 24,000 oz to tons

$$\frac{24{,}000\text{ oz}}{1}\cdot\frac{1\text{ lb}}{16\text{ oz}}\cdot\frac{1\text{ T}}{2000\text{ lb}}$$
$$= \frac{24{,}000}{1}\cdot\frac{1}{16}\cdot\frac{1}{2000}\text{ T}$$
$$= \frac{3}{4}\text{ or }0.75\text{ T}$$

For Exercises 55–62, the six problem-solving steps should be used, but are only shown for Exercises 55 and 56.

55. *Step 1*
The problem asks for the price per pound of strawberries.

Step 2
Convert ounces to pounds. Then divide the cost by the pounds.

Step 3
To estimate, round \$2.29 to \$2. Then, there are 16 oz in a pound, so 20 oz is a little more than 1 lb. Thus, $\$2 \div 1 = \$2$ per pound is our estimate.

Step 4

$$\frac{20\text{ oz}}{1}\cdot\frac{1\text{ lb}}{16\text{ oz}} = \frac{5}{4}\text{ lb} = 1.25\text{ lb}$$
$$\frac{\$2.29}{1.25\text{ lb}} = 1.832 \approx 1.83$$

Step 5
The strawberries are \$1.83 per pound (to the nearest cent).

Step 6
The answer, \$1.83, is close to our estimate of \$2.

57. Find the total number of feet needed. Then convert feet to yards. Then multiply to find the cost.

$$24\cdot 2 = 48\text{ ft}$$
$$\frac{48\text{ ft}}{1}\cdot\frac{1\text{ yd}}{3\text{ ft}} = 16\text{ yd}$$
$$(16\text{ yd})(\$8.75) = \$140$$

It will cost \$140 to equip all the stations.

59. **(a)** Convert seconds per foot to seconds per mile.

$$\frac{1\text{ sec}}{5\text{ ft}}\cdot\frac{5280\text{ ft}}{1\text{ mi}} = \frac{5280}{5}\text{ sec/mi} = 1056\text{ sec/mi}$$

It would take the cockroach 1056 seconds to travel 1 mile.

(b) $\dfrac{1056\ \cancel{sec}}{1\text{ mi}} \cdot \dfrac{1\text{ min}}{60\ \cancel{sec}} = \dfrac{\overset{88}{\cancel{1056}}}{\underset{5}{\cancel{60}}}\text{ min/mi}$

$= 17.6\text{ min/mi}$

It would take the cockroach 17.6 minutes to travel 1 mile.

61. **(a)** Find the total number of cups per week. Then convert cups to quarts.

$\dfrac{2}{3} \cdot 15 \cdot 5 = \dfrac{2}{\underset{1}{\cancel{3}}} \cdot \dfrac{\overset{5}{\cancel{15}}}{1} \cdot \dfrac{5}{1} = 50\text{ c}$

$\dfrac{\overset{25}{\cancel{50}}\ \cancel{c}}{1} \cdot \dfrac{1\ \cancel{pt}}{\underset{1}{\cancel{2}}\ \cancel{c}} \cdot \dfrac{1\text{ qt}}{2\ \cancel{pt}} = \dfrac{25}{2}\text{ qt} = 12\dfrac{1}{2}\text{ qt}$

The center needs $12\frac{1}{2}$ qt of milk per week.

(b) Convert quarts to gallons.

$\dfrac{12\frac{1}{2}\ \cancel{qt}}{1} \cdot \dfrac{1\text{ gal}}{4\ \cancel{qt}} = \dfrac{25}{2} \cdot \dfrac{1}{4}\text{ gal} = 3.125\text{ gal}$

The center should order 4 containers, because you can't buy part of a container.

63. **(a)** $\dfrac{34{,}646{,}437}{24{,}907} \approx 1391$

You would need to travel about 1391 times around Earth.

(b) $\dfrac{34{,}646{,}437}{2786} \approx 12{,}436$

You would need to drive about 12,436 times from Los Angeles to New York.

64. **(a)** $\dfrac{34{,}646{,}437}{24{,}907} \approx \dfrac{35{,}000{,}000}{25{,}000}$

$= \dfrac{35{,}000}{25} = \dfrac{7000}{5} = 1400$

The estimate is 1400 times around Earth.

(b) $\dfrac{34{,}646{,}437}{2786} \approx \dfrac{35{,}000{,}000}{2800}$

$= \dfrac{350{,}000}{28} = \dfrac{50{,}000}{4} = 12{,}500$

The estimate is 12,500 trips.

(c) They are very close and just as useful for making comparisons of such large numbers.

65. **(a)** $\dfrac{16.7\ \cancel{mi}}{1} \cdot \dfrac{5280\text{ ft}}{1\ \cancel{mi}} = (16.7)(5280)\text{ ft}$

$= 88{,}176\text{ ft tall}$

(b) $\dfrac{29{,}035\ \cancel{ft}}{1} \cdot \dfrac{1\text{ mi}}{5280\ \cancel{ft}} = \dfrac{29{,}035}{5280}\text{ mi}$

$= 5.5\text{ mi (rounded)}$

(c) $\dfrac{16.7\ \cancel{mi}}{5.5\ \cancel{mi}} \approx 3.0$, so Olympic Mons is about 3 times taller than Mount Everest.

66. **(a)** $\dfrac{36\text{ (Mars)}}{100\text{ (Earth)}} = \dfrac{x\text{ (Mars)}}{180\text{ (Earth)}}$ **OR** $\dfrac{9}{25} = \dfrac{x}{180}$

$25 \cdot x = 9 \cdot 180$

$\dfrac{25 \cdot x}{25} = \dfrac{1620}{25}$

$x = 64.8$

The person would weigh 65 lb (rounded) on Mars.

(b) $\dfrac{36\text{ (Mars)}}{100\text{ (Earth)}} = \dfrac{x\text{ (Mars)}}{9\text{ (Earth)}}$ **OR** $\dfrac{9}{25} = \dfrac{x}{9}$

$25 \cdot x = 9 \cdot 9$

$\dfrac{25 \cdot x}{25} = \dfrac{81}{25}$

$x = 3.24$

The baby would weigh 3 lb (rounded) on Mars.

(c) Answers will vary, but should be about $\frac{1}{3}$ of your weight on Earth.

8.2 The Metric System—Length

8.2 Section Exercises

1. *kilo* means <u>1000</u>, so 1 km = <u>1000</u> m

3. *milli* means <u>$\frac{1}{1000}$ or 0.001</u>, so

1 mm = <u>$\frac{1}{1000}$ or 0.001 m</u>

5. *centi* means <u>$\frac{1}{100}$ or 0.01</u>, so 1 cm = <u>$\frac{1}{100}$ or 0.01 m</u>

7. The width of your hand in centimeters • Answers will vary; about 8 to 10 cm.

9. The width of your thumb in millimeters • Answers will vary; about 15 to 25 mm.

11. The child was about 91 <u>cm</u> tall.

13. Ming-Na swam in the 200 <u>m</u> backstroke race.

15. Adriana drove 400 <u>km</u> on her vacation.

17. An aspirin tablet is 10 <u>mm</u> across.

19. A paper clip is about 3 <u>cm</u> long.

21. Dave's truck is 5 <u>m</u> long.

23. Some possible answers are: 35 mm film for cameras, track and field events, metric auto parts, and lead refills for mechanical pencils.

25. 7 m to cm

$\dfrac{7\ \cancel{m}}{1} \cdot \dfrac{100\text{ cm}}{1\ \cancel{m}} = 7 \cdot 100\text{ cm} = 700\text{ cm}$

27. 40 mm to m

$$\frac{40\ \cancel{mm}}{1} \cdot \frac{1\text{ m}}{1000\ \cancel{mm}} = \frac{40}{1000}\text{ m} = 0.040 \text{ or } 0.04\text{ m}$$

29. 9.4 km to m

$$\frac{9.4\ \cancel{km}}{1} \cdot \frac{1000\text{ m}}{1\ \cancel{km}} = (9.4)(1000)\text{ m} = 9400\text{ m}$$

31. 509 cm to m

$$\frac{509\ \cancel{cm}}{1} \cdot \frac{1\text{ m}}{100\ \cancel{cm}} = \frac{509}{100}\text{ m} = 5.09\text{ m}$$

33. 400 mm to cm
Count 1 place to the *left* on the conversion line.
$40_\wedge0.$ mm = 40.0 cm = 40 cm

35. 0.91 m to mm
Count 3 places to the *right* on the conversion line.
One zero is written in as a placeholder.
$0.910_\wedge$ m = 910 mm

37. 82 cm to m
Count 2 places to the *left* on the conversion line.
$_\wedge82.$ cm = 0.82 m

0.82 m is less than 1 m, so 82 cm is **less than** 1 m.
The difference in length is 1 m − 0.82 m =
0.18 m or 100 cm − 82 cm = 18 cm.

39. 5 mm
1 mm
5 mm to centimeters
Count one place to the *left* on the conversion line.
$_\wedge5.$ mm = 0.5 cm
Similarly, 1 mm = 0.1 cm.

41. 18 m to km
Count 3 places to the *left* on the conversion line.
One zero is written in as a placeholder.
$_\wedge018.$ m = 0.018 km

The north fork of the Roe River is just under 0.018 km long.

43. 1.64 m to centimeters and millimeters
Count two (three) places to the *right* on the conversion line.
$1.64_\wedge$ m = 164 cm;

$1.640_\wedge$ m = 1640 mm

The median height for U.S. females who are 20 to 29 years old is about 164 cm, or equivalently, 1640 mm.

45. 5.6 mm to km

$$\frac{5.6\ \cancel{mm}}{1} \cdot \frac{1\ \cancel{m}}{1000\ \cancel{mm}} \cdot \frac{1\text{ km}}{1000\ \cancel{m}} = \frac{5.6}{1{,}000{,}000}\text{ km}$$
$$= 0.0000056\text{ km}$$

8.3 The Metric System—Capacity and Weight (Mass)

8.3 Section Exercises

1. The glass held 250 <u>mL</u> of water. (Liquids are measured in mL or L.)

3. Dolores can make 10 <u>L</u> of soup in that pot. (Liquids are measured in mL or L.)

5. Our yellow Labrador dog grew up to weigh 40 <u>kg</u>. (Weight is measured in mg, g, or kg.)

7. Lori caught a small sunfish weighing 150 <u>g</u>. (Weight is measured in mg, g, or kg.)

9. Andre donated 500 <u>mL</u> of blood today. (Blood is a liquid, and liquids are measured in mL or L.)

11. The patient received a 250 <u>mg</u> tablet of medication each hour. (Weight is measured in mg, g, or kg.)

13. The gas can for the lawn mower holds 4 <u>L</u>. (Gasoline is a liquid, and liquids are measured in mL or L.)

15. Pam's backpack weighs 5 <u>kg</u> when it is full of books. (Weight is measured in mg, g, or kg.)

17. This is unreasonable (too much) since 4.1 liters would be about 4 quarts.

19. This is unreasonable (too much) since 5 kilograms of Epsom salts would be about 11 pounds with approximately a quart of water.

21. This is reasonable because 15 milliliters would be about 3 teaspoons.

23. This is reasonable because 350 milligrams would be a little more than 1 tablet, which is about 325 milligrams.

25. Some capacity examples are 2 L bottles of soda and shampoo bottles marked in mL; weight examples are grams of fat listed on cereal boxes and vitamin doses in milligrams.

27. The unit for your answer (g) is in the numerator. The unit being changed (kg) is in denominator, so it will divide out. The unit fraction is $\frac{1000\text{ g}}{1\text{ kg}}$.

29. 15 L to mL
Count 3 places to the *right* on the conversion line.
$15.000_\wedge$ L = 15,000 mL

31. 3000 mL to L
Count 3 places to the *left* on the conversion line.
3∧000. mL = 3 L

33. 925 mL to L

$$\frac{925\ \cancel{mL}}{1} \cdot \frac{1\ L}{1000\ \cancel{mL}} = \frac{925}{1000}\ L = 0.925\ L$$

35. 8 mL to L

$$\frac{8\ \cancel{mL}}{1} \cdot \frac{1\ L}{1000\ \cancel{mL}} = \frac{8}{1000}\ L = 0.008\ L$$

37. 4.15 L to mL
Count 3 places to the *right* on the conversion line.
4.150∧ L = 4150 mL

39. 8000 g to kg
Count 3 places to the *left* on the conversion line.
8∧000. g = 8 kg

41. 5.2 kg to g
Count 3 places to the *right* on the conversion line.
5.200∧ kg = 5200 g

43. 0.85 g to mg
Count 3 places to the *right* on the conversion line.
0.850∧ g = 850 mg

45. 30,000 mg to g
Count 3 places to the *left* on the conversion line.
30∧000. mg = 30 g

47. 598 mg to g
Count 3 places to the *left* on the conversion line.
∧598. mg = 0.598 g

49. 60 mL to L

$$\frac{60\ \cancel{mL}}{1} \cdot \frac{1\ L}{1000\ \cancel{mL}} = \frac{60}{1000}\ L = 0.060 \text{ or } 0.06\ L$$

51. 3 g to kg

$$\frac{3\ \cancel{g}}{1} \cdot \frac{1\ kg}{1000\ \cancel{g}} = \frac{3}{1000}\ kg = 0.003\ kg$$

53. 0.99 L to mL

$$\frac{0.99\ \cancel{L}}{1} \cdot \frac{1000\ mL}{1\ \cancel{L}} = 990\ mL$$

55. The masking tape is 19 <u>mm</u> wide. (Length is measured in mm, cm, m.)

57. Buy a 60 <u>mL</u> jar of acrylic paint for art class. (Paint is a liquid and liquids are measured in mL or L.)

59. My waist measurement is 65 <u>cm</u>. (Length is measured in mm, cm, m.)

61. A single postage stamp weighs 90 <u>mg</u>. (Weight is measured in mg, g, or kg.)

63. Convert 300 mL to L.
Count 3 places to the *left* on the conversion line.
∧300. mL = 0.3 L

Each day 0.3 L of sweat is released.

65. Convert 1.34 kg to g.
Count 3 places to the *right* on the conversion line.
1.340∧ kg = 1340 g

The average weight of a human brain is 1340 g.

67. Convert 900 mL to L.

$$\frac{900\ \cancel{mL}}{1} \cdot \frac{1\ L}{1000\ \cancel{mL}} = \frac{900}{1000}\ L$$
$$= 0.900 \text{ or } 0.9\ L$$

On average, we breathe in and out roughly 0.9 L of air every 10 seconds.

69. Convert 3000 g to kg and 4000 g to kg.

$$\frac{3000\ \cancel{g}}{1} \cdot \frac{1\ kg}{1000\ \cancel{g}} = \frac{3000}{1000}\ kg = 3\ kg$$
$$\frac{4000\ \cancel{g}}{1} \cdot \frac{1\ kg}{1000\ \cancel{g}} = \frac{4000}{1000}\ kg = 4\ kg$$

A small adult cat weighs from 3 kg to 4 kg.

71. There are 1000 milligrams in a gram so 1005 mg is *greater* than 1 g. The difference in weight is 1005 mg − 1000 mg = 5 mg or 1.005 g − 1 g = 0.005 g.

73. Convert 1 kg to g.

$$\frac{1\ \cancel{kg}}{1} \cdot \frac{1000\ g}{1\ \cancel{kg}} = 1000\ g$$

Divide the 1000 g by the weight of 1 nickel.

$$\frac{1000\ \cancel{g}}{5\ \cancel{g}} = 200$$

There are 200 nickels in 1 kg of nickels.

75. **(a)** 1 Mm = <u>1,000,000</u> m

(b) 3.5 Mm to m

$$\frac{3.5\ \cancel{Mm}}{1} \cdot \frac{1{,}000{,}000\ m}{1\ \cancel{Mm}} = 3{,}500{,}000\ m$$

76. **(a)** 1 Gm = <u>1,000,000,000</u> m

(b) 2500 m to Gm

$$\frac{2500\ \cancel{m}}{1} \cdot \frac{1\ Gm}{1{,}000{,}000{,}000\ \cancel{m}} = 0.0000025\ Gm$$

77. **(a)** 1 Tm = 1,000,000,000,000 m

(b) $\frac{1\ \text{Tm}}{1} \cdot \frac{1{,}000{,}000{,}000{,}000\ \text{m}}{1\ \text{Tm}} \cdot \frac{1\ \text{Gm}}{1{,}000{,}000{,}000\ \text{m}}$

$= \frac{1{,}000{,}000{,}000{,}000}{1{,}000{,}000{,}000}\ \text{Gm} = 1000\ \text{Gm}$

So 1 Tm = 1000 Gm.

$\frac{1\ \text{Tm}}{1} \cdot \frac{1{,}000{,}000{,}000{,}000\ \text{m}}{1\ \text{Tm}} \cdot \frac{1\ \text{Mm}}{1{,}000{,}000\ \text{m}}$

$= \frac{1{,}000{,}000{,}000{,}000}{1{,}000{,}000}\ \text{Mm} = 1{,}000{,}000\ \text{Mm}$

So 1 Tm = 1,000,000 Mm.

78. 1 MB = 1,000,000 bytes
1 GB = 1,000,000,000 bytes
$2^{20} = 1{,}048{,}576$
$2^{30} = 1{,}073{,}741{,}824$

8.4 Problem Solving with Metric Measurement

8.4 Section Exercises

1. Convert 850 g to kg.

$\frac{850\ \text{g}}{1} \cdot \frac{1\ \text{kg}}{1000\ \text{g}} = 0.85\ \text{kg}$

$\frac{\$0.98}{1\ \text{kg}} \cdot \frac{0.85\ \text{kg}}{1} = \$0.833 \approx \$0.83$

She will pay $0.83 (rounded) for the rice.

3. Convert 500 g to kg.

$\frac{500\ \text{g}}{1} \cdot \frac{1\ \text{kg}}{1000\ \text{g}} = 0.5\ \text{kg}$

Find the difference.
90 kg − 0.5 kg = 89.5 kg

The difference in the weights of the two dogs is 89.5 kg.

5. Write 5 L in terms of mL.

$5\ \text{L} = \frac{5\ \text{L}}{1} \cdot \frac{1000\ \text{mL}}{1\ \text{L}} = 5000\ \text{mL}$

$\frac{5000\ \text{mL}}{1} \cdot \frac{1\ \text{beat}}{70\ \text{mL}} \approx 71.42857\ \text{beats} \approx 71\ \text{beats}$

It takes 71 beats (rounded) to pass all the blood through the heart.

7. Multiply to find the total length in mm, then convert mm to cm.
60 mm • 30 = 1800 mm

$\frac{\overset{180}{1800}\ \text{mm}}{1} \cdot \frac{1\ \text{cm}}{\underset{1}{10}\ \text{mm}} = 180\ \text{cm}$

The total length is 180 cm.
Divide to find the cost per cm.

$\frac{\$3.29}{180\ \text{cm}} \approx \$0.0183/\text{cm} \approx \$0.02/\text{cm}$

The cost is $0.02/cm (rounded).

9. Convert centimeters to meters, then add the lengths of the two pieces.

2 m 8 cm	=	2.08 m
2 m 95 cm	=	+ 2.95 m
		5.03 m

Together the boards are 5.03 m.

11. Convert milliliters to liters.
85 mL = 0.085 L
Multiply by the number of students.
(0.085 L)(45) = 3.825 L
Four one-liter bottles should be ordered.

The amount of acid left over is

4 L − 3.825 L = 0.175 L.

0.175‸ L = 175 mL

13. Three cups a day for one week is 3 • 7 = 21 cups in one week.

$\frac{21\ \text{cups}}{1} \cdot \frac{90\ \text{mg}}{1\ \text{cup}} = 1890\ \text{mg}$

Convert mg to g.

$\frac{1890\ \text{mg}}{1} \cdot \frac{1\ \text{g}}{1000\ \text{mg}} = 1.89\ \text{g}$

15. **(a)** $\frac{10{,}000\ \text{m}}{1\ \text{sec}} \cdot \frac{1\ \text{km}}{1000\ \text{m}} = 10\ \text{km/sec}$

(b) $\frac{60\ \text{sec}}{1\ \text{min}} \cdot \frac{10\ \text{km}}{1\ \text{sec}} = 600\ \text{km/min}$

(c) $\frac{60\ \text{min}}{1\ \text{hr}} \cdot \frac{600\ \text{km}}{1\ \text{min}} = 36{,}000\ \text{km/hr}$

17. A box of 50 envelopes weighs 255 g. Subtract the packaging weight to find the net weight.

255 g
− 40 g
215 g

There are 50 envelopes. Divide to find the weight of one envelope.

$\frac{215\ \text{g}}{50\ \text{envelopes}} = \frac{4.3\ \text{g}}{\text{envelope}}$

Convert grams to milligrams.

$\frac{4.3\ \text{g}}{1} \cdot \frac{1000\ \text{mg}}{1\ \text{g}} = 4300\ \text{mg}$

18. Box of 1000 staples
Subtract the weight of the packaging to find the net weight.

$$\begin{array}{r} 350\text{ g} \\ -\ 20\text{ g} \\ \hline 330\text{ g} \end{array}$$

There are 1000 staples, so divide the net weight by 1000 or move the decimal point 3 places to the left to find the weight of one staple.

$330\text{ g} \div 1000 = 0.33\text{ g}$

The weight of one staple is 0.33 g or 330 mg.

19. Given that the weight of one sheet of paper is 3000 mg, find the weight of one sheet in grams.

$3000\text{ mg} = 3{\wedge}000.\text{ g} = 3\text{ g}$

To find the net weight of the ream of paper, multiply 500 sheets by 3 g per sheet.

$$\frac{500\text{ }\cancel{\text{sheets}}}{1} \cdot \frac{3\text{ g}}{1\text{ }\cancel{\text{sheet}}} = 1500\text{ g}$$

The net weight of the paper is 1500 g.
Since the weight of the packaging is 50 g, the total weight is $1500\text{ g} + 50\text{ g} = 1550\text{ g}$.

20. Box of 100 small paper clips
Use the metric conversion line to find the weight of one paper clip in grams.

$500\text{ mg} = {\wedge}500.\text{ g} = 0.5\text{ g}$

The net weight is
$(0.5\text{ g})(100) = 50\text{ g}$
The total weight is
$50\text{ g} + 5\text{ g} = 55\text{ g}$.

21. $$\frac{0.45\text{ }\cancel{\text{kg}}}{1{,}000{,}000\text{ seeds}} \cdot \frac{1{,}000{,}000\text{ mg}}{1\text{ }\cancel{\text{kg}}} = 0.45\text{ mg/seed}$$

Each seed weighs 0.45 mg.

8.5 Metric–English Conversions and Temperature

8.5 Section Exercises

1. 20 m to yards

$$\frac{20\text{ }\cancel{\text{m}}}{1} \cdot \frac{1.09\text{ yd}}{1\text{ }\cancel{\text{m}}} \approx 21.8\text{ yd}$$

3. 80 m to feet

$$\frac{80\text{ }\cancel{\text{m}}}{1} \cdot \frac{3.28\text{ ft}}{1\text{ }\cancel{\text{m}}} \approx 262.4\text{ ft}$$

5. 16 ft to meters

$$\frac{16\text{ }\cancel{\text{ft}}}{1} \cdot \frac{0.30\text{ m}}{1\text{ }\cancel{\text{ft}}} \approx 4.8\text{ m}$$

7. 150 g to ounces

$$\frac{150\text{ }\cancel{\text{g}}}{1} \cdot \frac{0.035\text{ oz}}{1\text{ }\cancel{\text{g}}} \approx 5.3\text{ oz}$$

9. 248 lb to kilograms

$$\frac{248\text{ }\cancel{\text{lb}}}{1} \cdot \frac{0.45\text{ kg}}{1\text{ }\cancel{\text{lb}}} \approx 111.6\text{ kg}$$

11. 28.6 L to quarts

$$\frac{28.6\text{ }\cancel{\text{L}}}{1} \cdot \frac{1.06\text{ qt}}{1\text{ }\cancel{\text{L}}} \approx 30.3\text{ qt}$$

13. **(a)** Convert 5 g to ounces.

$$\frac{5\text{ }\cancel{\text{g}}}{1} \cdot \frac{0.035\text{ oz}}{1\text{ }\cancel{\text{g}}} \approx 0.2\text{ oz}$$

About 0.2 oz of gold was used.

(b) The extra weight probably did not slow him down.

15. Convert 8.4 gal to liters.

$$\frac{8.4\text{ }\cancel{\text{gal}}}{1} \cdot \frac{3.78\text{ L}}{1\text{ }\cancel{\text{gal}}} \approx 31.8\text{ L}$$

The dishwasher uses about 31.8 L.

17. Convert 0.5 in. to centimeters.

$$\frac{0.5\text{ }\cancel{\text{in.}}}{1} \cdot \frac{2.54\text{ cm}}{1\text{ }\cancel{\text{in.}}} \approx 1.3$$

The dwarf gobie is about 1.3 cm long.

19. Convert lb to kg.

$$\frac{40\text{ }\cancel{\text{lb}}}{1} \cdot \frac{0.45\text{ kg}}{1\text{ }\cancel{\text{lb}}} \approx 18\text{ kg}$$

Convert in. to cm.

$$\frac{22\text{ }\cancel{\text{in.}}}{1} \cdot \frac{2.54\text{ cm}}{1\text{ }\cancel{\text{in.}}} \approx 56\text{ cm}$$

$$\frac{14\text{ }\cancel{\text{in.}}}{1} \cdot \frac{2.54\text{ cm}}{1\text{ }\cancel{\text{in.}}} \approx 36\text{ cm}$$

$$\frac{9\text{ }\cancel{\text{in.}}}{1} \cdot \frac{2.54\text{ cm}}{1\text{ }\cancel{\text{in.}}} \approx 23\text{ cm}$$

The luggage cannot exceed 18 kg in weight or measure more than 56 cm by 36 cm by 23 cm (all rounded).

21. Convert 0.09 oz to g.

$$\frac{0.09\ \cancel{oz}}{1} \cdot \frac{28.35\ g}{1\ \cancel{oz}} \approx 2.55\ g$$

Convert 2.5 g to oz.

$$\frac{2.5\ \cancel{g}}{1} \cdot \frac{0.035\ oz}{1\ \cancel{g}} \approx 0.0875\ oz$$

Converting ounces to grams, 0.09 oz ≈ 2.55 g, which rounds to 2.6 g. However, converting grams to ounces, 2.5 g ≈ 0.0875 oz, which does round to 0.09 oz.

23. Convert 3.5 kg to lb.

$$\frac{3.5\ \cancel{kg}}{1} \cdot \frac{2.20\ lb}{1\ \cancel{kg}} \approx 7.7\ lb$$

Since the minimum weight is 3.5 kg ≈ 7.7 lb, the 8-lb baby is heavy enough.

Convert 53 cm to in.

$$\frac{53\ \cancel{cm}}{1} \cdot \frac{0.39\ in.}{1\ \cancel{cm}} \approx 20.7\ in.$$

Since the minimum length is 53 cm ≈ 20.7 in., the 19.5-inch baby is not long enough to be in the carrier.

For Exercises 25–30, see the thermometer before Example 3.

25. A snowy day
Since a *very cold winter day* is –18 °C, the most reasonable choice is –8 °C (12 °C is a spring day and 28 °C is a summer day).

27. A high fever
Since *normal body temperature* is 37 °C, the most reasonable choice is 40 °C.

29. Oven temperature
Since *water boils* at 100 °C, the most reasonable choice is 150 °C.

31. 60 °F

$$C = \frac{5(60-32)}{9} = \frac{5 \cdot 28}{9} = \frac{140}{9} \approx 15.6 \approx 16$$

60 °F ≈ 16 °C

33. –4 °F

$$C = \frac{5(-4-32)}{9} = \frac{5(-36)}{9} = \frac{-5 \cdot 4 \cdot \overset{1}{\cancel{9}}}{\underset{1}{\cancel{9}}}$$

$$= -20$$

–4 °F = –20 °C

35. 8 °C

$$F = \frac{9 \cdot 8}{5} + 32 = \frac{72}{5} + 32 = 14.4 + 32 = 46.4$$

8 °C ≈ 46 °F

37. –5 °C

$$F = \frac{9(-5)}{5} + 32 = \frac{-9 \cdot \overset{1}{\cancel{5}}}{\underset{1}{\cancel{5}}} + 32 = -9 + 32 = 23$$

–5 °C = 23 °F

39. 136 °F

$$C = \frac{5(136-32)}{9} = \frac{5 \cdot 104}{9} = \frac{520}{9} \approx 57.8$$

136 °F ≈ 58 °C

–129 °F

$$C = \frac{5(-129-32)}{9} = \frac{5(-161)}{9} \approx -89$$

–129 °F ≈ –89 °C

41. 50 °F

$$C = \frac{5(50-32)}{9} = \frac{5(18)}{9} = \frac{5 \cdot 2 \cdot \overset{1}{\cancel{9}}}{\underset{1}{\cancel{9}}} = 10$$

50 °F = 10 °C

105 °F

$$C = \frac{5(105-32)}{9} = \frac{5(73)}{9} \approx 41$$

105 °F ≈ 41 °C

43. A drop of 20 Celsius degrees is more than a drop of 20 Fahrenheit degrees . There are 180 degrees between freezing and boiling on the Fahrenheit scale, but only 100 degrees on the Celsius scale, so each Celsius degree is a greater change in temperature.

45. **(a)** Since the comfort range of the boots is from 24 °C to 4 °C, you would wear these boots in pleasant weather—above freezing, but not hot.

(b) Change 24 °C to Fahrenheit.

$$F = \frac{9 \cdot C}{5} + 32 = \frac{9 \cdot 24}{5} + 32 = \frac{216}{5} + 32$$
$$= 43.2 + 32 = 75.2$$

Thus, 24 °C ≈ 75 °F.

Change 4 °C to Fahrenheit.

$$F = \frac{9 \cdot C}{5} + 32 = \frac{9 \cdot 4}{5} + 32 = \frac{36}{5} + 32$$
$$= 7.2 + 32 = 39.2$$

Thus, 4 °C ≈ 39 °F.

The boots are designed for Fahrenheit temperatures of about 75 °F to about 39 °F.

(c) The range of metric temperatures in January would depend on where you live. In Minnesota, it's 0°C to −40°C, and in California it's 24°C to 0°C.

47. Length of model plane

$$\frac{6\cancel{\text{ft}}}{1}\cdot\frac{0.30\text{ m}}{1\cancel{\text{ft}}}\approx 6(0.30\text{ m}) = 1.8\text{ m}$$

48. Weight of plane

$$\frac{11\cancel{\text{lb}}}{1}\cdot\frac{0.45\text{ kg}}{1\cancel{\text{lb}}}\approx 11(0.45\text{ kg}) = 4.95\text{ kg}$$
$$\approx 5.0\text{ kg}$$

49. Length of flight path

$$\frac{1888.3\cancel{\text{mi}}}{1}\cdot\frac{1.61\text{ km}}{1\cancel{\text{mi}}}\approx 1888.3(1.61\text{ km})$$
$$\approx 3040.2\text{ km}$$

50. Time of flight = 38 hr 23 min

51. Cruising altitude

$$\frac{1000\cancel{\text{ft}}}{1}\cdot\frac{0.30\text{ m}}{1\cancel{\text{ft}}}\approx 1000(0.30\text{ m}) = 300\text{ m}$$

52. Fuel at the start

$$\frac{1\cancel{\text{gal}}}{1}\cdot\frac{3.78\cancel{\text{L}}}{1\cancel{\text{gal}}}\cdot\frac{1000\text{ mL}}{1\cancel{\text{L}}}\approx(3.78)(1000\text{ mL})$$
$$= 3780\text{ mL}$$

Since it was "less than a gallon of fuel," there would be less than 3780 mL.

53. Fuel left after landing

$$\frac{\overset{1}{\cancel{2}}\,\cancel{\text{fl oz}}}{1}\cdot\frac{1\cancel{\text{qt}}}{\underset{16}{\cancel{32}}\,\cancel{\text{fl oz}}}\cdot\frac{0.95\cancel{\text{L}}}{1\cancel{\text{qt}}}\cdot\frac{1000\text{ mL}}{1\cancel{\text{L}}}$$
$$=\frac{950}{16}=59.375\approx 59.4\text{ mL}$$

54. $$\frac{1\cancel{\text{gal}}}{1}\cdot\frac{4\cancel{\text{qt}}}{1\cancel{\text{gal}}}\cdot\frac{32\text{ fl oz}}{1\cancel{\text{qt}}}=128\text{ fl oz}$$
$$\frac{2\text{ fl oz}}{128\text{ fl oz}}=\frac{1}{64}=0.015625\approx 1.6\%$$

Chapter 8 Review Exercises

1. 1 lb = <u>16</u> oz

2. <u>3</u> ft = 1 yd

3. 1 T = <u>2000</u> lb

4. <u>4</u> qt = 1 gal

5. 1 hr = <u>60</u> min

6. 1 c = <u>8</u> fl oz

7. <u>60</u> sec = 1 min

8. <u>5280</u> ft = 1 mi

9. <u>12</u> in. = 1 ft

10. 4 ft to inches

$$\frac{4\cancel{\text{ft}}}{1}\cdot\frac{12\text{ in.}}{1\cancel{\text{ft}}}=4\cdot 12\text{ in.}=48\text{ in.}$$

11. 6000 lb to tons

$$\frac{\overset{3}{\cancel{6000}}\,\cancel{\text{lb}}}{1}\cdot\frac{1\text{ T}}{\underset{1}{\cancel{2000}}\,\cancel{\text{lb}}}=3\text{ T}$$

12. 64 oz to pounds

$$\frac{\overset{4}{\cancel{64}}\,\cancel{\text{oz}}}{1}\cdot\frac{1\text{ lb}}{\underset{1}{\cancel{16}}\,\cancel{\text{oz}}}=4\text{ lb}$$

13. 18 hr to days

$$\frac{\overset{3}{\cancel{18}}\,\cancel{\text{hr}}}{1}\cdot\frac{1\text{ day}}{\underset{4}{\cancel{24}}\,\cancel{\text{hr}}}=\frac{3}{4}\text{ or }0.75\text{ day}$$

14. 150 min to hours

$$\frac{\overset{5}{\cancel{150}}\,\cancel{\text{min}}}{1}\cdot\frac{1\text{ hr}}{\underset{2}{\cancel{60}}\,\cancel{\text{min}}}=\frac{5}{2}\text{ hr}=2\frac{1}{2}\text{ or }2.5\text{ hr}$$

15. $1\frac{3}{4}$ lb to ounces

$$\frac{1\frac{3}{4}\cancel{\text{lb}}}{1}\cdot\frac{16\text{ oz}}{1\cancel{\text{lb}}}=\frac{7}{\underset{1}{\cancel{4}}}\cdot\frac{\overset{4}{\cancel{16}}}{1}\text{ oz}=28\text{ oz}$$

16. $6\frac{1}{2}$ ft to inches

$$\frac{6\frac{1}{2}\cancel{\text{ft}}}{1}\cdot\frac{12\text{ in.}}{1\cancel{\text{ft}}}=\frac{13}{\underset{1}{\cancel{2}}}\cdot\frac{\overset{6}{\cancel{12}}}{1}\text{ in.}=78\text{ in.}$$

17. 7 gal to cups

$$\frac{7\cancel{\text{gal}}}{1}\cdot\frac{4\cancel{\text{qt}}}{1\cancel{\text{gal}}}\cdot\frac{2\cancel{\text{pt}}}{1\cancel{\text{qt}}}\cdot\frac{2\text{ c}}{1\cancel{\text{pt}}}=7\cdot 4\cdot 2\cdot 2\text{ c}=112\text{ c}$$

18. 4 days to seconds

$$\frac{4\cancel{\text{days}}}{1}\cdot\frac{24\cancel{\text{hr}}}{1\cancel{\text{day}}}\cdot\frac{60\cancel{\text{min}}}{1\cancel{\text{hr}}}\cdot\frac{60\text{ sec}}{1\cancel{\text{min}}}$$
$$=4\cdot 24\cdot 60\cdot 60\text{ sec}$$
$$=345{,}600\text{ sec}$$

19. **(a)** 12,460 ft to yards

$$\frac{12{,}460\ \cancel{ft}}{1} \cdot \frac{1\ \text{yd}}{3\ \cancel{ft}} = \frac{12{,}460}{3}\ \text{yd}$$
$$= 4153\tfrac{1}{3}\ \text{yd (exact)}$$
or 4153.3 yd (rounded).

The average depth is $4153\frac{1}{3}$ yd (exact) or 4153.3 yd (rounded).

(b) 12,460 ft to miles

$$\frac{12{,}460\ \cancel{ft}}{1} \cdot \frac{1\ \text{mi}}{5280\ \cancel{ft}} \approx 2.36\ \text{mi} \approx 2.4\ \text{mi}$$

The average depth is 2.4 mi (rounded).

20. *Step 1*
The problem asks for the amount of money the company made.

Step 2
Convert pounds to tons. Then multiply to find the total amount.

Step 3
To estimate, round 123,260 lb to 100,000 lb. Then, there are 2000 lb in a ton, so 100,000 lb is 50 T. So $50 \cdot \$40 = \$2000$ is our estimate.

Step 4
$$\frac{123{,}260\ \cancel{lb}}{1} \cdot \frac{1\ \text{T}}{2000\ \cancel{lb}} = 61.63\ \text{T}$$

$61.63 \cdot \$40 = \$2465.20$

Step 5
The company made $2465.20.

Step 6
The answer, $2465.20, is close to our estimate of $2000.

21. My thumb is 20 <u>mm</u> wide.

22. Her waist measurement is 66 <u>cm</u>.

23. The two towns are 40 <u>km</u> apart.

24. A basketball court is 30 <u>m</u> long.

25. The height of the picnic bench is 45 <u>cm</u>.

26. The eraser on the end of my pencil is 5 <u>mm</u> long.

27. 5 m to cm
Count 2 places to the *right* on the metric conversion line.
5.00‸ m = 500 cm

28. 8.5 km to m

$$\frac{8.5\ \cancel{km}}{1} \cdot \frac{1000\ \text{m}}{1\ \cancel{km}} = 8500\ \text{m}$$

29. 85 mm to cm
Count 1 place to the *left* on the metric conversion line.
8‸5. mm = 8.5 cm

30. 370 cm to m

$$\frac{370\ \cancel{cm}}{1} \cdot \frac{1\ \text{m}}{100\ \cancel{cm}} = 3.7\ \text{m}$$

31. 70 m to km
Count 3 places to the *left* on the metric conversion line.
‸070. m = 0.07 km

32. 0.93 m to mm

$$\frac{0.93\ \cancel{m}}{1} \cdot \frac{1000\ \text{mm}}{1\ \cancel{m}} = 930\ \text{mm}$$

33. The eye dropper holds 1 <u>mL</u>.

34. I can heat 3 <u>L</u> of water in this pan.

35. Loretta's hammer weighed 650 <u>g</u>.

36. Yongshu's suitcase weighed 20 <u>kg</u> when it was packed.

37. My fish tank holds 80 <u>L</u> of water.

38. I'll buy the 500 <u>mL</u> bottle of mouthwash.

39. Mara took a 200 <u>mg</u> antibiotic pill.

40. This piece of chicken weighs 100 <u>g</u>.

41. 5000 mL to L
Count 3 places to the *left* on the metric conversion line.
5‸000. mL = 5 L

42. 8 L to mL
Count 3 places to the *right* on the metric conversion line.
8.000‸ L = 8000 mL

43. 4.58 g to mg

$$\frac{4.58\ \cancel{g}}{1} \cdot \frac{1000\ \text{mg}}{1\ \cancel{g}} = 4580\ \text{mg}$$

44. 0.7 kg to g

$$\frac{0.7\ \cancel{kg}}{1} \cdot \frac{1000\ \text{g}}{1\ \cancel{kg}} = 700\ \text{g}$$

45. 6 mg to g
Count 3 places to the *left* on the metric conversion line.
‸006. mg = 0.006 g

46. 35 mL to L
Count 3 places to the *left* on the metric conversion line.
$_\wedge 035.\ \text{mL} = 0.035\ \text{L}$

47. Convert milliliters to liters.

$$\frac{180\ \cancel{\text{mL}}}{1} \cdot \frac{1\ \text{L}}{1000\ \cancel{\text{mL}}} = \frac{180}{1000}\ \text{L} = 0.18\ \text{L}$$

Multiply 0.18 L by the number of servings.
$(0.18\ \text{L})(175) = 31.5\ \text{L}$

For 175 servings, 31.5 L of punch are needed.

48. Convert kilograms to grams
10 kg = 10,000 g
Divide by the number of people.

$$\frac{10{,}000\ \text{g}}{28\ \text{people}} \approx 357$$

Jason is allowing 357 g (rounded) of turkey for each person.

49. Convert grams to kilograms.
Using the metric conversion line,
4 kg 750 g = 4.75 kg.
Subtract the weight loss from his original weight.

$$\begin{array}{r l} 92.00 & \text{kg} \\ -\ 4.75 & \text{kg} \\ \hline 87.25 & \text{kg} \end{array}$$

Yerald weighs 87.25 kg.

50. Convert to kilograms.
Using the metric conversion line,
750 g = 0.75 kg.
Multiply by the price per kilogram.

$$\begin{array}{r l} \$1.49 & \leftarrow \textit{2 decimal places} \\ \times\ 0.75 & \leftarrow \textit{2 decimal places} \\ \hline 745 & \\ 1\ 043 & \\ \hline \$1.1175 & \leftarrow \textit{4 decimal places} \end{array}$$

Young-Mi paid $1.12 (rounded) for the onions.

51. 6 m to yards

$$\frac{6\ \cancel{\text{m}}}{1} \cdot \frac{1.09\ \text{yd}}{1\ \cancel{\text{m}}} \approx 6.5\ \text{yd}$$

52. 30 cm to inches

$$\frac{30\ \cancel{\text{cm}}}{1} \cdot \frac{0.39\ \text{in.}}{1\ \cancel{\text{cm}}} \approx 11.7\ \text{in.}$$

53. 108 km to miles

$$\frac{108\ \cancel{\text{km}}}{1} \cdot \frac{0.62\ \text{mi}}{1\ \cancel{\text{km}}} \approx 67.0\ \text{mi}$$

54. 800 miles to km

$$\frac{800\ \cancel{\text{mi}}}{1} \cdot \frac{1.61\ \text{km}}{1\ \cancel{\text{mi}}} \approx 1288\ \text{km}$$

55. 23 quarts to L

$$\frac{23\ \cancel{\text{qt}}}{1} \cdot \frac{0.95\ \text{L}}{1\ \cancel{\text{qt}}} \approx 21.9\ \text{L}$$

56. 41.5 L to quarts

$$\frac{41.5\ \cancel{\text{L}}}{1} \cdot \frac{1.06\ \text{qt}}{1\ \cancel{\text{L}}} \approx 44.0\ \text{qt}$$

57. Water freezes at <u>0 °C</u>.

58. Water boils at <u>100 °C</u>.

59. Normal body temperature is about <u>37 °C</u>.

60. Comfortable room temperature is about <u>20 °C</u>.

61. 77 °F

$$C = \frac{5(77-32)}{9} = \frac{5(\overset{5}{\cancel{45}})}{\underset{1}{\cancel{9}}} = 25$$

77 °F = 25 °C

62. 5 °F

$$C = \frac{5(5-32)}{9} = \frac{5(-27)}{9} = \frac{-5 \cdot 3 \cdot \overset{1}{\cancel{9}}}{\underset{1}{\cancel{9}}} = -15$$

5 °F = −15 °C

63. −2 °C

$$F = \frac{9(-2)}{5} + 32 = \frac{-18}{5} + 32 = -3.6 + 32 = 28.4$$

−2 °C ≈ 28 °F

64. 49 °C

$$F = \frac{9 \cdot 49}{5} + 32 = 88.2 + 32 = 120.2 \approx 120$$

49 °C ≈ 120 °F

65. **[8.3]** I added 1 <u>L</u> of oil to my car.

66. **[8.3]** The box of books weighed 15 <u>kg</u>.

67. **[8.2]** Larry's shoe is 30 <u>cm</u> long.

68. **[8.3]** Jan used 15 <u>mL</u> of shampoo on her hair.

69. **[8.2]** My fingernail is 10 <u>mm</u> wide.

70. **[8.2]** I walked 2 <u>km</u> to school.

71. **[8.3]** The tiny bird weighed 15 <u>g</u>.

72. **[8.2]** The new library building is 18 <u>m</u> wide.

73. **[8.3]** The cookie recipe uses 250 <u>mL</u> of milk.

74. **[8.3]** Renee's pet mouse weighs 30 g.

75. **[8.3]** One postage stamp weighs 90 mg.

76. **[8.3]** I bought 30 L of gas for my car.

77. **[8.2]** 10.5 cm to mm
Count 1 place to the *right* on the metric conversion line.
10.5∧ cm = 105 mm

78. **[8.1]** 45 min to hours

$$\frac{\overset{3}{\cancel{45}}\ \cancel{\text{min}}}{1} \cdot \frac{1\text{ hr}}{\underset{4}{\cancel{60}}\ \cancel{\text{min}}} = \frac{3}{4} \text{ or } 0.75 \text{ hr}$$

79. **[8.1]** 90 in. to feet

$$\frac{\overset{15}{\cancel{90}}\ \cancel{\text{in.}}}{1} \cdot \frac{1\text{ ft}}{\underset{2}{\cancel{12}}\ \cancel{\text{in.}}} = \frac{15}{2}\text{ ft} = 7\frac{1}{2} \text{ or } 7.5 \text{ ft}$$

80. **[8.2]** 1.3 m to cm
Count 2 places to the *right* on the metric conversion line.
1.30∧ m = 130 cm

81. **[8.5]** 25 °C to Fahrenheit

$$F = \frac{9 \cdot \overset{5}{\cancel{25}}}{\underset{1}{\cancel{5}}} + 32 = 45 + 32 = 77$$

25 °C = 77 °F

82. **[8.1]** $3\frac{1}{2}$ gal to quarts

$$\frac{3\frac{1}{2}\ \cancel{\text{gal}}}{1} \cdot \frac{4\text{ qt}}{1\ \cancel{\text{gal}}} = \frac{7}{\underset{1}{\cancel{2}}} \cdot \frac{\overset{2}{\cancel{4}}}{1}\text{ qt} = 14 \text{ qt}$$

83. **[8.3]** 700 mg to g
Count 3 places to the *left* on the metric conversion line.
∧700. mg = 0.7 g

84. **[8.3]** 0.81 L to mL
Count 3 places to the *right* on the metric conversion line.
0.810∧ L = 810 mL

85. **[8.1]** 5 lb to ounces

$$\frac{5\ \cancel{\text{lb}}}{1} \cdot \frac{16\text{ oz}}{1\ \cancel{\text{lb}}} = 5 \cdot 16 \text{ oz} = 80 \text{ oz}$$

86. **[8.3]** 60 kg to g

$$\frac{60\ \cancel{\text{kg}}}{1} \cdot \frac{1000\text{ g}}{1\ \cancel{\text{kg}}} = 60{,}000 \text{ g}$$

87. **[8.3]** 1.8 L to mL
Count 3 places to the *right* on the metric conversion line.
1.800∧ L = 1800 mL

88. **[8.5]** 30 °F to Celsius

$$C = \frac{5(30-32)}{9} = \frac{5(-2)}{9} = \frac{-10}{9} \approx -1.1$$

30 °F ≈ −1 °C

89. **[8.2]** 0.36 m to cm
Count 2 places to the *right* on the metric conversion line.
0.36∧ m = 36 cm

90. **[8.3]** 55 mL to L
Count 3 places to the *left* on the metric conversion line.
∧055. mL = 0.055 L

91. **[8.4]** Convert centimeters to meters.

$$\begin{aligned} 2\text{ m } 4\text{ cm} &= 2 + 0.04\text{ m} = & 2.04\text{ m} \\ 78\text{ cm} &= & -\ 0.78\text{ m} \\ & & \overline{1.26\text{ m}} \end{aligned}$$

The board is 1.26 m long.

92. **[8.1]** 3000 lb to tons

$$\frac{3000\ \cancel{\text{lb}}}{1} \cdot \frac{1\text{ T}}{2000\ \cancel{\text{lb}}} = \frac{3}{2} \text{ or } 1.5 \text{ T}$$

Multiply by 12 days.
(1.5 T)(12) = 18 T

18 T of cookies are sold in all.

93. **[8.5]** Convert ounces to grams.

$$\frac{4\ \cancel{\text{oz}}}{1} \cdot \frac{28.35\text{ g}}{1\ \cancel{\text{oz}}} \approx 113 \text{ g}$$

350 °F to Celsius

$$C = \frac{5(350-32)}{9} = \frac{5(318)}{9} = \frac{1590}{9} \approx 177\ °\text{C}$$

94. **[8.5]** Convert kilograms to pounds.

$$\frac{80.9\ \cancel{\text{kg}}}{1} \cdot \frac{2.20\text{ lb}}{1\ \cancel{\text{kg}}} \approx 178.0 \text{ lb}$$

Convert meters to feet.

$$\frac{1.83\ \cancel{\text{m}}}{1} \cdot \frac{3.28\text{ ft}}{1\ \cancel{\text{m}}} \approx 6.0 \text{ ft}$$

Jalo weighs about 178.0 lb and is about 6.0 ft.

95. **[8.5]** 44 ft to meters

$$\frac{44\ \cancel{\text{ft}}}{1} \cdot \frac{0.30\text{ m}}{1\ \cancel{\text{ft}}} \approx 13.2 \text{ m}$$

96. **[8.5]** 6.7 m to feet

$$\frac{6.7\ \cancel{m}}{1} \cdot \frac{3.28\ \text{ft}}{1\ \cancel{m}} \approx 22.0\ \text{ft}$$

97. **[8.5]** 4 yd to meters

$$\frac{4\ \cancel{yd}}{1} \cdot \frac{0.91\ \text{m}}{1\ \cancel{yd}} \approx 3.6\ \text{m}$$

98. **[8.5]** 102 cm to inches

$$\frac{102\ \cancel{cm}}{1} \cdot \frac{0.39\ \text{in.}}{1\ \cancel{cm}} \approx 39.8\ \text{in.}$$

99. **[8.5]** 220,000 kg to pounds

$$\frac{220{,}000\ \cancel{kg}}{1} \cdot \frac{2.20\ \text{lb}}{1\ \cancel{kg}} \approx 484{,}000\ \text{lb}$$

100. **[8.5]** 5 gal to liters

$$\frac{5\ \cancel{gal}}{1} \cdot \frac{3.78\ \text{L}}{1\ \cancel{gal}} \approx 18.9\ \text{L}$$

6 gal to liters

$$\frac{6\ \cancel{gal}}{1} \cdot \frac{3.78\ \text{L}}{1\ \cancel{gal}} \approx 22.7\ \text{L}$$

101. **[8.1]** **(a)** From Exercise 99:

$$\frac{484{,}000\ \cancel{lb}}{1} \cdot \frac{1\ \text{T}}{2000\ \cancel{lb}} = 242\ \text{T}$$

(b) From Exercise 97:

$$\frac{4\ \cancel{yd}}{1} \cdot \frac{3\ \cancel{ft}}{1\ \cancel{yd}} \cdot \frac{12\ \text{in.}}{1\ \cancel{ft}} = 144\ \text{in.}$$

102. **[8.2]** **(a)** From Exercise 98:

$$\frac{102\ \cancel{cm}}{1} \cdot \frac{1\ \text{m}}{100\ \cancel{cm}} = 1.02\ \text{m}$$

(b) From Exercise 96:

$$\frac{6.7\ \cancel{m}}{1} \cdot \frac{100\ \text{cm}}{1\ \cancel{m}} = 670\ \text{cm}$$

Chapter 8 Test

1. 9 gal to quarts

$$\frac{9\ \cancel{gal}}{1} \cdot \frac{4\ \text{qt}}{1\ \cancel{gal}} = 36\ \text{qt}$$

2. 45 ft to yards

$$\frac{45\ \cancel{ft}}{1} \cdot \frac{1\ \text{yd}}{3\ \cancel{ft}} = \frac{45}{3}\ \text{yd} = 15\ \text{yd}$$

3. 135 min to hours

$$\frac{\overset{27}{\cancel{135}}\ \cancel{min}}{1} \cdot \frac{1\ \text{hr}}{\underset{12}{\cancel{60}}\ \cancel{min}} = \frac{27}{12}\ \text{hr} = 2.25 \text{ or } 2\frac{1}{4}\ \text{hr}$$

4. 9 in. to feet

$$\frac{\overset{3}{\cancel{9}}\ \cancel{in.}}{1} \cdot \frac{1\ \text{ft}}{\underset{4}{\cancel{12}}\ \cancel{in.}} = \frac{3}{4} \text{ or } 0.75\ \text{ft}$$

5. $3\frac{1}{2}$ lb to ounces

$$\frac{3\frac{1}{2}\ \cancel{lb}}{1} \cdot \frac{16\ \text{oz}}{1\ \cancel{lb}} = \frac{7}{\underset{1}{\cancel{2}}} \cdot \frac{\overset{8}{\cancel{16}}}{1}\ \text{oz} = 56\ \text{oz}$$

6. 5 days to minutes

$$\frac{5\ \cancel{days}}{1} \cdot \frac{24\ \cancel{hr}}{1\ \cancel{day}} \cdot \frac{60\ \text{min}}{1\ \cancel{hr}} = \frac{5 \cdot 24 \cdot 60}{1}\ \text{min} = 7200\ \text{min}$$

7. My husband weighs 75 kg.

8. I hiked 5 km this morning.

9. She bought 125 mL of cough syrup.

10. This apple weighs 180 g.

11. This page is 21 cm wide.

12. My watch band is 10 mm wide.

13. I bought 10 L of soda for the picnic.

14. The bracelet is 16 cm long.

15. 250 cm to meters
Count 2 places to the *left* on the metric conversion line.
2∧50. cm = 2.5 m

16. 4.6 km to meters

$$\frac{4.6\ \cancel{km}}{1} \cdot \frac{1000\ \text{m}}{1\ \cancel{km}} = 4600\ \text{m}$$

17. 5 mm to centimeters
Count 1 place to the *left* on the metric conversion line.
∧5. mm = 0.5 cm

18. 325 mg to grams

$$\frac{325\ \cancel{mg}}{1} \cdot \frac{1\ \text{g}}{1000\ \cancel{mg}} = \frac{325}{1000}\ \text{g} = 0.325\ \text{g}$$

19. 16 L to milliliters
Count 3 places to the *right* on the metric conversion line.
16.000∧ L = 16,000 mL

20. 0.4 kg to grams

$$\frac{0.4\ \cancel{kg}}{1} \cdot \frac{1000\ \text{g}}{1\ \cancel{kg}} = 400\ \text{g}$$

21. 10.55 m to centimeters
Count 2 places to the *right* on the metric conversion line.
10.55‸ m = 1055 cm

22. 95 mL to liters
Count 3 places to the *left* on the metric conversion line.
‸095. mL = 0.095 L

23. Convert 460 in. to feet.

$$\frac{460\ \cancel{\text{in}}}{1} \cdot \frac{1\text{ ft}}{12\ \cancel{\text{in}}} = \frac{460}{12}\text{ ft} = \frac{115}{3}\text{ ft}$$

Divide by 12 months

$$\frac{\frac{115}{3}\text{ ft}}{12\text{ months}} = \frac{115}{3} \cdot \frac{1}{12} \approx 3.2\text{ ft/month}$$

It rains about 3.2 ft per month.

24. **(a)** Convert 2.9 g to mg.

$$\frac{2.9\ \cancel{\text{g}}}{1} = \frac{1000\text{ mg}}{1\ \cancel{\text{g}}} = 2900\text{ mg}$$

Find the difference.
2900 mg − 590 mg = 2310 mg
It has 2310 mg more sodium.

(b) 2900 mg is more than 2400 mg
2900 mg − 2400 mg = 500 mg

$$\frac{500\ \cancel{\text{mg}}}{1} \cdot \frac{1\text{ g}}{1000\ \cancel{\text{mg}}} = 0.5\text{ g}$$

The "Super Melt" has 500 mg or 0.5 g more sodium than the 2400 mg recommended daily amount.

25. The water is almost boiling.
On the Celsius scale, water boils at 100° so, 95 °C would be almost boiling.

26. The tomato plants may freeze tonight.
Water freezes at 0 °C.

27. 6 ft to meters

$$\frac{6\ \cancel{\text{ft}}}{1} \cdot \frac{0.30\text{ m}}{1\ \cancel{\text{ft}}} \approx 1.8\text{ m}$$

28. 125 lb to kilograms

$$\frac{125\ \cancel{\text{lb}}}{1} \cdot \frac{0.45\text{ kg}}{1\ \cancel{\text{lb}}} \approx 56.3\text{ kg}$$

29. 50 L to gallons

$$\frac{50\ \cancel{\text{L}}}{1} \cdot \frac{0.26\text{ gal}}{1\ \cancel{\text{L}}} \approx 13\text{ gal}$$

30. 8.1 km to miles

$$\frac{8.1\ \cancel{\text{km}}}{1} \cdot \frac{0.62\text{ mi}}{1\ \cancel{\text{km}}} \approx 5.0\text{ mi}$$

31. 74 °F to Celsius

$$C = \frac{5(74-32)}{9} = \frac{5(\overset{14}{\cancel{42}})}{\underset{3}{\cancel{9}}} = \frac{70}{3} \approx 23\text{ °C}$$

74 °F ≈ 23 °C

32. −12 °C to Fahrenheit

$$F = \frac{9(-12)}{5} + 32 = \frac{-108}{5} + 32 = -21.6 + 32$$
$$= 10.4$$

−12 °C ≈ 10 °F

33. Convert 1 m 20 cm to meters.

$$1\text{ m} + 20\text{ cm} = 1\text{ m} + \frac{20\ \cancel{\text{cm}}}{1} \cdot \frac{1\text{ m}}{100\ \cancel{\text{cm}}}$$
$$= 1\text{ m} + 0.2\text{ m} = 1.2\text{ m}$$

Multiply to find the number of meters for 5 pillows.
5(1.2 m) = 6 m

Convert 6 m to yards.

$$\frac{6\ \cancel{\text{m}}}{1} \cdot \frac{1.09\text{ yd}}{1\ \cancel{\text{m}}} \approx 6.54\text{ yd}$$

Multiply by \$3.98 to find the cost.

$$\frac{6.54\ \cancel{\text{yd}}}{1} \cdot \frac{\$3.98}{\cancel{\text{yd}}} \approx \$26.03$$

It will cost about \$26.03.

34. Possible answers: Use same system as rest of the world; easier system for children to learn; less use of fractional numbers; compete internationally.

Cumulative Review Exercises (Chapters 1–8)

1. **(a)** In words, 603,005,040,000 is six hundred three billion, five million, forty thousand.

(b) In words, 9.040 is nine and forty thousandths.

2. **(a)** Written in digits, eighty and eight hundredths is 80.08.

(b) Written in digits, two hundred million, sixty-five thousand, four is 200,065,004.

3. **(a) 0.9802 to the nearest tenth**

Underline the tenths place: 0.$\underline{9}$802

The next digit is 5 or more. Add 1 to 9. Write 0 and regroup the 1 to the ones place. Drop all digits to the right of the underlined place. **1.0**

(b) 495 to the nearest ten

Underline the tens place: 4$\underline{9}$5

The next digit is 5 or more. Add 1 to 9. Write 0 and regroup the 1 to the hundreds place. Change 5 to 0. **500**

(c) 306,472,000 to the nearest million

Underline the millions place: 30$\underline{6}$,472,000

The next digit is 4 or less. Leave 6. Change all digits to the right of the underlined place to zero.

306,000,000

4. $\text{Mean} = \dfrac{\text{sum of all values}}{\text{number of values}}$

$$= \frac{(22+18+40+18 + 20+21+45+25)}{8}$$

$$= \frac{209}{8} = 26.125 \text{ hours}$$

The mean is 26.1 hours (rounded).

Median: arrange the data from smallest to largest

$$18, 18, 20, \underline{21}, \underline{22}, 25, 40, 45$$

Since there is an even number of values, the median is the average of the middle values.

$$\text{Median} = \frac{21+22}{2} = 21.5 \text{ hours}$$

5. $(-2)^3 - 3^2$

$$= (-2)(-2)(-2) - (3)(3)$$
$$= -8 - 9$$
$$= -8 + (-9)$$
$$= -17$$

6. $2\frac{2}{5} - \frac{3}{4} = \frac{12}{5} - \frac{3}{4} = \frac{48}{20} - \frac{15}{20} = \frac{48-15}{20}$

$$= \frac{33}{20} \text{ or } 1\frac{13}{20}$$

7. $\dfrac{-4}{0.16}$ Different signs, quotient is negative.

$$0.16\overline{)4.\,0\,0} \rightarrow 16\overline{)4\,0\,0}$$

$$\begin{array}{r} 25 \\ 16\overline{)400} \\ \underline{32} \\ 80 \\ \underline{80} \\ 0 \end{array}$$

Answer: -25

8. $\dfrac{7}{6x^2} \cdot \dfrac{9x}{14y} = \dfrac{\overset{1}{\cancel{7}} \cdot \overset{1}{\cancel{3}} \cdot 3 \cdot \overset{1}{\cancel{x}}}{2 \cdot \underset{1}{\cancel{3}} \cdot \underset{1}{\cancel{x}} \cdot x \cdot 2 \cdot \underset{1}{\cancel{7}} \cdot y} = \dfrac{3}{4xy}$

9. $(-0.003)(-0.05)$ Same signs, product is positive

$$\begin{array}{rcl} 0.003 & \leftarrow & \textit{3 decimal places} \\ \times\, 0.05 & \leftarrow & \textit{2 decimal places} \\ \hline 0.00015 & \leftarrow & \textit{5 decimal places} \end{array}$$

Answer: 0.00015

10. $0.083 - 42$
$$= 0.083 + (-42)$$

Find the absolute values.

$$|0.083| = 0.083; |-42| = 42$$

-42 is larger in absolute value so the answer will be negative.

$$\text{Subtract: } \begin{array}{r} \overset{1}{4}\,\overset{1}{\cancel{2}}.\overset{9}{\cancel{10}}\,\overset{9}{\cancel{10}}\,\overset{10}{\cancel{0}} \\ -\;0.0\,8\,3 \\ \hline 41.9\,1\,7 \end{array}$$

Answer: -41.917

11. $\dfrac{3}{c} - \dfrac{5}{6} = \dfrac{3 \cdot 6}{c \cdot 6} - \dfrac{5 \cdot c}{6 \cdot c} = \dfrac{18}{6c} - \dfrac{5c}{6c} = \dfrac{18-5c}{6c}$

12. $-15 - 15$
$$= -15 + (-15)$$
$$= -30$$

13. $10 - 3(6-7)$
$$= 10 - 3(-1)$$
$$= 10 - (-3)$$
$$= 10 + 3$$
$$= 13$$

14. $\dfrac{-3(-4)}{27-3^3} = \dfrac{12}{27-27} = \dfrac{12}{0}$, which is undefined.

15. $\dfrac{3}{5} \text{ of } (-400) = \dfrac{3}{5} \cdot \dfrac{-400}{1} = -\dfrac{3 \cdot \overset{1}{\cancel{5}} \cdot 80}{\underset{1}{\cancel{5}} \cdot 1}$

$$= -\frac{240}{1} = -240$$

16. $3\frac{7}{12} - 4 = \dfrac{43}{12} - \dfrac{4}{1} = \dfrac{43}{12} - \dfrac{48}{12}$

$$= \frac{43-48}{12} = \frac{-5}{12} = -\frac{5}{12}$$

17. $\dfrac{9y^2}{8x} \div \dfrac{y}{6x^2} = \dfrac{9y^2}{8x} \cdot \dfrac{6x^2}{y}$

$$= \frac{3 \cdot 3 \cdot \overset{1}{\cancel{y}} \cdot y \cdot \overset{1}{\cancel{2}} \cdot 3 \cdot \overset{1}{\cancel{x}} \cdot x}{\underset{1}{\cancel{2}} \cdot 4 \cdot \underset{1}{\cancel{x}} \cdot \underset{1}{\cancel{y}}}$$

$$= \frac{27xy}{4}$$

18. $\dfrac{2}{3} + \dfrac{n}{m} = \dfrac{2 \cdot m}{3 \cdot m} + \dfrac{n \cdot 3}{m \cdot 3} = \dfrac{2m}{3m} + \dfrac{3n}{3m}$

$$= \frac{2m+3n}{3m}$$

19. $1\frac{1}{6}+1\frac{2}{3}=\frac{7}{6}+\frac{5}{3}=\frac{7}{6}+\frac{10}{6}=\frac{7+10}{6}$
$=\frac{17}{6}$ or $2\frac{5}{6}$

20. $\frac{\frac{14}{15}}{-6}=\frac{14}{15}\div(-6)=\frac{14}{15}\div\left(-\frac{6}{1}\right)$
$=\frac{14}{15}\cdot\left(-\frac{1}{6}\right)$
$=-\frac{\overset{1}{\cancel{2}}\cdot 7\cdot 1}{15\cdot \underset{1}{\cancel{2}}\cdot 3}$
$=-\frac{7}{45}$

21. $\frac{12\div(2-5)+12(-1)}{2^3-(-4)^2}$

Numerator: $12\div(2-5)+12(-1)$
$=12\div(2+(-5))+12(-1)$
$=12\div(-3)+(-12)$
$=-4+(-12)$
$=-16$

Denominator: $2^3-(-4)^2$
$=8-16$
$=8+(-16)$
$=-8$

Last step is division: $\frac{-16}{-8}=2$

22. $(-0.8)^2\div(0.8-1)$
$=(-0.8)\cdot(-0.8)\div(0.8+(-1))$
$=0.64\div(-0.2)$
$=-3.2$

Last division step:

$$\begin{array}{r} 3.2 \\ 0.2_\wedge\overline{)0.6_\wedge 4} \\ \underline{6} \\ 0\ 4 \\ \underline{4} \\ 0 \end{array}$$

23. $\left(-\frac{1}{3}\right)^2-\frac{1}{4}\left(\frac{4}{9}\right)$
$=\left(-\frac{1}{3}\right)\left(-\frac{1}{3}\right)-\frac{1\cdot\overset{1}{\cancel{4}}}{\underset{1}{\cancel{4}}\cdot 9}$
$=\frac{1}{9}-\frac{1}{9}$
$=0$

24. $-3k+4$ Replace k with -6.
$=-3\cdot(-6)+4$
$=18+4$
$=22$

25. $4k-5n$ Replace k with -6 and n with 2.
$=4(-6)-5(2)$
$=-24-10$
$=-24+(-10)$
$=-34$

26. $3m^3n$ Replace m with -1 and n with 2.
$=3\cdot m\cdot m\cdot m\cdot n$
$=(3)(-1)(-1)(-1)(2)$
$=-6$

27. $3p-3p^2-4p$
$=3p+(-3p^2)+(-4p)$ Combine like terms.
$=-1p+(-3p^2)$
$=-3p^2-p$

28. $-5(x+2)-4$
$=-5x-10-4$
$=-5x+(-10)+(-4)$
$=-5x+(-14)$ or $-5x-14$

29. $7(-2r^3)$
$=(7(-2))r^3$
$=-14r^3$

30.
$$\begin{aligned} 2b+2 &= -5+5 \\ 2b+2 &= 0 \\ \underline{-2} &\quad \underline{-2} \\ 2b+0 &= -2 \\ \frac{2b}{2} &= \frac{-2}{2} \\ b &= -1 \end{aligned}$$

The solution is -1.

31.
$$\begin{aligned} 12.92-a &= 4.87 \\ 12.92-1a &= 4.87 \\ \underline{-12.92} &\quad \underline{-12.92} \\ 0-1a &= -8.05 \\ \frac{-1a}{-1} &= \frac{-8.05}{-1} \\ a &= 8.05 \end{aligned}$$

The solution is 8.05.

32.
$$\begin{aligned} 7(t+6) &= 42 \\ 7t+42 &= 42 \\ \underline{-42} &\quad \underline{-42} \\ 7t+0 &= 0 \\ \frac{7t}{7} &= \frac{0}{7} \\ t &= 0 \end{aligned}$$

The solution is 0.

33.
$$\begin{aligned} -5n &= n - 12 \\ -5n &= 1n - 12 \\ \underline{-1n} \quad & \underline{-1n} \\ -6n &= 0 - 12 \\ \frac{-6n}{-6} &= \frac{-12}{-6} \\ n &= 2 \end{aligned}$$

The solution is 2.

34.
$$\begin{aligned} 2 &= \frac{1}{4}w - 3 \\ \underline{+3} \quad & \underline{\quad +3} \\ 5 &= \frac{1}{4}w + 0 \\ \frac{4}{1}(5) &= \frac{4}{1}\left(\frac{1}{4}w\right) \\ 20 &= w \end{aligned}$$

The solution is 20.

35.
$$\begin{aligned} \frac{1.5}{45} &= \frac{x}{12} \\ 45 \cdot x &= 1.5 \cdot 12 \\ \frac{45x}{45} &= \frac{18}{45} \\ x &= 0.4 \end{aligned}$$

The solution is 0.4.

36.
$$\begin{aligned} 4y - 3 &= 7y + 12 \\ \underline{-4y} \quad & \quad \underline{-4y} \\ 0 - 3 &= 3y + 12 \\ -3 &= 3y + 12 \\ \underline{-12} \quad & \quad \underline{-12} \\ -15 &= 3y + 0 \\ \frac{-15}{3} &= \frac{3y}{3} \\ -5 &= y \end{aligned}$$

The solution is −5.

37.
$$\begin{aligned} 3(k - 6) - 4 &= -2(k + 1) \\ 3k - 18 - 4 &= -2k - 2 \\ 3k - 22 &= -2k - 2 \\ \underline{+2k} \quad & \quad \underline{+2k} \\ 5k - 22 &= 0 - 2 \\ 5k - 22 &= -2 \\ \underline{+22} \quad & \quad \underline{+22} \\ 5k + 0 &= 20 \\ \frac{5k}{5} &= \frac{20}{5} \\ k &= 4 \end{aligned}$$

The solution is 4.

38. Let n represent the unknown number.

$$\begin{aligned} 11n - 8n &= -9 \\ 3n &= -9 \\ \frac{3n}{3} &= \frac{-9}{3} \\ n &= -3 \end{aligned}$$

The number is −3.

39. Let n represent the unknown number.

$$\begin{aligned} 2n - 8 &= n + 7 \\ 2n - 8 &= 1n + 7 \\ \underline{-1n} \quad & \quad \underline{-1n} \\ n - 8 &= 0 + 7 \\ n - 8 &= 7 \\ \underline{+8} \quad & \quad \underline{+8} \\ n + 0 &= 15 \\ n &= 15 \end{aligned}$$

The number is 15.

40. *Step 1*
Unknown: length
Known: perimeter is 124 cm; width is 25 cm

Step 2
Let l represent the length.

Step 3
$$\begin{aligned} P &= 2l + 2w \\ 124 \text{ cm} &= 2l + 2(25) \text{ cm} \end{aligned}$$

Step 4
$$\begin{aligned} 124 \text{ cm} &= 2l + 50 \text{ cm} \\ \underline{-50 \text{ cm}} \quad & \quad \underline{-50 \text{ cm}} \\ 74 \text{ cm} &= 2l + 0 \\ \frac{74 \text{ cm}}{2} &= \frac{2l}{2} \\ 37 \text{ cm} &= l \end{aligned}$$

Step 5
The length is 37 cm.

Step 6
Check:

$2(37)\text{ cm} + 2(25)\text{ cm} = 74\text{ cm} + 50\text{ cm} = 124\text{ cm}$

41. *Step 1*
Unknown: length of each piece
Known: total length is 90 ft; one piece is 6 ft shorter

Step 2
Let p be the length of the longer piece. Then $p - 6$ is the length of the shorter piece.

Step 3
$p + (p - 6) = 90$

Step 4

$$\begin{aligned} 2p - 6 &= 90 \\ +6 &\quad +6 \\ \frac{2p}{2} &= \frac{96}{2} \\ p &= 48 \text{ ft} \end{aligned}$$

Step 5
The long piece is 48 ft, and the short piece is 48 – 6 = 42 ft.

Step 6
Check: 42 ft is 6 feet shorter than 48 ft and 42 ft + 48 ft = 90 ft.

42. $P = 4s$

$P = 4 \cdot 2\frac{1}{4}$ ft

$P = \frac{4}{1} \cdot \frac{9}{4}$ ft

$P = \frac{\overset{1}{\cancel{4}}}{1} \cdot \frac{9}{\underset{1}{\cancel{4}}}$ ft

$P = 9$ ft

$A = s^2$

$A = s \cdot s$

$A = 2\frac{1}{4}$ ft $\cdot 2\frac{1}{4}$ ft

$A = \frac{9}{4}$ ft $\cdot \frac{9}{4}$ ft

$A = \frac{81}{16}$ ft$^2 = 5\frac{1}{16}$ ft^2

$A \approx 5.1$ ft^2

43. $C = \pi \cdot d$

$C \approx 3.14 \cdot 9$ mm

$C = 28.26$ mm

$C \approx 28.3$ mm

$r = \frac{9 \text{ mm}}{2} = 4.5$ mm

$A = \pi r^2$

$A = \pi \cdot r \cdot r$

$A \approx 3.14 \cdot 4.5$ mm $\cdot 4.5$ mm

$A = 63.585$ mm^2

$A \approx 63.6$ mm^2

44. $P = 2.4$ cm $+ 1.5$ cm $+ 1.4$ cm $+ 1.15$ cm

$P = 6.45$ cm

$P \approx 6.5$ cm

Use subtraction to find the third side of the triangle.
2.4 cm – 1.4 cm = 1.0 cm

$A =$ triangle area + rectangle area

$A = \frac{1}{2}bh + lw$

$A = \frac{1}{2}(1 \text{ cm})(1.15 \text{ cm}) + (1.4 \text{ cm})(1.15 \text{ cm})$

$A = 0.575$ cm$^2 + 1.61$ cm^2

$A = 2.185$ cm^2

$A \approx 2.2$ cm^2

45. The triangle is a right triangle.

$\text{Leg} = \sqrt{(\text{hypotenuse})^2 - (\text{leg})^2}$

$y = \sqrt{(20)^2 - (16)^2}$

$y = \sqrt{400 - 256}$

$y = \sqrt{144}$

$y = 12$ yd

$A = \frac{1}{2}bh$

$A = \frac{1}{2}(16 \text{ yd})(12 \text{ yd})$

$A = 96$ yd^2

46. $V = l \cdot w \cdot h$

$V = 6$ cm $\cdot 4$ cm $\cdot 4$ cm

$V = 96$ cm^3

$SA = 2lw + 2lh + 2wh$

$SA = 2 \cdot 6$ cm $\cdot 4$ cm $+ 2 \cdot 6$ cm $\cdot 4$ cm $+ 2 \cdot 4$ cm $\cdot 4$ cm

$SA = 48$ cm$^2 + 48$ cm$^2 + 32$ cm^2

$SA = 128$ cm^2

47. Redraw the triangles so that you can see which sides correspond.

$\frac{7}{x} = \frac{8.5}{17}$ **OR** $\frac{7}{x} = \frac{1}{2}$

$1x = 7 \cdot 2$

$x = 14$ in.

Note that the lengths of the sides are approximate, that is, they do not satisfy the Pythagorean theorem.

48. $4\frac{1}{2}$ ft to inches

$\frac{9 \text{ ft}}{2} \cdot \frac{12 \text{ in.}}{1 \text{ ft}} = \frac{9 \cancel{\text{ft}}}{\underset{1}{\cancel{2}}} \cdot \frac{\overset{6}{\cancel{12}} \text{ in.}}{1 \cancel{\text{ft}}} = 54$ in.

49. 72 hours to days

$\frac{72 \text{ hr}}{1} \cdot \frac{1 \text{ day}}{24 \text{ hr}} = \frac{\overset{3}{\cancel{72}} \cancel{\text{hr}}}{1} \cdot \frac{1 \text{ day}}{\underset{1}{\cancel{24}} \cancel{\text{hr}}} = 3$ days

50. 3.7 kg to grams

From kg to g is 3 places to the right.

3.7 kg = 3.700 kg = 3700 g

51. 60 cm to meters

From cm to m is 2 places to the left.

60 cm = 060. cm = 0.60 m or 0.6 m

52. 7 mL to liters

From mL to L is 3 places to the left.

7 mL = 0007. mL = 0.007 L

53. −20 °C to Fahrenheit

$$F = \frac{9C}{5} + 32$$

$$F = \frac{9(-\overset{4}{\cancel{20}})}{\underset{1}{\cancel{5}}} + 32 = -36 + 32$$

$$= -4$$

−20 °C = −4 °F

54. Part: 35 credits earned
Whole: 60 credits needed
Percent: unknown (p)

$$\text{Percent} \cdot \text{Whole} = \text{Part}$$
$$p \cdot 60 = 35$$
$$\frac{60p}{60} = \frac{35}{60}$$
$$p = 0.58\overline{3}$$

She has 58% (rounded) of the necessary credits.

55. $15\frac{1}{2}$ ounces of Brand T for
\$2.99 − \$0.30 (coupon) = \$2.69

$$\frac{\$2.69}{15\frac{1}{2}\text{ ounces}} \approx \$0.174 \text{ per ounce } (*)$$

14 ounces of Brand F for \$2.49

$$\frac{\$2.49}{14\text{ ounces}} \approx \$0.178 \text{ per ounce}$$

18 ounces of Brand H for
\$3.89 − \$0.40 (coupon) = \$3.49

$$\frac{\$3.49}{18\text{ ounces}} \approx \$0.194 \text{ per ounce}$$

The best buy is Brand T at 15.5 ounces for \$2.99 with a \$0.30 coupon.

56. **(a)** Set up a proportion.

$$\frac{1\text{ cm}}{12\text{ km}} = \frac{7.8\text{ cm}}{x}$$
$$x \cdot 1 = (7.8)(12)$$
$$x = 93.6$$

The distance is 93.6 kilometers.

(b) $\frac{93.6\cancel{\text{km}}}{1} \cdot \frac{0.62\text{ mi}}{1\cancel{\text{km}}} \approx 58.0\text{ mi}$

The distance is 58.0 mi (rounded).

57. Find the total grams, then convert grams to kilograms.
450 g + 48 • 115 g = 450 g + 5520 g = 5970 g

$$\frac{5970\cancel{g}}{1} \cdot \frac{1\text{ kg}}{1000\cancel{g}} = 5.97\text{ kg}$$

The carton would weigh 5.97 kg.

58. First add the amount of canvas material that was used.

$$1\frac{2}{3} = 1\frac{8}{12}$$
$$+1\frac{3}{4} = 1\frac{9}{12}$$
$$2\frac{17}{12} = 3\frac{5}{12}\text{ yd}$$

Subtract $3\frac{5}{12}$ from the amount of canvas material that Steven bought.

$$4\frac{1}{2} = 4\frac{6}{12}$$
$$-3\frac{5}{12} = 3\frac{5}{12}$$
$$1\frac{1}{12}\text{ yd}$$

There will be $1\frac{1}{12}$ yd left.

59. Convert 650 g to kg.
‸650. g = 0.650 kg

Multiply to find the cost.
(0.650)(\$14.98) ≈ \$9.74

Mark paid \$9.74 (rounded).

60. *amount of discount* = *rate of discount* • *cost of item*
= (0.10)(\$189.94)
≈ \$18.99

The amount of discount is \$18.99.
The sale price is \$189.94 − \$18.99 = \$170.95.

61. Set up a proportion.

$$\frac{6\text{ rows}}{5\text{ cm}} = \frac{x}{100\text{ cm}}$$
$$5 \cdot x = 6 \cdot 100$$
$$\frac{5 \cdot x}{5} = \frac{600}{5}$$
$$x = 120$$

Akuba will knit 120 rows.

62. \$3500 at $7\frac{1}{2}\%$ for 6 months

$$I = p \cdot r \cdot t \qquad 7.5\% = 0.075$$
$$= (3500)(0.075)\left(\frac{6}{12}\right)$$
$$= 131.25$$

The interest is \$131.25.

$$\begin{aligned} \text{amount due} &= \text{principal} + \text{interest} \\ &= \$3500 + \$131.25 \\ &= \$3631.25 \end{aligned}$$

The total amount due is \$3631.25.

63. How many \$17 blankets can be purchased from a \$400 donation? Use division.

$$\begin{array}{r} 23 \\ 17\overline{)400} \\ \underline{34} \\ 60 \\ \underline{51} \\ 9 \end{array}$$

23 blankets can be purchased. \$9 will be left over.

64. $\dfrac{\$6}{\text{40 applications}} = \$0.15/\text{application}$

65. decrease = 8 − 5 = 3 days

$$\frac{3\text{ days}}{8\text{ days}} = \frac{x}{100}$$
$$8 \cdot x = 3 \cdot 100$$
$$\frac{8 \cdot x}{8} = \frac{300}{8}$$
$$x = 37.5$$

The length of a hospital stay has decreased by 37.5%.

CHAPTER 9 GRAPHS

9.1 Problem Solving with Tables and Pictographs

9.1 Section Exercises

1. **(a)** Look in the points column for Wilt Chamberlain. He scored 31,419 points.

   (b) Look down the points column and find the number(s) greater than 31,419, then read across to the player's name.

   Since 32,292 is greater than 31,419, Michael Jordan scored more points than Wilt Chamberlain.

3. **(a)** Look down the games column and find the largest number, which is 1072. Then read across to find the player's name.

   Michael Jordan has been in the greatest number of games.

   (b) The smallest number in the games column is 527.

   Allen Iverson has been in the fewest number of games.

5. Look down the points column and find the largest and smallest number. Then find the difference.

   $$32{,}292 - 14{,}436 = 17{,}856$$

   The difference is 17,856 points.

7. Round answers to the nearest tenth to match other numbers in column.

   Jerry West: $\dfrac{25{,}192}{932} \approx 27.0$

   Allen Iverson: $\dfrac{14{,}436}{535} \approx 27.0$

   Bob Pettit: $\dfrac{20{,}880}{792} \approx 26.4$

9. The asterisks next to O'Neal's and Iverson's names means that they played in the entire 2003-2004 season.

11. **(a)** Look down the 140-pounds column and across the aerobic dance activity row.

    The person will burn 255 calories.

    (b) Look down the 140-pounds column for the largest number. Then read across to the activity. Moderate jogging burns the most calories.

13. **(a)** Look down the 110-pounds column for numbers greater than or equal to 200. Then read across to the activities. Moderate jogging, aerobic dance, and racquetball are activities that burn at least 200 calories.

    (b) Use the same method as in part (a), except use the 170-pound column. Moderate jogging, moderate bicycling, aerobic dance, racquetball, and tennis are activities that burn at least 200 calories.

15. 15 minutes is $\frac{1}{2}$ of 30 minutes.
    60 minutes is 2 times 30 minutes.
    Look down the 140-pound column.

    $$\tfrac{1}{2}\cdot 180 + 2\cdot 140 = 90 + 280 = 370$$

    The person will burn 370 calories.

For Exercises 17–20, other proportions are possible.

17. **(a)** $\dfrac{322 \text{ calories}}{110 \text{ pounds}} = \dfrac{x \text{ calories}}{125 \text{ pounds}}$

    $$110\cdot x = 322\cdot 125$$
    $$\frac{110x}{110} = \frac{40{,}250}{110}$$
    $$x \approx 366$$

    The person would burn approximately 366 calories.

    (b) $\dfrac{210 \text{ calories}}{110 \text{ pounds}} = \dfrac{x \text{ calories}}{125 \text{ pounds}}$

    $$110\cdot x = 210\cdot 125$$
    $$\frac{110x}{110} = \frac{26{,}250}{110}$$
    $$x \approx 239$$

    The person would burn approximately 239 calories.

19. Aerobic dance: $\dfrac{255 \text{ calories}}{140 \text{ pounds}} = \dfrac{x \text{ calories}}{158 \text{ pounds}}$

    $$140\cdot x = 255\cdot 158$$
    $$\frac{140x}{140} = \frac{40{,}290}{140}$$
    $$x \approx 288$$

    15 min is $\frac{1}{2}$ of 30 min: $\frac{1}{2}\cdot 288 \approx 144$

    Walking: $\dfrac{140 \text{ calories}}{140 \text{ pounds}} = \dfrac{x \text{ calories}}{158 \text{ pounds}}$

    $$140\cdot x = 140\cdot 158$$
    $$x = 158$$

    20 min is $\frac{2}{3}$ of 30 min: $\frac{2}{3}\cdot 158 \approx 105$

    The difference is $144 - 105 = 39$ calories.

For Exercises 21–28, each symbol represents 10 million passenger arrivals and departures.

21. **(a)** Chicago is represented by 7 symbols.

    $7(10 \text{ million}) = 70$ million or 70,000,000

    (b) Houston is represented by $3\frac{1}{2}$ symbols (or 3.5).

    $3.5(10 \text{ million}) = 35$ million or 35,000,000

23. Atlanta: $8(10 \text{ million}) = 80$ million

Chicago: $7(10 \text{ million}) = 70$ million

Los Angeles: $5.5(10 \text{ million}) = 55$ million

The total number is
80 million + 70 million + 55 million =
205 million or 205,000,000.

25. Los Angeles: $5.5(10 \text{ million}) = 55$ million

Chicago: $7(10 \text{ million}) = 70$ million

The difference is $70 - 55 = 15$ million or 15,000,000.

27. There are $8 + 7 + 5.5 + 3.5 + 2.5 = 26.5$ symbols.

$26.5(10 \text{ million}) = 265$ million

The total number is 265 million or 265,000,000.

29. Answers will vary. One possibility: choose Southwest because it has the best on-time performance.

30. Answers will vary. Possibilities include planning more time between each flight, or doing some or all of your business via conference calls or e-mail.

31. Answers will vary. One possibility: choose Airtran because it has the fewest luggage problems.

32. Answers will vary. Possibilities include buying heavy-duty luggage or shipping the golf clubs via a delivery service.

33. Answers will vary. Possibilities include a lot of bad weather, maintenance problems, new computer system.

34. Answers will vary. Possibilities include availability of nonstop flights, convenience of departure times, type and size of aircraft, availability of low-cost fares.

9.2 Reading and Constructing Circle Graphs

9.2 Section Exercises

1. **(a)** The total cost of adding the family room is
$12,100 + $9800 + $2000 + $900 + $1800
+ $3000 + $2400 = $32,000

(b) The largest single expense in adding the family room is carpentry at $12,100.

3. **(a)**
$$\frac{\text{cost of materials}}{\text{total remodeling cost}} = \frac{\$9800}{\$32{,}000} = \frac{9800 \div 200}{32{,}000 \div 200} = \frac{49}{160}$$

(b)
$$\frac{\text{windows}}{\text{electrical}} = \frac{\$3000}{\$2000} = \frac{3}{2}$$

5.
$$\frac{\text{carpentry} + \text{windows} + \text{window coverings}}{\text{total remodeling cost}} = \frac{\$12{,}100 + \$3000 + \$900}{\$32{,}000} = \frac{\$16{,}000}{\$32{,}000} = \frac{1}{2}$$

7. **(a)** The smallest number and the smallest sector represent the reason "Don't know."

(b) Find the second to the smallest sector.

"Quicker" was the reason given by the second-fewest number of people.

9. Total people $= 1740 + 1200 + 1140 + 180 + 1020 + 720 = 6000$

"Quicker" to total:

$$\frac{720}{6000} = \frac{720 \div 240}{6000 \div 240} = \frac{3}{25}$$

11. "Less work/No clean up" to "Atmosphere":

$$\frac{1020}{1200} = \frac{1020 \div 60}{1200 \div 60} = \frac{17}{20}$$

13. "Wanted food they couldn't cook at home" to "Don't know":

$$\frac{1740}{180} = \frac{1740 \div 60}{180 \div 60} = \frac{29}{3}$$

15. The percent for restrooms is 30%.

$$\begin{aligned} \text{percent} \cdot \text{whole} &= \text{part} \\ (0.30)(\$1{,}740{,}000) &= n \\ \$522{,}000 &= n \end{aligned}$$

The amount spent on restrooms is $522,000.

17. The percent for doors and thresholds is 10%.

$$\begin{aligned} (0.10)(\$1{,}740{,}000) &= n \\ \$174{,}000 &= n \end{aligned}$$

The amount spent on doors and thresholds is $174,000.

19. The percent for walkways and curbs is 15%.

$$\begin{aligned} (0.15)(\$1{,}740{,}000) &= n \\ \$261{,}000 &= n \end{aligned}$$

The amount spent on walkways and curbs is $261,000.

21. To determine how many people prefer onions for their hot dog topping, use the percent equation.

$$\begin{aligned} \text{percent} \cdot \text{whole} &= \text{part} \\ (0.05)(3200) &= n \\ 160 &= n \end{aligned}$$

160 people favored onions.

23. The most popular topping was mustard (30%).

$$(0.30)(3200) = n$$
$$960 = n$$

960 people favored mustard.

25. Chili = 12% of 3200

$$(0.12)(3200) = n$$
$$384 = n$$

Relish = 10% of 3200

$$(0.10)(3200) = n$$
$$320 = n$$

There were 384 – 320 = 64 more people who chose chili than relish.

27. First find the percent of the total that is to be represented by each item. Next, multiply the percent by 360° to find the size of each sector. Finally, use a protractor to draw each sector.

29. **(a)** 25% of total is rent.

$$\text{Degrees of a circle} = 25\% \text{ of } 360° = (0.25)(360) = 90°$$

(b) $\text{Percent for food} = \dfrac{72°}{360°} = 0.20 = 20\%$

(c) $\text{Percent for clothing} = \dfrac{\$546}{\$5460} = 0.10 = 10\%$

$$\text{Degrees of a circle} = 10\% \text{ of } 360° = (0.10)(360°) = 36°$$

(d) Since the dollar amount, $546, is the same as part (c), the percent of total is 10% and the number of degrees in the circle is 36°.

(e) $\text{Percent for tuition and fees} = \dfrac{\$819}{\$5460} = 0.15 = 15\%$

$$\text{Degrees of a circle} = 15\% \text{ of } 360° = (0.15)(360°) = 54°$$

(f) $\text{Percent for savings} = \dfrac{\$273}{\$5460} = 0.05 = 5\%$

$$\text{Degrees of a circle} = 5\% \text{ of } 360° = (0.05)(360°) = 18°$$

(g) Since the dollar amount, $819, is the same as part (e), the percent of total is 15% and the number of degrees in the circle is 54°.

(h)

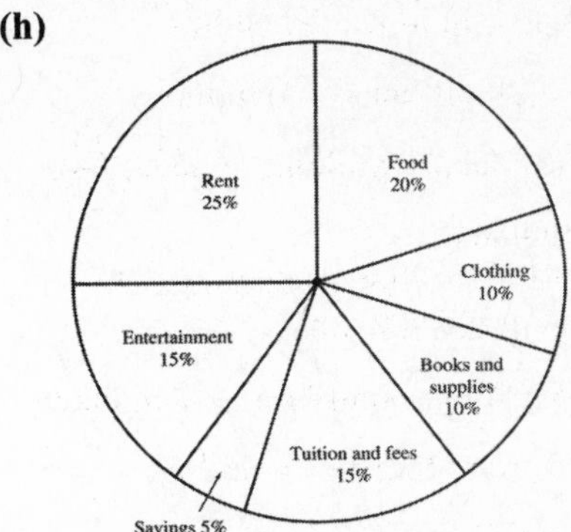

For part (a) – (e), the "Number of Americans" (in the second column) is given in the numerator of the fraction.

31. **(a)** Total sales = $12,500 + $40,000 + $60,000 + $50,000 + $37,500 = $200,000

(b) Adventure classes = $12,500

$$\text{percent of total} = \frac{12{,}500}{200{,}000} = 0.0625 = 6.25\%$$

Grocery/provision sales = $40,000

$$\text{percent of total} = \frac{40{,}000}{200{,}000} = 0.2 = 20\%$$

Equipment rentals = $60,000

$$\text{percent of total} = \frac{60{,}000}{200{,}000} = 0.3 = 30\%$$

Rafting tours = $50,000

$$\text{percent of total} = \frac{50{,}000}{200{,}000} = 0.25 = 25\%$$

Equipment sales = $37,500

$$\text{percent of total} = \frac{37{,}500}{200{,}000} = 0.1875 = 18.75\%$$

(c) Adventure classes:

$$\text{number of degrees} = (0.0625)(360°) = 22.5°$$

Grocery/provision sales:

$$\text{number of degrees} = (0.2)(360°) = 72°$$

Equipment rentals:

$$\text{number of degrees} = (0.3)(360°) = 108°$$

Rafting tours:

$$\text{number of degrees} = (0.25)(360°) = 90°$$

Equipment sales:

$$\text{number of degrees} = (0.1875)(360°) = 67.5°$$

(d)

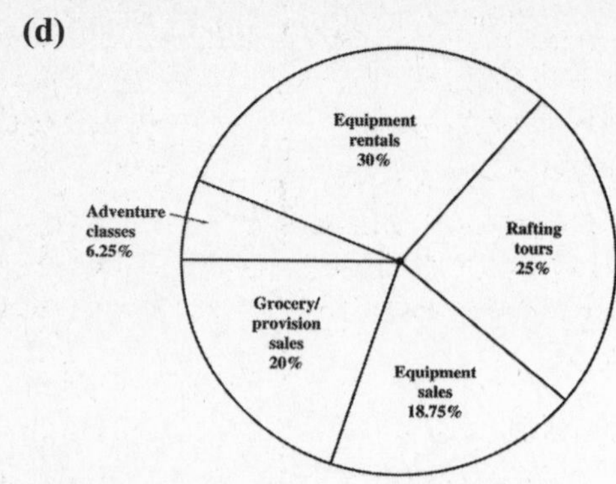

9.3 Bar Graphs and Line Graphs

9.3 Section Exercises

1. Look for the longest bar. The top reason is, "I can shop during off hours." 74% gave this reason.

3. 57% say they find better prices on-line.

   $0.57(600) = 342$

   342 people gave this answer.

5. $\frac{1}{2} = 50\%$

   $\frac{1}{2}$ of the people said, "I can compare products more easily."

   $\frac{3}{4} = 75\% \approx 74\%$

   Nearly $\frac{3}{4}$ of the people said, "I can shop during off hours."

7. May had the greatest number of unemployed workers. The total was 10,000 unemployed workers.

9. Unemployed workers in February of 2005 = 7000
   Unemployed workers in February of 2004 = 5500

   $$7000 - 5500 = 1500$$

   There were 1500 more workers unemployed in February of 2005.

11. The number of unemployed workers increased from 5500 in February of 2004 to 8000 in April of 2005. The increase was 8000 – 5500 = 2500 workers.

   $$\begin{array}{cccc} \text{percent of} & \text{February 2004 unemployed workers} & = & \text{amount of increase} \\ p \quad \cdot & 5500 & = & 2500 \\ & \dfrac{p \cdot 5500}{5500} & = & \dfrac{2500}{5500} \\ & p & = & \dfrac{5}{11} \approx 0.45 = 45\% \end{array}$$

   The percent of increase was 45% (rounded).

13. 150,000 gallons of supreme unleaded gasoline were sold in 2001.

15. The greatest difference occurred in 2001. The difference was 400,000 – 150,000 = 250,000 gallons.

17. 700,000 gallons of supreme unleaded gasoline were sold in 2005. 150,000 gallons of supreme unleaded gasoline were sold in 2001.

   $$700{,}000 - 150{,}000 = 550{,}000$$

   There was an increase of 550,000 gallons of supreme unleaded gasoline sales.

   $$\begin{array}{cccc} \text{percent of} & \text{supreme unleaded gasoline sales in 2001} & = & \text{amount of increase} \\ p \quad \cdot & 150{,}000 & = & 550{,}000 \\ & \dfrac{p \cdot 150{,}000}{150{,}000} & = & \dfrac{550{,}000}{150{,}000} \\ & p & = & \dfrac{11}{3} \approx 3.67 = 367\% \end{array}$$

   The percent of increase was 367% (rounded).

19. The number of PCs shipped in 1990 was 24.1 million or 24,100,000.

21. The increase in the number of PCs shipped in 2005 from the number shipped in 1985 was 197.4 – 11.8 = 185.6 million or 185,600,000.

23. The amount of increase in shipments from 1995 to 2000 was 144.6 – 62.3 = 82.3 million or 82,300,000 PCs.

   $$\begin{array}{cccc} \text{percent of} & \text{1995 shipments} & = & \text{amount of increase} \\ p \quad \cdot & 62.3 & = & 82.3 \\ & \dfrac{p \cdot 62.3}{62.3} & = & \dfrac{82.3}{62.3} \\ & p & \approx & 1.32 = 132\% \end{array}$$

   The percent of increase was 132% (rounded).

25. **(a)** Chain Store A sold 3000 • 1000 = 3,000,000 CDs in 2001.

   (b) Chain Store B sold 1500 • 1000 = 1,500,000 CDs in 2001.

27. **(a)** Chain Store A sold 2500 • 1000 = 2,500,000 CDs in 2004.

   (b) Chain Store A sold 3000 • 1000 = 3,000,000 CDs in 2005.

29. Answers will vary. Possibilities include: Both stores had decreased sales from 2001 to 2002 and increased sales from 2003 to 2005; Store B had lower sales than Store A in 2001-02 but higher sales than Store A in 2003-2005.

31. On the blue line graph (Sales), the lowest point corresponds to the year 2003 and the amount $25,000.

33. **For 2002:** Profit $= \$10{,}000$, Sales $= \$35{,}000$

$$\text{Percent the profit is of sales} = \frac{\$10{,}000}{\$35{,}000} \approx 0.29 = 29\%$$

For 2003: Profit $= \$5{,}000$, Sales $= \$25{,}000$

$$\text{Percent the profit is of sales} = \frac{\$5{,}000}{\$25{,}000} = 0.20 = 20\%$$

For 2004: Profit $= \$5{,}000$, Sales $= \$30{,}000$

$$\text{Percent the profit is of sales} = \frac{\$5{,}000}{\$30{,}000} \approx 0.17 = 17\%$$

For 2005: Profit $= \$15{,}000$, Sales $= \$40{,}000$

$$\text{Percent the profit is of sales} = \frac{\$15{,}000}{\$40{,}000} = 0.375 \approx 38\%$$

35. Answers will vary. Possibilities include: The decrease in sales may have resulted from poor service or greater competition; the increase in sales may have been a result of more advertising or better service.

37. Shipments have increased at a rapid rate since 1985.

38. Answers will vary. Some possibilities are: lower prices; more uses and applications for students, home use, and businesses; improved technology.

39. See exercise 23 for more detail.

(a) 1985 to 1990:

$$11.8p = 24.1 - 11.8$$
$$p = \frac{12.3}{11.8} \approx 1.04 = \mathbf{104\%}$$

(b) 1990 to 1995:

$$24.1p = 62.3 - 24.1$$
$$p = \frac{38.2}{24.1} \approx 1.59 = \mathbf{159\%}$$

(c) 1995 to 2000:

$$62.3p = 144.6 - 62.3$$
$$p = \frac{82.3}{62.3} \approx 1.32 = \mathbf{132\%}$$

(d) 2000 to 2005:

$$144.6p = 197.4 - 144.6$$
$$p = \frac{52.8}{144.6} \approx 0.37 = \mathbf{37\%}$$

40. Since 1990, the percent of increase for each 5-year period has been dropping.

41. Answers will vary. Some possibilities are: more people will already own a computer and not want to buy another; some new invention will replace computers.

42. **(a)** The sales for Chain Store A were 3,000,000 in 2001 and in 2005. Since these values are the same, there is no percent of increase or decrease (0%).

(b) Chain Store B:

$$\text{percent of} \cdot \text{2001 sales} = \text{amount of increase}$$
$$p \cdot 1{,}500{,}000 = 4{,}000{,}000 - 1{,}500{,}000$$
$$\frac{p \cdot 1{,}500{,}000}{1{,}500{,}000} = \frac{2{,}500{,}000}{1{,}500{,}000}$$
$$p = \frac{5}{3} \approx 1.67 = 167\%$$

From 2001 to 2005, the percent of increase for Chain Store B was 167% (rounded).

43. **(a)** Answers will vary; perhaps 3,500,000 CDs in 2006.

(b) Answers will vary; perhaps 4,500,000 CDs in 2006.

Answers will vary; one possibility is predicting a continuing increase based on the increases from 2003 to 2005.

44. **(a)** Most people will probably pick Store B because of its greater sales and more consistent upward trend.

(b) Answers will vary. Some possibilities include: age and physical condition of the store, annual expenses, sales of other products, annual profit.

9.4 The Rectangular Coordinate System

9.4 Section Exercises

1.

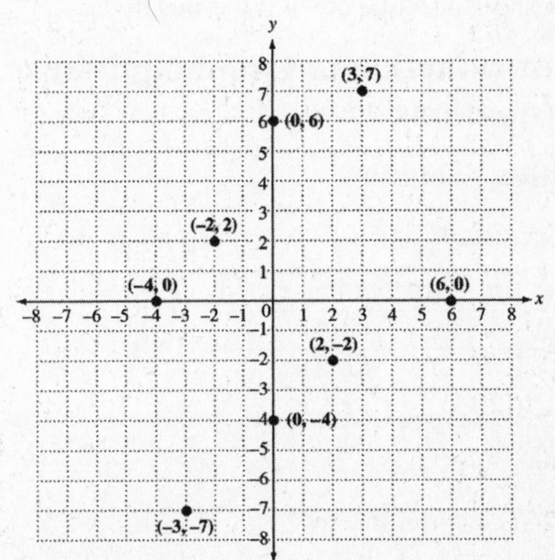

3.

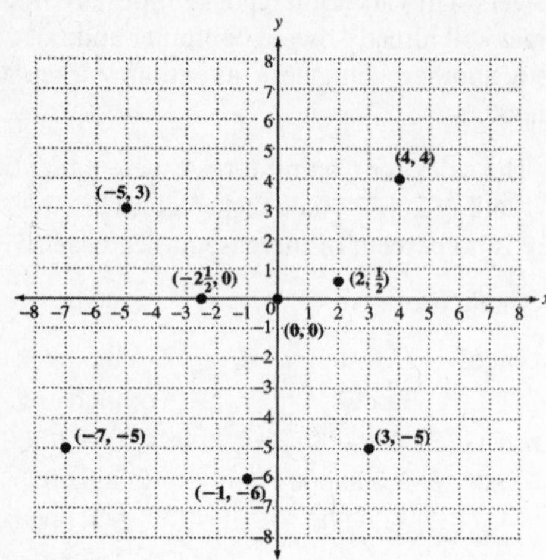

5. A is $(3, 4)$; B is $(5, -5)$; C is $(-4, -2)$; D is approximately $(4, \frac{1}{2})$; E is $(0, -7)$; F is $(-5, 5)$; G is $(-2, 0)$; H is $(0, 0)$.

7. $(-3, -7)$ is in Quadrant III.
$(0, 4)$ is on the y-axis, so it is not in a quadrant.
$(10, -16)$ is in Quadrant IV.
$(-9, 5)$ is in Quadrant II.

9. **(a)** Any *positive* number, because points in Quadrant II have the pattern $(-, +)$.

(b) Any *negative* number, because points in Quadrant IV have the pattern $(+, -)$.

(c) 0, because points not in a quadrant have the form $(0, \pm)$ or $(\pm, 0)$.

(d) Any *negative* number, because points in Quadrant III have the pattern $(-, -)$.

(e) Any *positive* number, because points in Quadrant I have the pattern $(+, +)$.

11. Starting at the origin, move left or right along the x-axis to the number a; then move up if b is positive or move down if b is negative.

9.5 Introduction to Graphing Linear Equations

9.5 Section Exercises

1. $x + y = 4$

x	y	Ordered Pair (x, y)
0	4	$(0, 4)$
1	3	$(1, 3)$
2	2	$(2, 2)$

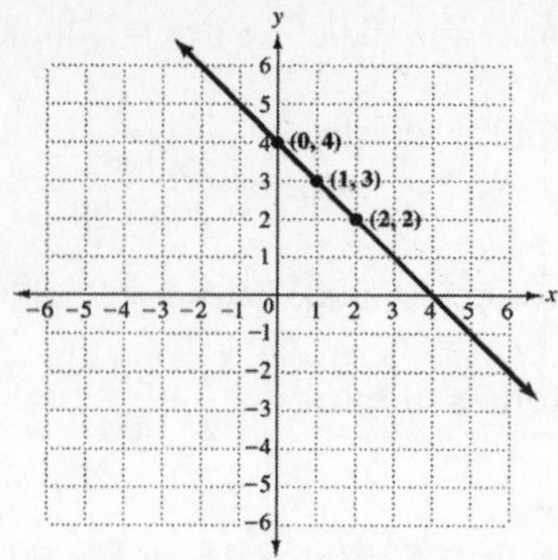

Two other possible solutions are $(3, 1)$ and $(4, 0)$. All points on the line are solutions.

3. $x + y = -1$

x	y	Ordered Pair (x, y)
0	-1	$(0, -1)$
1	-2	$(1, -2)$
2	-3	$(2, -3)$

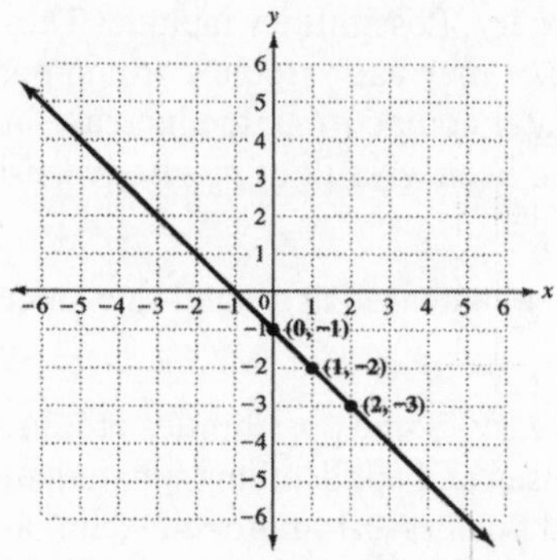

Two other possible solutions are $(-1, 0)$ and $(-2, 1)$. All points on the line are solutions.

5. The line in Exercise 1, $x + y = 4$, crosses the y-axis at 4 or $(0, 4)$.

The line in Exercise 3, $x + y = -1$, crosses the y-axis at -1 or $(0, -1)$.

Based on these examples:

The line $x + y = -6$ will cross the y-axis at -6 or $(0, -6)$.

The line $x + y = 99$ will cross the y-axis at 99 or $(0, 99)$.

7. $y = x - 2$

x	$x - 2 = y$	Ordered Pair (x, y)
1	$1 - 2 = -1$	$(1, -1)$
2	$2 - 2 = 0$	$(2, 0)$
3	$3 - 2 = 1$	$(3, 1)$

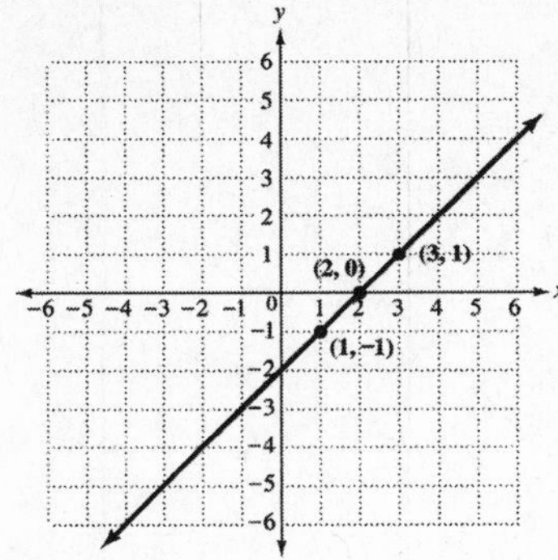

9. $y = x + 2$

x	$x + 2 = y$	(x, y)
0	$0 + 2 = 2$	$(0, 2)$
−1	$-1 + 2 = 1$	$(-1, 1)$
−2	$-2 + 2 = 0$	$(-2, 0)$

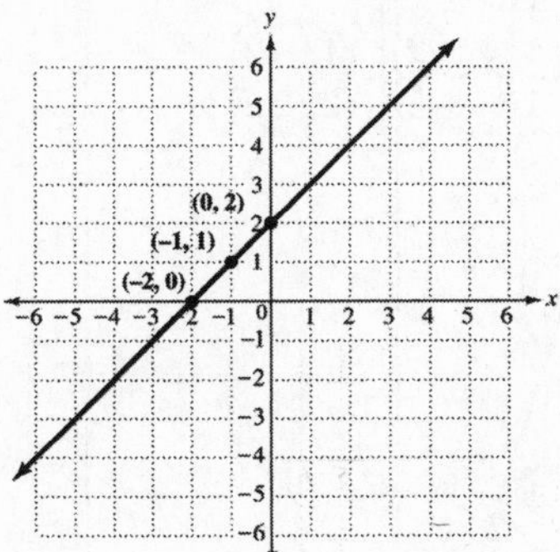

11. $y = -3x$

x	$-3 \cdot x = y$	(x, y)
0	$-3(0) = 0$	$(0, 0)$
1	$-3(1) = -3$	$(1, -3)$
2	$-3(2) = -6$	$(2, -6)$

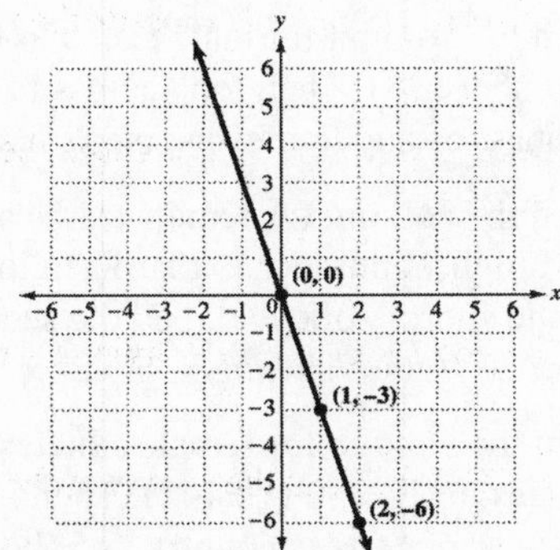

13. **(a)** The graph of $y = x - 2$ in Exercise 7 has a positive slope since the line rises. As the value of x increases, the value of y increases.

(b) The graph of $y = x + 2$ in Exercise 9 has a positive slope since the line rises. As the value of x decreases, the value of y decreases.

(c) The graph of $y = -3x$ in Exercise 11 has a negative slope since the line falls. As the value of x increases, the value of y decreases.

15. $y = \frac{1}{3}x$

x	$\frac{1}{3} \cdot x = y$	(x, y)
0	$\frac{1}{3}(0) = 0$	$(0, 0)$
3	$\frac{1}{3}(3) = 1$	$(3, 1)$
6	$\frac{1}{3}(6) = 2$	$(6, 2)$

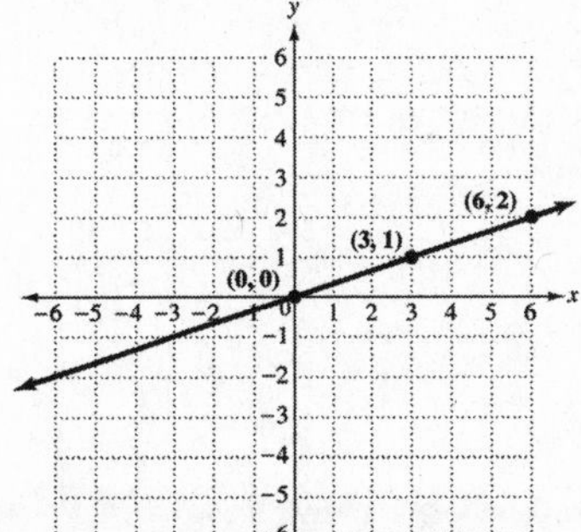

17. $y = x$

x	$y = x$	(x, y)
−1	−1	$(-1, -1)$
−2	−2	$(-2, -2)$
−3	−3	$(-3, -3)$

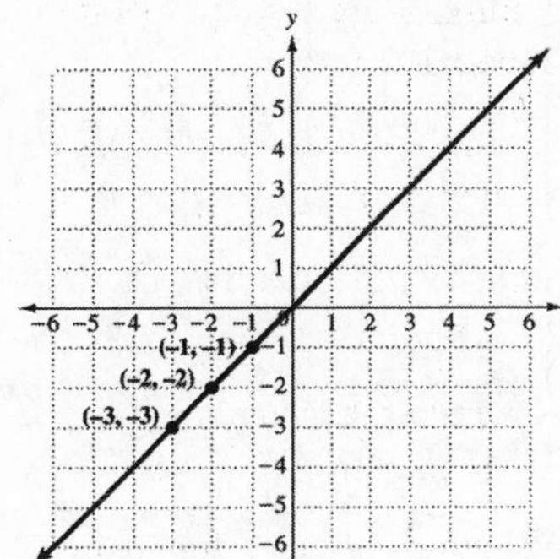

19. $y = -2x + 3$

x	$-2 \cdot x + 3 = y$	(x, y)
0	$-2(0) + 3 = 3$	$(0, 3)$
1	$-2(1) + 3 = 1$	$(1, 1)$
2	$-2(2) + 3 = -1$	$(2, -1)$

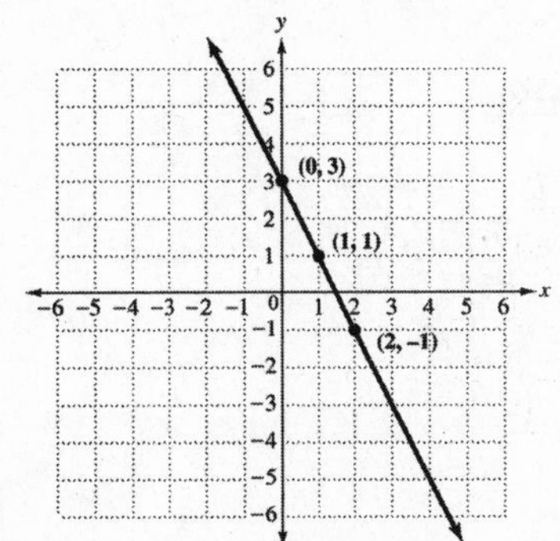

21. $x + y = -3$

One possible table:

x	y	(x, y)
0	-3	$(0, -3)$
1	-4	$(1, -4)$
2	-5	$(2, -5)$

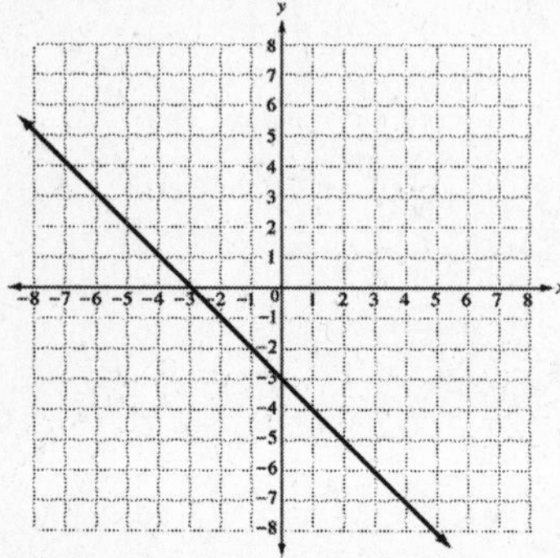

The line has a negative slope (falls).

23. $y = \frac{1}{4}x$

One possible table:

x	$\frac{1}{4} \cdot x = y$	(x, y)
0	$\frac{1}{4}(0) = 0$	$(0, 0)$
4	$\frac{1}{4}(4) = 1$	$(4, 1)$
8	$\frac{1}{4}(8) = 2$	$(8, 2)$

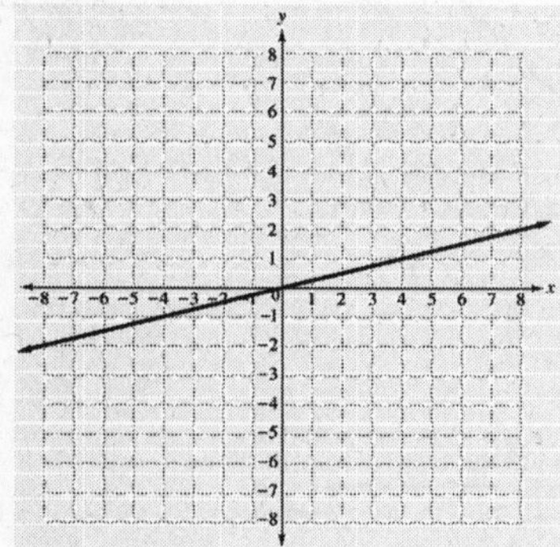

The line has a positive slope (rises).

25. $y = x - 5$

One possible table:

x	$x - 5 = y$	(x, y)
2	$2 - 5 = -3$	$(2, -3)$
3	$3 - 5 = -2$	$(3, -2)$
4	$4 - 5 = -1$	$(4, -1)$

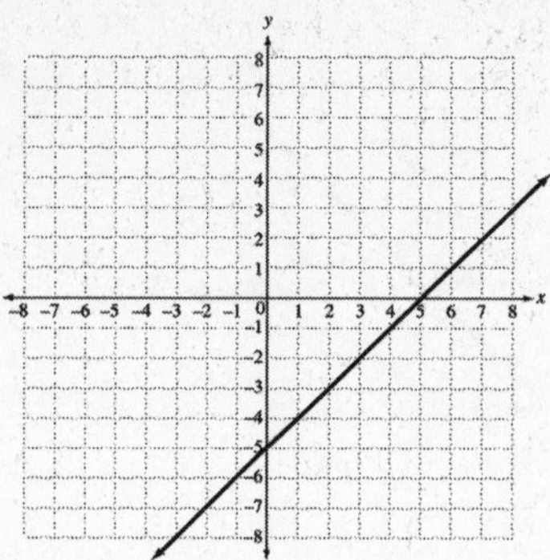

The line has a positive slope (rises).

27. $y = -3x + 1$

One possible table:

x	$-3 \cdot x + 1 = y$	(x, y)
0	$-3(0) + 1 = 1$	$(0, 1)$
1	$-3(1) + 1 = -2$	$(1, -2)$
2	$-3(2) + 1 = -5$	$(2, -5)$

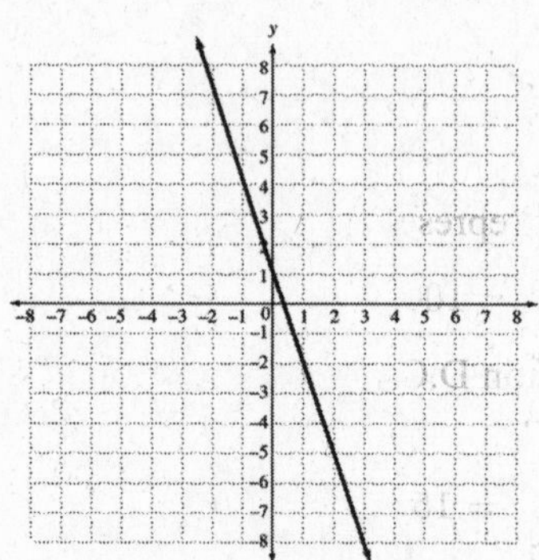

The line has a negative slope (falls).

Chapter 9 Review Exercises

1. **(a)** Look down the column for male teams to find the smallest number (28). Then look across to find the sport. Gymnastics had the fewest men's teams.

   (b) Look down the column for female teams and find the second highest number (923). Then look across to find the sport. Volleyball had the second greatest number of women's teams.

2. **(a)** Look down the column for female athletes for the number closest to 10,000 (10,141). Then look across to find the sport. Cross country had about 10,000 female athletes.

   (b) Look down the column for male athletes for the number closest to 7000 (7197). Then look across to find the sport. Golf had about 7000 male athletes.

3. **(a)** From the table, basketball had 15,141 male athletes and cross country had 10,271 male athletes.

$$15{,}141 - 10{,}271 = 4870$$

4870 more men participated in basketball than cross country.

(b) From the table, gymnastics had 1311 female athletes and basketball had 13,392 female athletes.

$$13{,}392 - 1311 = 12{,}081$$

12,081 fewer women participated in gymnastics than basketball.

4. Average squad size for men's volleyball

$$= \frac{1052}{74} \approx 14.2$$

Average squad size for women's volleyball

$$= \frac{12{,}284}{923} \approx 13.3$$

For Exercises 5–8, each symbol represents 10 inches of snowfall.

5. **(a)** Juneau is represented by 10 symbols.

$10(10 \text{ inches}) = 100 \text{ inches}$

(b) Washington D.C. is represented by $1\frac{1}{2}$ or 1.5 symbols.

$1.5(10 \text{ inches}) = 15 \text{ inches}$

6. **(a)** Minneapolis is represented by 5 symbols.

$5(10 \text{ inches}) = 50 \text{ inches}$

(b) Cleveland is represented by $5\frac{1}{2}$ or 5.5 symbols.

$5.5(10 \text{ inches}) = 55 \text{ inches}$

7. **(a)** Buffalo: $9(10 \text{ inches}) = 90 \text{ inches}$
Cleveland: $5.5(10 \text{ inches}) = 55 \text{ inches}$

The difference in average yearly snowfall is 90 inches − 55 inches = 35 inches.

(b) Memphis: $\frac{1}{2}(10 \text{ inches}) = 5 \text{ inches}$
Minneapolis: $5(10 \text{ inches}) = 50 \text{ inches}$

The difference in average yearly snowfall is 50 inches − 5 inches = 45 inches.

8. greatest amount: Juneau, 100 inches
least amount: Memphis, 5 inches

The difference in average yearly snowfall between Juneau and Memphis is

100 inches − 5 inches = 95 inches.

9. **(a)** Lodging is the largest sector in the circle graph: \$560

(b) Food is the second largest sector in the circle graph: \$400

(c) $\text{Total cost} = \$560 + \$400 + \$300 + \$280 + \$160 = \$1700$

10. $$\frac{\text{lodging}}{\text{food}} = \frac{\$560}{\$400} = \frac{560}{400} = \frac{7}{5}$$

11. $$\frac{\text{gasoline}}{\text{total cost}} = \frac{\$300}{\$1700} = \frac{300}{1700} = \frac{3}{17}$$

12. $$\frac{\text{sightseeing}}{\text{total cost}} = \frac{\$280}{\$1700} = \frac{280 \div 20}{1700 \div 20} = \frac{14}{85}$$

13. $$\frac{\text{gasoline}}{\text{other}} = \frac{\$300}{\$160} = \frac{300 \div 20}{160 \div 20} = \frac{15}{8}$$

14. The most popular project is painting and wallpapering since it has the longest bar. The percent is 63%.

15. The project selected the least is construction work at 33%.

16. 43% of the homeowners in the survey selected carpentry projects.

$$\text{percent} \cdot \text{whole} = \text{part}$$
$$(0.43)(341) = n$$
$$n = 146.63 \approx 147$$

About 147 homeowners selected carpentry.

17. 54% of the homeowners in the survey selected landscaping or gardening projects.

$$\text{percent} \cdot \text{whole} = \text{part}$$
$$(0.54)(341) = n$$
$$n = 184.14 \approx 184$$

About 184 homeowners selected landscaping or gardening projects.

18. **(a)** $\frac{1}{2} = 0.5 = 50\%$

51% (about $\frac{1}{2}$) of the homeowners selected interior decorating.

(b) $\frac{1}{3} = 0.\overline{3} = 33.\overline{3}\%$

33% (about $\frac{1}{3}$) of the homeowners selected construction work.

19. Answers will vary. Possibilities include: painting and wallpapering are easier to do, take less time, or cost less than construction work.

20. In 2005, the greatest amount of water in the lake occurred in March when there were 8,000,000 acre-feet of water.

21. In 2004, the least amount of water in the lake occurred in June when there were 2,000,000 acre-feet of water.

22. In June of 2005, there were 5,000,000 acre-feet of water in the lake.

23. In January of 2004, there were 6,000,000 acre-feet of water in the lake.

24. March 2005: 8,000,000 acre-feet
June 2005: 5,000,000 acre-feet

This is a 8,000,000 − 5,000,000 = 3,000,000 acre-feet decrease.

percent of March 2005 amount = amount of decrease

$$p \cdot 8{,}000{,}000 = 3{,}000{,}000$$

$$\frac{p \cdot 8{,}000{,}000}{8{,}000{,}000} = \frac{3{,}000{,}000}{8{,}000{,}000}$$

$$p = \frac{3}{8} = 0.375 = 37.5\%$$

In 2005, the percent of decrease from March to June was 37.5%.

25. April 2004: 5,000,000 acre-feet
June 2004: 2,000,000 acre-feet

This is a 5,000,000 − 2,000,000 = 3,000,000 acre-feet decrease.

percent of April 2004 amount = amount of decrease

$$p \cdot 5{,}000{,}000 = 3{,}000{,}000$$

$$\frac{p \cdot 5{,}000{,}000}{5{,}000{,}000} = \frac{3{,}000{,}000}{5{,}000{,}000}$$

$$p = \frac{3}{5} = 0.6 = 60\%$$

In 2004, the percent of decrease from April to June was 60%.

26. In 2002, Center A sold \$50,000,000 worth of floor covering.

27. In 2004, Center A sold \$20,000,000 worth of floor covering.

28. In 2003, Center B sold \$20,000,000 worth of floor covering.

29. In 2005, Center B sold \$40,000,000 worth of floor covering.

30. Sales decreased for two years and then moved up slightly. Answers will vary. Perhaps there is less new construction, remodeling, and home improvement in the area near Center A. Also, better product selection and service may have reversed the decline in sales.

31. Sales are increasing. Answers will vary. New construction may have increased in the area near Center B, or greater advertising may attract more attention.

32. Degrees for Plumbing and electrical changes

$$= 10\% \text{ of } 360°$$
$$= (0.10)(360°) = 36°$$

33. Percent for Work stations (total = \$22,400)

$$= \frac{\$7840}{\$22{,}400} = 0.35 = 35\%$$

$$\text{Degrees} = (0.35)(360°) = 126°$$

34. Percent for Small appliances

$$= \frac{\$4480}{\$22{,}400} = 0.20 = 20\%$$

$$\text{Degrees} = (0.20)(360°) = 72°$$

35. Percent for Interior decoration

$$= \frac{\$5600}{\$22{,}400} = 0.25 = 25\%$$

$$\text{Degrees} = (0.25)(360°) = 90°$$

36. The unknowns are dollar amount and percent of total.

$$\text{Percent} = \frac{36°}{360°} = 0.10 \text{ or } 10\%$$

$$\text{Dollar amount} = (0.10)(\$22{,}400) = \$2240$$

37.

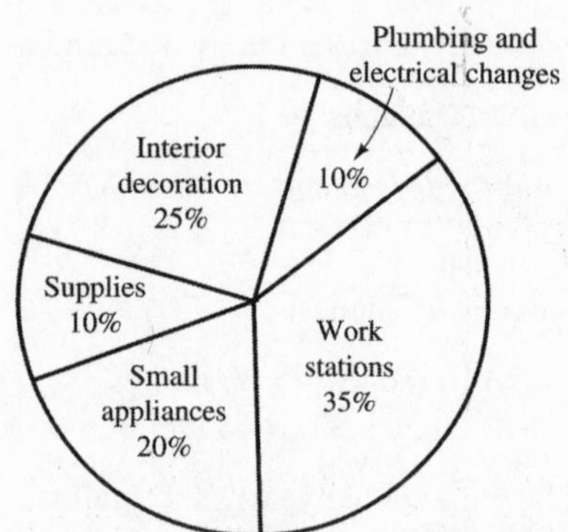

38.

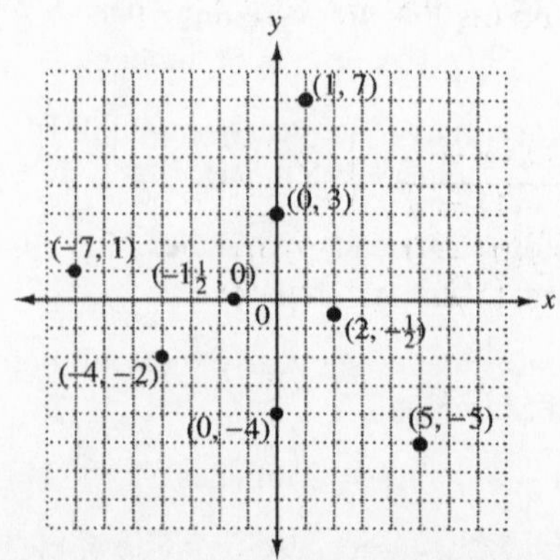

39. A is $(0, 6)$; B is approximately $\left(-2, 2\frac{1}{2}\right)$;
C is $(0, 0)$; D is $(-6, -6)$; E is $(4, 3)$;
F is approximately $\left(3\frac{1}{2}, 0\right)$; G is $(2, -4)$.

40. $x + y = -2$

x	y	(x, y)
0	-2	$(0, -2)$
1	-3	$(1, -3)$
2	-4	$(2, -4)$

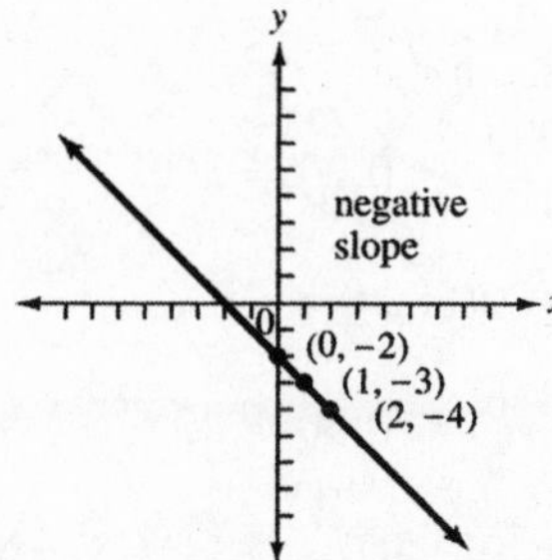

The graph of $x + y = -2$ has a *negative* slope. Two other solutions of $x + y = -2$ are $(-2, 0)$ and $(-1, -1)$.
All points on the line are solutions.

41. $y = x + 3$

x	$x + 3 = y$	(x, y)
0	$0 + 3 = 3$	$(0, 3)$
1	$1 + 3 = 4$	$(1, 4)$
2	$2 + 3 = 5$	$(2, 5)$

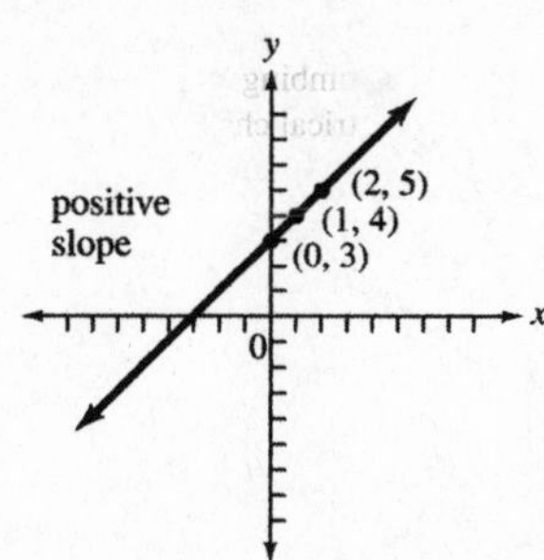

The graph of $y = x + 3$ has a *positive* slope. Two other solutions for $y = x + 3$ are $(-1, 2)$ and $(-2, 1)$.
All points on the line are solutions.

42. $y = -4x$

x	$-4 \cdot x = y$	(x, y)
-1	$-4(-1) = 4$	$(-1, 4)$
0	$-4(0) = 0$	$(0, 0)$
1	$-4(1) = -4$	$(1, -4)$

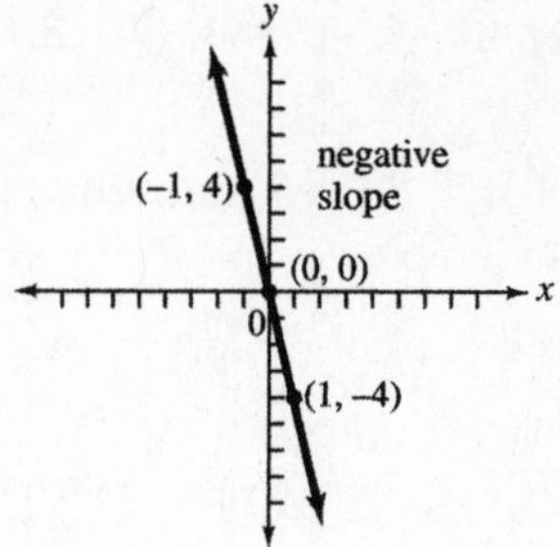

The graph of $y = -4x$ has a *negative* slope. Two other solutions of $y = -4x$ are $\left(-\frac{1}{2}, 2\right)$ and $\left(\frac{1}{2}, -2\right)$.
All points on the line are solutions.

Chapter 9 Test

1. **(a)** Look down the calcium column and find the largest number (371). Then look across to find the food. Sardines have the greatest amount of calcium.

(b) Look down the calcium column and find the smallest number (23). Then look across to find the food. Cream cheese has the least amount of calcium.

2. Set up a proportion.

$$\frac{95 \text{ calories}}{1 \text{ oz}} = \frac{x \text{ calories}}{1.75 \text{ oz}}$$
$$1 \cdot x = 95 \cdot 1.75$$
$$x \approx 166$$

There are approximately 166 calories in 1.75 ounces of Swiss cheese.

3. Set up a proportion.

$$\frac{345 \text{ mg}}{8 \text{ oz}} = \frac{x \text{ mg}}{6 \text{ oz}}$$
$$8 \cdot x = 345 \cdot 6$$
$$\frac{8x}{8} = \frac{2070}{8}$$
$$x \approx 259$$

There are approximately 259 mg of calcium in 6 oz of fruit-flavored yogurt.

For Exercises 4–6, each symbol represents 10 species.

4. **(a)** Birds are represented by 8 symbols, more than any other group.

(b) $8(10 \text{ species}) = 80$ species

5. Fish: $7(10 \text{ species}) = 70$ species
Reptiles: $1.5(10 \text{ species}) = 15$ species

There are $70 - 15 = 55$ more species of fish endangered than reptiles.

6. There are $6.5 + 8 + 1.5 + 7 + 3.5 = 26.5$ symbols total.

There are $26.5(10) = 265$ endangered species shown in the pictograph.

7. Look for the largest sector (or largest percent).

29% of \$2,800,000
$= 0.29(\$2{,}800{,}000) = \$812{,}000$

Television has the largest budget of \$812,000.

8. Look for the smallest sector (or smallest percent).

3% of \$2,800,000
$= 0.03(\$2{,}800{,}000) = \$84{,}000$

Miscellaneous has the smallest budget of \$84,000.

9. 11% of \$2,800,000
$= 0.11(\$2{,}800{,}000) = \$308{,}000$

\$308,000 is budgeted for internet advertising.

10. 23% of \$2,800,000
$= 0.23(\$2{,}800{,}000) = \$644{,}000$

\$644,000 is budgeted for newspaper ads.

11. In 2005, expenses exceeded income by

$\$21{,}000 - \$17{,}000 = \$4000.$

12. From 2003 to 2004, expenses increased by

$\$18{,}000 - \$13{,}000 = \$5000$

percent of	2003 expenses	=	amount of increase
p	$\cdot\ 13{,}000$	=	$5{,}000$

$$\frac{p \cdot 13{,}000}{13{,}000} = \frac{5{,}000}{13{,}000}$$

$$p = \frac{5}{13} \approx 0.38 = 38\%$$

The amount of increase in the student's expenses was \$5000 and the percent of increase was 38%.

13. In 2005, the student's income declined. Explanations will vary. Some possibilities are: laid off from work, changed jobs, was ill, cut down on hours worked.

14. 5500 students were enrolled at College A in 2003.

3000 students were enrolled at College B in 2004.

15. College B had a higher enrollment in 2006.

College A = 4500 students

College B = 5500 students

College B had $5500 - 4500 = 1000$ more students.

16. Explanations will vary. For example, College B may have added new courses or lowered tuition or added child care.

17. $\text{Percent} = \dfrac{\$168{,}000}{\$480{,}000} = 0.35 = 35\%$

$35\% \text{ of } 360° = 0.35(360°) = 126°$

18. $\text{Percent} = \dfrac{\$24{,}000}{\$480{,}000} = 0.05 = 5\%$

$5\% \text{ of } 360° = 0.05(360°) = 18°$

19. $\text{Percent} = \dfrac{\$96{,}000}{\$480{,}000} = 0.20 = 20\%$

$20\% \text{ of } 360° = 0.20(360°) = 72°$

20. $\text{Percent} = \dfrac{\$144{,}000}{\$480{,}000} = 0.30 = 30\%$

$30\% \text{ of } 360° = 0.30(360°) = 108°$

21. $\text{Percent} = \dfrac{36°}{360°} = 0.10 = 10\%$

$10\% \text{ of } \$480{,}000 = 0.10(\$480{,}000) = \$48{,}000$

22.

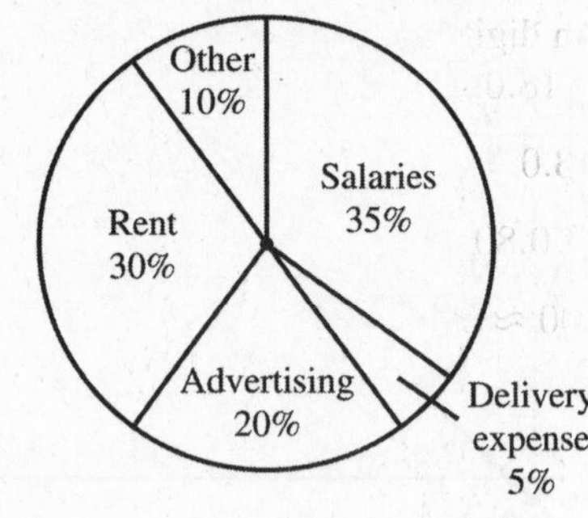

23.–26.

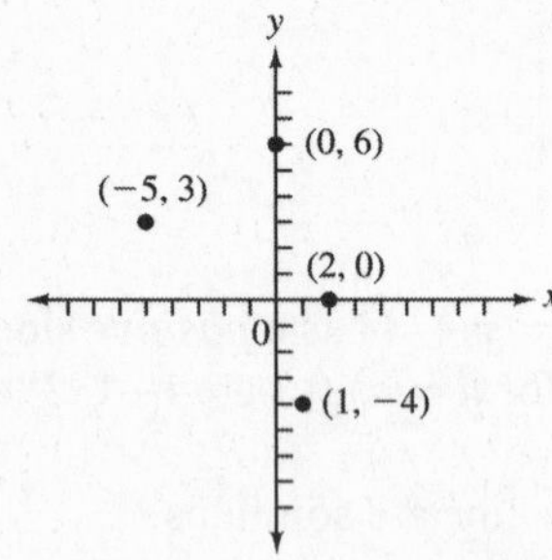

27. Point A is $(0, 0)$; no quadrant

28. Point B is $(-5, -4)$; quadrant III

29. Point C is $(3, 3)$; quadrant I

30. Point D is $(-2, 4)$; quadrant II

31. $y = x - 4$

x	$x - 4 = y$	(x, y)
0	$0 - 4 = -4$	$(0, -4)$
1	$1 - 4 = -3$	$(1, -3)$
2	$2 - 4 = -2$	$(2, -2)$

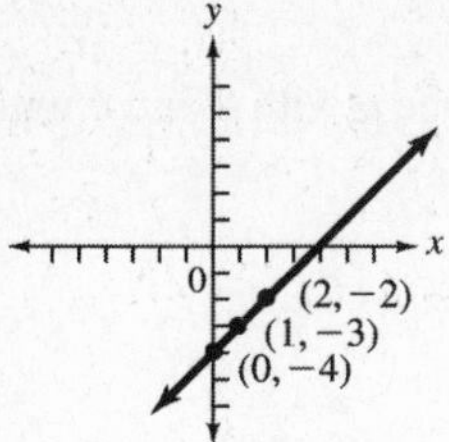

32. **(a)** Answers will vary; all points on the line are solutions. Some possibilities are $(3, -1)$ and $(4, 0)$.

(b) Since the graph of $y = x - 4$ rises, its slope is positive.

Cumulative Review Exercises (Chapters 1–9)

1. **(a)** In words, 0.0602 is six hundred two ten-thousandths.

(b) In words, 300,000,560 is three hundred million, five hundred sixty.

2. **(a)** Written in digits, seventy billion, five million, forty-three is 70,005,000,043.

(b) Written in digits, eighteen and nine hundredths is 18.09.

3. **(a)** $3.\underline{0}49 \approx 3.0$

(b) $0.7\underline{9}82 \approx 0.80$

(c) $6\underline{8},592,000 \approx 69,000,000$

4. Mean =

$$\frac{23 + 29 + 18 + 23 + 36 + 62 + 23 + 19 + 27 + 30}{10}$$

$$= \frac{290}{10} = 29 \text{ years}$$

Order the data:

$$18, 19, 23, 23, \underline{23}, \underline{27}, 29, 30, 36, 62$$

Since there is an even number of data points, the median is the average of the middle data points.

$$\text{Median} = \frac{23 + 27}{2} = \frac{50}{2} = 25 \text{ years}$$

Mode: 23 occurs the most times (3), so the mode is 23 years.

5. $\frac{4}{5} + 2\frac{1}{3} = \frac{4}{5} + \frac{7}{3} = \frac{12}{15} + \frac{35}{15} = \frac{12 + 35}{15}$

$= \frac{47}{15}$ or $3\frac{2}{15}$

6. $\frac{0.8}{-3.2}$ different signs, negative quotient

$$3.2\overline{)0.8} \rightarrow 32\overline{)8.00}$$

$$\begin{array}{r} 0.25 \\ 32\overline{)8.00} \\ \underline{6\;4} \\ 1\;60 \\ \underline{1\;60} \\ 0 \end{array}$$

Answer: -0.25

7. $5^2 + (-4)^3 = 25 + (-64) = -39$

8. $(0.002)(-0.05)$: different signs, negative product

$$\begin{array}{rl} 0.002 & \leftarrow 3 \text{ decimal places} \\ \times\ 0.05 & \leftarrow 2 \text{ decimal places} \\ \hline 0.00010 & \leftarrow 5 \text{ decimal places} \end{array}$$

Answer: -0.00010 or -0.0001

9. $\frac{4a}{9} \cdot \frac{6b}{2a^3} = \frac{\cancel{2}\cdot 2\cdot \cancel{a}\cdot 2\cdot \cancel{3}\cdot b}{\cancel{3}\cdot 3\cdot \cancel{2}\cdot \cancel{a}\cdot a\cdot a} = \frac{4b}{3a^2}$

10. $1\frac{1}{4} - 3\frac{5}{6} = \frac{5}{4} - \frac{23}{6} = \frac{15}{12} - \frac{46}{12} = \frac{15 - 46}{12}$

$= \frac{-31}{12} = -\frac{31}{12}$ or $-2\frac{7}{12}$

11. $-13 + 2.993$

$|-13| = 13, |2.993| = 2.993$

Since -13 has the larger absolute value, the sum will be negative.

Subtract:

$$\begin{array}{r} 13.000 \\ -\ \ 2.993 \\ \hline 10.007 \end{array}$$

Answer: -10.007

12. $\frac{2}{7} - \frac{8}{x} = \frac{2\cdot x}{7\cdot x} - \frac{8\cdot 7}{x\cdot 7} = \frac{2x}{7x} - \frac{56}{7x} = \frac{2x - 56}{7x}$

13. $-3 - 33 = -3 + (-33) = -36$

14. $\frac{10m}{3n^2} \div \frac{2m^2}{5n} = \frac{10m}{3n^2} \cdot \frac{5n}{2m^2}$

$= \frac{\cancel{2}\cdot 5\cdot \cancel{m}\cdot 5\cdot \cancel{n}}{3\cdot \cancel{n}\cdot n\cdot \cancel{2}\cdot \cancel{m}\cdot m}$

$= \frac{25}{3mn}$

15. $\frac{-3}{-\frac{9}{10}} = -\frac{3}{1} \div \left(-\frac{9}{10}\right) = -\frac{\cancel{3}}{1} \cdot \left(-\frac{10}{\cancel{9}}\right)$

$= \frac{10}{3}$ or $3\frac{1}{3}$

16. $\frac{3(-7)}{2^4 - 16} = \frac{-21}{16 - 16} = \frac{-21}{0}$, which is undefined.

17. $10-2\frac{5}{8}=\frac{10}{1}-\frac{21}{8}=\frac{80}{8}-\frac{21}{8}$
$=\frac{59}{8}$ or $7\frac{3}{8}$

18. $\frac{3}{w}+\frac{x}{6}=\frac{3\cdot 6}{w\cdot 6}+\frac{x\cdot w}{6\cdot w}$
$=\frac{18}{6w}+\frac{wx}{6w}=\frac{18+wx}{6w}$

19. $8+4(2-5)=8+4(-3)=8+(-12)=-4$

20. $\frac{7}{8}$ of $960=\frac{7}{\cancel{8}_1}\cdot\frac{\cancel{960}^{120}}{1}=840$

21. $\frac{(-4)^2+8(0-2)}{8\div 2(-3+5)-10}$

Numerator:
$(-4)^2+8(0-2)$
$=16+8(-2)$
$=16+(-16)$
$=0$

Denominator:
$8\div 2(-3+5)-10$
$=8\div 2(2)-10$
$=4(2)-10$
$=8-10$
$=-2$

The last step is division: $\frac{0}{-2}=0$

22. $0.5-0.25(3.2)^2$
$=0.5-0.25(10.24)$
$=0.5-2.56$
$=-2.06$

23. $6\left(-\frac{1}{2}\right)^3+\frac{2}{3}\left(\frac{3}{5}\right)=\frac{6}{1}\cdot\left(-\frac{1}{8}\right)+\frac{2\cdot\cancel{3}^1}{\cancel{3}_1\cdot 5}$
$=-\frac{\cancel{2}^1\cdot 3\cdot 1}{1\cdot\cancel{2}_1\cdot 4}+\frac{2}{5}$
$=-\frac{3}{4}+\frac{2}{5}$
$=-\frac{15}{20}+\frac{8}{20}$
$=\frac{-15+8}{20}$
$=\frac{-7}{20}$ or $-\frac{7}{20}$

24. $20+4c$ Replace c with -5.
$=20+4(-5)$
$=20+(-20)$
$=0$

25. $7b-4c$ Replace b with 3 and c with -5.
$=7(3)-4(-5)$
$=21-(-20)$
$=21+20$
$=41$

26. $-2ac^2$
$=-2\cdot a\cdot c\cdot c$ Replace a with 2 and c with -5.
$=-2\cdot 2\cdot(-5)\cdot(-5)$
$=-4(-5)(-5)$
$=20(-5)$
$=-100$

27. $-4x+x^2-x=-4x+x^2-1x$
$=-4x+x^2+(-1x)$
$=-5x+x^2$ or x^2-5x

28. $3(y-4)+2y=3y-12+2y$
$=5y-12$

29. $-5(8h^2)=(-5\cdot 8)h^2$
$=-40h^2$

30.
$$\begin{aligned}2x-3&=-20+3\\2x-3&=-17\\+3\quad&\quad+3\\2x+0&=-14\\\frac{2x}{2}&=\frac{-14}{2}\\x&=-7\end{aligned}$$

The solution is -7.

31.
$$\begin{aligned}-12&=3(y+2)\\-12&=3y+6\\-6\quad&\quad-6\\-18&=3y+0\\\frac{-18}{3}&=\frac{3y}{3}\\-6&=y\end{aligned}$$

The solution is -6.

32.
$$\begin{aligned}6x&=14-x\\6x&=14-1x\\+1x\quad&\quad+1x\\7x&=14+0\\\frac{7x}{7}&=\frac{14}{7}\\x&=2\end{aligned}$$

The solution is 2.

33.
$$\begin{aligned}-8&=\frac{2}{3}m+2\\-2\quad&\quad-2\\-10&=\frac{2}{3}m+0\\\frac{3}{2}(-10)&=\frac{3}{2}\left(\frac{2}{3}m\right)\\-15&=m\end{aligned}$$

The solution is -15.

34. $\dfrac{2}{13.5} = \dfrac{2.4}{n}$

$2 \cdot n = 13.5 \cdot 2.4$ Cross products

$\dfrac{2n}{2} = \dfrac{32.4}{2}$

$n = 16.2$

The solution is 16.2.

35. $-20 = -w$

$\dfrac{-20}{-1} = \dfrac{-1w}{-1}$

$20 = w$

The solution is 20.

36.
$$\begin{aligned} 3.4x - 6 &= 8 + 1.4x \\ +6 \quad & \quad +6 \\ \hline 3.4x + 0 &= 14 + 1.4x \\ 3.4x &= 14 + 1.4x \\ -1.4x \quad & \quad -1.4x \\ \hline 2.0x &= 14 + 0 \\ \frac{2x}{2} &= \frac{14}{2} \\ x &= 7 \end{aligned}$$

The solution is 7.

37.
$$\begin{aligned} 2(h-1) &= -3(h+12) - 11 \\ 2h - 2 &= -3h - 36 - 11 \\ 2h - 2 &= -3h + (-36) + (-11) \\ 2h - 2 &= -3h + (-47) \\ +3h \quad & \quad +3h \\ \hline 5h - 2 &= 0 + (-47) \\ 5h - 2 &= -47 \\ +2 \quad & \quad +2 \\ \hline 5h + 0 &= -45 \\ \frac{5h}{5} &= \frac{-45}{5} \\ h &= -9 \end{aligned}$$

The solution is –9.

38. Let n represent the number.

$\underbrace{\text{If five times a number is subtracted from 12,}}_{12 - 5n} \quad \underbrace{\text{the result is}}_{=} \quad \underbrace{\text{the number.}}_{n}$

$$\begin{aligned} 12 - 5n &= n \\ +5n \quad & \quad +5n \\ \hline 12 + 0 &= 6n \\ 12 &= 6n \\ \frac{12}{6} &= \frac{6n}{6} \\ 2 &= n \end{aligned}$$

The number is 2.

39. Let n represent the number.

$\underbrace{\text{When } -8}_{-8} \quad \underbrace{\text{is added to}}_{+} \quad \underbrace{\text{twice a number,}}_{2n} \quad \underbrace{\text{the result is}}_{=} \quad \underset{\downarrow}{-28.} \; -28$

$$\begin{aligned} -8 + 2n &= -28 \\ +8 \quad & \quad +8 \\ \hline 0 + 2n &= -20 \\ \frac{2n}{2} &= \frac{-20}{2} \\ n &= -10 \end{aligned}$$

The number is –10.

40. *Step 1*
Unknown: amount received by each person.
Known: total prize is \$1800, one person gets \$500 more

Step 2
Let m represent the amount of prize money received by one person. Then $m + 500$ represents the amount of prize money received by the other person.

Step 3
$m + (m + 500) = 1800$

Step 4
$$\begin{aligned} m + m + 500 &= 1800 \\ 2m + 500 &= 1800 \\ -500 \quad & \quad -500 \\ \hline 2m &= 1300 \\ \frac{2m}{2} &= \frac{1300}{2} \\ m &= 650 \end{aligned}$$

Step 5
One person receives \$650 and the other receives \$650 + \$500 = \$1150.

Step 6
\$1150 is \$500 more than \$650 and the sum of \$650 and \$1150 is \$1800.

41. *Step 1*
Unknown: length and width
Known: perimeter is 280 feet, length is 3 times width

Step 2
Let w represent the width. Then $3w$ represents the length.

Step 3
$2L + 2W = P$

Step 4

$$2(3w) + 2w = 280$$
$$6w + 2w = 280$$
$$8w = 280$$
$$\frac{8w}{8} = \frac{280}{8}$$
$$w = 35$$

Step 5
The width is 35 feet and the length is $3(35) = 105$ feet.

Step 6
105 feet is 3 times 35 feet and
$35 + 35 + 105 + 105 = 280$ feet.

42. $P = 10 \text{ ft} + 8 \text{ ft} + 10 \text{ ft} + 8 \text{ ft} = 36 \text{ ft}$
$A = b \cdot h = 10 \text{ ft} \cdot 7 \text{ ft} = 70 \text{ ft}^2$

43. $P = 48 \text{ m} + 46 \text{ m} + 22 \text{ m} + 22 \text{ m} + 26 \text{ m} + 24 \text{ m}$
$= 188 \text{ m}$

To find the area, draw a horizontal line to cut the figure into two pieces: a rectangle and a square.

$$A = \text{Area rectangle} + \text{Area square}$$
$$= l \cdot w + s \cdot s$$
$$= 48 \text{ m} \cdot 24 \text{ m} + 22 \text{ m} \cdot 22 \text{ m}$$
$$= 1152 \text{ m}^2 + 484 \text{ m}^2$$
$$= 1636 \text{ m}^2$$

44. The figure is a right triangle with unknown hypotenuse.

$$\text{hypotenuse} = \sqrt{(\text{leg})^2 + (\text{leg})^2}$$
$$x = \sqrt{(7)^2 + (24)^2}$$
$$x = \sqrt{49 + 576}$$
$$x = \sqrt{625}$$
$$x = 25 \text{ mi}$$

$$P = 24 \text{ mi} + 7 \text{ mi} + 25 \text{ mi}$$
$$P = 56 \text{ mi}$$

$$A = \frac{1}{2} \cdot b \cdot h$$
$$A = 0.5(7 \text{ mi})(24 \text{ mi})$$
$$A = 84 \text{ mi}^2$$

45. $C = \pi \cdot d$
$C \approx 3.14 \cdot 6 \text{ ft}$
$C = 18.84 \text{ ft}$
$C \approx 18.8 \text{ ft}$

$A = \pi \cdot r \cdot r$
$A \approx 3.14 \cdot 3 \text{ ft} \cdot 3 \text{ ft}$
$A = 28.26 \text{ ft}^2$
$A \approx 28.3 \text{ ft}^2$

46. $V = \pi r^2 h$
$V = \pi \cdot r \cdot r \cdot h$
$V \approx 3.14 \cdot 10 \text{ m} \cdot 10 \text{ m} \cdot 12 \text{ m}$
$V = 3768 \text{ m}^3$

47. $V = l \cdot w \cdot h$

$$V = 4 \text{ yd} \cdot 3 \text{ yd} \cdot 2\frac{1}{2} \text{ yd}$$
$$V = \frac{\overset{2}{\cancel{4}}}{1} \text{ yd} \cdot \frac{3}{1} \text{ yd} \cdot \frac{5}{\underset{1}{\cancel{2}}} \text{ yd}$$
$$V = 30 \text{ yd}^3$$

48. $2\frac{1}{4}$ hr to minutes

$$2\frac{1}{4} \text{ hr} = \frac{9 \cancel{\text{hr}}}{\underset{1}{\cancel{4}}} \cdot \frac{\overset{15}{\cancel{60}} \text{ min}}{1 \cancel{\text{hr}}}$$
$$= 135 \text{ minutes}$$

49. 54 in. to feet

$$54 \text{ in.} = \frac{\overset{9}{\cancel{54}} \cancel{\text{in.}}}{1} \cdot \frac{1 \text{ ft}}{\underset{2}{\cancel{12}} \cancel{\text{in.}}}$$
$$= \frac{9}{2} \text{ ft} = 4\frac{1}{2} \text{ or } 4.5 \text{ ft}$$

50. 1.85 L to milliliters

$$1.85 \text{ L} = \frac{1.85 \cancel{\text{L}}}{1} \cdot \frac{1000 \text{ mL}}{1 \cancel{\text{L}}} = 1850 \text{ mL}$$

Or use the metric conversion line. mL is 3 places to the right of L, so move the decimal point 3 places right:

1.850 L = 1850 mL

51. 35 mm to centimeters

$$35 \text{ mm} = \frac{35 \cancel{\text{mm}}}{1} \cdot \frac{1 \text{ cm}}{10 \cancel{\text{mm}}} = \frac{35}{10} \text{ cm} = 3.5 \text{ cm}$$

Or use the metric conversion line. cm is one place to the left of mm, so move the decimal point 1 place left:

3 5. mm = 3.5 cm

52. 10 g to kilograms

$$10 \text{ g} = \frac{10 \cancel{\text{g}}}{1} \cdot \frac{1 \text{ kg}}{1000 \cancel{\text{g}}} = \frac{10}{1000} \text{ kg}$$
$$= \frac{1}{100} \text{ kg} = 0.01 \text{ kg}$$

Or use the metric conversion line. kg is 3 places to the left of g, so move the decimal 3 places left:

10. g = 010. g = 0.01 kg

53. 25°F to Celsius

$$C = \frac{5(F-32)}{9}$$
$$C = \frac{5(25-32)}{9}$$
$$C = \frac{5(-7)}{9}$$
$$C = \frac{-35}{9} = -3.\overline{8} \approx -4°$$

25°F ≈ −4°C

54. If there are 19 nonsmokers out of 25 adults, how many (n) are nonsmokers out of 732 employees?

$$\frac{19}{25} = \frac{n}{732}$$
$$25 \cdot n = 19 \cdot 732$$
$$\frac{25n}{25} = \frac{13{,}908}{25}$$
$$n = 556.32$$

About 556 employees would be nonsmokers.

55. Original price • Discount rate = Discount amount

$129 • 0.15 = Discount amount

$19.35 = Discount amount

Sale price = Original price − Discount amount

= $129 − $19.35

= $109.65

Tax amount:

Price • Tax Rate = Tax amount

$109.65 • 0.065 = Tax amount

$7.13 ≈ Tax amount

Total cost = $109.65 + $7.13 = $116.78

Her total cost for the phone was $116.78 (rounded).

56. Part: 167 points
Whole: 180 points
Percent: unknown (p)

percent • whole = part

$$p \cdot 180 = 167$$
$$\frac{180p}{180} = \frac{167}{180}$$
$$p \approx 0.928 = 92.8\%$$

She earned 92.8% (rounded) of the points.

57. **(a)** $30,000 grant to purchase $957 computers.

```
       3 1
957)3 0, 0 0 0
    2 8 7 1
      1 2 9 0
        9 5 7
        3 3 3
```

With the grant money, Century College could purchase 31 computers and have $333 left over.

(b) $333 left over to purchase $19 calculators.

```
    1 7
19)3 3 3
   1 9
   1 4 3
   1 3 3
     1 0
```

With the $333, Century College could purchase 17 calculators and have $10 left over.

58. Purchased: $2\frac{1}{4}$ pounds

Used: $5\left(\frac{1}{6}\text{ pound}\right) = \frac{5}{1}\left(\frac{1}{6}\text{ pound}\right) = \frac{5}{6}$ pound

$$\text{Amount left} = 2\frac{1}{4} - \frac{5}{6}$$
$$= \frac{9}{4} - \frac{5}{6}$$
$$= \frac{27}{12} - \frac{10}{12}$$
$$= \frac{17}{12} \text{ or } 1\frac{5}{12} \text{ pounds}$$

There are $1\frac{5}{12}$ pounds of meat left.

59.

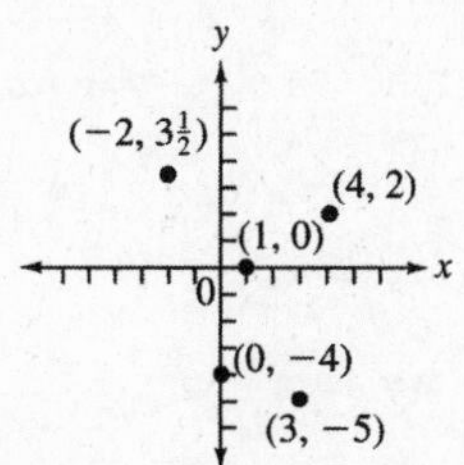

Second quadrant point: $\left(-2, 3\frac{1}{2}\right)$

Third quadrant points: none

60. $y = x + 6$

x	$x + 6 = y$	(x, y)
0	$0 + 6 = 6$	$(0, 6)$
-1	$-1 + 6 = 5$	$(-1, 5)$
-2	$-2 + 6 = 4$	$(-2, 4)$

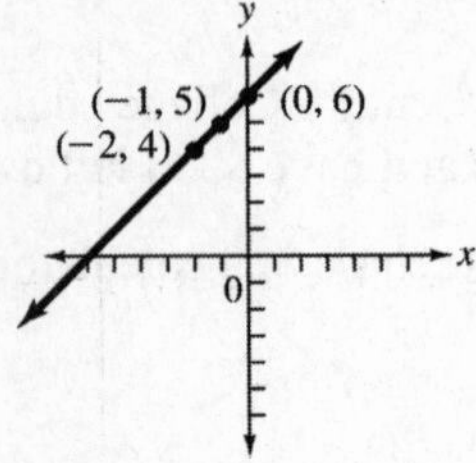

The line has a *positive* slope. All points on the line are solutions, so there are many possibilities such as $(-3, 3)$ and $(-4, 2)$.

CHAPTER 10 EXPONENTS AND POLYNOMIALS

10.1 The Product Rule and Power Rules for Exponents

10.1 Section Exercises

1. $xy^2 = x^1y^2$

The understood exponent for x is 1.

3. $3^3 = 3 \cdot 3 \cdot 3 = 27$, so the statement $3^3 = 9$ is *false*.

5. $(a^2)^3 = a^{2(3)} = a^6$, so the statement $(a^2)^3 = a^5$ is *false*.

7. $\underbrace{(-2)(-2)(-2)(-2)(-2)}_{\text{5 factors of } -2} = (-2)^5$

9. $\left(\frac{1}{2}\right)\left(\frac{1}{2}\right)\left(\frac{1}{2}\right)\left(\frac{1}{2}\right)\left(\frac{1}{2}\right)\left(\frac{1}{2}\right) = \left(\frac{1}{2}\right)^6$

11. $(-8p)(-8p) = (-8p)^2$

13. In $(-3)^4$, -3 is the base.

$$(-3)^4 = (-3)(-3)(-3)(-3) = 81$$

In -3^4, 3 is the base.

$$-3^4 = -(3 \cdot 3 \cdot 3 \cdot 3) = -81$$

15. In the exponential expression 3^5, the base is 3 and the exponent is 5.

$$3^5 = 3 \cdot 3 \cdot 3 \cdot 3 \cdot 3 = 243$$

17. In the expression $(-3)^5$, the base is -3 and the exponent is 5.

$$(-3)^5 = (-3)(-3)(-3)(-3)(-3) = -243$$

19. In the expression $(-6x)^4$, the base is $-6x$ and the exponent is 4.

21. In the expression $-6x^4$, -6 is not part of the base. The base is x and the exponent is 4.

23. The product rule does not apply to $5^2 + 5^3$ because the expression is a sum, not a product. The product rule would apply if we had $5^2 \cdot 5^3$.

$$5^2 + 5^3 = 25 + 125 = 150$$

25. $5^2 \cdot 5^6 = 5^{2+6} = 5^8$

27. $4^2 \cdot 4^7 \cdot 4^3 = 4^{2+7+3} = 4^{12}$

29. $(-7)^3(-7)^6 = (-7)^{3+6} = (-7)^9$

31. $t^3 \cdot t \cdot t^{13} = t^{3+1+13} = t^{17}$

33. $$\begin{aligned}(-8r^4)(7r^3) &= -8 \cdot 7 \cdot r^4 \cdot r^3 \\ &= -56r^{4+3} \\ &= -56r^7\end{aligned}$$

35. $$\begin{aligned}(-6p^5)(-7p^5) &= (-6)(-7)p^5 \cdot p^5 \\ &= 42p^{5+5} \\ &= 42p^{10}\end{aligned}$$

37. $5x^4 + 9x^4 = (5+9)x^4 = 14x^4$

$(5x^4)(9x^4) = (5 \cdot 9)x^{4+4} = 45x^8$

39. $-7a^2 + 2a^2 + 10a^2 = (-7+2+10)a^2 = 5a^2$

$$\begin{aligned}(-7a^2)(2a^2)(10a^2) &= (-7 \cdot 2 \cdot 10)a^{2+2+2} \\ &= -140a^6\end{aligned}$$

41. $$\begin{aligned}(4^3)^2 &= 4^{3 \cdot 2} \quad \textit{Power rule (a)} \\ &= 4^6\end{aligned}$$

43. $(t^4)^5 = t^{4 \cdot 5} = t^{20}$ *Power rule (a)*

45. $(7r)^3 = 7^3r^3$ *Power rule (b)*

47. $(5xy)^5 = 5^5x^5y^5$ *Power rule (b)*

49. $8(qr)^3 = 8q^3r^3$ *Power rule (b)*

51. $\left(\frac{1}{2}\right)^3 = \frac{1^3}{2^3} = \frac{1}{2^3}$ *Power rule (c)*

53. $\left(\frac{a}{b}\right)^3 (b \neq 0) = \frac{a^3}{b^3}$ *Power rule (c)*

55. $\left(\frac{9}{5}\right)^8 = \frac{9^8}{5^8}$ *Power rule (c)*

57. $$\begin{aligned}(-2x^2y)^3 &= (-2)^3(x^2)^3y^3 \\ &= (-2)^3x^6y^3\end{aligned}$$

59. $$\begin{aligned}(3a^3b^2)^2 &= 3^2(a^3)^2(b^2)^2 \\ &= 3^2a^6b^4\end{aligned}$$

61. Use the formula for the area of a rectangle, $A = LW$, with $L = 4x^3$ and $W = 3x^2$.

$$\begin{aligned}A &= (4x^3)(3x^2) \\ &= 4 \cdot 3 \cdot x^3 \cdot x^2 \\ &= 12x^5\end{aligned}$$

10.2 Integer Exponents and the Quotient Rule

10.2 Section Exercises

1. $(-2)^{-3} = \frac{1}{(-2)^3}$ is negative, because $(-2)^3$ is a negative number raised to an odd exponent, which is negative, and the quotient of a positive number and a negative number is a negative number.

3. $-2^4 = -(2^4)$ is negative, because 2^4 is positive.

5. $(-2)^6$ is positive, because the exponent is even.

7. $1-5^0=1-1=0$

9. $(-4)^0=1$ *Definition of zero exponent*

11. $-9^0=-(9^0)=-(1)=-1$

13. $(-2)^0-2^0=1-1=0$

15. $\frac{0^{10}}{10^0}=\frac{0}{1}=0$

17. $7^0+9^0=1+1=2$

19. $b^0=1 \ (b\neq 0)$

21. $15x^0=15(1)=15 \ (x\neq 0)$

23. $4^{-3}=\frac{1}{4^3}$ *Definition of negative exponent*

$=\frac{1}{64}$

25. $5^{-1}+3^{-1}=\frac{1}{5}+\frac{1}{3}$

$=\frac{3}{15}+\frac{5}{15}=\frac{8}{15}$

27. $x^{-4}=\frac{1}{x^4}$

29. $\frac{25}{25}=1$

30. $\frac{25}{25}=\frac{5^2}{5^2}$

31. $\frac{5^2}{5^2}=5^{2-2}=5^0$

32. $5^0=1$; This supports the definition of an exponent of 0.

33. $\frac{6^7}{6^2}=6^{7-2}=6^5$

35. $\frac{10^4}{10}=\frac{10^4}{10^1}=10^{4-1}=10^3$

37. $\frac{y^2}{y^6}=y^{2-6}=y^{-4}=\frac{1}{y^4}$

39. $\frac{c^6}{c^5}=c^{6-5}=c^1$ or c

41. $\frac{5^3}{5^7}=5^{3-7}=5^{-4}=\frac{1}{5^4}$

43. $\frac{m^7}{m^8}=m^{7-8}=m^{-1}=\frac{1}{m^1}$ or $\frac{1}{m}$

45. $\frac{3^{-4}}{3^{-8}}=3^{-4-(-8)}=3^{-4+8}=3^4$

47. $\frac{a^{-2}}{a^{-5}}=a^{-2-(-5)}=a^{-2+5}=a^3$

49. $\frac{2^{-10}}{2^{-2}}=2^{-10-(-2)}=2^{-10+2}=2^{-8}=\frac{1}{2^8}$

51. $\frac{r^{-12}}{r^{-8}}=r^{-12-(-8)}=r^{-12+8}=r^{-4}=\frac{1}{r^4}$

53. $\frac{10^6}{10^{-4}}=10^{6-(-4)}=10^{6+4}=10^{10}$

55. $\frac{10^{-2}}{10^3}=10^{-2-3}=10^{-5}=\frac{1}{10^5}$

57. $\frac{10^3}{10^{-8}}=10^{3-(-8)}=10^{3+8}=10^{11}$

59. $\frac{10^{-4}}{10^4}=10^{-4-4}=10^{-8}=\frac{1}{10^8}$

61. $10^6(10^{-2})=10^{6+(-2)}=10^4$

63. $a^{-4}(a^3)=a^{-4+3}=a^{-1}=\frac{1}{a^1}$ or $\frac{1}{a}$

65. $(2^{-4})(2^{-4})=2^{-4+(-4)}=2^{-8}=\frac{1}{2^8}$

67. $(x^{-10})(x^{-1})=x^{-10+(-1)}=x^{-11}=\frac{1}{x^{11}}$

69. $10^8\cdot 10^{-2}\cdot 10^{-4}=10^{8+(-2)+(-4)}=10^2$

71. $(y^{-3})(y^5)(y^{-4})=y^{-3+5+(-4)}=y^{-2}=\frac{1}{y^2}$

73. $m^{-6}\cdot m^3\cdot m^{12}=m^{-6+3+12}=m^9$

75. $(10^{-1})(10^{-2})(10^{-3})=10^{-1+(-2)+(-3)}$

$=10^{-6}=\frac{1}{10^6}$

Summary Exercises on Exponents

1. $(-3)^2=(-3)(-3)=9$

3. $15^0=1$

5. $p^0=1$

7. $-3^2=-(3\cdot 3)=-9$

9. $-2^3=-(2\cdot 2\cdot 2)=-8$

11. $6^{-1}+2^{-2}=\frac{1}{6^1}+\frac{1}{2^2}$

$=\frac{1}{6}+\frac{1}{4}$

$=\frac{4}{24}+\frac{6}{24}$

$=\frac{4+6}{24}=\frac{10}{24}=\frac{5}{12}$

13. $2^{-5}=\frac{1}{2^5}=\frac{1}{2\cdot 2\cdot 2\cdot 2\cdot 2}=\frac{1}{32}$

15. $(-2)^3=(-2)(-2)(-2)=-8$

17. $(6^4)(6^4)=6^{4+4}=6^8$

19. $10^2(10^{-10})=10^{2+(-10)}=10^{-8}=\frac{1}{10^8}$

21. $(y^{-12})(y^6)(y^6) = y^{-12+6+6} = y^0 = 1$

23. $(-5)^2(-5)^8 = (-5)^{2+8} = (-5)^{10}$

25. $(m^6)^4 = m^{6\cdot4} = m^{24}$

27. $(10^{-2})(10^{-2})(10^{-2}) = 10^{-2+(-2)+(-2)}$
$= 10^{-6} = \dfrac{1}{10^6}$

29. $4x^4 \cdot 4x^4 = 4\cdot4x^{4+4} = 16x^8$

31. $(10^4)^2 = 10^{4\cdot2} = 10^8$

33. $\left(\dfrac{x}{y}\right)^6 = \dfrac{x^6}{y^6}$

35. $\dfrac{7^{12}}{7^{11}} = 7^{12-11} = 7^1$ or 7

37. $-8a^3(2a) = -8\cdot2a^{3+1} = -16a^4$

39. $(5xy)^3 = 5^3x^3y^3$
$= 5\cdot5\cdot5x^3y^3$ or $125x^3y^3$

41. $\dfrac{10^{-3}}{10^4} = 10^{-3-4} = 10^{-7} = \dfrac{1}{10^7}$

43. $(b^3)^5 = b^{3\cdot5} = b^{15}$

45. $10^{-8}(10^5) = 10^{-8+5} = 10^{-3} = \dfrac{1}{10^3}$

47. $\dfrac{10^5}{10^{-8}} = 10^{5-(-8)} = 10^{5+8} = 10^{13}$

49. $10(ab)^4 = 10(a)^4(b)^4 = 10a^4b^4$

10.3 An Application of Exponents: Scientific Notation

10.3 Section Exercises

1. $0.0021 = 2.1 \times 10^{-3}$

Move the decimal point 3 places. The *original* number was "small" (between 0 and 1), so the exponent is *negative* 3.

$317.83 = 3.1783 \times 10^2$

Move the decimal point 2 places. The *original* number was "large" (10 or more), so the exponent is *positive* 2.

3. $831{,}200{,}000 = 8.312 \times 10^8$

Move the decimal point 8 places. The *original* number was "large" (10 or more), so the exponent is *positive* 8.

$319{,}000{,}000 = 3.19 \times 10^8$

Move the decimal point 8 places. The *original* number was "large" (10 or more), so the exponent is *positive* 8.

5. 4.56×10^3 is written in scientific notation because 4.56 is between 1 and 10, and 10^3 is a power of 10.

7. 5,600,000 is not written in scientific notation. It can be written in scientific notation as 5.6×10^6.

9. 0.004 is not written in scientific notation because $|0.004| = 0.004$ is not between 1 and 10. It can be written in scientific notation as 4×10^{-3}.

11. 0.8×10^2 is not written in scientific notation because $|0.8| = 0.8$ is not greater than or equal to 1 and less than 10. It can be written in scientific notation as 8×10^1.

13. To write a number in scientific notation ($a \times 10^n$) move the decimal point to the right of the first nonzero digit. The number of places moved is the absolute value of the exponent n. If the original number is "large," then n is positive. If the original number is "small," then n is negative.

15. 5,876,000,000

Move the decimal point to the right of the first nonzero digit and count the number of places the decimal point was moved.

5.876,000,000 *9 places*

Because moving the decimal point to the *left* made the number *smaller*, we must multiply by a *positive* power of 10 so that the product 5.876×10^n will equal the larger number. Thus, $n = 9$, and

$$5{,}876{,}000{,}000 = 5.876 \times 10^9.$$

17. 82,350

Move the decimal point left 4 places so it is to the right of the first nonzero digit.

8.2350 *4 places*

Since the number got smaller, multiply by a positive power of 10.

$$82{,}350 = 8.2350 \times 10^4 = 8.235 \times 10^4$$

(Note that the final zero need not be written.)

19. 0.000007

Move the decimal point to the right of the first nonzero digit.

0.000007 *6 places*

Since moving the decimal point to the *right* made the number *larger*, we must multiply by a *negative* power of 10 so that the product 7×10^n will equal the smaller number. Thus, $n = -6$, and

$$0.000007 = 7 \times 10^{-6}.$$

21. 0.00203

To move the decimal point to the right of the first nonzero digit, we move it 3 places. Since 2.03 is larger than 0.00203, the exponent on 10 must be negative.

$$0.00203 = 2.03 \times 10^{-3}$$

23. 7.5×10^5

Since the exponent is positive, make 7.5 larger by moving the decimal point 5 places to the right.

$$7.5 \times 10^5 = 750{,}000$$

25. 5.677×10^{12}

Since the exponent is positive, make 5.677 larger by moving the decimal point 12 places to the right. We need to add 9 zeros.

$$5.677 \times 10^{12} = 5{,}677{,}000{,}000{,}000$$

27. 6.21×10^0

Because the exponent is 0, the decimal point should not be moved.

$$6.21 \times 10^0 = 6.21$$

We know this result is correct because $10^0 = 1$.

29. 7.8×10^{-4}

Since the exponent is negative, make 7.8 smaller by moving the decimal point 4 places to the left.

$$7.8 \times 10^{-4} = 0.00078$$

31. 5.134×10^{-9}

Since the exponent is negative, make 5.134 smaller by moving the decimal point 9 places to the left.

$$5.134 \times 10^{-9} = 0.000\,000\,005\,134$$

33. $(2 \times 10^8)(3 \times 10^3)$

$= (2 \times 3)(10^8 \times 10^3)$ *Commutative and associative properties*

$= 6 \times 10^{11}$ *Product rule for exponents*

$= 600{,}000{,}000{,}000$

35. $(5 \times 10^4)(3 \times 10^2)$
$= (5 \times 3)(10^4 \times 10^2)$
$= 15 \times 10^6$
$= 1.5 \times 10^7$
$= 15{,}000{,}000$

37. $(3.15 \times 10^{-4})(2.04 \times 10^8)$
$= (3.15 \times 2.04)(10^{-4} \times 10^8)$

$= 6.426 \times 10^4$ *Scientific notation*

$= 64{,}260$ *Without exponents*

39. $\dfrac{9 \times 10^{-5}}{3 \times 10^{-1}} = \dfrac{9}{3} \times \dfrac{10^{-5}}{10^{-1}}$
$= 3 \times 10^{-5-(-1)}$
$= 3 \times 10^{-4}$
$= 0.0003$

41. $\dfrac{8 \times 10^3}{2 \times 10^2} = \dfrac{8}{2} \times \dfrac{10^3}{10^2}$
$= 4 \times 10^1$
$= 40$

43. $\dfrac{(2.6 \times 10^{-3})(7.0 \times 10^{-1})}{(2 \times 10^2)(3.5 \times 10^{-3})}$

$= \dfrac{(2.6 \times 7.0)(10^{-3} \times 10^{-1})}{(2 \times 3.5)(10^2 \times 10^{-3})}$

$= \dfrac{18.2 \times 10^{-4}}{7 \times 10^{-1}}$

$= \dfrac{18.2}{7} \times \dfrac{10^{-4}}{10^{-1}}$

$= 2.6 \times 10^{-3}$
$= 0.0026$

45. $3{,}700{,}000 = 3.7 \times 10^6$

$\dfrac{1.304 \times 10^9 \text{ people}}{3.7 \times 10^6 \text{ mi}^2} = \dfrac{1.304}{3.7} \times \dfrac{10^9}{10^6}$
$\approx 0.352 \times 10^3$
$= 3.52 \times 10^2$ (rounded)
$= 352$ people per square mile

47. **(a)** 1 million $= 1 \times 10^6$

$(1 \times 10^{-10} \text{ m})(1 \times 10^6) = (1 \times 1)(10^{-10} \times 10^6)$
$= 1 \times 10^{-4}$
$= 0.0001$ m

(b) 1 billion $= 1 \times 10^9$

$$\begin{aligned}(1 \times 10^{-10}\text{ m})(1 \times 10^9) &= (1 \times 1)(10^{-10} \times 10^9)\\ &= 1 \times 10^{-1}\\ &= 0.1\text{ m}\end{aligned}$$

49. $$\begin{aligned}\frac{2 \times 10^{10}}{2.88 \times 10^8} &= \frac{2}{2.88} \times \frac{10^{10}}{10^8}\\ &\approx 0.694 \times 10^2\\ &= 6.94 \times 10^1\end{aligned}$$

or about 69 items

10.4 Adding and Subtracting Polynomials

10.4 Section Exercises

1. In the term $7x^5$, the coefficient is 7 and the exponent is 5.

3. The degree of the term $-4x^8$ is 8, the exponent.

5. When $x^2 + 10$ is evaluated for $x = 4$, the result is

$$4^2 + 10 = 16 + 10 = 26.$$

7. The polynomial $6x^4$ has one term. The coefficient of this term is 6.

9. The polynomial t^4 has one term. Since $t^4 = 1 \cdot t^4$, the coefficient of this term is 1.

11. The polynomial $-19r^2 - r$ has two terms. The coefficient of r^2 is -19 and the coefficient of r is -1.

13. $x + 8x^2 - 8x^3 = 1x + 8x^2 - 8x^3$ has 3 terms, and the coefficients are 1, 8, and -8.

15. $-3m^5 + 5m^5 = (-3 + 5)m^5 = 2m^5$

17. $$\begin{aligned}2r^5 + (-3r^5) &= [2 + (-3)]r^5\\ &= -1r^5 = -r^5\end{aligned}$$

19. The polynomial $0.2m^5 - 0.5m^2$ cannot be simplified. The two terms are unlike because the exponents on the variables are different, so they cannot be combined. The polynomial is already written in descending powers.

21. $$\begin{aligned}-3x^5 + 2x^5 - 4x^5 &= (-3 + 2 - 4)x^5\\ &= -5x^5\end{aligned}$$

23. $$\begin{aligned}-4p^7 + 8p^7 + 5p^9 &= (-4 + 8)p^7 + 5p^9\\ &= 4p^7 + 5p^9\end{aligned}$$

In descending powers of the variable, this polynomial is written $5p^9 + 4p^7$.

25. $$\begin{aligned}&-4y^2 + 3y^2 - 2y^2 + y^2\\ &= (-4 + 3 - 2 + 1)y^2\\ &= -2y^2\end{aligned}$$

27. $6x^4 - 9x$

This polynomial has no like terms, so it is already simplified. It is already written in descending powers of the variable x. The highest degree of any nonzero term is 4, so the degree of the polynomial is 4. There are two terms, so this is a *binomial.*

29. $5m^4 - 3m^2 + 6m^5 - 7m^3$

This polynomial is already simplified (no like terms). In descending powers, it is $6m^5 + 5m^4 - 7m^3 - 3m^2$. The degree is 5 (the largest exponent on the variable m). The polynomial has four terms, so it is neither a monomial, nor a binomial, nor a trinomial.

31. $$\begin{aligned}&\frac{5}{3}x^4 - \frac{2}{3}x^4 + \frac{1}{3}x^2 - 4\\ &= \left(\frac{5}{3} - \frac{2}{3}\right)x^4 + \frac{1}{3}x^2 - 4\\ &= x^4 + \frac{1}{3}x^2 - 4\end{aligned}$$

The resulting polynomial is a *trinomial* of degree 4.

33. $$\begin{aligned}&0.8x^4 - 0.3x^4 - 0.5x^4 + 7x\\ &= (0.8 - 0.3 - 0.5)x^4 + 7x\\ &= 0x^4 + 7x = 7x\end{aligned}$$

Since $7x$ can be written as $7x^1$, the degree of the polynomial is 1. The simplified polynomial has one term, so it is a *monomial.*

35. (a) $$\begin{aligned}-2x + 3 &= -2(2) + 3 \quad \textit{Let } x = 2.\\ &= -4 + 3\\ &= -1\end{aligned}$$

(b) $$\begin{aligned}-2x + 3 &= -2(-1) + 3 \quad \textit{Let } x = -1.\\ &= 2 + 3\\ &= 5\end{aligned}$$

37. (a) $$\begin{aligned}&2x^2 + 5x + 1\\ &= 2(2)^2 + 5(2) + 1 \quad \textit{Let } x = 2.\\ &= 2(4) + 10 + 1\\ &= 8 + 10 + 1\\ &= 18 + 1\\ &= 19\end{aligned}$$

(b) $$\begin{aligned}&2x^2 + 5x + 1\\ &= 2(-1)^2 + 5(-1) + 1 \quad \textit{Let } x = -1.\\ &= 2(1) - 5 + 1\\ &= 2 - 5 + 1\\ &= -3 + 1\\ &= -2\end{aligned}$$

39. **(a)** $2x^5 - 4x^4 + 5x^3 - x^2$
$= 2(2)^5 - 4(2)^4 + 5(2)^3 - (2)^2$ *Let x = 2.*
$= 2(32) - 4(16) + 5(8) - 4$
$= 64 - 64 + 40 - 4$
$= 36$

(b) $2x^5 - 4x^4 + 5x^3 - x^2$
$= 2(-1)^5 - 4(-1)^4 + 5(-1)^3 - (-1)^2$ *Let x = −1.*
$= 2(-1) - 4(1) + 5(-1) - 1$
$= -2 - 4 - 5 - 1$
$= -12$

41. **(a)** $-4x^5 + x^2$
$= -4(2)^5 + (2)^2$ *Let x = 2.*
$= -4(32) + 4$
$= -128 + 4$
$= -124$

(b) $-4x^5 + x^2$
$= -4(-1)^5 + (-1)^2$ *Let x = −1.*
$= -4(-1) + 1$
$= 4 + 1$
$= 5$

43. $3m^2 + 5m$ and $2m^2 - 2m$

$$\begin{array}{l} 3m^2 + 5m \\ \underline{2m^2 - 2m} \\ 5m^2 + 3m \end{array}$$

45. $-6x^3 - 4x + 1$ and $5x + 5$

$$\begin{array}{r} -6x^3 - 4x + 1 \\ \underline{5x + 5} \\ -6x^3 + 1x + 6 = -6x^3 + x + 6 \end{array}$$

47. $3w^3 - 2w^2 + 8w$ and $2w^2 - 8w + 5$

$$\begin{array}{l} 3w^3 - 2w^2 + 8w \\ \underline{\quad\quad 2w^2 - 8w + 5} \\ 3w^3 + 0w^2 + 0w + 5 = 3w^3 + 5 \end{array}$$

49. $(12x^4 - x^2) - (8x^4 + 3x^2)$
$= (12x^4 - x^2) + (-8x^4 - 3x^2)$
$= (12x^4 - 8x^4) + (-x^2 - 3x^2)$
$= 4x^4 - 4x^2$

51. $(2r^2 + 3r - 12) + (6r^2 + 2r)$
$= (2r^2 + 6r^2) + (3r + 2r) - 12$
$= 8r^2 + 5r - 12$

53. $(8m^2 - 7m) - (3m^2 + 7m - 6)$
$= (8m^2 - 7m) + (-3m^2 - 7m + 6)$
$= (8m^2 - 3m^2) + (-7m - 7m) + 6$
$= 5m^2 - 14m + 6$

55. $(16x^3 - x^2 + 3x) + (-12x^3 + 3x^2 + 2x)$
$= (16x^3 - 12x^3) + (-x^2 + 3x^2) + (3x + 2x)$
$= 4x^3 + 2x^2 + 5x$

57. $(12m^3 - 8m^2 + 6m + 7) - (5m^2 - 4)$
$= (12m^3 - 8m^2 + 6m + 7) + (-5m^2 + 4)$
$= 12m^3 + (-8m^2 - 5m^2) + 6m + (7 + 4)$
$= 12m^3 - 13m^2 + 6m + 11$

59. $(-2x^2 - 6x + 4) - (9x^2 - 3x + 7)$
$= (-2x^2 - 6x + 4) + (-9x^2 + 3x - 7)$
$= (-2x^2 - 9x^2) + (-6x + 3x) + (4 - 7)$
$= -11x^2 - 3x - 3$

61. Use the formula for the perimeter of a rectangle, $P = 2L + 2W$, with length $L = 4x^2 + 3x + 1$ and width $W = x + 2$.

$P = 2L + 2W$
$= 2(4x^2 + 3x + 1) + 2(x + 2)$
$= 8x^2 + 6x + 2 + 2x + 4$
$= 8x^2 + 8x + 6$

The perimeter of the rectangle is $8x^2 + 8x + 6$.

63. Use the formula for the perimeter of a triangle, $P = a + b + c$, with $a = 3t^2 + 2t + 7$, $b = 5t^2 + 2$, and $c = 6t + 4$.

$P = (3t^2 + 2t + 7) + (5t^2 + 2) + (6t + 4)$
$= (3t^2 + 5t^2) + (2t + 6t) + (7 + 2 + 4)$
$= 8t^2 + 8t + 13$

The perimeter of the triangle is $8t^2 + 8t + 13$.

65. $D = 100t - 13t^2$ *Replace t with 1.*
$D = 100(1) - 13(1)^2$
$D = 100 - 13$
$D = 87$ ft

66. From the previous exercise, $D = 100t - 13t^2$.

t (sec)	D (ft)	(t, D)
0	$100(0) - 13(0)^2 = 0$	$(0, 0)$
0.5	$100(0.5) - 13(0.5)^2$ $= 46.75 \approx 47$	$(0.5, 47)$
1	87 (See Exercise 65)	$(1, 87)$
1.5	$100(1.5) - 13(1.5)^2$ $= 120.75 \approx 121$	$(1.5, 121)$
2	$100(2) - 13(2)^2 = 148$	$(2, 148)$
2.5	$100(2.5) - 13(2.5)^2$ $= 168.75 \approx 169$	$(2.5, 169)$
3	$100(3) - 13(3)^2 = 183$	$(3, 183)$
3.5	$100(3.5) - 13(3.5)^2$ $= 190.75 \approx 191$	$(3.5, 191)$

67.–68.

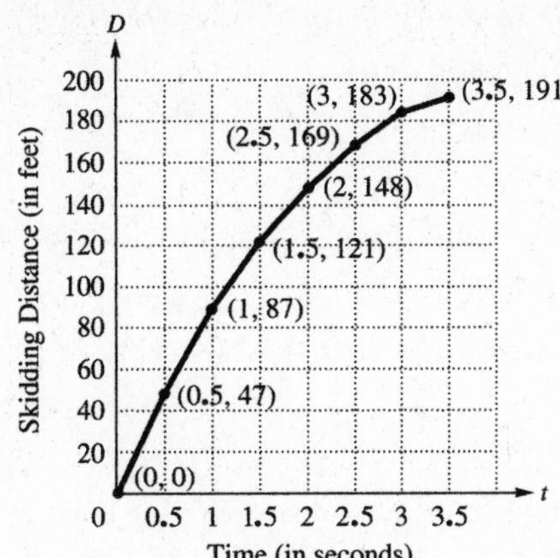

Answers will vary. This graph is a curve instead of the straight lines graphed for linear equations. Also, the rate of change in the distance is not consistent. From 0.5 sec to 1 sec, the distance increases 40 ft, but from 3 sec to 3.5 sec, the distance only increases 8 ft.

10.5 Multiplying Polynomials: An Introduction

10.5 Section Exercises

1. $5y(8y^2-3)$

$= 5y(8y^2)+5y(-3)$ Distributive property

$= 40y^3-15y$ Multiply monomials.

3. $-2m(3m+2)$

$= -2m(3m)+(-2m)(2)$ Distributive property

$= -6m^2-4m$ Multiply monomials.

5. $4x^2(6x^2-3x+2)$

$= 4x^2(6x^2)+4x^2(-3x)+4x^2(2)$

$= 24x^4-12x^3+8x^2$

7. $-3k^3(2k^3-3k^2-k+1)$

$= -3k^3(2k^3)+(-3k^3)(-3k^2)$
$+(-3k^3)(-k)+(-3k^3)(1)$

$= -6k^6+9k^5+3k^4-3k^3$

In Exercises 9–22, we can multiply the polynomials horizontally or vertically. The following solutions illustrate these two methods.

9. Multiply each term of the second polynomial by each term of the first. Then combine like terms.

$(n-2)(n+3)$

$= n(n)+n(3)+(-2)(n)+(-2)(3)$

$= n^2+3n+(-2n)+(-6)$

$= n^2+n-6$

11. Multiply each term of the second polynomial by each term of the first. Then combine like terms.

$(4r+1)(2r-3)$

$= 4r(2r)+4r(-3)+1(2r)+1(-3)$

$= 8r^2+(-12r)+2r+(-3)$

$= 8r^2-10r-3$

13. $(6x+1)(2x^2+4x+1)$

$= (6x)(2x^2)+(6x)(4x)+(6x)(1)$
$+(1)(2x^2)+(1)(4x)+(1)(1)$

$= 12x^3+24x^2+6x+2x^2+4x+1$

$= 12x^3+26x^2+10x+1$

15. $(4m+3)(5m^3-4m^2+m-5)$

Multiply vertically.

$$\begin{array}{rrrrr} & 5m^3 & -\;4m^2 & +\;m & -\;5 \\ & & & 4m & +\;3 \\ \hline & 15m^3 & -\;12m^2 & +\;3m & -\;15 \\ 20m^4 & -\;16m^3 & +\;4m^2 & -\;20m & \\ \hline 20m^4 & -\;m^3 & -\;8m^2 & -\;17m & -\;15 \end{array}$$

17. $(5x^2+2x+1)(x^2-3x+5)$

Multiply vertically.

$$\begin{array}{rrrrr} & & 5x^2 & +\;2x & +\;1 \\ & & x^2 & -\;3x & +\;5 \\ \hline & & 25x^2 & +\;10x & +\;5 \\ & -\;15x^3 & -\;6x^2 & -\;3x & \\ 5x^4 & +\;2x^3 & +\;x^2 & & \\ \hline 5x^4 & -\;13x^3 & +\;20x^2 & +\;7x & +\;5 \end{array}$$

19. $(-x-3)(-x-4)$

$= -x(-x)+(-x)(-4)+(-3)(-x)+(-3)(-4)$

$= x^2+4x+3x+12$

$= x^2+7x+12$

21. $(2x+1)(x^2+3x)$

$= 2x(x^2)+2x(3x)+1(x^2)+1(3x)$

$= 2x^3+6x^2+x^2+3x$

$= 2x^3+7x^2+3x$

23. $(a+4)(a+5)$

$= a(a)+a(5)+4(a)+4(5)$

$= a^2+5a+4a+20$

$= a^2+9a+20$ **Choice D**

$(a-4)(a+5)$

$= a(a)+a(5)+(-4)(a)+(-4)(5)$

$= a^2+5a+(-4a)+(-20)$

$= a^2+a-20$ **Choice B**

continued

$(a+4)(a-5)$
$= a(a) + a(-5) + 4(a) + 4(-5)$
$= a^2 + (-5a) + 4a + (-20)$
$= a^2 - a - 20$ **Choice A**

$(a-4)(a-5)$
$= a(a) + a(-5) + (-4)(a) + (-4)(-5)$
$= a^2 + (-5a) + (-4a) + 20$
$= a^2 - 9a + 20$ **Choice C**

25. Area = Width • length
$= 10(3x+6)$
$= 10(3x) + 10(6)$
$= (30x + 60)\ \text{yd}^2$

26. Area = 600
$30x + 60 = 600$ Subtract 60.
$30x = 540$ Divide by 30.
$\frac{30x}{30} = \frac{540}{30}$
$x = 18$

27. With $x = 18$, $3x + 6 = 3(18) + 6 = 60$.

The rectangle measures 10 yd by 60 yd.

28. ($3.50 per yd^2)(600 yd^2) = $2100

29. Perimeter = 2(length) + 2(width)
$= 2(60) + 2(10)$
$= 120 + 20$
$= 140$ yd

30. ($9.00 per yd)(140 yd) = $1260

31. $(x+4)(x-4) = x(x) + x(-4) + 4(x) + 4(-4)$
$= x^2 + (-4x) + 4x + (-16)$
$= x^2 - 16$

$(y+3)(y-3) = y(y) + y(-3) + 3(y) + 3(-3)$
$= y^2 + (-3y) + 3y + (-9)$
$= y^2 - 9$

$(r+7)(r-7) = r(r) + r(-7) + 7(r) + 7(-7)$
$= r^2 + (-7r) + 7r + (-49)$
$= r^2 - 49$

Each product is the difference of the square of the first term and the square of the last term from the binomial.

Chapter 10 Review Exercises

1. $4^3 \cdot 4^8 = 4^{3+8} = 4^{11}$

2. $(-5)^6(-5)^5 = (-5)^{6+5} = (-5)^{11}$

3. $(-8x^4)(9x^3) = (-8)(9)(x^4)(x^3)$
$= -72x^{4+3} = -72x^7$

4. $(2x^2)(5x^3)(x^9) = (2)(5)(x^2)(x^3)(x^9)$
$= 10x^{2+3+9} = 10x^{14}$

5. $(19x)^5 = 19^5x^5$

6. $(-4y)^7 = (-4)^7y^7$

7. $5(pt)^4 = 5p^4t^4$

8. $\left(\frac{7}{5}\right)^6 = \frac{7^6}{5^6}$

9. $(3x^2y^3)^3$
$= 3^3(x^2)^3(y^3)^3$
$= 3^3x^{2\cdot3}y^{3\cdot3}$
$= 3^3x^6y^9$

10. The product rule for exponents does not apply because it is the *sum* of 7^2 and 7^4, not their *product*.

11. $5^0 + 8^0 = 1 + 1 = 2$

12. $2^{-5} = \frac{1}{2^5} = \frac{1}{32}$

13. $10w^0 = 10 \cdot 1 = 10$

14. $4^{-2} + 4^{-1} = \frac{1}{4^2} + \frac{1}{4^1}$
$= \frac{1}{16} + \frac{1}{4}$
$= \frac{1}{16} + \frac{4}{16} = \frac{5}{16}$

15. $\frac{6^{-3}}{6^{-5}} = 6^{-3-(-5)} = 6^{-3+5} = 6^2$

16. $\frac{x^{-7}}{x^{-9}} = x^{-7-(-9)} = x^{-7+9} = x^2$

17. $\frac{p^{-8}}{p^4} = p^{-8-4} = p^{-12} = \frac{1}{p^{12}}$

18. $\frac{r^{-2}}{r^{-6}} = r^{-2-(-6)} = r^{-2+6} = r^4$

19. $\frac{5^4}{5^5} = 5^{4-5} = 5^{-1} = \frac{1}{5}$

20. $10^3(10^{-10}) = 10^{3+(-10)} = 10^{-7} = \frac{1}{10^7}$

21. $n^{-2} \cdot n^3 \cdot n^{-4} = n^{-2+3+(-4)} = n^{-3} = \frac{1}{n^3}$

22. $(10^4)(10^{-1}) = 10^{4+(-1)} = 10^3$

23. $(2^{-4})(2^{-4}) = 2^{-4+(-4)} = 2^{-8} = \frac{1}{2^8}$

24. $h^{10} \cdot h^{-4} \cdot h^{-3} = h^{10+(-4)+(-3)} = h^3$

25. $\frac{10^{-4}}{10^9} = 10^{-4-9} = 10^{-13} = \frac{1}{10^{13}}$

26. $\dfrac{x^2}{x^{-4}} = x^{2-(-4)} = x^{2+4} = x^6$

27. $48{,}000{,}000 = 4.8 \times 10^7$

Move the decimal point 7 places. The *original* number was "large" (10 or more), so the exponent is *positive* 7.

28. $28{,}988{,}000{,}000 = 2.8988 \times 10^{10}$

Move the decimal point 10 places. The *original* number was "large" (10 or more), so the exponent is *positive* 10.

29. $0.000065 = 6.5 \times 10^{-5}$

Move the decimal point 5 places. The *original* number was "small" (between 0 and 1), so the exponent is *negative* 5.

30. $0.0000000824 = 8.24 \times 10^{-8}$

Move the decimal point 8 places. The *original* number was "small" (between 0 and 1), so the exponent is *negative* 8.

31. $2.4 \times 10^4 = 24{,}000$

Since the exponent is positive, move the decimal point 4 places to the *right*.

32. $7.83 \times 10^7 = 78{,}300{,}000$

Since the exponent is positive, move the decimal point 7 places to the *right*.

33. $8.97 \times 10^{-7} = 0.000\,000\,897$

Since the exponent is negative, move the decimal point 7 places to the *left*.

34. $9.95 \times 10^{-12} = 0.000\,000\,000\,009\,95$

Since the exponent is negative, move the decimal point 12 places to the *left*.

35.
$$\begin{aligned}&(2 \times 10^{-3})(4 \times 10^5)\\ &= (2 \times 4)(10^{-3} \times 10^5)\\ &= 8 \times 10^{-3+5} = 8 \times 10^2\\ &= 800\end{aligned}$$

36.
$$\begin{aligned}\frac{8 \times 10^4}{2 \times 10^{-2}} &= \frac{8}{2} \times \frac{10^4}{10^{-2}} = 4 \times 10^{4-(-2)}\\ &= 4 \times 10^6 = 4{,}000{,}000\end{aligned}$$

37.
$$\begin{aligned}\frac{12 \times 10^{-8}}{4 \times 10^{-3}} &= \frac{12}{4} \times \frac{10^{-8}}{10^{-3}}\\ &= 3 \times 10^{-8-(-3)}\\ &= 3 \times 10^{-5}\\ &= 0.00003\end{aligned}$$

38.
$$\begin{aligned}&\frac{(2.5 \times 10^5)(4.8 \times 10^{-4})}{(7.5 \times 10^8)(1.6 \times 10^{-5})}\\ &= \frac{(2.5 \times 4.8)(10^5 \times 10^{-4})}{(7.5 \times 1.6)(10^8 \times 10^{-5})}\\ &= \frac{12 \times 10^1}{12 \times 10^3}\\ &= \frac{12}{12} \times \frac{10^1}{10^3}\\ &= 1 \times 10^{-2}\\ &= 0.01\end{aligned}$$

39.
$$\begin{aligned}\frac{2.88 \times 10^8 \text{ people}}{3.54 \times 10^6 \text{ mi}^2} &= \frac{2.88}{3.54} \times \frac{10^8}{10^6}\\ &\approx 0.81 \times 10^2\\ &= 8.1 \times 10^1 \text{ (rounded)}\\ &= 81 \text{ people per square mile}\end{aligned}$$

40.
$$\begin{aligned}(36 \text{ sec})(3.0 \times 10^5 \text{ km per sec}) &= (36 \times 3.0)(10^5)\\ &= 108 \times 10^5\\ &= 1.08 \times 10^7\\ &= 10{,}800{,}000 \text{ km}\end{aligned}$$

41.
$$\begin{aligned}9m^2 + 11m^2 + 2m^2 &= (9 + 11 + 2)m^2\\ &= 22m^2\end{aligned}$$

The degree is 2.

To determine if the polynomial is a monomial, binomial, or trinomial, count the number of terms in the final expression.

There is one term, so this is a *monomial*.

42.
$$\begin{aligned}&-4p + p^3 - p^2 + 8p + 2\\ &= p^3 - p^2 + (-4 + 8)p + 2\\ &= p^3 - p^2 + 4p + 2\end{aligned}$$

The degree is 3.

To determine if the polynomial is a monomial, binomial, or trinomial, count the number of terms in the final expression. Since there are four terms, it is none of these.

43. $12a^5 - 9a^4 + 8a^3 + 2a^2 - a + 3$ cannot be simplified further and is already written in descending powers of the variable.

The degree is 5.

This polynomial has 6 terms, so it is none of the names listed.

44.
$$\begin{aligned}&-7y^5 - 8y^4 - y^5 + y^4 + 9y\\ &= -7y^5 - 1y^5 - 8y^4 + 1y^4 + 9y\\ &= (-7 - 1)y^5 + (-8 + 1)y^4 + 9y\\ &= -8y^5 - 7y^4 + 9y\end{aligned}$$

The degree is 5.

There are three terms, so the polynomial is a *trinomial*.

45. $(-2a^3 + 5a^2) + (-3a^3 - a^2)$

$$\begin{array}{r} -2a^3 + 5a^2 \\ -3a^3 - a^2 \\ \hline -5a^3 + 4a^2 \end{array}$$

46. $(-4r^3 - 8r^2) + (6r + 4r^3)$

$$\begin{array}{l} -4r^3 - 8r^2 \\ 4r^3 \qquad + 6r \\ \hline 0r^3 - 8r^2 + 6r = -8r^2 + 6r \end{array}$$

47. Subtract.

$$\begin{array}{r} 6y^2 - 8y + 2 \\ -5y^2 + 2y - 7 \\ \hline \end{array}$$

Change all signs in the second row and then add.

$$\begin{array}{r} 6y^2 - 8y + 2 \\ 5y^2 - 2y + 7 \\ \hline 11y^2 - 10y + 9 \end{array}$$

48. $(2m^3 - 8m^2 + 4) + (8m^3 + 2m^2 - 7)$
$= (2m^3 + 8m^3) + (-8m^2 + 2m^2) + (4 - 7)$
$= 10m^3 - 6m^2 - 3$

49. $(-5y^2 + 3y + 11) + (4y^3 - 7y + 15)$
$= 4y^3 - 5y^2 + (3y - 7y) + (11 + 15)$
$= 4y^3 - 5y^2 - 4y + 26$

50. $(6p^2 - p - 8) - (-4p^2 + 2p + 3)$
$= (6p^2 - p - 8) + (4p^2 - 2p - 3)$
$= (6p^2 + 4p^2) + (-p - 2p) + (-8 - 3)$
$= 10p^2 - 3p - 11$

51. $(12r^4 - 7r^3 + 2r^2) - (5r^4 - 3r^3 + 2r^2 + 1)$
$= (12r^4 - 7r^3 + 2r^2) + (-5r^4 + 3r^3 - 2r^2 - 1)$
$= (12r^4 - 5r^4) + (-7r^3 + 3r^3) + (2r^2 - 2r^2) - 1$
$= 7r^4 - 4r^3 - 1$

52. $5x(2x + 14)$
$= 5x(2x) + 5x(14)$
$= 10x^2 + 70x$

53. $-3p^3(2p^2 - 5p)$
$= -3p^3(2p^2) + (-3p^3)(-5p)$
$= -6p^5 + 15p^4$

54. $(3r - 2)(2r^2 + 4r - 3)$
Multiply vertically.

$$\begin{array}{rrrr} & 2r^2 & +4r & -3 \\ & & 3r & -2 \\ \hline & -4r^2 & -8r & +6 \\ 6r^3 & +12r^2 & -9r & \\ \hline 6r^3 & +8r^2 & -17r & +6 \end{array}$$

55. $(2y + 3)(4y^2 - 6y + 9)$
Multiply vertically.

$$\begin{array}{rrrr} & 4y^2 & -6y & +9 \\ & & 2y & +3 \\ \hline & 12y^2 & -18y & +27 \\ 8y^3 & -12y^2 & +18y & \\ \hline 8y^3 & +0y^2 & +0y & +27 \end{array} = 8y^3 + 27$$

56. $(5p^2 + 3p)(p^3 - p^2 + 5)$
$= 5p^2(p^3) + 5p^2(-p^2) + 5p^2(5)$
$\quad + 3p(p^3) + 3p(-p^2) + 3p(5)$
$= 5p^5 - 5p^4 + 25p^2 + 3p^4 - 3p^3 + 15p$
$= 5p^5 - 2p^4 - 3p^3 + 25p^2 + 15p$

57. $(3k - 6)(2k + 1)$
$= (3k)(2k) + (3k)(1) + (-6)(2k) + (-6)(1)$
$= 6k^2 + 3k - 12k - 6$
$= 6k^2 - 9k - 6$

58. $(6p - 3q)(2p - 7q)$
$= 6p(2p) + 6p(-7q) + (-3q)(2p) + (-3q)(-7q)$
$= 12p^2 + (-42pq) + (-6pq) + (21q^2)$
$= 12p^2 - 48pq + 21q^2$

59.

$$\begin{array}{rrrrr} & & m^2 & +m & -9 \\ & & 2m^2 & +3m & -1 \\ \hline & & -m^2 & -m & +9 \\ & 3m^3 & +3m^2 & -27m & \\ 2m^4 & +2m^3 & -18m^2 & & \\ \hline 2m^4 & +5m^3 & -16m^2 & -28m & +9 \end{array}$$

60. **[10.2]** $19^0 - 3^0 = 1 - 1 = 0$

61. **[10.1]** $(3p)^4 = 3^4p^4$ or $81p^4$

62. **[10.2]** $7^{-2} = \frac{1}{7^2}$ or $\frac{1}{49}$

63. **[10.5]** $-m^5(8m^2 + 10m + 6)$
$= -m^5(8m^2) + (-m^5)(10m) + (-m^5)(6)$
$= -8m^7 + (-10m^6) + (-6m^5)$
$= -8m^7 - 10m^6 - 6m^5$

64. **[10.2]** $2^{-1} + 4^{-1} = \frac{1}{2^1} + \frac{1}{4^1}$
$= \frac{1}{2} + \frac{1}{4}$
$= \frac{2}{4} + \frac{1}{4} = \frac{3}{4}$

65. **[10.5]** $(a+2)(a^2-4a+1)$
Multiply vertically.

$$\begin{array}{rrrr} & a^2 & -\ 4a & +\ 1 \\ & & a & +\ 2 \\ \hline & 2a^2 & -\ 8a & +\ 2 \\ a^3 & -\ 4a^2 & +\ a & \\ \hline a^3 & -\ 2a^2 & -\ 7a & +\ 2 \end{array}$$

66. **[10.4]** $(5y^3-8y^2+7)-(-3y^3+y^2+2)$
$= (5y^3-8y^2+7)+(3y^3-y^2-2)$
$= (5y^3+3y^3)+(-8y^2-y^2)+(7-2)$
$= 8y^3-9y^2+5$

67. **[10.5]** Perimeter $= 2 \cdot \text{length} + 2 \cdot \text{width}$
$= 2(2x-3)+2(x+2)$
$= 2(2x)+2(-3)+2(x)+2(2)$
$= 4x+(-6)+2x+4$
$= 6x-2$

Area $= \text{length} \cdot \text{width}$
$= (2x-3)(x+2)$
$= 2x(x)+2x(2)+(-3)(x)+(-3)(2)$
$= 2x^2+4x+(-3x)+(-6)$
$= 2x^2+x-6$

Chapter 10 Test

1. $5^{-4} = \dfrac{1}{5^4} = \dfrac{1}{625}$

2. $(-3)^0+4^0 = 1+1 = 2$

3. $4^{-1}+3^{-1} = \dfrac{1}{4^1}+\dfrac{1}{3^1} = \dfrac{3}{12}+\dfrac{4}{12} = \dfrac{7}{12}$

4. $6^{-3} \cdot 6^4 = 6^{-3+4} = 6^1$ or 6

5. $12(xy)^3 = 12x^3y^3$

6. $\dfrac{10^5}{10^9} = 10^{5-9} = 10^{-4} = \dfrac{1}{10^4}$

7. $r^{-4} \cdot r^{-4} \cdot r^3 = r^{-4+(-4)+3} = r^{-5} = \dfrac{1}{r^5}$

8. $(7x^2)(-3x^5) = 7(-3)(x^2 \cdot x^5)$
$= -21x^{2+5}$
$= -21x^7$

9. $\dfrac{m^{-5}}{m^{-8}} = m^{-5-(-8)} = m^{-5+8} = m^3$

10. $(3a^4b)^2 = 3^2(a^4)^2b^2$
$= 3^2a^{4 \cdot 2}b^2$
$= 3^2a^8b^2$ or $9a^8b^2$

11. $3^4+3^2 = 81+9 = 90$
Note that this is a *sum*, not a *product*, so we cannot use the product rule for exponents.

12. $300{,}000{,}000{,}000 = 3.0 \times 10^{11}$

Move the decimal point 11 places. The *original* number was "large" (10 or more), so the exponent is *positive* 11.

13. $0.00000557 = 5.57 \times 10^{-6}$

Move the decimal point 6 places. The *original* number was "small" (between 0 and 1), so the exponent is *negative* 6.

14. $2.9 \times 10^7 = 29{,}000{,}000$

Since the exponent is positive, move the decimal point 7 places to the *right*.

15. $6.07 \times 10^{-8} = 0.000\,000\,060\,7$

Since the exponent is negative, move the decimal point 8 places to the *left*.

16. $(330{,}000)(5.98 \times 10^{24}\text{ kg})$
$= (3.3 \times 10^5)(5.98 \times 10^{24})$
$= (3.3 \times 5.98)(10^5 \times 10^{24})$
$= 19.734 \times 10^{29}$
$\approx 1.97 \times 10^{30}$ kg

17. $5x^2+8x-12x^2 = (5-12)x^2+8x$
$= -7x^2+8x$

The degree is 2 (the largest exponent on the variable x). The polynomial has two terms, so it is a binomial.

18. $13n^3-n^2+n^4+3n^4-9n^2$
$= (1+3)n^4+13n^3+(-1-9)n^2$
$= 4n^4+13n^3-10n^2$

The degree is 4 (the largest exponent on the variable n). The polynomial has three terms, so it is a trinomial.

19. $(5t^4-3t^2+3)-(t^4-t^2+3)$
$= (5t^4-3t^2+3)+(-t^4+t^2-3)$
$= (5t^4-t^4)+(-3t^2+t^2)+(3-3)$
$= 4t^4-2t^2$

20. $(2y^4-8y+8)+(-3y^2+2y-8)$
$= 2y^4-3y^2+(-8y+2y)+(8-8)$
$= 2y^4-3y^2-6y$

21. $(9t^3-4t^2+2)-(9t^3+8t^2-6)$
$= (9t^3-4t^2+2)+(-9t^3-8t^2+6)$
$= (9t^3-9t^3)+(-4t^2-8t^2)+(2+6)$
$= -12t^2+8$

22. $-4x^3(3x^2 - 5x)$
$= -4x^3(3x^2) + (-4x^3)(-5x)$
$= -12x^5 + 20x^4$

23. $(y+2)(3y-1)$
$= y(3y) + y(-1) + 2(3y) + 2(-1)$
$= 3y^2 + (-y) + 6y + (-2)$
$= 3y^2 + 5y - 2$

24. $(2r-3)(r^2+2r-5)$

Multiply vertically.

$$\begin{array}{r} r^2 + 2r - 5 \\ 2r - 3 \\ \hline -3r^2 - 6r + 15 \\ 2r^3 + 4r^2 - 10r \quad\quad \\ \hline 2r^3 + r^2 - 16r + 15 \end{array}$$

25. Use the formula for the perimeter of a square, $P = 4s$, with $s = 3x + 9$.

$$\begin{aligned} P &= 4(3x+9) \\ &= 4(3x) + 4(9) \\ &= 12x + 36 \end{aligned}$$

Use the formula for the area of a square, $A = s^2$, with $s = 3x + 9$.

$$\begin{aligned} A &= (3x+9)^2 \\ &= (3x+9)(3x+9) \\ &= (3x)(3x) + (3x)(9) + (9)(3x) + (9)(9) \\ &= 9x^2 + 27x + 27x + 81 \\ &= 9x^2 + 54x + 81 \end{aligned}$$

Cumulative Review Exercises (Chapters 1–10)

1. **(a)** In words, 10.035 is written as ten and thirty-five thousandths.

(b) In words, 410,000,351,109 is written as four hundred ten billion, three hundred fifty-one thousand, one hundred nine.

2. **(a)** Using digits, three hundred million, six thousand, eighty is written as 300,006,080.

(b) Using digits, fifty-five ten-thousandths is written as 0.0055.

3. **(a)** 0.8029 to the nearest hundredth

2 is less than 5, so drop all digits to the right of 0 in the hundredths place. **0.80**

(b) 340,519,000 to the nearest million

Since the number in the hundred-thousands place is 5 or greater, round 519,000 up to 1,000,000. **341,000,000**

(c) 14.973 to the nearest tenth

Since the number in the tenths place, 9, is 5 or greater, round 0.973 up to 1. **15.0**

4. $\text{Mean} = \dfrac{\text{sum of all values}}{\text{number of values}}$

$$= \frac{(41+65+37+90+41+65+48+41+59+40)}{10}$$

$$= \frac{527}{10} = 52.7$$

The mean is \$52.70.

Arrange in increasing order:

$$37, 40, 41, 41, 41, 48, 59, 65, 65, 90$$

The median is the average of the 5th and 6th values.

$$\frac{41+48}{2} = \frac{89}{2} = 44.5$$

The median is \$44.50.

\$41 occurs the greatest number of times (3), so the mode is \$41.

5. $-12 + 7.829 = -4.171$

6. $7 + 5(3-8) = 7 + 5(-5) = 7 - 25 = -18$

7. $1\frac{2}{3} + \frac{3}{5} = \frac{5}{3} + \frac{3}{5} = \frac{25}{15} + \frac{9}{15} = \frac{34}{15} = 2\frac{4}{15}$

8. $$\begin{aligned} \frac{(-6)^2 + 9(0-4)}{10 \div 5(-7+5) - 10} &= \frac{36 + 9(-4)}{10 \div 5(-2) - 10} \\ &= \frac{36-36}{2(-2)-10} \\ &= \frac{0}{-4-10} \\ &= \frac{0}{-14} \\ &= 0 \end{aligned}$$

9. $$\begin{aligned} \frac{8x^2}{9} \cdot \frac{12w}{2x^3} &= \frac{\overset{4}{\cancel{8}}x^2}{\underset{3}{\cancel{9}}} \cdot \frac{\overset{4}{\cancel{12}}w}{\underset{1}{\cancel{2}}x^3} \\ &= \frac{16x^2w}{3x^3} \\ &= \frac{16w}{3x} \end{aligned}$$

10. $\dfrac{-0.7}{5.6} = \dfrac{-0.7}{8(0.7)} = -\dfrac{1}{8} = -0.125$

11. $\dfrac{10^5}{10^6} = 10^{5-6} = 10^{-1} = \dfrac{1}{10}$

12. $2\frac{1}{4} - 2\frac{5}{6} = \frac{9}{4} - \frac{17}{6} = \frac{27}{12} - \frac{34}{12} = -\frac{7}{12}$

13. $(-6)^2 + (-2)^3 = 36 + (-8) = 28$

14. $(-4y^3)(3y^4) = -4 \cdot 3y^{3+4} = -12y^7$

15. $\frac{5b}{2a^2} \div \frac{3b^2}{10a} = \frac{5b}{\cancel{2}a^2} \cdot \frac{\overset{5}{\cancel{10}}a}{3b^2}$

$= \frac{25ab}{3a^2b^2}$

$= \frac{25}{3ab}$

16. $\frac{r}{8} + \frac{6}{t} = \frac{r \cdot t}{8 \cdot t} + \frac{6 \cdot 8}{t \cdot 8} = \frac{rt}{8t} + \frac{48}{8t} = \frac{rt + 48}{8t}$

17. $\frac{-4(6)}{3^3 - 27} = \frac{-24}{27 - 27} = \frac{-24}{0}$, which is undefined.

18. $\frac{5}{6}$ of $900 = \frac{5}{6} \cdot 900$

$= \frac{5 \cdot 900}{6}$

$= \frac{5 \cdot 6 \cdot 150}{6}$

$= 5 \cdot 150$

$= 750$

19. $(-0.003)(0.04) = (-3 \times 10^{-3})(4 \times 10^{-2})$

$= (-3 \times 4)(10^{-3+(-2)})$

$= -12 \times 10^{-5}$

$= -0.00012$

20. $(2a^3b^2)^4 = 2^4(a^3)^4(b^2)^4$

$= 2^4 \cdot a^{3 \cdot 4}b^{2 \cdot 4}$

$= 2^4a^{12}b^8$ or $16a^{12}b^8$

21. $\frac{4}{9} - \frac{6}{m} = \frac{4 \cdot m}{9 \cdot m} - \frac{6 \cdot 9}{m \cdot 9} = \frac{4m}{9m} - \frac{54}{9m} = \frac{4m - 54}{9m}$

22. $t^5 \cdot t^{-2} \cdot t^{-4} = t^{5-2-4} = t^{-1} = \frac{1}{t}$

23. $\frac{-\frac{15}{16}}{-6} = -\frac{15}{16} \div (-6) = -\frac{15}{16}\left(-\frac{1}{6}\right)$

$= \frac{\overset{1}{\cancel{3}} \cdot 5}{16 \cdot 2 \cdot \underset{1}{\cancel{3}}} = \frac{5}{32}$

24. $\left(-\frac{1}{2}\right)^4 + \overset{2}{\cancel{6}}\left(\frac{2}{\underset{1}{\cancel{3}}}\right) = \frac{1}{16} + 4$

$= \frac{1}{16} + \frac{64}{16}$

$= \frac{65}{16}$ or $4\frac{1}{16}$

25. $\frac{n^{-3}}{n^{-4}} = n^{-3-(-4)} = n^{-3+4} = n^1$ or n

26. $-8 - 88 = -8 + (-88) = -96$

27. $8^0 + 2^{-1} = 1 + \frac{1}{2} = 1\frac{1}{2}$

28. $9 - 7\frac{4}{5} = \frac{45}{5} - \frac{39}{5} = \frac{6}{5}$ or $1\frac{1}{5}$

29. When $x = 4$ and $y = -1$,

$-3xy^3 = -3(4)(-1)^3$

$= -3(4)(-1)$

$= 12$

30. When $y = -1$ and $w = -2$,

$15y - 6w = 15(-1) - 6(-2)$

$= -15 - (-12)$

$= -3$

31. When $x = 4$ and $y = -1$,

$x^2 - 2xy + 6 = 4^2 - 2(4)(-1) + 6$

$= 16 + 8 + 6$

$= 30$

32. $4 + h = 3h - 6$ Subtract 4.

$h = 3h - 10$ Subtract $3h$.

$-2h = -10$ Divide by -2.

$\frac{-2h}{-2} = \frac{-10}{-2}$

$h = 5$

The solution is 5.

33. $-1.65 = 0.5x + 2.3$ Subtract 2.3.

$-3.95 = 0.5x$ Divide by 0.5.

$\frac{-3.95}{0.5} = \frac{0.5x}{0.5}$

$x = -7.9$

The solution is -7.9.

34. $3(a - 5) = -3 + a$ Distributive property

$3a - 15 = -3 + a$ Add 15.

$3a = 12 + a$ Subtract a.

$2a = 12$ Divide by 2.

$\frac{2a}{2} = \frac{12}{2}$

$a = 6$

The solution is 6.

35. Let n represent the number.

$20 - 3n = 2n$ Add $3n$.

$20 = 5n$ Divide by 5.

$\frac{20}{5} = \frac{5n}{5}$

$n = 4$

The number is 4.

36. *Step 2*
Let b represent the number of elephant breaths. Then $16b$ represents the number of mouse breaths.

Step 3
$b + 16b = 170$

Step 4
$$17b = 170$$
$$\frac{17b}{17} = \frac{170}{17}$$
$$b = 10$$

Step 5
An elephant takes 10 breaths. A mouse takes 16 times that, or 160 breaths.

Step 6
Check: 16 times 10 is 160 and the sum of 10 and 160 is 170.

37. $$\frac{78 \text{ runs}}{234 \text{ innings}} = \frac{n \text{ runs}}{9 \text{ innings}}$$
$$234 \cdot n = 78 \cdot 9$$
$$\frac{234n}{234} = \frac{702}{234}$$
$$n = 3$$

At this rate, he will give up 3 runs in a 9-inning game.

38. percent • whole = part
$$p \cdot 1.8 = 2.61$$
$$\frac{1.8p}{1.8} = \frac{2.61}{1.8}$$
$$p = 1.45 \quad (145\%)$$

2.61 inches is 145% of 1.8 inches.

39. Tax = $(\$64.95)(0.075) \approx \$4.87$
Total = $\$64.95 + \$4.87 = \$69.82$

The total cost of the camera was $69.82 (rounded).

40. $$\frac{5.87 \times 10^{12} \text{ miles}}{1 \text{ year}} \cdot \frac{1 \text{ year}}{365 \text{ days}}$$
$$= \frac{5.87}{365} \times 10^{12}$$
$$\approx 0.0161 \times 10^{12}$$
$$= 1.61 \times 10^{10}$$

Light will travel 1.61×10^{10} (rounded) or 16,100,000,000 miles in one day.

41. **(a)** Size 6 shoes correspond to a foot length of $5\frac{1}{8}$ in.

(b) $5\frac{7}{16} > 5\frac{1}{8}$ $\left(= 5\frac{2}{16}\right)$, so size 6 is too small. She should order size 7.

(c) A size 6 is $5\frac{1}{8}$ in. and a size 5 is $4\frac{13}{16}$ in.

$$5\frac{1}{8} = 4\frac{9}{8} = 4\frac{18}{16}$$
$$-\,4\frac{13}{16} = 4\frac{13}{16}$$
$$\frac{5}{16} \text{ in.}$$

The difference in length between a size 5 and a size 6 shoe is $\frac{5}{16}$ inch.

42. Perimeter $= 4 \text{ m} + 6 \text{ m} + 3.2 \text{ m} + 3.2 \text{ m} + 6 \text{ m}$
$= 22.4 \text{ m}$

Area = area of rectangle + area of triangle
$$= l \cdot w + \frac{1}{2} \cdot b \cdot h$$
$$= (6 \text{ m})(4 \text{ m}) + \frac{1}{2}(4 \text{ m})(3.8 \text{ m})$$
$$= 24 \text{ m}^2 + 7.6 \text{ m}^2$$
$$= 31.6 \text{ m}^2$$

43. Circumference $= \pi \cdot \text{diameter} = \pi(9 \text{ cm})$
$\approx 3.14(9)$
$\approx 28.3 \text{ cm}$

Area $= \pi \cdot (\text{radius})^2 = \pi\left(\frac{9}{2} \text{ cm}\right)^2$
$\approx 3.14(4.5)^2$
$\approx 63.6 \text{ cm}^2$

WHOLE NUMBERS COMPUTATION: PRETEST

Adding Whole Numbers

1.
```
  1
 368
+ 22
----
 390
```

3.
```
  111
   85
+2968
-----
 3053
```

5. 714 + 3728 + 9 + 683,775
```
    2 12
      714
    3 728
        9
+ 683,775
---------
  688,226
```

Subtracting Whole Numbers

1.
```
 312
 426
- 76
----
 350
```

3.
```
   9 15 9
 2 10 5 10 12
  30,602
 - 5 708
 -------
  24,894
```

5. 679,420 − 88,033
```
          11
 517    3 1 10
   679,420
 -  88,033
 ---------
   591,387
```

Multiplying Whole Numbers

1. $3 \times 3 \times 0 \times 6 = 0$ because 0 times any number is 0.

3. (520)(3000)
```
      520
  × 3 000
---------
1,560,000
```

5. Multiply 359 and 48.
```
  23
  47
  359
× 48
-----
 2872   ←   8 × 359
1436    ←   4 × 359
------
17,232
```

Dividing Whole Numbers

1.
```
    2 3
3 ) 6 9
    6
    ---
      9
      9
    ---
      0
```

3. $\frac{25,036}{4}$
```
     6 2 5 9
4 ) 2 5,0 3 6
    2 4
    ---
      1 0
        8
      ---
        2 3
        2 0
        ---
          3 6
          3 6
          ---
            0
```

5.
```
        3 4
52 ) 1 7 6 8
     1 5 6
     -----
       2 0 8
       2 0 8
       -----
           0
```

7.
```
        6 0
38 ) 2 3 0 0
     2 2 8
     -----
         2 0
           0
         ---
         2 0
```

Answer: 60 **R**20

CHAPTER R WHOLE NUMBERS REVIEW

R.1 Adding Whole Numbers

R.1 Section Exercises

1. (a)
$$\begin{array}{rl} 5 & \\ 7 & 5+7=12 \\ 6 & 12+6=18 \\ +5 & 18+5=23 \\ \hline 23 & \end{array}$$

(b)
$$\begin{array}{rl} 9 & \\ 2 & 9+2=11 \\ 1 & 11+1=12 \\ 3 & 12+3=15 \\ +4 & 15+4=19 \\ \hline 19 & \end{array}$$

3. (a) $3213 + 5715 = 8928$

(b) $38{,}204 + 21{,}020 = 59{,}224$

5.
$$\begin{array}{r} \overset{1}{6}7 \\ +83 \\ \hline 150 \end{array}$$

7.
$$\begin{array}{r} 7\overset{1}{4}6 \\ +905 \\ \hline 1651 \end{array}$$

9.
$$\begin{array}{r} \overset{1}{7}\overset{1}{9}8 \\ +206 \\ \hline 1004 \end{array}$$

11.
$$\begin{array}{r} \overset{1}{7}\overset{1}{9}\overset{1}{6}8 \\ +1285 \\ \hline 9253 \end{array}$$

13.
$$\begin{array}{r} \overset{1}{7}\,\overset{1}{8}\overset{1}{9}6 \\ +\ 3\,728 \\ \hline 11{,}624 \end{array}$$

15.
$$\begin{array}{r} \overset{1}{3}\,\overset{1}{7}05 \\ 3\,916 \\ +9\,037 \\ \hline 16{,}658 \end{array}$$

17.
$$\begin{array}{r} \overset{1\,1}{}32 \\ +4\,977 \\ \hline 5009 \end{array}$$

19.
$$\begin{array}{r} 3\,\overset{1}{0}\overset{1}{7}7 \\ 8 \\ +\ \ 421 \\ \hline 3506 \end{array}$$

21.
$$\begin{array}{r} 9\,\overset{2}{0}\overset{2}{5}6 \\ 78 \\ 6\,089 \\ +\ \ \ 731 \\ \hline 15{,}954 \end{array}$$

23.
$$\begin{array}{r} \overset{1\,1\,2}{}18 \\ 708 \\ 9\,286 \\ +\ \ \ 636 \\ \hline 10{,}648 \end{array}$$

25. Add up to check addition.
$$\begin{array}{rl} 769 & \\ \hline 179 & \\ 214 & \\ +376 & \\ \hline 759 & \text{incorrect; should be 769} \end{array}$$

27. Add up to check addition.
$$\begin{array}{rl} 5420 & \\ \hline 4713 & \\ 28 & \\ 615 & \\ +64 & \\ \hline 5420 & \text{correct} \end{array}$$

29. The shortest route between Southtown and Rena is through Thomasville.
$$\begin{array}{rl} 21 & \textit{Southtown to Thomasville} \\ +12 & \textit{Thomasville to Rena} \\ \hline 33 & \textit{miles} \end{array}$$

31. The shortest route between Thomasville and Murphy is through Rena and Austin.
$$\begin{array}{rl} 12 & \textit{Thomasville to Rena} \\ 15 & \textit{Rena to Austin} \\ +11 & \textit{Austin to Murphy} \\ \hline 38 & \textit{miles} \end{array}$$

33.
$$\begin{array}{rl} \$\overset{2}{7}9 & \textit{auto tune-up} \\ 24 & \textit{tire rotation} \\ +19 & \textit{oil change} \\ \hline \$122 & \textit{total cost} \end{array}$$

35.
$$\begin{array}{rl} 413 & \textit{women} \\ +286 & \textit{men} \\ \hline 699 & \textit{total people} \end{array}$$

37.

```
   1 1 1
   33,871,648
 + 20,851,820
   54,723,468
```

The total population of the two states is 54,723,468 people.

39. The perimeter is the sum of all of the sides in the figure.

```
   3
   98
   49
   98
 + 49
  294
```

294 in. is the perimeter of the figure.

41. The perimeter is the sum of all of the sides in the figure.

```
  11
  286
  308
  114
  708
```

708 ft is the perimeter of the figure.

R.2 Subtracting Whole Numbers

R.2 Section Exercises

1.

```
   89                 1
 - 27  Given         27   Check by
   63              + 63   addition.
                     90
```

$90 \neq 89$, so 63 is incorrect. Rework.

```
   89
 - 27
   62   is correct.
```

3.

```
   382              261   Check by
 - 261  Given     + 131   addition.
   131              392
```

$392 \neq 382$, so 131 is incorrect. Rework.

```
   382
 - 261
   121   is correct.
```

5.

```
  216
   36
 - 28
    8
```

7.

```
  713
   83
 - 58
   25
```

9.

```
  315
   45
 - 29
   16
```

11.

```
   611
   719
 - 658
    61
```

13.

```
   611
   771
 - 252
   519
```

15.

```
    7 1511
   9 8 6 1
 -   6 8 4
   9 1 7 7
```

17.

```
    81718
   9 9 8 8
 - 2 3 9 9
   7 5 8 9
```

19.

```
    217 121215
   3 8, 3 3 5
 - 2 9, 4 7 6
     8  8 5 9
```

21.

```
  310
   40
 - 37
    3
```

23.

```
  510
   60
 - 37
   23
```

25.

```
       9
   5 101110
   6 0 2 0
 - 4 0 7 8
   1 9 4 2
```

27.

```
       9
   7141013
   8 5 0 3
 - 2 8 1 6
   5 6 8 7
```

29.

```
           9
   710 6 1015
   8 0, 7 0 5
 - 6 1, 6 6 7
   1 9, 0 3 8
```

31.

```
         9  9
     5  10 10 10
   6 6, 0  0  0
 -         4  4  4
   6 5, 5  5  6
```

33.
$$\begin{array}{r} \overset{\ \ 9\ \ 9}{\overset{1\,10\ 10\,17\,10}{\cancel{2}\,\cancel{0},\cancel{0}\,\cancel{8}\,\cancel{0}}} \\ -\quad 9\ 6 \\ \hline 19,\ 9\ 8\ 4 \end{array}$$

35.
$$\begin{array}{rl} 3070 & \\ -\ 576 & \textit{Subtraction} \\ \hline 2596 & \textit{problem} \end{array} \qquad \begin{array}{rl} \overset{111}{576} & \textit{Check by} \\ +\ 2596 & \textit{addition.} \\ \hline 3172 & \end{array}$$

$3172 \neq 3070$, so 2596 is incorrect. Rework.

$$\begin{array}{r} \overset{\ \ 9}{\overset{2\,10\,16\,10}{\cancel{3}\ \cancel{0}\ \cancel{7}\ \cancel{0}}} \\ -\ 5\ 7\ 6 \\ \hline 2\ 4\ 9\ 4 \end{array}$$ is correct.

37.
$$\begin{array}{rl} 27{,}600 & \\ -\ 807 & \textit{Subtraction} \\ \hline 26{,}793 & \textit{problem} \end{array} \qquad \begin{array}{rl} \overset{1\ 11}{807} & \textit{Check by} \\ +\ 26{,}793 & \textit{addition.} \\ \hline 27{,}600 & \end{array}$$

Matches: 26,793 is correct.

39.
$$\begin{array}{rl} 103 & \textit{calories man burns} \\ -\ 88 & \textit{calories woman burns} \\ \hline 15 & \textit{fewer calories woman burned} \end{array}$$

The woman burned 15 fewer calories.

41.
$$\begin{array}{rl} 254 & \textit{number of passengers} \\ -\ 133 & \textit{passengers departing in Atlanta} \\ \hline 121 & \textit{passengers left on the plane} \end{array}$$

There were 121 passengers left on the plane.

43. **(a)** From the table, the occupation with the highest earnings is electrical engineer at \$64,910 and the occupation with the lowest earnings is photographer at \$22,350.

(b)
$$\begin{array}{rl} \$6\overset{8}{4},\overset{1}{\cancel{9}}\overset{1}{\cancel{1}}0 & \textit{electrical engineer earnings} \\ -\ 22{,}350 & \textit{photographer earnings} \\ \hline \$42{,}560 & \textit{difference} \end{array}$$

45.
$$\begin{array}{rl} \overset{\ \ 7\,11}{1\,\cancel{8}\,\cancel{1}\,5} & \textit{CN Tower height} \\ -\ 1\ 4\ 5\ 0 & \textit{Sears Tower height} \\ \hline 3\ 6\ 5 & \textit{difference} \end{array}$$

There is a difference in height of 365 feet.

R.3 Multiplying Whole Numbers

R.3 Section Exercises

1. $3 \times 1 \times 3 = (3 \times 1) \times 3 = 3 \times 3 = 9$

3. $9 \times 1 \times 7 = (9 \times 1) \times 7 = 9 \times 7 = 63$

5. $9 \cdot 5 \cdot 0 = (9 \cdot 5) \cdot 0 = 45 \cdot 0 = 0$
The product of any number and 0 is 0.

7. $4 \cdot 1 \cdot 6 = (4 \cdot 1) \cdot 6 = 4 \cdot 6 = 24$

9. $(2)(3)(6) = [(2)(3)](6) = (6)(6) = 36$

11.
$$\begin{array}{r} \overset{3}{35} \\ \times\ 7 \\ \hline 245 \end{array}$$

$7 \cdot 5 = 35$ Write 5, regroup 3 tens.
$7 \cdot 3 = 21$ Add 3 to get 24. Write 24.

13.
$$\begin{array}{r} \overset{4}{28} \\ \times\ 6 \\ \hline 168 \end{array}$$

$6 \cdot 8 = 48$ Write 8, regroup 4 tens.
$6 \cdot 2 = 12$ Add 4 to get 16. Write 16.

15.
$$\begin{array}{r} \overset{141}{3182} \\ \times\ \ 6 \\ \hline 19{,}092 \end{array}$$

$6 \cdot 2 = 12$ Write 2, regroup 1 ten.
$6 \cdot 8 = 48$ Add 1 to get 49. Write 9, regroup 4.
$6 \cdot 1 = 6$ Add 4 to get 10. Write 0, regroup 1.
$6 \cdot 3 = 18$ Add 1 to get 19. Write 19.

17.
$$\begin{array}{r} \overset{46\ 1}{36{,}921} \\ \times\quad 7 \\ \hline 258{,}447 \end{array}$$

$7 \cdot 1 = 7$ Write 7.
$7 \cdot 2 = 14$ Write 4, regroup 1.
$7 \cdot 9 = 63$ Add 1 to get 64. Write 4, regroup 6.
$7 \cdot 6 = 42$ Add 6 to get 48. Write 8, regroup 4.
$7 \cdot 3 = 21$ Add 4 to get 25. Write 25.

19.
$$\begin{array}{r} 125 \\ \times\ 100 \\ \hline \end{array} \qquad \begin{array}{r} 125 \\ \times\ 1 \\ \hline 125 \end{array} \qquad \begin{array}{r} 125 \\ \times\ 100 \\ \hline 12{,}500 \end{array}$$ *Attach* 00.

21.
$$\begin{array}{r} 1485 \\ \times\ \ 30 \\ \hline \end{array} \qquad \begin{array}{r} 1485 \\ \times\ 3 \\ \hline 4455 \end{array} \qquad \begin{array}{r} 1485 \\ \times\ \ 30 \\ \hline 44{,}550 \end{array}$$ *Attach* 0.

23.
$$\begin{array}{r} 900 \\ \times\ 300 \\ \hline \end{array} \qquad \begin{array}{r} 9 \\ \times\ 3 \\ \hline 27 \end{array} \qquad \begin{array}{r} 900 \\ \times\ 300 \\ \hline 270{,}000 \end{array}$$ *Attach* 0000.

25.
$$\begin{array}{r} 43{,}000 \\ \times\ 2000 \\ \hline \end{array} \qquad \begin{array}{r} 43 \\ \times\ 2 \\ \hline 86 \end{array} \qquad \begin{array}{r} 43{,}000 \\ \times\ 2000 \\ \hline 86{,}000{,}000 \end{array}$$ *Attach* 000000.

27.
$$\begin{array}{rl} 68 & \\ \times\ \ 22 & \\ \hline 136 & \leftarrow 2 \times 68 \\ 136\ \ & \leftarrow 2 \times 68 \\ \hline 1496 & \end{array}$$

29.
$$\begin{array}{r l} 83 & \\ \times\ 45 & \\ \hline 415 & \leftarrow 5 \times 83 \\ 332 & \leftarrow 4 \times 83 \\ \hline 3735 & \end{array}$$

31. (32)(475)
$$\begin{array}{r l} 475 & \\ \times\ 32 & \\ \hline 950 & \leftarrow 2 \times 475 \\ 1425 & \leftarrow 3 \times 475 \\ \hline 15{,}200 & \end{array}$$

33. (729)(45)
$$\begin{array}{r l} 729 & \\ \times\ 45 & \\ \hline 3645 & \leftarrow 5 \times 729 \\ 2916 & \leftarrow 4 \times 729 \\ \hline 32{,}805 & \end{array}$$

35.
$$\begin{array}{r l} 538 & \\ \times\ 342 & \\ \hline 1076 & \leftarrow 2 \times 538 \\ 2152 & \leftarrow 4 \times 538 \\ 1614 & \leftarrow 3 \times 538 \\ \hline 183{,}996 & \end{array}$$

37.
$$\begin{array}{r l} 8162 & \\ \times\ 407 & \\ \hline 57134 & \leftarrow 7 \times 8162 \\ 326480 & \leftarrow 40 \times 8162 \\ \hline 3{,}321{,}934 & \end{array}$$

39.
$$\begin{array}{r l} 6310 & \\ \times\ 3008 & \\ \hline 50480 & \leftarrow 8 \times 6310 \\ 1893000 & \leftarrow 300 \times 6310 \\ \hline 18{,}980{,}480 & \end{array}$$

41.
$$\begin{array}{r l} 18 & \textit{inches per day} \\ \times\ 14 & \textit{days} \\ \hline 72 & \\ 18 & \\ \hline 252 & \textit{inches in two weeks} \end{array}$$

$$\begin{array}{r l} 18 & \textit{inches per day} \\ \times\ 30 & \textit{days} \\ \hline 540 & \textit{inches in 30 days} \end{array}$$

43.
$$\begin{array}{r l} 48 & \textit{flats} \\ \times\ 12 & \textit{tomato plants per flat} \\ \hline 96 & \\ 48 & \\ \hline 576 & \textit{tomato plants} \end{array}$$

The total number of tomato plants is 576.

45.
$$\begin{array}{r l} 55 & \textit{miles per gallon} \\ \times\ 11 & \textit{gallons} \\ \hline 55 & \\ 55 & \\ \hline 605 & \textit{miles} \end{array}$$

The Prius can travel 605 miles on 11 gallons of gas.

47.
$$\begin{array}{r l} 2695 & \textit{Reno to Atlantic Ocean} \\ -\ 255 & \textit{Reno to Pacific Ocean} \\ \hline 2440 & \textit{difference} \end{array}$$

It is 2440 miles farther from Reno to the Atlantic Ocean than it is from Reno to the Pacific Ocean.

$$\begin{array}{r l} 2695 & \textit{miles per trip} \\ \times\ 6 & \textit{trips (3 round trips)} \\ \hline 16{,}170 & \textit{miles} \end{array}$$

You will earn 16,170 frequent flier miles.

49.
$$\begin{array}{r l} \overset{916}{14\not{0}\not{6}} & \textit{calories per high-fat meal} \\ -\ 348 & \textit{calories per low-fat meal} \\ \hline 1058 & \textit{calories difference} \end{array}$$

$$\begin{array}{r l} \overset{45}{1058} & \textit{more calories per meal} \\ \times\ 7 & \textit{meals} \\ \hline 7406 & \textit{more calories} \end{array}$$

There are 7406 more calories in seven high-fat meals than in seven low-fat meals.

R.4 Dividing Whole Numbers

R.4 Section Exercises

1. $\frac{12}{12} = 1$; $12\overline{)12}$ or $12 \div 12$

3. $24 \div 0$ is undefined; $\frac{24}{0}$ or $0\overline{)24}$

5. $\frac{0}{4} = 0$; $4\overline{)0}$ or $0 \div 4$

7. $0 \div 12 = 0$; $\frac{0}{12}$ or $12\overline{)0}$

9. $0\overline{)21}$ is undefined; $\frac{21}{0}$ or $21 \div 0$

11. $\begin{array}{r} 21 \\ 4\overline{)84} \end{array}$

The dividend is 84, the divisor is 4, and the quotient is 21.

Check: $4 \times 21 = 84$

13. $\begin{array}{r} 213 \\ 3\overline{)639} \end{array}$

The dividend is 639, the divisor is 3, and the quotient is 213.

Check: $3 \times 213 = 639$

15. $\begin{array}{r} 1\ 5\ 2\ 2\ \textbf{R5} \\ 6\overline{)9\,{}^{3}1\,{}^{1}3\,{}^{1}7} \end{array}$

Check: $6 \times 1522 + 5 = 9132 + 5 = 9137$

17. $\begin{array}{r} 30\ 9\,\textbf{R4} \\ 6\overline{)185\,{}^{5}8} \end{array}$

Check: $6 \times 309 + 4 = 1854 + 4 = 1858$

19. $4024 \div 4$ $\quad \begin{array}{r} 1006 \\ 4\overline{)4024} \end{array}$

Check: $4 \times 1006 = 4024$

21. $15{,}019 \div 3$ $\quad \begin{array}{r} 5\,00\ 6\,\textbf{R1} \\ 3\overline{)15{,}01\,{}^{1}9} \end{array}$

Check: $3 \times 5006 + 1 = 15{,}018 + 1 = 15{,}019$

23. $\frac{26{,}684}{4}$ $\quad \begin{array}{r} 6\ 6\ 71 \\ 4\overline{)26,\,{}^{2}6\,{}^{2}84} \end{array}$

Check: $4 \times 6671 = 26{,}684$

25. $\frac{74{,}751}{6}$ $\quad \begin{array}{r} 1\ 2,\ 4\ 5\ 8\ \textbf{R3} \\ 6\overline{)7\,{}^{1}4,\,{}^{2}7\,{}^{3}5\,{}^{5}1} \end{array}$

Check: $6 \times 12{,}458 + 3 = 74{,}748 + 3 = 74{,}751$

27. $\frac{71{,}776}{7}$ $\quad \begin{array}{r} 10,\ 2\ 5\ 3\ \textbf{R5} \\ 7\overline{)71,\,{}^{1}7\,{}^{3}7\,{}^{2}6} \end{array}$

Check: $7 \times 10{,}253 + 5 = 71{,}771 + 5 = 71{,}776$

29. $\frac{128{,}645}{7}$ $\quad \begin{array}{r} 1\ 8,\ 3\ 7\ 7\ \textbf{R6} \\ 7\overline{)12\,{}^{5}8,\,{}^{2}6\,{}^{5}4\,{}^{5}5} \end{array}$

Check: $7 \times 18{,}377 + 6 = 128{,}639 + 6 = 128{,}645$

31. $\begin{array}{r} 67\ \textbf{R2} \\ 7\overline{)4692} \end{array}$

Check: $7 \times 67 + 2 = 469 + 2 = 471$ *incorrect*

Rework:

$\begin{array}{r} 6\ 70\ \textbf{R2} \\ 7\overline{)46\,{}^{4}92} \end{array}$

Check: $7 \times 670 + 2 = 4690 + 2 = 4692$ *correct*

33. $\begin{array}{r} 3\ 568\ \textbf{R2} \\ 6\overline{)21{,}409} \end{array}$

Check: $6 \times 3568 + 2 = 21{,}408 + 2 = 21{,}410$ *incorrect*

Rework:

$\begin{array}{r} 3\ 5\ 6\ 8\ \textbf{R1} \\ 6\overline{)21,\,{}^{3}4\,{}^{4}0\,{}^{4}9} \end{array}$

Check: $6 \times 3568 + 1 = 21{,}408 + 1 = 21{,}409$

35. $\begin{array}{r} 3\ 003\ \textbf{R5} \\ 6\overline{)18{,}023} \end{array}$

Check: $6 \times 3003 + 5 = 18{,}018 + 5 = 18{,}023$ *correct*

37. $\begin{array}{r} 11{,}523\ \textbf{R2} \\ 6\overline{)69{,}140} \end{array}$

Check: $6 \times 11{,}523 + 2 = 69{,}138 + 2 = 69{,}140$ *correct*

39. $\begin{array}{r} 2\ 3 \\ 8\overline{)18\,{}^{2}4} \end{array} \quad \begin{array}{r} 1\ 4 \\ 8\overline{)11\,{}^{3}2} \end{array} \quad \begin{array}{r} 1\ 9 \\ 8\overline{)15\,{}^{7}2} \end{array}$

Kaci earns $23/hour. Her workers earn $14/hour and $19/hour.

41. $\begin{array}{r} 1\ 6,\ 600 \\ 6\overline{)9\,{}^{3}9,\,{}^{3}600} \end{array}$

Each van costs $16,600.

43. $\begin{array}{r} 3\ 7\ 8 \\ 5\overline{)18\,{}^{3}9\,{}^{4}0} \end{array}$ (378 $5 tickets)

$\begin{array}{r} 2\ 7\ 0 \\ 7\overline{)18\,{}^{4}9\ 0} \end{array}$ (270 $7 tickets)

$\begin{array}{r} 210 \\ 9\overline{)1890} \end{array}$ (210 $9 tickets)

45. 300 × $5 = $1500 (too little for the $1890 budget)

300 × $7 = $2100 (enough to cover the $1890 budget)

$$\begin{array}{r} \overset{10}{\ } \\ 1\ \not{0}\ 10 \\ \$\not{2}\ 1\ \not{0}\ 0 \\ -\ 1\ 8\ 9\ 0 \\ \hline \$\ 2\ 1\ 0 \end{array}$$

Selling 300 $7 tickets would cover the $1890 budget with $210 extra.

47. **(a)** Circle the numbers that end in 0, 2, 4, 6, or 8; that is, circle 358 and 190.

(b) Find the sum of the digits for each number.

736: $7+3+6=16$
10,404: $1+0+4+0+4=9$
5603: $5+6+0+3=14$
78: $7+8=15$

Since the sums 16 and 14 *are not* divisible by 3, the numbers 736 and 5603 *are not* divisible by 3. Since the sums 9 and 15 *are* divisible by 3, the numbers 10,404 and 78 *are* divisible by 3 and should be circled.

(c) Circle the numbers that end in 0 or 5; that is, circle 13,740 and 985.

49. 30 ends in 0, so it is divisible by 2, 5, and 10. The sum of its digits, 3, is divisible by 3, so 30 is divisible by 3.

51. 184 ends in 4, so it is divisible by 2, but not divisible by 5 or 10. The sum of its digits, 13, is not divisible by 3, so 184 is not divisible by 3.

53. 445 ends in 5, so it is divisible by 5, but not divisible by 2 or 10. The sum of its digits, 13, is not divisible by 3, so 445 is not divisible by 3.

55. 903 ends in 3, so it is not divisible by 2, 5, or 10. The sum of its digits, 12, is divisible by 3, so 903 is divisible by 3.

57. 5166 ends in 6, so it is divisible by 2, but not divisible by 5 or 10. The sum of its digits, 18, is divisible by 3, so 5166 is divisible by 3.

59. 21,763 ends in 3, so it is not divisible by 2, 5, or 10. The sum of its digits, 19, is not divisible by 3, so 21,763 is not divisible by 3.

R.5 Long Division

R.5 Section Exercises

1. Because 24 is closer to 20 than to 30, use 2 as a trial divisor.

$$\begin{array}{r} 3 \\ 24\overline{)7\,6\,8} \end{array}$$

3 goes over the 6 because $\frac{76}{24}$ is about 3.

3. Because 18 is closer to 20 than to 10, use 2 as a trial divisor.

$$\begin{array}{r} 2 \\ 18\overline{)4\,5\,0\,0} \end{array}$$

2 goes over the 5 because $\frac{45}{18}$ is about 2.

5. Because 86 is closer to 90 than to 80, use 9 as a trial divisor.

$$\begin{array}{r} 1 \\ 86\overline{)1\,0,\,3\,2\,7} \end{array}$$

1 goes over the 3 because $\frac{103}{86}$ is about 1.

7. Because 52 is closer to 50 than to 60, use 5 as a trial divisor.

$$\begin{array}{r} 7 \\ 52\overline{)3\,8,\,0\,2\,5} \end{array}$$

7 goes over the 0 because $\frac{380}{52}$ is about 7.

9. Because 77 is closer to 80 than to 70, use 8 as a trial divisor.

$$\begin{array}{r} 3 \\ 77\overline{)2\,4\,9,\,8\,2\,6} \end{array}$$

3 goes over the 9 because $\frac{249}{77}$ is about 3.

11. Because 420 is closer to 400 than to 500, use 4 as a trial divisor.

$$\begin{array}{r} 1 \\ 420\overline{)4\,7\,0,\,8\,0\,0} \end{array}$$

1 goes over the first 0 because $\frac{470}{420}$ is about 1.

13.
$$\begin{array}{r} 6\,4\ \textbf{R3} \\ 29\overline{)1\,8\,5\,9} \\ \underline{1\,7\,4} \\ 1\,1\,9 \\ \underline{1\,1\,6} \\ 3 \end{array}$$

Check:
$$\begin{array}{r} 64 \\ \times\ \ 29 \\ \hline 576 \\ 128 \\ \hline 1856 \\ +\ \ \ 3 \\ \hline 1859 \end{array}$$

15.
$$\begin{array}{r} 2\,3\,6\ \textbf{R29} \\ 47\overline{)1\,1,\,1\,2\,1} \\ \underline{9\,4} \\ 1\,7\,2 \\ \underline{1\,4\,1} \\ 3\,1\,1 \\ \underline{2\,8\,2} \\ 2\,9 \end{array}$$

Check:
$$\begin{array}{r} 236 \\ \times\ \ 47 \\ \hline 1\,652 \\ 9\,44 \\ \hline 11{,}092 \\ +\ \ \ 29 \\ \hline 11{,}121 \end{array}$$

17.
$$\begin{array}{r} 2\,4\,0\,7\ \textbf{R1} \\ 26\overline{)6\,2,\,5\,8\,3} \\ \underline{5\,2} \\ 1\,0\,5 \\ \underline{1\,0\,4} \\ 1\,8\,3 \\ \underline{1\,8\,2} \\ 1 \end{array}$$

Check:
$$\begin{array}{r} 2407 \\ \times\ \ 26 \\ \hline 14442 \\ 4814 \\ \hline 62{,}582 \\ +\ \ \ 1 \\ \hline 62{,}583 \end{array}$$

19.
```
    1 2 3 9 R15
63)7 8, 0 7 2
   6 3
   1 5 0
   1 2 6
     2 4 7
     1 8 9
       5 8 2
       5 6 7
         1 5
```
Check:
```
   1239
 ×   63
   3717
  7434
 78,057
 +   15
 78,072
```

21. 150)4 9 9, 7 6 0 Drop 1 zero.
```
    3 3 3 1 R11
15)4 9, 9 7 6
   4 5
     4 9
     4 5
       4 7
       4 5
         2 6
         1 5
         1 1
```
Check:
```
   3331
 ×   15
  16655
  3331
 49,965
 +   11
 49,976
```

Note: If you get a nonzero remainder when dropping zeros, you must add the same number of zeros to the remainder after you divide. Hence, the answer is 3331 **R**110.

23. 400)3 4 0, 0 0 0 Drop 2 zeros.
```
   8 5 0
4)3 4 0 0
  3 2
    2 0
    2 0
      0
```
Check:
```
   850
 ×   4
  3400
```

25.
```
    1 0 6 R17
56)5 9 4 3
```
Check:
```
   106
 ×  56
   636
  530
  5936
 +  17
  5953 incorrect
```

Rework.
```
    1 0 6 R7
56)5 9 4 3
   5 6
     3 4 3
     3 3 6
         7
```
Check:
```
   106
 ×  56
   636
  530
  5936
 +   7
  5943
```

The correct answer is 106 **R**7.

27.
```
     6 5 8 R9
600)3 9 4, 8 0 0
```
Check:
```
     658
 ×   600
 394,800
 +     9
 394,809 incorrect
```

From the check, we can see that the correct answer is 658.

29.
```
     6 2 R3
410)2 5, 4 2 0
```
Check:
```
     410
 ×    62
     820
   24 60
  25,420
 +     3
  25,423 incorrect
```

From the check, we can see that the correct answer is 62.

31.
```
    4 5 0 R65
72)3 2, 4 6 5
```
Check:
```
     450
 ×    72
     900
   3150
   32400
 +    65
  32,465 correct
```

33.
```
    4
    59 hours per week         38 hours per week
 ×  50 weeks per year      ×  50 weeks per year
  2950                      1900
```

2950 − 1900 = 1050 more hours were worked per year in 1900 than today.

Alternatively, we see that there are 59 − 38 = 21 more hours of work per week in 1900 than today, so we get 50 × 21 = 1050 more hours per year.

35.
```
  32
  375 dollars per night        32 dollars per night
 ×  4 number of nights      ×  4 number of nights
 1500                         128
```

```
    9
  4 1̸0 10
  1 5̸ 0̸ 0̸
 −  1 2 8
  1 3 7 2
```
The amount saved by staying four nights at the Motel 6 rather than the Ritz-Carlton is $1372.

Alternatively, the difference is $375 − $32 = $343 per night, so for 4 nights, the total savings is 4 × $343 = $1372.

37. Divide. $11{,}088 \div 36 = 308$
Judy's monthly payment is \$308.

39.
$$\begin{array}{rl} 42 & \textit{circuits per hour} \\ \times\ 8 & \textit{hours per day} \\ \hline 336 & \textit{circuits per day} \\ \times\ 5 & \textit{days per week} \\ \hline 1680 & \textit{circuits per week} \end{array}$$

He can assemble 1680 circuits in a 40-hr workweek.

41. First subtract.

$$\begin{array}{rl} \$7588 & \textit{money raised} \\ -\ \ 838 & \textit{expenses} \\ \hline \$6750 & \textit{remaining money} \end{array}$$

Then divide.

$$\text{Number of teams} \rightarrow 18\overline{)6\,7\,5\,0} \leftarrow \text{Remaining money}$$

$$\begin{array}{r} 375 \\ 18\overline{)6750} \\ 54 \\ \hline 135 \\ 126 \\ \hline 90 \\ 90 \\ \hline 0 \end{array}$$

Each team received \$375.

Chapter R Review Exercises

1.
$$\begin{array}{r} \overset{1}{7}4 \\ +\ 18 \\ \hline 92 \end{array}$$

2.
$$\begin{array}{r} \overset{1}{3}5 \\ +\ 78 \\ \hline 113 \end{array}$$

3.
$$\begin{array}{r} \overset{1}{}\overset{1}{8}07 \\ 4606 \\ +\ \ \ 51 \\ \hline 5464 \end{array}$$

4.
$$\begin{array}{r} 8\,2\overset{1}{1}5 \\ 9 \\ +\ 7\,433 \\ \hline 15{,}657 \end{array}$$

5.
$$\begin{array}{r} \overset{1}{\not2}\ \overset{12}{\not3}\ \overset{18}{\not8} \\ -\ 1\ 9\ 9 \\ \hline 3\ 9 \end{array} \qquad \text{Check: } \begin{array}{r} \overset{11}{199} \\ +\ 39 \\ \hline 238 \end{array}$$

6.
$$\begin{array}{r} \overset{4}{\not5}\ \overset{16}{\not7}\ \overset{13}{\not3} \\ -\ 3\ 8\ 9 \\ \hline 1\ 8\ 4 \end{array} \qquad \text{Check: } \begin{array}{r} \overset{11}{389} \\ +\ 184 \\ \hline 573 \end{array}$$

7.
$$\begin{array}{r} \overset{1}{2}\ \overset{11}{2}\ \overset{10}{\not1}\ \overset{10}{\not0} \\ -\ 1\ 9\ 8\ 6 \\ \hline 2\ 2\ 4 \end{array} \qquad \text{Check: } \begin{array}{r} \overset{111}{1986} \\ +\ 224 \\ \hline 2210 \end{array}$$

8.
$$\begin{array}{r} 9\overset{8}{\not9},\overset{16}{\not7}\ \overset{\overset{9}{10}}{\not0}\ \overset{14}{\not4} \\ -\ 73,8\ 3\ 8 \\ \hline 25,8\ 6\ 6 \end{array} \qquad \text{Check: } \begin{array}{r} \overset{1\ 11}{25{,}866} \\ +\ 73{,}838 \\ \hline 99{,}704 \end{array}$$

9. $2 \times 4 \times 6$
$= (2 \times 4) \times 6$
$= 8 \times 6 = 48$

10. $9 \times 1 \times 5$
$= (9 \times 1) \times 5$
$= 9 \times 5 = 45$

11. $(6)(1)(8)$
$= [(6)(1)](8)$
$= (6)(8) = 48$

12. $7 \cdot 7 \cdot 0 = 0$ Any number times 0 equals 0.

13.
$$\begin{array}{r} \overset{1}{4}3 \\ \times\ 4 \\ \hline 172 \end{array}$$

14.
$$\begin{array}{r} \overset{5}{7}81 \\ \times\ 7 \\ \hline 5467 \end{array}$$

15.
$$\begin{array}{r} \overset{2}{5}\,\overset{2}{4}40 \\ \times\ 6 \\ \hline 32{,}640 \end{array}$$

16.
$$\begin{array}{r} \overset{1}{9}3{,}1\overset{2}{0}5 \\ \times\ \ \ \ \ 5 \\ \hline 465{,}525 \end{array}$$

17.
$$\begin{array}{r} 320 \\ \times\ 60 \\ \hline \end{array} \qquad \begin{array}{r} 32 \\ \times\ 6 \\ \hline 192 \end{array} \qquad \begin{array}{r} 320 \\ \times\ 60 \\ \hline 19{,}200 \end{array} \quad \textit{Attach } 00.$$

18.
$$\begin{array}{r} 280 \\ \times\ 90 \\ \hline \end{array} \qquad \begin{array}{r} 28 \\ \times\ 9 \\ \hline 252 \end{array} \qquad \begin{array}{r} 280 \\ \times\ 90 \\ \hline 25{,}200 \end{array} \quad \textit{Attach } 00.$$

19.
$$\begin{array}{r} 517 \\ \times\ 400 \\ \hline \end{array} \qquad \begin{array}{r} 517 \\ \times\ 4 \\ \hline 2068 \end{array} \qquad \begin{array}{r} 517 \\ \times\ 400 \\ \hline 206{,}800 \end{array} \quad \textit{Attach } 00.$$

20.
$$\begin{array}{r} 16{,}000 \\ \times\ 8000 \\ \hline \end{array} \qquad \begin{array}{r} 16 \\ \times\ 8 \\ \hline 128 \end{array} \qquad \begin{array}{r} 16{,}000 \\ \times\ 8\,000 \\ \hline 128{,}000{,}000 \end{array} \quad \textit{Attach } 000000.$$

21.
$$\begin{array}{r} \overset{3}{3}4 \\ \times\ 18 \\ \hline 272 \\ 34 \\ \hline 612 \end{array}$$

22.
$$\begin{array}{r} \overset{1}{5}2 \\ \times\ 36 \\ \hline 312 \\ 156 \\ \hline 1872 \end{array}$$

23.
$$\begin{array}{r} \overset{1}{6}\overset{1}{5}5 \\ \times\ 21 \\ \hline 655 \\ 1310 \\ \hline 13{,}755 \end{array}$$

24.
$$\begin{array}{r} \overset{6}{3}\overset{1}{9}2 \\ \times\ 77 \\ \hline 2744 \\ 2744 \\ \hline 30{,}184 \end{array}$$

25. $42 \div 7 = 6$

26. $18 \div 18 = 1$

27. $\frac{125}{0}$ is undefined.

28. $\frac{0}{35} = 0$

29. $4\overline{)43^32}$ quotient $10\ 8$ Check: $\begin{array}{r} 108 \\ \times 4 \\ \hline 432 \end{array}$

30. $9\overline{)21^36}$ quotient $2\ 4$ Check: $\begin{array}{r} 24 \\ \times 9 \\ \hline 216 \end{array}$

31.
$$\begin{array}{r} 352 \\ 76\overline{)26{,}752} \\ 228 \\ \hline 395 \\ 380 \\ \hline 152 \\ 152 \\ \hline 0 \end{array}$$

Check:
$$\begin{array}{r} 352 \\ \times\ 76 \\ \hline 2112 \\ 2464 \\ \hline 26{,}752 \end{array}$$

32. $2704 \div 18$
$$\begin{array}{r} 150\ \textbf{R}4 \\ 18\overline{)2704} \\ 18 \\ \hline 90 \\ 90 \\ \hline 4 \end{array}$$

Check:
$$\begin{array}{r} 150 \\ \times\ 18 \\ \hline 1200 \\ 150 \\ \hline 2700 \\ +\ \ 4 \\ \hline 2704 \end{array}$$

33. **[R.3]**
$$\begin{array}{rl} 52 & \textit{cards per deck} \\ \times 2 & \textit{decks} \\ \hline 104 & \textit{total cards} \end{array} \qquad \begin{array}{rl} 52 & \textit{cards per deck} \\ \times 6 & \textit{decks} \\ \hline 312 & \textit{total cards} \end{array}$$

34. **[R.3]**
$$\begin{array}{rl} 238 & \textit{cartons} \\ \times\ 12 & \textit{textbooks per carton} \\ \hline 476 & \\ 238 & \\ \hline 2856 & \textit{total textbooks} \end{array}$$

35. **[R.3]**
$$\begin{array}{rl} \$47 & \textit{rent per day} \\ \times\ \ 4 & \textit{number of days} \\ \hline \$188 & \textit{total rental charge} \\ +\ 24 & \textit{gas charge} \\ \hline \$212 & \textit{total amount spent} \end{array}$$

Raoul was over his budget by $212 − 200 = $12.

36. **[R.2]**
$$\begin{array}{r} \overset{1014}{\overset{2\not{0}\ \not{4}\ \ 12}{\$\not{3}\not{1}\ \not{5},\not{2}80}} \\ -\ 8\ 7{,}340 \\ \hline \$22\ 7{,}940 \end{array}$$

They must raise $227,940.

37. **[R.3]**
$$\begin{array}{rl} 2000 & \textit{hours per home} \\ \times 5 & \textit{homes} \\ \hline 10{,}000 & \textit{total hours} \end{array}$$

$$\begin{array}{rl} 2000 & \textit{hours per home} \\ \times 12 & \textit{homes} \\ \hline 24{,}000 & \textit{total hours} \end{array}$$

38. **[R.3]**
$$\begin{array}{rl} 80 & \textit{miles per hour} \\ \times 3 & \textit{hours} \\ \hline 240 & \textit{total miles} \end{array}$$

$$\begin{array}{rl} 80 & \textit{miles per hour} \\ \times 7 & \textit{hours} \\ \hline 560 & \textit{total miles} \end{array}$$

39. **[R.5]** To find the total acres fertilized, divide the total pounds by the pounds needed per acre.

Drop one 0 from the divisor and the dividend.

$5750 \div 250 = 575 \div 25$

$$\begin{array}{r} 23 \\ 25\overline{)575} \\ \underline{50} \\ 75 \\ \underline{75} \\ 0 \end{array}$$

23 acres can be fertilized.

40. **[R.5]** To find the number of homes that can be fenced, divide the total number of feet by the number of feet needed for each home.

Drop one 0 from the divisor and the dividend.

$5760 \div 180 = 576 \div 18$

$$\begin{array}{r} 32 \\ 18\overline{)576} \\ \underline{54} \\ 36 \\ \underline{36} \\ 0 \end{array}$$

32 homes can be fenced.

41. **[R.2]** To find the new account balance, first subtract the amount of the check from the old balance.

$\$382 - 135 = \$247$

Next, add the amount of the deposit.

$\$247 + \$563 = \$810$

She has a new balance of \$810.

42. **[R.3]** Total cost = cost of four T-shirts

+ cost of three sweatshirts

$= (4 \times \text{cost per T-shirt})$

$+ (3 \times \text{cost per sweatshirt})$

$= (4 \times 14) + (3 \times 29)$

$= 56 + 87 = 143$

The total cost is \$143.

43. **[R.3]**

Total cost = cost of 18 adult admissions

+ cost of 6 child admissions

+ cost of 15 senior admissions

$= (18 \times 18) + (6 \times 14) + (15 \times 17)$

$= 324 + 84 + 255$

$= 663$

The total cost is \$663.

44. **[R.3]** To find the total monthly collections, add the amount collected from the 56 daily customers to the amount collected from the 23 weekend customers. To find the amount collected from the 56 daily customers, multiply the amount collected from one daily customer by 56. To find the amount collected from the 23 weekend customers, multiply the amount collected from one weekend customer by 23.

Total monthly collections

= (amount collected from 56 daily customers)

+ (amount collected from 23 weekend customers)

$= (56 \times 18) + (23 \times 11)$

$= 1008 + 253 = 1261$

The total monthly collection is \$1261.

Chapter R Test

1. addends; sum or total

2. factors; product

3. difference; quotient

4.
$$\begin{array}{r} {\scriptstyle 121} \\ 984 \\ 65 \\ +7561 \\ \hline 8610 \end{array}$$

5.
$$\begin{array}{r} {\scriptstyle 11\ 12} \\ 17{,}063 \\ 7 \\ 12 \\ 1\,505 \\ 93{,}710 \\ +\quad 333 \\ \hline 112{,}630 \end{array}$$

6.
$$\begin{array}{r} {\scriptstyle 9\ \ 9} \\ {\scriptstyle 6\ \cancel{10}\ \cancel{10}\ 12} \\ 1\cancel{7},\cancel{0}\ \cancel{0}\ \cancel{2} \\ -\qquad 5\ 4 \\ \hline 16,9\ 4\ 8 \end{array}$$

7.
$$\begin{array}{r} {\scriptstyle 9} \\ {\scriptstyle 4\ \cancel{10}\ 15\ 12} \\ \cancel{5}\ \cancel{0}\ \cancel{6}\ \cancel{2} \\ -1\ 9\ 7\ 8 \\ \hline 3\ 0\ 8\ 4 \end{array}$$

8. $5 \times 7 \times 4$

$(5 \times 7) \times 4 = 35 \times 4 = 140$

or

$5 \times (7 \times 4) = 5 \times 28 = 140$

9. $57 \cdot 3000$
$$\begin{array}{r} 57 \\ \times\ 3 \\ \hline 171 \end{array}$$

$57 \cdot 3000 = 171{,}000$ *Attach* 000.

10. (85)(21)

$$\begin{array}{r} 85 \\ \times\ 21 \\ \hline 85 \\ 170 \\ \hline 1785 \end{array}$$

11.

$$\begin{array}{r} 7381 \\ \times\ \ 603 \\ \hline 22143 \\ 4428\,60 \\ \hline 4{,}450{,}743 \end{array}$$

Check:

$$\begin{array}{r} 7381 \\ 603\overline{)4{,}450{,}743} \\ 4221 \\ \hline 2297 \\ 1809 \\ \hline 4884 \\ 4824 \\ \hline 603 \\ 603 \\ \hline 0 \end{array}$$

12.

$$\begin{array}{r} 206 \\ 6\overline{)123^{3}6} \end{array}$$

Check: $6 \times 206 = 1236$

13. $\frac{791}{0}$ is undefined.
Division by zero is not possible.

14. $38{,}472 \div 84$

$$\begin{array}{r} 458 \\ 84\overline{)38{,}472} \\ 336 \\ \hline 487 \\ 420 \\ \hline 672 \\ 672 \\ \hline 0 \end{array}$$

Check:

$$\begin{array}{r} 458 \\ \times\ \ 84 \\ \hline 1832 \\ 3664 \\ \hline 38{,}472 \end{array}$$

15. $280\overline{)44{,}800}$ Drop 1 zero.

$$\begin{array}{r} 160 \\ 28\overline{)4480} \\ 28 \\ \hline 168 \\ 168 \\ \hline 0 \end{array}$$

16.

$$\begin{array}{rl} \$11 & \textit{cost per beaker} \\ \times\ 48 & \textit{beakers} \\ \hline \$528 & \textit{total cost} \end{array}$$

17.

$$\begin{array}{rl} {}^{2\ 14\ 10} & \\ \$\cancel{3}\,\cancel{5}\,\cancel{0} & \textit{most expensive phone} \\ -\ 5\ 9 & \textit{least expensive phone} \\ \hline \$2\ 9\ 1 & \textit{difference} \end{array}$$

The difference in price between the most expensive and the least expensive phone is $291.

18. Drop 2 zeros.

$31{,}500 \div 900 = 315 \div 9$

$$\begin{array}{r} 35 \\ 9\overline{)31^{4}5} \end{array}$$

It will take 35 hours.

19. Add the amounts paid.

$$\begin{array}{r} {}^{21} \\ \$690 \\ 185 \\ +\ 68 \\ \hline \$943 \end{array}$$

Subtract from the balance.

$$\begin{array}{r} {}^{10\,10} \\ \$\cancel{1}\,\cancel{1}\,\cancel{0}\,8 \\ -\ 9\ 4\ 3 \\ \hline \$1\ 6\ 5 \end{array}$$

Her new balance is $165.

20.

$$\begin{array}{rl} 118 & \textit{self-cleaning ovens per hour} \\ \times\ 4 & \textit{hours} \\ \hline 472 & \textit{self-cleaning ovens} \end{array}$$

$$\begin{array}{rl} 139 & \textit{standard ovens per hour} \\ \times\ 4 & \textit{hours} \\ \hline 556 & \textit{standard ovens} \end{array}$$

$$\begin{array}{rl} 472 & \textit{self-cleaning ovens} \\ +\ 556 & \textit{standard ovens} \\ \hline 1028 & \textit{total ovens} \end{array}$$

The total number of ovens assembled in the 8-hour period is 1028.

21. Add the rent collected.

$$\begin{array}{r} {}^{11} \\ \$785 \\ 800 \\ 815 \\ 725 \\ \hline \$3125 \end{array}$$

Subtract expenses.

$$\begin{array}{r} {}^{012} \\ \$3\cancel{1}\cancel{2}5 \\ -\ 1085 \\ \hline \$2040 \end{array}$$

remains after expenses.

22. A number is divisible by 2 if it ends in 0, 2, 4, 6, or 8.

Examples: 312 is divisible by 2.
415 is not divisible by 2.

A number is divisible by 5 if it ends in 0 or 5.

Examples: 75 is divisible by 5.
93 is not divisible by 5.

A number is divisible by 10 if it ends in 0.

Examples: 470 is divisible by 10.
293 is not divisible by 10.

APPENDIX B INDUCTIVE AND DEDUCTIVE REASONING

Appendix B Exercises

1. $2, 9, 16, 23, 30, \ldots$

Inspect the sequence and note that 7 is added to a term to obtain the next term. So the term immediately following 30 is $30 + 7 = 37$.

3. $0, 10, 8, 18, 16, \ldots$

Inspect the sequence and note that 10 is added, then 2 is subtracted, then 10 is added, then 2 is subtracted, and so on. So the term immediately following 16 is $16 + 10 = 26$.

5. $1, 2, 4, 8, \ldots$

Inspect the sequence and note that 2 is multiplied times a term to obtain the next term. So, the term immediately following 8 is $(8)(2) = 16$.

7. $1, 3, 9, 27, 81, \ldots$

Inspect the sequence and note that 3 is multiplied times a term to obtain the next term. So, the term immediately following 81 is $(81)(3) = 243$.

9. $1, 4, 9, 16, 25, \ldots$

Inspect the sequence and note that the pattern is add 3, add 5, add 7, etc., or 1^1, 2^2, 3^2, etc. So, the term immediately following 25 is $6^2 = 36$.

11. The first three shapes are unique. The fourth shape is the same as the first shape except that it is reversed, and reversing the position of the second shape gives the fifth shape. So, the next shape will be the reverse of the third shape.

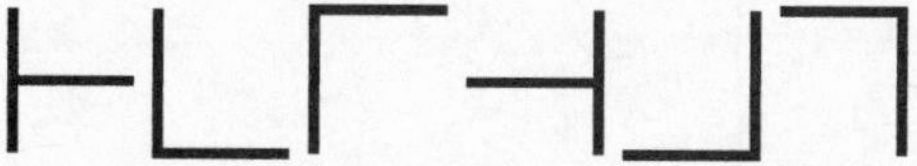

13. The first three shapes are the same except different images. Since the third shape is the reverse of the first shape, the fourth shape should be a reverse of the second shape.

15. All animals are wild.
All lions are animals.
$\therefore$ All lions are wild.

The statement "All animals are wild" is shown by a large circle that represents all creatures that are wild with a smaller circle inside that represents animals.

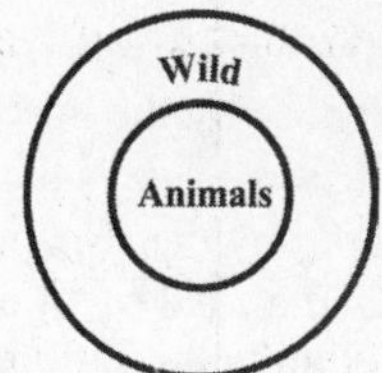

The statement "All lions are animals" is represented by adding a third circle representing lions inside the circle representing animals.

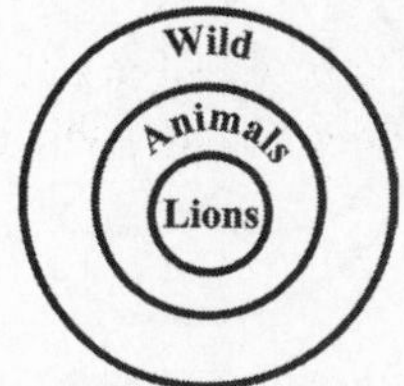

Since the circle representing lions is completely inside the circle representing creatures that are wild, it follows that:

All lions are wild.

The conclusion follows from the premises.

17. All teachers are serious.
All mathematicians are serious.
$\therefore$ All mathematicians are teachers.

The statement "All teachers are serious" is represented by a large circle representing serious and a smaller circle inside the large circle representing teachers.

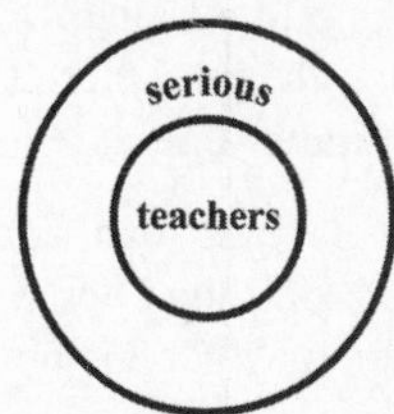

The statement "All mathematicians are serious" is represented by another circle that represents mathematicians inside the larger circle that represents serious. (The circle for teachers would overlap the one for mathematicians.)

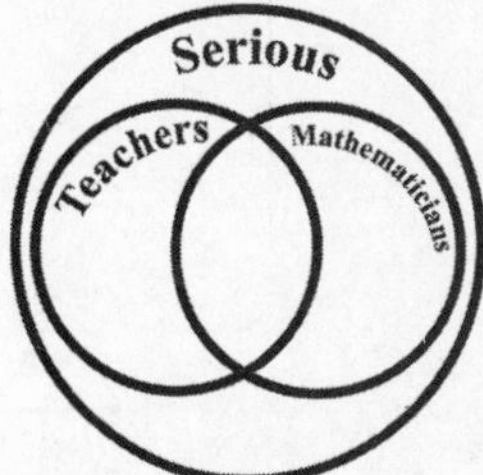

Since the circle representing mathematicians is not completely inside the circle representing teachers, the conclusion does *not* follow from the premises.

19. Two intersecting circles represent the days of television watching by the husband and wife. Since they watched 18 days together, place an 18 in the area shared by the two smaller circles. Since the wife watched a total of 25 days, place a $25 - 18 = 7$ in the other region of the circle labeled Wife.

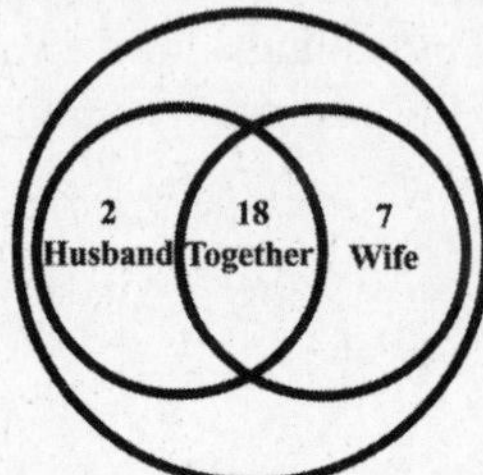

Since the husband watched a total of 20 days, place a $20 - 18 = 2$ in the other region labeled Husband. The 2, 18, and 7 give a total of 27 days. This results in $30 - 27 = 3$ days for the region outside the intersecting circles. This represents 3 days of neither one watching television.

21. Tom, Dick, Mary, and Joan

One is a secretary, one is a computer operator, one is a receptionist, and one is a mail clerk. Find the one who is a computer operator.

1. Tom and Joan eat dinner with the computer operator (fact a), so neither Tom nor Joan is the computer operator.
2. Mary works on the same floor as the computer operator and the mail clerk (fact c), so Mary is not the computer operator.

We know that if Tom, Joan, and Mary are not the computer operator, then Dick must be the computer operator. Fact b is not needed.